# Praise for the earlier editions of
# *Programming in Scala*

*Programming in Scala* is probably one of the best programming books I've ever read. I like the writing style, the brevity, and the thorough explanations. The book seems to answer every question as it enters my mind—it's always one step ahead of me. The authors don't just give you some code and take things for granted. They give you the meat so you really understand what's going on. I really like that.

   - Ken Egervari, Chief Software Architect

*Programming in Scala* is clearly written, thorough, and easy to follow. It has great examples and useful tips throughout. It has enabled our organization to ramp up on the Scala language quickly and efficiently. This book is great for any programmer who is trying to wrap their head around the flexibility and elegance of the Scala language.

   - Larry Morroni, Owner, Morroni Technologies, Inc.

The *Programming in Scala* book serves as an excellent tutorial to the Scala language. Working through the book, it flows well with each chapter building on concepts and examples described in earlier ones. The book takes care to explain the language constructs in depth, often providing examples of how the language differs from Java. As well as the main language, there is also some coverage of libraries such as containers and actors.

I have found the book really easy to work through, and it is probably one of the better written technical books I have read recently. I really would recommend this book to any programmer wanting to find out more about the Scala language.

   - Matthew Todd

i

I am amazed by the effort undertaken by the authors of *Programming in Scala*. This book is an invaluable guide to what I like to call Scala the Platform: a vehicle to better coding, a constant inspiration for scalable software design and implementation. If only I had Scala in its present mature state and this book on my desk back in 2003, when co-designing and implementing parts of the Athens 2004 Olympic Games Portal infrastructure!

To all readers: No matter what your programming background is, I feel you will find programming in Scala liberating and this book will be a loyal friend in the journey.

- Christos KK Loverdos, Software Consultant, Researcher

*Programming in Scala* is a superb in-depth introduction to Scala, and it's also an excellent reference. I'd say that it occupies a prominent place on my bookshelf, except that I'm still carrying it around with me nearly everywhere I go.

- Brian Clapper, President, ArdenTex, Inc.

Great book, well written with thoughtful examples. I would recommend it to both seasoned programmers and newbies.

- Howard Lovatt

The book *Programming in Scala* is not only about *how*, but more importantly, *why* to develop programs in this new programming language. The book's pragmatic approach in introducing the power of combining object-oriented and functional programming leaves the reader without any doubts as to what Scala really is.

- Dr. Ervin Varga, CEO/founder, EXPRO I.T. Consulting

This is a great introduction to functional programming for OO programmers. Learning about FP was my main goal, but I also got acquainted with some nice Scala surprises like case classes and pattern matching. Scala is an intriguing language and this book covers it well.

There's always a fine line to walk in a language introduction book between giving too much or not enough information. I find *Programming in Scala* to achieve a perfect balance.

- Jeff Heon, Programmer Analyst

I bought an early electronic version of the *Programming in Scala* book, by Odersky, Spoon, and Venners, and I was immediately a fan. In addition to the fact that it contains the most comprehensive information about the language, there are a few key features of the electronic format that impressed me. I have never seen links used as well in a PDF, not just for bookmarks, but also providing active links from the table of contents and index. I don't know why more authors don't use this feature, because it's really a joy for the reader. Another feature which I was impressed with was links to the forums ("Discuss") and a way to send comments ("Suggest") to the authors via email. The comments feature by itself isn't all that uncommon, but the simple inclusion of a page number in what is generated to send to the authors is valuable for both the authors and readers. I contributed more comments than I would have if the process would have been more arduous.

Read *Programming in Scala* for the content, but if you're reading the electronic version, definitely take advantage of the digital features that the authors took the care to build in!

- Dianne Marsh, Founder/Software Consultant, SRT Solutions

Lucidity and technical completeness are hallmarks of any well-written book, and I congratulate Martin Odersky, Lex Spoon, and Bill Venners on a job indeed very well done! The *Programming in Scala* book starts by setting a strong foundation with the basic concepts and ramps up the user to an intermediate level & beyond. This book is certainly a must buy for anyone aspiring to learn Scala.

- Jagan Nambi, Enterprise Architecture, GMAC Financial Services

*Programming in Scala* is a pleasure to read. This is one of those well-written technical books that provide deep and comprehensive coverage of the subject in an exceptionally concise and elegant manner.

The book is organized in a very natural and logical way. It is equally well suited for a curious technologist who just wants to stay on top of the current trends and a professional seeking deep understanding of the language core features and its design rationales. I highly recommend it to all interested in functional programming in general. For Scala developers, this book is unconditionally a must-read.

- Igor Khlystov, Software Architect/Lead Programmer, Greystone Inc.

The book *Programming in Scala* outright oozes the huge amount of hard work that has gone into it. I've never read a tutorial-style book before that accomplishes to be introductory yet comprehensive: in their (misguided) attempt to be approachable and not "confuse" the reader, most tutorials silently ignore aspects of a subject that are too advanced for the current discussion. This leaves a very bad taste, as one can never be sure as to the understanding one has achieved. There is always some residual "magic" that hasn't been explained and cannot be judged at all by the reader. This book never does that, it never takes anything for granted: every detail is either sufficiently explained or a reference to a later explanation is given. Indeed, the text is extensively cross-referenced and indexed, so that forming a complete picture of a complex topic is relatively easy.

- Gerald Loeffler, Enterprise Java Architect

*Programming in Scala* by Martin Odersky, Lex Spoon, and Bill Venners: in times where good programming books are rare, this excellent introduction for intermediate programmers really stands out. You'll find everything here you need to learn this promising language.

- Christian Neukirchen

# Programming in Scala

Third Edition

# Programming in Scala

## Third Edition

Martin Odersky, Lex Spoon, Bill Venners

artima

ARTIMA PRESS
WALNUT CREEK, CALIFORNIA

Programming in Scala
Third Edition

Martin Odersky is the creator of the Scala language and a professor at EPFL in
Lausanne, Switzerland. Lex Spoon worked on Scala for two years as a post-doc
with Martin Odersky. Bill Venners is president of Artima, Inc.

Artima Press is an imprint of Artima, Inc.
2070 N Broadway Unit 305, Walnut Creek, California 94597

First edition published as PrePrint™ eBook 2007
First edition published 2008
Second edition published as PrePrint™ eBook 2010
Second edition published 2010
Third edition published as PrePrint™ eBook 2016
Third edition published 2016
Build date of this impression May 26, 2017
Produced in the United States of America

21 20 19 18 17   3 4 5 6 7
ISBN-10: 0-9815316-8-7
ISBN-13: 978-0-9815316-8-7

Library of Congress Control Number: 2014936642

Text printed on acid-free, SFI (Sustainable Forestry Initiative)-certified paper.

*to Nastaran - M.O.*
*to Fay - L.S.*
*to Siew - B.V.*

# Overview

# Contents

Contents

# Contents

Contents

# List of Figures

List of Figures

# List of Tables

# List of Listings

# Foreword

You've chosen a great time to pick up this book! Scala adoption keeps accelerating, our community is thriving, and job ads abound. Whether you're programming for fun or profit (or both), Scala's promise of joy and productivity is proving hard to resist. To me, the true joy of programming comes from tackling interesting challenges with simple, sophisticated solutions. Scala's mission is not just to make this possible, but enjoyable, and this book will show you how.

I first experimented with Scala 2.5, and was immediately drawn to its syntactic and conceptual regularity. When I ran into the irregularity that type parameters couldn't have type parameters themselves, I (timidly) walked up to Martin Odersky at a conference in 2006 and proposed an internship to remove that restriction. My contribution was accepted, bringing support for type constructor polymorphism to Scala 2.7 and up. Since then, I've worked on most other parts of the compiler. In 2012 I went from post-doc in Martin's lab to Scala team lead at Typesafe, as Scala, with version 2.10, graduated from its pragmatic academic roots to a robust language for the enterprise.

Scala 2.10 was a turning point from fast-paced, feature-rich releases based on academic research, towards a focus on simplification and increased adoption in the enterprise. We shifted our attention to issues that won't be written up in dissertations, such as binary compatibility between major releases. To balance stability with our desire to keep evolving and refining the Scala platform, we're working towards a smaller core library, which we aim to stabilize while evolving the platform as a whole. To enable this, my first project as Scala tech lead was to begin modularizing the Scala standard library in 2.11.

To reduce the rate of change, Typesafe also decided to alternate changing the library and the compiler. This edition of *Programming in Scala* covers Scala 2.12, which will be a compiler release sporting a new back-end and optimizer to make the most of Java 8's new features. For interoperability with Java and to enjoy the same benefits from JVM optimizations, Scala compiles functions to the same bytecode as the Java 8 compiler. Similarly, Scala traits now compile to Java interfaces with default methods. Both compilation schemes reduce the magic that older Scala compilers had to perform, aligning us more closely with the Java platform, while improving both compile-time and run-time performance, with a smoother binary compatibility story to boot!

These improvement to the Java 8 platform are very exciting for Scala, and it's very rewarding to see Java align with the trend Scala has been setting for over a decade! There's no doubt that Scala provides a much better functional programming experience, with immutability by default, a uniform treatment of expressions (there's hardly a return statement in sight in this book), pattern matching, definition-site variance (Java's use-site variance make function subtyping quite awkward), and so on! To be blunt, there's more to functional programming than nice syntax for lambdas.

As stewards of the language, our goal is to develop the core language as much as to foster the ecosystem. Scala is successful because of the many excellent libraries, outstanding IDEs and tools, and the friendly and ever helpful members of our community. I've thoroughly enjoyed my first decade of Scala—as an implementer of the language, it's such a thrill and inspiration to meet programmers having fun with Scala across so many domains.

I love programming in Scala, and I hope you will too. On behalf of the Scala community, welcome!

Adriaan Moors
San Francisco, CA
January 14, 2016

# Acknowledgments

Many people have contributed to this book and to the material it covers. We are grateful to all of them.

Scala itself has been a collective effort of many people. The design and the implementation of version 1.0 was helped by Philippe Altherr, Vincent Cremet, Gilles Dubochet, Burak Emir, Stéphane Micheloud, Nikolay Mihaylov, Michel Schinz, Erik Stenman, and Matthias Zenger. Phil Bagwell, Antonio Cunei, Iulian Dragos, Gilles Dubochet, Miguel Garcia, Philipp Haller, Sean McDirmid, Ingo Maier, Donna Malayeri, Adriaan Moors, Hubert Plociniczak, Paul Phillips, Aleksandar Prokopec, Tiark Rompf, Lukas Rytz, and Geoffrey Washburn joined in the effort to develop the second and current version of the language and tools.

Gilad Bracha, Nathan Bronson, Caoyuan, Aemon Cannon, Craig Chambers, Chris Conrad, Erik Ernst, Matthias Felleisen, Mark Harrah, Shriram Krishnamurti, Gary Leavens, David MacIver, Sebastian Maneth, Rickard Nilsson, Erik Meijer, Lalit Pant, David Pollak, Jon Pretty, Klaus Ostermann, Jorge Ortiz, Didier Rémy, Miles Sabin, Vijay Saraswat, Daniel Spiewak, James Strachan, Don Syme, Erik Torreborre, Mads Torgersen, Philip Wadler, Jamie Webb, John Williams, Kevin Wright, and Jason Zaugg have shaped the design of the language by graciously sharing their ideas with us in lively and inspiring discussions, by contributing important pieces of code to the open source effort, as well as through comments on previous versions of this document. The contributors to the Scala mailing list have also given very useful feedback that helped us improve the language and its tools.

George Berger has worked tremendously to make the build process and the web presence for the book work smoothly. As a result this project has been delightfully free of technical snafus.

Many people gave us valuable feedback on early versions of the text. Thanks goes to Eric Armstrong, George Berger, Alex Blewitt, Gilad Bracha,

William Cook, Bruce Eckel, Stéphane Micheloud, Todd Millstein, David Pollak, Frank Sommers, Philip Wadler, and Matthias Zenger. Thanks also to the Silicon Valley Patterns group for their very helpful review: Dave Astels, Tracy Bialik, John Brewer, Andrew Chase, Bradford Cross, Raoul Duke, John P. Eurich, Steven Ganz, Phil Goodwin, Ralph Jocham, Yan-Fa Li, Tao Ma, Jeffery Miller, Suresh Pai, Russ Rufer, Dave W. Smith, Scott Turnquest, Walter Vannini, Darlene Wallach, and Jonathan Andrew Wolter. And we'd like to thank Dewayne Johnson and Kim Leedy for their help with the cover art, and Frank Sommers for his work on the index.

We'd also like to extend a special thanks to all of our readers who contributed comments. Your comments were very helpful to us in shaping this into an even better book. We couldn't print the names of everyone who contributed comments, but here are the names of readers who submitted at least five comments during the eBook PrePrint™ stage by clicking on the Suggest link, sorted first by the highest total number of comments submitted, then alphabetically. Thanks goes to: David Biesack, Donn Stephan, Mats Henricson, Rob Dickens, Blair Zajac, Tony Sloane, Nigel Harrison, Javier Diaz Soto, William Heelan, Justin Forder, Gregor Purdy, Colin Perkins, Bjarte S. Karlsen, Ervin Varga, Eric Willigers, Mark Hayes, Martin Elwin, Calum MacLean, Jonathan Wolter, Les Pruszynski, Seth Tisue, Andrei Formiga, Dmitry Grigoriev, George Berger, Howard Lovatt, John P. Eurich, Marius Scurtescu, Jeff Ervin, Jamie Webb, Kurt Zoglmann, Dean Wampler, Nikolaj Lindberg, Peter McLain, Arkadiusz Stryjski, Shanky Surana, Craig Bordelon, Alexandre Patry, Filip Moens, Fred Janon, Jeff Heon, Boris Lorbeer, Jim Menard, Tim Azzopardi, Thomas Jung, Walter Chang, Jeroen Dijkmeijer, Casey Bowman, Martin Smith, Richard Dallaway, Antony Stubbs, Lars Westergren, Maarten Hazewinkel, Matt Russell, Remigiusz Michalowski, Andrew Tolopko, Curtis Stanford, Joshua Cough, Zemian Deng, Christopher Rodrigues Macias, Juan Miguel Garcia Lopez, Michel Schinz, Peter Moore, Randolph Kahle, Vladimir Kelman, Daniel Gronau, Dirk Detering, Hiroaki Nakamura, Ole Hougaard, Bhaskar Maddala, David Bernard, Derek Mahar, George Kollias, Kristian Nordal, Normen Mueller, Rafael Ferreira, Binil Thomas, John Nilsson, Jorge Ortiz, Marcus Schulte, Vadim Gerassimov, Cameron Taggart, Jon-Anders Teigen, Silvestre Zabala, Will McQueen, and Sam Owen.

We would also like to thank those who submitted comments and errata after the first two editions were published, including Felix Siegrist, Lothar Meyer-Lerbs, Diethard Michaelis, Roshan Dawrani, Donn Stephan,

William Uther, Francisco Reverbel, Jim Balter, and Freek de Bruijn, Ambrose Laing, Sekhar Prabhala, Levon Saldamli, Andrew Bursavich, Hjalmar Peters, Thomas Fehr, Alain O'Dea, Rob Dickens, Tim Taylor, Christian Sternagel, Michel Parisien, Joel Neely, Brian McKeon, Thomas Fehr, Joseph Elliott, Gabriel da Silva Ribeiro, Thomas Fehr, Pablo Ripolles, Douglas Gaylor, Kevin Squire, Harry-Anton Talvik, Christopher Simpkins, Martin Witmann-Funk, Jim Balter, Peter Foster, Craig Bordelon, Heinz-Peter Gumm, Peter Chapin, Kevin Wright, Ananthan Srinivasan, Omar Kilani, Donn Stephan, Guenther Waffler.

Lex would like to thank Aaron Abrams, Jason Adams, Henry and Emily Crutcher, Joey Gibson, Gunnar Hillert, Matthew Link, Toby Reyelts, Jason Snape, John and Melinda Weathers, and all of the Atlanta Scala Enthusiasts for many helpful discussions about the language design, its mathematical underpinnings, and how to present Scala to working engineers.

A special thanks to Dave Briccetti and Adriaan Moors for reviewing the third edition, and to Marconi Lanna for not only reviewing, but providing motivation for the third edition by giving a talk entitled "What's new since *Programming in Scala*."

Bill would like to thank Gary Cornell, Greg Doench, Andy Hunt, Mike Leonard, Tyler Ortman, Bill Pollock, Dave Thomas, and Adam Wright for providing insight and advice on book publishing. Bill would also like to thank Dick Wall for collaborating on Escalate's *Stairway to Scala* course, which is in great part based on this book. Our many years of experience teaching *Stairway to Scala* has helped make this book better. Lastly, Bill would like to thank Darlene Gruendl and Samantha Woolf for their help in getting the third edition completed.

# Introduction

This book is a tutorial for the Scala programming language, written by people directly involved in the development of Scala. Our goal is that by reading this book, you can learn everything you need to be a productive Scala programmer. All examples in this book compile with Scala version 2.11.7, except for those marked 2.12, which compile with 2.12.0-M3.

## Who should read this book

The main target audience for this book is programmers who want to learn to program in Scala. If you want to do your next software project in Scala, then this is the book for you. In addition, the book should be interesting to programmers wishing to expand their horizons by learning new concepts. If you're a Java programmer, for example, reading this book will expose you to many concepts from functional programming as well as advanced object-oriented ideas. We believe learning about Scala, and the ideas behind it, can help you become a better programmer in general.

General programming knowledge is assumed. While Scala is a fine first programming language, this is not the book to use to learn programming.

On the other hand, no specific knowledge of programming languages is required. Even though most people use Scala on the Java platform, this book does not presume you know anything about Java. However, we expect many readers to be familiar with Java, and so we sometimes compare Scala to Java to help such readers understand the differences.

## How to use this book

Because the main purpose of this book is to serve as a tutorial, the recommended way to read this book is in chapter order, from front to back. We have tried hard to introduce one topic at a time, and explain new topics only in terms of topics we've already introduced. Thus, if you skip to the back to get an early peek at something, you may find it explained in terms of concepts you don't quite understand. To the extent you read the chapters in order, we think you'll find it quite straightforward to gain competency in Scala, one step at a time.

If you see a term you do not know, be sure to check the glossary and the index. Many readers will skim parts of the book, and that is just fine. The glossary and index can help you backtrack whenever you skim over something too quickly.

After you have read the book once, it should also serve as a language reference. There is a formal specification of the Scala language, but the language specification tries for precision at the expense of readability. Although this book doesn't cover every detail of Scala, it is quite comprehensive and should serve as an approachable language reference as you become more adept at programming in Scala.

# How to learn Scala

You will learn a lot about Scala simply by reading this book from cover to cover. You can learn Scala faster and more thoroughly, though, if you do a few extra things.

First of all, you can take advantage of the many program examples included in the book. Typing them in yourself is a way to force your mind through each line of code. Trying variations is a way to make them more fun and to make sure you really understand how they work.

Second, keep in touch with the numerous online forums. That way, you and other Scala enthusiasts can help each other. There are numerous mailing lists, discussion forums, a chat room, a wiki, and multiple Scala-specific article feeds. Take some time to find ones that fit your information needs. You will spend a lot less time stuck on little problems, so you can spend your time on deeper, more important questions.

Finally, once you have read enough, take on a programming project of your own. Work on a small program from scratch or develop an add-in to a larger program. You can only go so far by reading.

# EBook features

This book is available in both paper and PDF eBook form. The eBook is not simply an electronic copy of the paper version of the book. While the content is the same as in the paper version, the eBook has been carefully designed and optimized for reading on a computer screen.

The first thing to notice is that most references within the eBook are hyperlinked. If you select a reference to a chapter, figure, or glossary entry, your PDF viewer should take you immediately to the selected item so that you do not have to flip around to find it.

Additionally, at the bottom of each page in the eBook are a number of navigation links. The Cover, Overview, and Contents links take you to the front matter of the book. The Glossary and Index links take you to reference parts of the book. Finally, the Discuss link takes you to an online forum where you discuss questions with other readers, the authors, and the larger Scala community. If you find a typo, or something you think could be explained better, please click on the Suggest link, which will take you to an online web application where you can give the authors feedback.

Although the same pages appear in the eBook as in the printed book, blank pages are removed and the remaining pages renumbered. The pages are numbered differently so that it is easier for you to determine PDF page numbers when printing only a portion of the eBook. The pages in the eBook are, therefore, numbered exactly as your PDF viewer will number them.

## Typographic conventions

The first time a *term* is used, it is italicized. Small code examples, such as x + 1, are written inline with a mono-spaced font. Larger code examples are put into mono-spaced quotation blocks like this:

```
def hello() = {
  println("Hello, world!")
}
```

When interactive shells are shown, responses from the shell are shown in a lighter font:

```
scala> 3 + 4
res0: Int = 7
```

# Content overview

- Chapter 1, "A Scalable Language," gives an overview of Scala's design as well as the reasoning, and history, behind it.

- Chapter 2, "First Steps in Scala," shows you how to do a number of basic programming tasks in Scala, without going into great detail about how they work. The goal of this chapter is to get your fingers started typing and running Scala code.

- Chapter 3, "Next Steps in Scala," shows you several more basic programming tasks that will help you get up to speed quickly in Scala. After completing this chapter, you should be able to start using Scala for simple scripting tasks.

- Chapter 4, "Classes and Objects," starts the in-depth coverage of Scala with a description of its basic object-oriented building blocks and instructions on how to compile and run a Scala application.

- Chapter 5, "Basic Types and Operations," covers Scala's basic types, their literals, the operations you can perform on them, how precedence and associativity works, and what rich wrappers are.

- Chapter 6, "Functional Objects," dives more deeply into the object-oriented features of Scala, using functional (*i.e.*, immutable) rational numbers as an example.

- Chapter 7, "Built-in Control Structures," shows you how to use Scala's built-in control structures: `if`, `while`, `for`, `try`, and `match`.

- Chapter 8, "Functions and Closures," provides in-depth coverage of functions, the basic building block of functional languages.

- Chapter 9, "Control Abstraction," shows how to augment Scala's basic control structures by defining your own control abstractions.

- Chapter 10, "Composition and Inheritance," discusses more of Scala's support for object-oriented programming. The topics are not as fundamental as those in Chapter 4, but they frequently arise in practice.

- Chapter 11, "Scala's Hierarchy," explains Scala's inheritance hierarchy and discusses its universal methods and bottom types.

- Chapter 12, "Traits," covers Scala's mechanism for mixin composition. The chapter shows how traits work, describes common uses, and explains how traits improve on traditional multiple inheritance.

- Chapter 13, "Packages and Imports," discusses issues with programming in the large, including top-level packages, import statements, and access control modifiers like `protected` and `private`.

- Chapter 14, "Assertions and Tests," shows Scala's assertion mechanism and gives a tour of several tools available for writing tests in Scala, focusing on ScalaTest in particular.

- Chapter 15, "Case Classes and Pattern Matching," introduces twin constructs that support you when writing regular, non-encapsulated data structures. Case classes and pattern matching are particularly helpful for tree-like recursive data.

- Chapter 16, "Working with Lists," explains in detail lists, which are probably the most commonly used data structure in Scala programs.

- Chapter 17, "Working with Other Collections," shows you how to use the basic Scala collections, such as lists, arrays, tuples, sets, and maps.

- Chapter 18, "Mutable Objects," explains mutable objects and the syntax Scala provides to express them. The chapter concludes with a case study on discrete event simulation, which shows some mutable objects in action.

- Chapter 19, "Type Parameterization," explains some of the techniques for information hiding introduced in Chapter 13 by means of a concrete example: the design of a class for purely functional queues. The chapter builds up to a description of variance of type parameters and how it interacts with information hiding.

- Chapter 20, "Abstract Members," describes all kinds of abstract members that Scala supports; not only methods, but also fields and types, can be declared abstract.

- Chapter 21, "Implicit Conversions and Parameters," covers two constructs that can help you omit tedious details from source code, letting the compiler supply them instead.

- Chapter 22, "Implementing Lists," describes the implementation of class List. It is important to understand how lists work in Scala, and furthermore the implementation demonstrates the use of several of Scala's features.

- Chapter 23, "For Expressions Revisited," shows how for expressions are translated to invocations of map, flatMap, filter, and foreach.

- Chapter 24, "Collections in Depth," gives a detailed tour of the collections library.

- Chapter 25, "The Architecture of Scala Collections," shows how the collection library is built and how you can implement your own collections.

- Chapter 26, "Extractors," shows how to pattern match against arbitrary classes, not just case classes.

- Chapter 27, "Annotations," shows how to work with language extension via annotation. The chapter describes several standard annotations and shows you how to make your own.

- Chapter 28, "Working with XML," explains how to process XML in Scala. The chapter shows you idioms for generating XML, parsing it, and processing it once it is parsed.

- Chapter 29, "Modular Programming Using Objects," shows how you can use Scala's objects as a modules system.

- Chapter 30, "Object Equality," points out several issues to consider when writing an equals method. There are several pitfalls to avoid.

- Chapter 31, "Combining Scala and Java," discusses issues that arise when combining Scala and Java together in the same project, and suggests ways to deal with them.

- Chapter 32, "Futures and Concurrency," shows you how to use Scala's Future. Although you can use the Java platform's concurrency primitives and libraries for Scala programs, futures can help you avoid the deadlocks and race conditions that plague the traditional "threads and locks" approach to concurrency.

- Chapter 33, "Combinator Parsing," shows how to build parsers using Scala's library of parser combinators.

- Chapter 34, "GUI Programming," gives a quick tour of a Scala library that simplifies GUI programming with Swing.

- Chapter 35, "The SCells Spreadsheet," ties everything together by showing a complete spreadsheet application written in Scala.

## Resources

At http://www.scala-lang.org, the main website for Scala, you'll find the latest Scala release and links to documentation and community resources. For a more condensed page of links to Scala resources, visit this book's website: http://booksites.artima.com/programming_in_scala_3ed. To interact with other readers of this book, check out the Programming in Scala Forum, at: http://www.artima.com/forums/forum.jsp?forum=282.

## Source code

You can download a ZIP file containing the source code of this book, which is released under the Apache 2.0 open source license, from the book's website: http://booksites.artima.com/programming_in_scala_3ed.

## Errata

Although this book has been heavily reviewed and checked, errors will inevitably slip through. For a (hopefully short) list of errata for this book, visit http://booksites.artima.com/programming_in_scala_3ed/errata. If you find an error, please report it at the above URL, so that we can fix it in a future printing or edition of this book.

# Programming in Scala

## Third Edition

```
println("Hello, reader!")
```

# Chapter 1

# A Scalable Language

The name Scala stands for "scalable language." The language is so named because it was designed to grow with the demands of its users. You can apply Scala to a wide range of programming tasks, from writing small scripts to building large systems.[1]

Scala is easy to get into. It runs on the standard Java platform and interoperates seamlessly with all Java libraries. It's quite a good language for writing scripts that pull together Java components. But it can apply its strengths even more when used for building large systems and frameworks of reusable components.

Technically, Scala is a blend of object-oriented and functional programming concepts in a statically typed language. The fusion of object-oriented and functional programming shows up in many different aspects of Scala; it is probably more pervasive than in any other widely used language. The two programming styles have complementary strengths when it comes to scalability. Scala's functional programming constructs make it easy to build interesting things quickly from simple parts. Its object-oriented constructs make it easy to structure larger systems and adapt them to new demands. The combination of both styles in Scala makes it possible to express new kinds of programming patterns and component abstractions. It also leads to a legible and concise programming style. And because it is so malleable, programming in Scala can be a lot of fun.

This initial chapter answers the question, "Why Scala?" It gives a high-level view of Scala's design and the reasoning behind it. After reading the chapter you should have a basic feel for what Scala is and what kinds of

---

[1]Scala is pronounced *skah-lah*.

tasks it might help you accomplish. Although this book is a Scala tutorial, this chapter isn't really part of the tutorial. If you're eager to start writing some Scala code, you should jump ahead to Chapter 2.

## 1.1 A language that grows on you

Programs of different sizes tend to require different programming constructs. Consider, for example, the following small Scala program:

```
var capital = Map("US" -> "Washington", "France" -> "Paris")
capital += ("Japan" -> "Tokyo")
println(capital("France"))
```

This program sets up a map from countries to their capitals, modifies the map by adding a new binding ("Japan" -> "Tokyo"), and prints the capital associated with the country France.[2] The notation in this example is high level, to the point, and not cluttered with extraneous semicolons or type annotations. Indeed, the feel is that of a modern "scripting" language like Perl, Python, or Ruby. One common characteristic of these languages, which is relevant for the example above, is that they each support an "associative map" construct in the syntax of the language.

Associative maps are very useful because they help keep programs legible and concise, but sometimes you might not agree with their "one size fits all" philosophy because you need to control the properties of the maps you use in your program in a more fine-grained way. Scala gives you this fine-grained control if you need it, because maps in Scala are not language syntax. They are library abstractions that you can extend and adapt.

In the above program, you'll get a default Map implementation, but you can easily change that. You could for example specify a particular implementation, such as a HashMap or a TreeMap, or invoke the par method to obtain a ParMap that executes operations in parallel. You could specify a default value for the map, or you could override any other method of the map you create. In each case, you can use the same easy access syntax for maps as in the example above.

---

[2]Please bear with us if you don't understand all the details of this program. They will be explained in the next two chapters.

This example shows that Scala can give you both convenience and flexibility. Scala has a set of convenient constructs that help you get started quickly and let you program in a pleasantly concise style. At the same time, you have the assurance that you will not outgrow the language. You can always tailor the program to your requirements, because everything is based on library modules that you can select and adapt as needed.

**Growing new types**

Eric Raymond introduced the cathedral and bazaar as two metaphors of software development.[3] The cathedral is a near-perfect building that takes a long time to build. Once built, it stays unchanged for a long time. The bazaar, by contrast, is adapted and extended each day by the people working in it. In Raymond's work the bazaar is a metaphor for open-source software development. Guy Steele noted in a talk on "growing a language" that the same distinction can be applied to language design.[4] Scala is much more like a bazaar than a cathedral, in the sense that it is designed to be extended and adapted by the people programming in it. Instead of providing all constructs you might ever need in one "perfectly complete" language, Scala puts the tools for building such constructs into your hands.

Here's an example. Many applications need a type of integer that can become arbitrarily large without overflow or "wrap-around" of arithmetic operations. Scala defines such a type in library class `scala.BigInt`. Here is the definition of a method using that type, which calculates the factorial of a passed integer value:[5]

```
def factorial(x: BigInt): BigInt =
  if (x == 0) 1 else x * factorial(x - 1)
```

Now, if you call `factorial(30)` you would get:

```
265252859812191058636308480000000
```

`BigInt` looks like a built-in type because you can use integer literals and operators such as * and – with values of that type. Yet it is just a class that

---

[3]Raymond, *The Cathedral and the Bazaar*. [Ray99]

[4]Steele, "Growing a language." [Ste99]

[5]`factorial(x)`, or x! in mathematical notation, is the result of computing $1 * 2 * \ldots * x$, with 0! defined to be 1.

happens to be defined in Scala's standard library.[6] If the class were missing, it would be straightforward for any Scala programmer to write an implementation, for instance, by wrapping Java's class `java.math.BigInteger` (in fact that's how Scala's `BigInt` class is implemented).

Of course, you could also use Java's class directly. But the result is not nearly as pleasant, because although Java allows you to create new types, those types don't feel much like native language support:

```
import java.math.BigInteger

def factorial(x: BigInteger): BigInteger =
  if (x == BigInteger.ZERO)
    BigInteger.ONE
  else
    x.multiply(factorial(x.subtract(BigInteger.ONE)))
```

BigInt is representative of many other number-like types—big decimals, complex numbers, rational numbers, confidence intervals, polynomials—the list goes on. Some programming languages implement some of these types natively. For instance, Lisp, Haskell, and Python implement big integers; Fortran and Python implement complex numbers. But any language that attempted to implement all of these abstractions at the same time would simply become too big to be manageable. What's more, even if such a language were to exist, some applications would surely benefit from other number-like types that were not supplied. So the approach of attempting to provide everything in one language doesn't scale very well. Instead, Scala allows users to grow and adapt the language in the directions they need by defining easy-to-use libraries that *feel* like native language support.

**Growing new control constructs**

The previous example demonstrates that Scala lets you add new types that can be used as conveniently as built-in types. The same extension principle also applies to control structures. This kind of extensibility is illustrated by Akka, a Scala API for "actor-based" concurrent programming.

---

[6]Scala comes with a standard library, some of which will be covered in this book. For more information, you can also consult the library's Scaladoc documentation, which is available in the distribution and online at http://www.scala-lang.org.

As multicore processors continue to proliferate in the coming years, achieving acceptable performance may increasingly require that you exploit more parallelism in your applications. Often, this will mean rewriting your code so that computations are distributed over several concurrent threads. Unfortunately, creating dependable multi-threaded applications has proven challenging in practice. Java's threading model is built around shared memory and locking, a model that is often difficult to reason about, especially as systems scale up in size and complexity. It is hard to be sure you don't have a race condition or deadlock lurking—something that didn't show up during testing, but might just show up in production. An arguably safer alternative is a message passing architecture, such as the "actors" approach used by the Erlang programming language.

Java comes with a rich, thread-based concurrency library. Scala programs can use it like any other Java API. However, Akka is an additional Scala library that implements an actor model similar to Erlang's.

Actors are concurrency abstractions that can be implemented on top of threads. They communicate by sending messages to each other. An actor can perform two basic operations, message send and receive. The send operation, denoted by an exclamation point (!), sends a message to an actor. Here's an example in which the actor is named `recipient`:

```
recipient ! msg
```

A send is asynchronous; that is, the sending actor can proceed immediately, without waiting for the message to be received and processed. Every actor has a *mailbox* in which incoming messages are queued. An actor handles messages that have arrived in its mailbox via a `receive` block:

```
def receive = {
  case Msg1 => ... // handle Msg1
  case Msg2 => ... // handle Msg2
  // ...
}
```

A receive block consists of a number of cases that each query the mailbox with a message pattern. The first message in the mailbox that matches any of the cases is selected, and the corresponding action is performed on it. Once the mailbox does not contain any messages, the actor suspends and waits for further incoming messages.

7

As an example, here is a simple Akka actor implementing a checksum calculator service:

```scala
class ChecksumActor extends Actor {
  var sum = 0
  def receive = {
    case Data(byte) => sum += byte
    case GetChecksum(requester) =>
      val checksum = ~(sum & 0xFF) + 1
      requester ! checksum
  }
}
```

This actor first defines a local variable named sum with initial value zero. It defines a receive block that will handle messages. If it receives a Data message, it adds the contained byte to the sum variable. If it receives a GetChecksum message, it calculates a checksum from the current value of sum and sends the result back to the requester using the message send requester ! sum. The requester field is embedded in the GetChecksum message; it usually refers to the actor that made the request.

We don't expect you to fully understand the actor example at this point. Rather, what's significant about this example for the topic of scalability is that neither the receive block nor message send (!) are built-in operations in Scala. Even though the receive block may look and act very much like a built-in control construct, it is in fact a method defined in Akka's actors library. Likewise, even though '!' looks like a built-in operator, it too is just a method defined in the Akka actors library. Both of these constructs are completely independent of the Scala programming language.

The receive block and send (!) syntax look in Scala much like they look in Erlang, but in Erlang, these constructs are built into the language. Akka also implements most of Erlang's other concurrent programming constructs, such as monitoring failed actors and time-outs. All in all, the actor model has turned out to be a very pleasant means for expressing concurrent and distributed computations. Even though they must be defined in a library, actors can feel like an integral part of the Scala language.

This example illustrates that you can "grow" the Scala language in new directions even as specialized as concurrent programming. To be sure, you need good architects and programmers to do this. But the crucial thing is that it is feasible—you can design and implement abstractions in Scala that address radically new application domains, yet still feel like native language support when used.

## 1.2    What makes Scala scalable?

Scalability is influenced by many factors, ranging from syntax details to component abstraction constructs. If we were forced to name just one aspect of Scala that helps scalability, though, we'd pick its combination of object-oriented and functional programming (well, we cheated, that's really two aspects, but they are intertwined).

Scala goes further than all other well-known languages in fusing object-oriented and functional programming into a uniform language design. For instance, where other languages might have objects and functions as two different concepts, in Scala a function value *is* an object. Function types are classes that can be inherited by subclasses. This might seem nothing more than an academic nicety, but it has deep consequences for scalability. In fact the actor concept shown previously could not have been implemented without this unification of functions and objects. This section gives an overview of Scala's way of blending object-oriented and functional concepts.

### Scala is object-oriented

Object-oriented programming has been immensely successful. Starting from Simula in the mid-60s and Smalltalk in the 70s, it is now available in more languages than not. In some domains, objects have taken over completely. While there is not a precise definition of what object-oriented means, there is clearly something about objects that appeals to programmers.

In principle, the motivation for object-oriented programming is very simple: all but the most trivial programs need some sort of structure. The most straightforward way to do this is to put data and operations into some form of containers. The great idea of object-oriented programming is to make these containers fully general, so that they can contain operations as well as data, and that they are themselves values that can be stored in other containers, or passed as parameters to operations. Such containers are called objects. Alan Kay, the inventor of Smalltalk, remarked that in this way the simplest object has the same construction principle as a full computer: it combines data with operations under a formalized interface.[7] So objects have a lot to do with language scalability: the same techniques apply to the construction of small as well as large programs.

Even though object-oriented programming has been mainstream for a long time, there are relatively few languages that have followed Smalltalk in pushing this construction principle to its logical conclusion. For instance, many languages admit values that are not objects, such as the primitive values in Java. Or they allow static fields and methods that are not members of any object. These deviations from the pure idea of object-oriented programming look quite harmless at first, but they have an annoying tendency to complicate things and limit scalability.

By contrast, Scala is an object-oriented language in pure form: every value is an object and every operation is a method call. For example, when you say 1 + 2 in Scala, you are actually invoking a method named + defined in class Int. You can define methods with operator-like names that clients of your API can then use in operator notation. This is how the designer of Akka's actors API enabled you to use expressions such as requester ! sum shown in the previous example: '!' is a method of the Actor class.

Scala is more advanced than most other languages when it comes to composing objects. An example is Scala's *traits*. Traits are like interfaces in Java, but they can also have method implementations and even fields.[8] Objects are constructed by *mixin composition*, which takes the members of a class and adds the members of a number of traits to them. In this way, different aspects of classes can be encapsulated in different traits. This looks a bit like multiple inheritance, but differs when it comes to the details. Unlike a class, a trait can add some new functionality to an unspecified superclass. This makes traits more "pluggable" than classes. In particular, it avoids the classical "diamond inheritance" problems of multiple inheritance, which arise when the same class is inherited via several different paths.

## Scala is functional

In addition to being a pure object-oriented language, Scala is also a full-blown functional language. The ideas of functional programming are older than (electronic) computers. Their foundation was laid in Alonzo Church's

---

[7]Kay, "The Early History of Smalltalk." [Kay96]

[8]Starting with Java 8, interfaces can have default method implementations, but these do not offer all the features of Scala's traits.

lambda calculus, which he developed in the 1930s. The first functional programming language was Lisp, which dates from the late 50s. Other popular functional languages are Scheme, SML, Erlang, Haskell, OCaml, and F#. For a long time, functional programming has been a bit on the sidelines—popular in academia, but not that widely used in industry. However, in recent years, there has been an increased interest in functional programming languages and techniques.

Functional programming is guided by two main ideas. The first idea is that functions are first-class values. In a functional language, a function is a value of the same status as, say, an integer or a string. You can pass functions as arguments to other functions, return them as results from functions, or store them in variables. You can also define a function inside another function, just as you can define an integer value inside a function. And you can define functions without giving them a name, sprinkling your code with function literals as easily as you might write integer literals like 42.

Functions that are first-class values provide a convenient means for abstracting over operations and creating new control structures. This generalization of functions provides great expressiveness, which often leads to very legible and concise programs. It also plays an important role for scalability. As an example, the ScalaTest testing library offers an `eventually` construct that takes a function as an argument. It is used like this:

```
val xs = 1 to 3
val it = xs.iterator
eventually { it.next() shouldBe 3 }
```

The code inside `eventually`—the assertion, `it.next() shouldBe 3`—is wrapped in a function that is passed unexecuted to the `eventually` method. For a configured amount of time, `eventually` will repeatedly execute the function until the assertion succeeds.

In most traditional languages, by contrast, functions are not values. Languages that do have function values often relegate them to second-class status. For example, the function pointers of C and C++ do not have the same status as non-functional values in those languages: Function pointers can only refer to global functions, they do not allow you to define first-class nested functions that refer to some values in their environment. Nor do they allow you to define unnamed function literals.

The second main idea of functional programming is that the operations of a program should map input values to output values rather than change data in place. To see the difference, consider the implementation of strings in Ruby and Java. In Ruby, a string is an array of characters. Characters in a string can be changed individually. For instance you can change a semicolon character in a string to a period inside the same string object. In Java and Scala, on the other hand, a string is a sequence of characters in the mathematical sense. Replacing a character in a string using an expression like `s.replace(';', '.')` yields a new string object, which is different from s. Another way of expressing this is that strings are immutable in Java whereas they are mutable in Ruby. So looking at just strings, Java is a functional language, whereas Ruby is not. Immutable data structures are one of the cornerstones of functional programming. The Scala libraries define many more immutable data types on top of those found in the Java APIs. For instance, Scala has immutable lists, tuples, maps, and sets.

Another way of stating this second idea of functional programming is that methods should not have any *side effects*. They should communicate with their environment only by taking arguments and returning results. For instance, the `replace` method in Java's `String` class fits this description. It takes a string and two characters and yields a new string where all occurrences of one character are replaced by the other. There is no other effect of calling `replace`. Methods like `replace` are called *referentially transparent*, which means that for any given input the method call could be replaced by its result without affecting the program's semantics.

Functional languages encourage immutable data structures and referentially transparent methods. Some functional languages even require them. Scala gives you a choice. When you want to, you can write in an *imperative* style, which is what programming with mutable data and side effects is called. But Scala generally makes it easy to avoid imperative constructs when you want because good functional alternatives exist.

## 1.3  Why Scala?

Is Scala for you? You will have to see and decide for yourself. We have found that there are actually many reasons besides scalability to like programming in Scala. Four of the most important aspects will be discussed in this section: compatibility, brevity, high-level abstractions, and advanced static typing.

## Scala is compatible

Scala doesn't require you to leap backwards off the Java platform to step forward from the Java language. It allows you to add value to existing code—to build on what you already have—because it was designed for seamless interoperability with Java.[9] Scala programs compile to JVM bytecodes. Their run-time performance is usually on par with Java programs. Scala code can call Java methods, access Java fields, inherit from Java classes, and implement Java interfaces. None of this requires special syntax, explicit interface descriptions, or glue code. In fact, almost all Scala code makes heavy use of Java libraries, often without programmers being aware of this fact.

Another aspect of full interoperability is that Scala heavily re-uses Java types. Scala's `Ints` are represented as Java primitive integers of type `int`, `Floats` are represented as `floats`, `Booleans` as `booleans`, and so on. Scala arrays are mapped to Java arrays. Scala also re-uses many of the standard Java library types. For instance, the type of a string literal `"abc"` in Scala is `java.lang.String`, and a thrown exception must be a subclass of `java.lang.Throwable`.

Scala not only re-uses Java's types, but also "dresses them up" to make them nicer. For instance, Scala's strings support methods like `toInt` or `toFloat`, which convert the string to an integer or floating-point number. So you can write `str.toInt` instead of `Integer.parseInt(str)`. How can this be achieved without breaking interoperability? Java's `String` class certainly has no `toInt` method! In fact, Scala has a very general solution to solve this tension between advanced library design and interoperability. Scala lets you define *implicit conversions*, which are always applied when types would not normally match up, or when non-existing members are selected. In the case above, when looking for a `toInt` method on a string, the Scala compiler will find no such member of class `String`, but it will find an implicit conversion that converts a Java `String` to an instance of the Scala class `StringOps`, which does define such a member. The conversion will then be applied implicitly before performing the `toInt` operation.

Scala code can also be invoked from Java code. This is sometimes a bit more subtle, because Scala is a richer language than Java, so some of Scala's

---

[9]Originally, there was an implementation of Scala that ran on the .NET platform, but it is no longer active. More recently, an implementation of Scala that runs on JavaScript, Scala.js, has become increasingly popular.

more advanced features need to be encoded before they can be mapped to Java. Chapter 31 explains the details.

**Scala is concise**

Scala programs tend to be short. Scala programmers have reported reductions in number of lines of up to a factor of ten compared to Java. These might be extreme cases. A more conservative estimate would be that a typical Scala program should have about half the number of lines of the same program written in Java. Fewer lines of code mean not only less typing, but also less effort at reading and understanding programs and fewer possibilities of defects. There are several factors that contribute to this reduction in lines of code.

First, Scala's syntax avoids some of the boilerplate that burdens Java programs. For instance, semicolons are optional in Scala and are usually left out. There are also several other areas where Scala's syntax is less noisy. As an example, compare how you write classes and constructors in Java and Scala. In Java, a class with a constructor often looks like this:

```
// this is Java
class MyClass {

    private int index;
    private String name;

    public MyClass(int index, String name) {
        this.index = index;
        this.name = name;
    }

}
```

In Scala, you would likely write this instead:

```
class MyClass(index: Int, name: String)
```

Given this code, the Scala compiler will produce a class that has two private instance variables, an Int named index and a String named name, and a constructor that takes initial values for those variables as parameters. The code of this constructor will initialize the two instance variables with the

14

values passed as parameters. In short, you get essentially the same function-ality as the more verbose Java version.[10] The Scala class is quicker to write, easier to read, and most importantly, less error prone than the Java class.

Scala's type inference is another factor that contributes to its concise-ness. Repetitive type information can be left out, so programs become less cluttered and more readable.

But probably the most important key to compact code is code you don't have to write because it is done in a library for you. Scala gives you many tools to define powerful libraries that let you capture and factor out common behavior. For instance, different aspects of library classes can be separated out into traits, which can then be mixed together in flexible ways. Or, li-brary methods can be parameterized with operations, which lets you define constructs that are, in effect, your own control structures. Together, these constructs allow the definition of libraries that are both high-level and flexi-ble to use.

**Scala is high-level**

Programmers are constantly grappling with complexity. To program pro-ductively, you must understand the code on which you are working. Overly complex code has been the downfall of many a software project. Unfortu-nately, important software usually has complex requirements. Such com-plexity can't be avoided; it must instead be managed.

Scala helps you manage complexity by letting you raise the level of ab-straction in the interfaces you design and use. As an example, imagine you have a String variable name, and you want to find out whether or not that String contains an upper case character. Prior to Java 8, you might have written a loop, like this:

```
boolean nameHasUpperCase = false;  // this is Java
for (int i = 0; i < name.length(); ++i) {
    if (Character.isUpperCase(name.charAt(i))) {
        nameHasUpperCase = true;
        break;
    }
}
```

---

[10]The only real difference is that the instance variables produced in the Scala case will be final. You'll learn how to make them non-final in Section 10.6.

15

Whereas in Scala, you could write this:

```
val nameHasUpperCase = name.exists(_.isUpper)
```

The Java code treats strings as low-level entities that are stepped through character by character in a loop. The Scala code treats the same strings as higher-level sequences of characters that can be queried with *predicates*. Clearly the Scala code is much shorter and—for trained eyes—easier to understand than the Java code. So the Scala code weighs less heavily on the total complexity budget. It also gives you less opportunity to make mistakes.

The predicate _.isUpper is an example of a function literal in Scala.[11] It describes a function that takes a character argument (represented by the underscore character) and tests whether it is an upper case letter.[12]

Java 8 introduced support for *lambdas* and *streams*, which enable you to perform a similar operation in Java. Here's what it might look like:

```
boolean nameHasUpperCase =    // This is Java 8
    name.chars().anyMatch(
        (int ch) -> Character.isUpperCase((char) ch)
    );
```

Although a great improvement over earlier versions of Java, the Java 8 code is still more verbose than the equivalent Scala code. This extra "heaviness" of Java code, as well as Java's long tradition of loops, may encourage many Java programmers in need of new methods like exists to just write out loops and live with the increased complexity in their code.

On the other hand, function literals in Scala are really lightweight, so they are used frequently. As you get to know Scala better you'll find more and more opportunities to define and use your own control abstractions. You'll find that this helps avoid code duplication and thus keeps your programs shorter and clearer.

**Scala is statically typed**

A static type system classifies variables and expressions according to the kinds of values they hold and compute. Scala stands out as a language with a very advanced static type system. Starting from a system of nested class

---

[11]A function literal can be called a *predicate* if its result type is Boolean.

[12]This use of the underscore as a placeholder for arguments is described in Section 8.5.

types much like Java's, it allows you to parameterize types with *generics*, to combine types using *intersections*, and to hide details of types using *abstract types*.[13] These give a strong foundation for building and composing your own types, so that you can design interfaces that are at the same time safe and flexible to use.

If you like dynamic languages, such as Perl, Python, Ruby, or Groovy, you might find it a bit strange that Scala's static type system is listed as one of its strong points. After all, the absence of a static type system has been cited by some as a major advantage of dynamic languages. The most common arguments against static types are that they make programs too verbose, prevent programmers from expressing themselves as they wish, and make impossible certain patterns of dynamic modifications of software systems. However, often these arguments do not go against the idea of static types in general, but against specific type systems, which are perceived to be too verbose or too inflexible. For instance, Alan Kay, the inventor of the Smalltalk language, once remarked: "I'm not against types, but I don't know of any type systems that aren't a complete pain, so I still like dynamic typing."[14]

We hope to convince you in this book that Scala's type system is far from being a "complete pain." In fact, it addresses nicely two of the usual concerns about static typing: Verbosity is avoided through type inference, and flexibility is gained through pattern matching and several new ways to write and compose types. With these impediments out of the way, the classical benefits of static type systems can be better appreciated. Among the most important of these benefits are verifiable properties of program abstractions, safe refactorings, and better documentation.

***Verifiable properties.*** Static type systems can prove the absence of certain run-time errors. For instance, they can prove properties like: Booleans are never added to integers; private variables are not accessed from outside their class; functions are applied to the right number of arguments; only strings are ever added to a set of strings.

Other kinds of errors are not detected by today's static type systems. For instance, they will usually not detect non-terminating functions, array bounds violations, or divisions by zero. They will also not detect that your

---

[13]Generics are discussed in Chapter 19; intersections (*e.g.*, A with B with C) in Chapter 12; and abstract types in Chapter 20.

[14]Kay, in an email on the meaning of object-oriented programming. [Kay03]

program does not conform to its specification (assuming there is a spec, that is!). Static type systems have therefore been dismissed by some as not being very useful. The argument goes that since such type systems can only detect simple errors, whereas unit tests provide more extensive coverage, why bother with static types at all? We believe that these arguments miss the point. Although a static type system certainly cannot *replace* unit testing, it can reduce the number of unit tests needed by taking care of some properties that would otherwise need to be tested. Likewise, unit testing cannot replace static typing. After all, as Edsger Dijkstra said, testing can only prove the presence of errors, never their absence.[15] So the guarantees that static typing gives may be simple, but they are real guarantees of a form no amount of testing can deliver.

***Safe refactorings.*** A static type system provides a safety net that lets you make changes to a codebase with a high degree of confidence. Consider for instance a refactoring that adds an additional parameter to a method. In a statically typed language you can do the change, re-compile your system, and simply fix all lines that cause a type error. Once you have finished with this, you are sure to have found all places that need to be changed. The same holds for many other simple refactorings, like changing a method name or moving methods from one class to another. In all cases a static type check will provide enough assurance that the new system works just like the old.

***Documentation.*** Static types are program documentation that is checked by the compiler for correctness. Unlike a normal comment, a type annotation can never be out of date (at least not if the source file that contains it has recently passed a compiler). Furthermore, compilers and integrated development environments (IDEs) can make use of type annotations to provide better context help. For instance, an IDE can display all the members available for a selection by determining the static type of the expression on which the selection is made and looking up all members of that type.

Even though static types are generally useful for program documentation, they can sometimes be annoying when they clutter the program. Typically, useful documentation is what readers of a program cannot easily derive by themselves. In a method definition like:

---

[15]Dijkstra, "Notes on Structured Programming." [Dij70]

```
def f(x: String) = ...
```

it's useful to know that f's argument should be a String. On the other hand, at least one of the two annotations in the following example is annoying:

```
val x: HashMap[Int, String] = new HashMap[Int, String]()
```

Clearly, it should be enough to say just once that x is a HashMap with Ints as keys and Strings as values; there's no need to repeat the same phrase twice.

Scala has a very sophisticated type inference system that lets you omit almost all type information that's usually considered annoying. In the previous example, the following two less annoying alternatives would work just as well:

```
val x = new HashMap[Int, String]()
val x: Map[Int, String] = new HashMap()
```

Type inference in Scala can go quite far. In fact, it's not uncommon for user code to have no explicit types at all. Therefore, Scala programs often look a bit like programs written in a dynamically typed scripting language. This holds particularly for client application code, which glues together pre-written library components. It's less true for the library components themselves, because these often employ fairly sophisticated types to allow flexible usage patterns. This is only natural. After all, the type signatures of the members that make up the interface of a reusable component should be explicitly given, because they constitute an essential part of the contract between the component and its clients.

## 1.4   Scala's roots

Scala's design has been influenced by many programming languages and ideas in programming language research. In fact, only a few features of Scala are genuinely new; most have been already applied in some form in other languages. Scala's innovations come primarily from how its constructs are put together. In this section, we list the main influences on Scala's design. The list cannot be exhaustive—there are simply too many smart ideas around in programming language design to enumerate them all here.

At the surface level, Scala adopts a large part of the syntax of Java and C#, which in turn borrowed most of their syntactic conventions from C and

C++. Expressions, statements, and blocks are mostly as in Java, as is the syntax of classes, packages and imports.[16] Besides syntax, Scala adopts other elements of Java, such as its basic types, its class libraries, and its execution model.

Scala also owes much to other languages. Its uniform object model was pioneered by Smalltalk and taken up subsequently by Ruby. Its idea of universal nesting (almost every construct in Scala can be nested inside any other construct) is also present in Algol, Simula, and, more recently, in Beta and gbeta. Its uniform access principle for method invocation and field selection comes from Eiffel. Its approach to functional programming is quite similar in spirit to the ML family of languages, which has SML, OCaml, and F# as prominent members. Many higher-order functions in Scala's standard library are also present in ML or Haskell. Scala's implicit parameters were motivated by Haskell's type classes; they achieve analogous results in a more classical object-oriented setting. Scala's main actor-based concurrency library, Akka, was heavily inspired by Erlang.

Scala is not the first language to emphasize scalability and extensibility. The historic root of extensible languages that can span different application areas is Peter Landin's 1966 paper, "The Next 700 Programming Languages."[17] (The language described in this paper, Iswim, stands beside Lisp as one of the pioneering functional languages.) The specific idea of treating an infix operator as a function can be traced back to Iswim and Smalltalk. Another important idea is to permit a function literal (or block) as a parameter, which enables libraries to define control structures. Again, this goes back to Iswim and Smalltalk. Smalltalk and Lisp both have a flexible syntax that has been applied extensively for building internal domain-specific languages. C++ is another scalable language that can be adapted and extended through operator overloading and its template system; compared to Scala it is built on a lower-level, more systems-oriented core.

---

[16] The major deviation from Java concerns the syntax for type annotations: it's "variable: Type" instead of "Type variable" in Java. Scala's postfix type syntax resembles Pascal, Modula-2, or Eiffel. The main reason for this deviation has to do with type inference, which often lets you omit the type of a variable or the return type of a method. Using the "variable: Type" syntax is easy—just leave out the colon and the type. But in C-style "Type variable" syntax you cannot simply leave off the type; there would be no marker to start the definition anymore. You'd need some alternative keyword to be a placeholder for a missing type (C# 3.0, which does some type inference, uses var for this purpose). Such an alternative keyword feels more ad-hoc and less regular than Scala's approach.

[17]Landin, "The Next 700 Programming Languages." [Lan66]

20

Scala is also not the first language to integrate functional and object-oriented programming, although it probably goes furthest in this direction. Other languages that have integrated some elements of functional programming into object-oriented programming (OOP) include Ruby, Smalltalk, and Python. On the Java platform, Pizza, Nice, Multi-Java—and Java 8 itself—have all extended a Java-like core with functional ideas. There are also primarily functional languages that have acquired an object system; examples are OCaml, F#, and PLT-Scheme.

Scala has also contributed some innovations to the field of programming languages. For instance, its abstract types provide a more object-oriented alternative to generic types, its traits allow for flexible component assembly, and its extractors provide a representation-independent way to do pattern matching. These innovations have been presented in papers at programming language conferences in recent years.[18]

## 1.5 Conclusion

In this chapter, we gave you a glimpse of what Scala is and how it might help you in your programming. To be sure, Scala is not a silver bullet that will magically make you more productive. To advance, you will need to apply Scala artfully, and that will require some learning and practice. If you're coming to Scala from Java, the most challenging aspects of learning Scala may involve Scala's type system (which is richer than Java's) and its support for functional programming. The goal of this book is to guide you gently up Scala's learning curve, one step at a time. We think you'll find it a rewarding intellectual experience that will expand your horizons and make you think differently about program design. Hopefully, you will also gain pleasure and inspiration from programming in Scala.

In the next chapter, we'll get you started writing some Scala code.

---

[18]For more information, see [Ode03], [Ode05], and [Emi07] in the bibliography.

Chapter 2

# First Steps in Scala

It's time to write some Scala code. Before we start on the in-depth Scala tutorial, we put in two chapters that will give you the big picture of Scala, and most importantly, get you writing code. We encourage you to actually try out all the code examples presented in this chapter and the next as you go. The best way to start learning Scala is to program in it.

To run the examples in this chapter, you should have a standard Scala installation. To get one, go to `http://www.scala-lang.org/downloads` and follow the directions for your platform. You can also use a Scala plug-in for Eclipse, IntelliJ, or NetBeans. For the steps in this chapter, we'll assume you're using the Scala distribution from scala-lang.org.[1]

If you are a veteran programmer new to Scala, the next two chapters should give you enough understanding to enable you to start writing useful programs in Scala. If you are less experienced, some of the material may seem a bit mysterious to you. But don't worry. To get you up to speed quickly, we had to leave out some details. Everything will be explained in a less "fire hose" fashion in later chapters. In addition, we inserted quite a few footnotes in these next two chapters to point you to later sections of the book where you'll find more detailed explanations.

## Step 1. Learn to use the Scala interpreter

The easiest way to get started with Scala is by using the Scala interpreter, an interactive "shell" for writing Scala expressions and programs. The inter-

---

[1] We tested the examples in this book with Scala version 2.11.7.

preter, which is called scala, will evaluate expressions you type and print the resulting value. You use it by typing scala at a command prompt:[2]

```
$ scala
Welcome to Scala version 2.11.7
Type in expressions to have them evaluated.
Type :help for more information.

scala>
```

After you type an expression, such as 1 + 2, and hit enter:

```
scala> 1 + 2
```

The interpreter will print:

```
res0: Int = 3
```

This line includes:

- an automatically generated or user-defined name to refer to the computed value (res0, which means result 0),

- a colon (:), followed by the type of the expression (Int),

- an equals sign (=),

- the value resulting from evaluating the expression (3).

The type Int names the class Int in the package scala. Packages in Scala are similar to packages in Java: They partition the global namespace and provide a mechanism for information hiding.[3] Values of class Int correspond to Java's int values. More generally, all of Java's primitive types have corresponding classes in the scala package. For example, scala.Boolean corresponds to Java's boolean. scala.Float corresponds to Java's float. And when you compile your Scala code to Java bytecodes, the Scala compiler will use Java's primitive types where possible to give you the performance benefits of the primitive types.

---

[2]If you're using Windows, you'll need to type the scala command into the "Command Prompt" DOS box.

[3]If you're not familiar with Java packages, you can think of them as providing a full name for classes. Because Int is a member of package scala, "Int" is the class's simple name, and "scala.Int" is its full name. The details of packages are explained in Chapter 13.

The resX identifier may be used in later lines. For instance, since res0 was set to 3 previously, res0 * 3 will be 9:

```
scala> res0 * 3
res1: Int = 9
```

To print the necessary, but not sufficient, Hello, world! greeting, type:

```
scala> println("Hello, world!")
Hello, world!
```

The println function prints the passed string to the standard output, similar to System.out.println in Java.

## Step 2. Define some variables

Scala has two kinds of variables, vals and vars. A val is similar to a final variable in Java. Once initialized, a val can never be reassigned. A var, by contrast, is similar to a non-final variable in Java. A var can be reassigned throughout its lifetime. Here's a val definition:

```
scala> val msg = "Hello, world!"
msg: String = Hello, world!
```

This statement introduces msg as a name for the string "Hello, world!". The type of msg is java.lang.String, because Scala strings are implemented by Java's String class.

If you're used to declaring variables in Java, you'll notice one striking difference here: neither java.lang.String nor String appear anywhere in the val definition. This example illustrates *type inference*, Scala's ability to figure out types you leave off. In this case, because you initialized msg with a string literal, Scala inferred the type of msg to be String. When the Scala interpreter (or compiler) can infer types, it is often best to let it do so rather than fill the code with unnecessary, explicit type annotations. You can, however, specify a type explicitly if you wish, and sometimes you probably should. An explicit type annotation can both ensure the Scala compiler infers the type you intend, as well as serve as useful documentation for future readers of the code. In contrast to Java, where you specify a variable's type before its name, in Scala you specify a variable's type after its name, separated by a colon. For example:

25

```
scala> val msg2: java.lang.String = "Hello again, world!"
msg2: String = Hello again, world!
```

Or, since java.lang types are visible with their simple names[4] in Scala programs, simply:

```
scala> val msg3: String = "Hello yet again, world!"
msg3: String = Hello yet again, world!
```

Going back to the original msg, now that it is defined, you can use it as you'd expect, for example:

```
scala> println(msg)
Hello, world!
```

What you can't do with msg, given that it is a val, not a var, is reassign it.[5] For example, see how the interpreter complains when you attempt the following:

```
scala> msg = "Goodbye cruel world!"
<console>:8: error: reassignment to val
       msg = "Goodbye cruel world!"
           ^
```

If reassignment is what you want, you'll need to use a var, as in:

```
scala> var greeting = "Hello, world!"
greeting: String = Hello, world!
```

Since greeting is a var not a val, you can reassign it later. If you are feeling grouchy later, for example, you could change your greeting to:

```
scala> greeting = "Leave me alone, world!"
greeting: String = Leave me alone, world!
```

To enter something into the interpreter that spans multiple lines, just keep typing after the first line. If the code you typed so far is not complete, the interpreter will respond with a vertical bar on the next line.

---

[4]The simple name of java.lang.String is String.

[5]In the interpreter, however, you can *define* a new val with a name that was already used before. This mechanism is explained in Section 7.7.

```
scala> val multiLine =
     |   "This is the next line."
multiLine: String = This is the next line.
```

If you realize you have typed something wrong, but the interpreter is still waiting for more input, you can escape by pressing enter twice:

```
scala> val oops =
     |
     |
You typed two blank lines.  Starting a new command.
scala>
```

In the rest of the book, we'll leave out the vertical bars to make the code easier to read (and easier to copy and paste from the PDF eBook into the interpreter).

## Step 3. Define some functions

Now that you've worked with Scala variables, you'll probably want to write some functions. Here's how you do that in Scala:

```
scala> def max(x: Int, y: Int): Int = {
         if (x > y) x
         else y
       }
max: (x: Int, y: Int)Int
```

Function definitions start with def. The function's name, in this case max, is followed by a comma-separated list of parameters in parentheses. A type annotation must follow every function parameter, preceded by a colon, because the Scala compiler (and interpreter, but from now on we'll just say compiler) does not infer function parameter types. In this example, the function named max takes two parameters, x and y, both of type Int. After the close parenthesis of max's parameter list you'll find another ": Int" type annotation. This one defines the *result type* of the max function itself.[6] Following the

---

[6]In Java, the type of the value returned from a method is its return type. In Scala, that same concept is called *result type*.

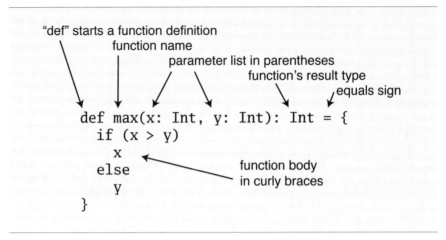

"def" starts a function definition
function name
parameter list in parentheses
function's result type
equals sign

```
def max(x: Int, y: Int): Int = {
    if (x > y)
        x
    else                    function body
        y                   in curly braces
}
```

Figure 2.1 · The basic form of a function definition in Scala.

function's result type is an equals sign and pair of curly braces that contain
the body of the function. In this case, the body contains a single if expression, which selects either x or y, whichever is greater, as the result of the
max function. As demonstrated here, Scala's if expression can result in a
value, similar to Java's ternary operator. For example, the Scala expression
"if (x > y) x else y" behaves similarly to "(x > y) ? x : y" in Java. The
equals sign that precedes the body of a function hints that in the functional
world view, a function defines an expression that results in a value. The basic
structure of a function is illustrated in Figure 2.1.

Sometimes the Scala compiler will require you to specify the result type
of a function. If the function is *recursive*,[7] for example, you must explicitly
specify the function's result type. In the case of max, however, you may leave
the result type off and the compiler will infer it.[8] Also, if a function consists
of just one statement, you can optionally leave off the curly braces. Thus,
you could alternatively write the max function like this:

```
scala> def max(x: Int, y: Int) = if (x > y) x else y
max: (x: Int, y: Int)Int
```

---

[7]A function is recursive if it calls itself.

[8]Nevertheless, it is often a good idea to indicate function result types explicitly, even
when the compiler doesn't require it. Such type annotations can make the code easier to read,
because the reader need not study the function body to figure out the inferred result type.

Once you have defined a function, you can call it by name, as in:

```
scala> max(3, 5)
res4: Int = 5
```

Here's the definition of a function that takes no parameters and returns no interesting result:

```
scala> def greet() = println("Hello, world!")
greet: ()Unit
```

When you define the `greet()` function, the interpreter will respond with `greet: ()Unit`. "greet" is, of course, the name of the function. The empty parentheses indicate the function takes no parameters. And `Unit` is `greet`'s result type. A result type of `Unit` indicates the function returns no interesting value. Scala's `Unit` type is similar to Java's `void` type; in fact, every void-returning method in Java is mapped to a `Unit`-returning method in Scala. Methods with the result type of `Unit`, therefore, are only executed for their side effects. In the case of `greet()`, the side effect is a friendly greeting printed to the standard output.

In the next step, you'll place Scala code in a file and run it as a script. If you wish to exit the interpreter, you can do so by entering `:quit` or `:q`.

```
scala> :quit
$
```

## Step 4. Write some Scala scripts

Although Scala is designed to help programmers build very large-scale systems, it also scales down nicely to scripting. A script is just a sequence of statements in a file that will be executed sequentially. Put this into a file named `hello.scala`:

```
println("Hello, world, from a script!")
```

then run:[9]

---

[9]You can run scripts without typing "scala" on Unix and Windows using a "pound-bang" syntax, which is shown in Appendix A.

```
$ scala hello.scala
```

And you should get yet another greeting:

```
Hello, world, from a script!
```

Command line arguments to a Scala script are available via a Scala array named `args`. In Scala, arrays are zero based, and you access an element by specifying an index in parentheses. So the first element in a Scala array named `steps` is `steps(0)`, not `steps[0]`, as in Java. To try this out, type the following into a new file named `helloarg.scala`:

```
// Say hello to the first argument
println("Hello, " + args(0) + "!")
```

then run:

```
$ scala helloarg.scala planet
```

In this command, `"planet"` is passed as a command line argument, which is accessed in the script as `args(0)`. Thus, you should see:

```
Hello, planet!
```

Note that this script included a comment. The Scala compiler will ignore characters between `//` and the next end of line and any characters between `/*` and `*/`. This example also shows `Strings` being concatenated with the + operator. This works as you'd expect. The expression `"Hello, " + "world!"` will result in the string `"Hello, world!"`.

## Step 5. Loop with `while`; decide with `if`

To try out a `while`, type the following into a file named `printargs.scala`:

```
var i = 0
while (i < args.length) {
  println(args(i))
  i += 1
}
```

> **Note**
> Although the examples in this section help explain while loops, they do
> not demonstrate the best Scala style. In the next section, you'll see better
> approaches that avoid iterating through arrays with indexes.

This script starts with a variable definition, var i = 0. Type inference
gives i the type scala.Int, because that is the type of its initial value, 0.
The while construct on the next line causes the *block* (the code between
the curly braces) to be repeatedly executed until the boolean expression
i < args.length is false. args.length gives the length of the args array.
The block contains two statements, each indented two spaces, the recommended indentation style for Scala. The first statement, println(args(i)),
prints out the ith command line argument. The second statement, i += 1, increments i by one. Note that Java's ++i and i++ don't work in Scala. To
increment in Scala, you need to say either i = i + 1 or i += 1. Run this script
with the following command:

```
$ scala printargs.scala Scala is fun
```

And you should see:

```
Scala
is
fun
```

For even more fun, type the following code into a new file with the name
echoargs.scala:

```
var i = 0
while (i < args.length) {
  if (i != 0)
    print(" ")
  print(args(i))
  i += 1
}
println()
```

In this version, you've replaced the println call with a print call, so that
all the arguments will be printed out on the same line. To make this readable,
you've inserted a single space before each argument except the first via the

if (i != 0) construct. Since i != 0 will be `false` the first time through the `while` loop, no space will get printed before the initial argument. Lastly, you've added one more `println` to the end, to get a line return after printing out all the arguments. Your output will be very pretty indeed. If you run this script with the following command:

```
$ scala echoargs.scala Scala is even more fun
```

You'll get:

```
Scala is even more fun
```

Note that in Scala, as in Java, you must put the boolean expression for a `while` or an `if` in parentheses. (In other words, you can't say in Scala things like if i < 10 as you can in a language such as Ruby. You must say if (i < 10) in Scala.) Another similarity to Java is that if a block has only one statement, you can optionally leave off the curly braces, as demonstrated by the `if` statement in `echoargs.scala`. And although you haven't seen any of them, Scala does use semicolons to separate statements as in Java, except that in Scala the semicolons are very often optional, giving some welcome relief to your right little finger. If you had been in a more verbose mood, therefore, you could have written the `echoargs.scala` script as follows:

```
var i = 0;
while (i < args.length) {
  if (i != 0) {
    print(" ");
  }
  print(args(i));
  i += 1;
}
println();
```

## Step 6. Iterate with `foreach` and `for`

Although you may not have realized it, when you wrote the `while` loops in the previous step, you were programming in an *imperative* style. In the imperative style, which is the style you normally use with languages like Java,

32

C++, and C, you give one imperative command at a time, iterate with loops, and often mutate state shared between different functions. Scala enables you to program imperatively, but as you get to know Scala better, you'll likely often find yourself programming in a more *functional* style. In fact, one of the main aims of this book is to help you become as comfortable with the functional style as you are with imperative style.

One of the main characteristics of a functional language is that functions are first class constructs, and that's very true in Scala. For example, another (far more concise) way to print each command line argument is:

```
args.foreach(arg => println(arg))
```

In this code, you call the foreach method on args and pass in a function. In this case, you're passing in a *function literal* that takes one parameter named arg. The body of the function is println(arg). If you type the above code into a new file named pa.scala and execute with the command:

```
$ scala pa.scala Concise is nice
```

You should see:

```
Concise
is
nice
```

In the previous example, the Scala interpreter infers the type of arg to be String, since String is the element type of the array on which you're calling foreach. If you'd prefer to be more explicit, you can mention the type name. But when you do, you'll need to wrap the argument portion in parentheses (which is the normal form of the syntax anyway):

```
args.foreach((arg: String) => println(arg))
```

Running this script has the same behavior as the previous one.

If you're in the mood for more conciseness instead of more explicitness, you can take advantage of a special shorthand in Scala. If a function literal consists of one statement that takes a single argument, you need not explicitly name and specify the argument.[10] Thus, the following code also works:

---

[10]This shorthand, called a *partially applied function*, is described in Section 8.6.

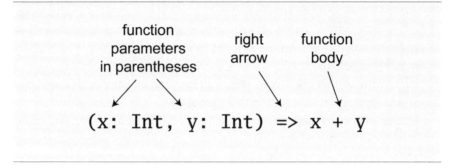

Figure 2.2 · The syntax of a function literal in Scala.

```
args.foreach(println)
```

To summarize, the syntax for a function literal is a list of named parameters, in parentheses, a right arrow, and then the body of the function. This syntax is illustrated in Figure 2.2.

Now, by this point you may be wondering what happened to those trusty for loops you have been accustomed to using in imperative languages, such as Java or C. In an effort to guide you in a functional direction, only a functional relative of the imperative for (called a *for expression*) is available in Scala. While you won't see their full power and expressiveness until you reach (or peek ahead to) Section 7.3, we'll give you a glimpse here. In a new file named forargs.scala, type the following:

```
for (arg <- args)
  println(arg)
```

The parentheses after the "for" contain arg <- args.[11] To the right of the <- symbol is the familiar args array. To the left of <- is "arg", the name of a val, not a var. (Because it is always a val, you just write "arg" by itself, not "val arg".) Although arg may seem to be a var, because it will get a new value on each iteration, it really is a val: arg can't be reassigned inside the body of the for expression. Instead, for each element of the args array, a *new* arg val will be created and initialized to the element value, and the body of the for will be executed.

If you run the forargs.scala script with the command:

---

[11] You can say "in" for the <- symbol. You'd read for (arg <- args), therefore, as "for arg in args."

```
$ scala forargs.scala for arg in args
```

You'll see:

```
for
arg
in
args
```

Scala's for expression can do much more than this, but this example is enough to get you started. We'll show you more about for in Section 7.3 and Chapter 23.

## Conclusion

In this chapter, you learned some Scala basics and, hopefully, took advantage of the opportunity to write a bit of Scala code. In the next chapter, we'll continue this introductory overview and get into more advanced topics.

Chapter 3

# Next Steps in Scala

This chapter continues the previous chapter's introduction to Scala. In this chapter, we'll introduce some more advanced features. When you complete this chapter, you should have enough knowledge to enable you to start writing useful scripts in Scala. As with the previous chapter, we recommend you try out these examples as you go. The best way to get a feel for Scala is to start writing Scala code.

## Step 7. Parameterize arrays with types

In Scala, you can instantiate objects, or class instances, using new. When you instantiate an object in Scala, you can *parameterize* it with values and types. Parameterization means "configuring" an instance when you create it. You parameterize an instance with values by passing objects to a constructor in parentheses. For example, the following Scala code instantiates a new java.math.BigInteger and parameterizes it with the value "12345":

```
val big = new java.math.BigInteger("12345")
```

You parameterize an instance with types by specifying one or more types in square brackets. An example is shown in Listing 3.1. In this example, greetStrings is a value of type Array[String] (an "array of string") that is initialized to length 3 by parameterizing it with the value 3 in the first line of code. If you run the code in Listing 3.1 as a script, you'll see yet another Hello, world! greeting. Note that when you parameterize an instance with both a type and a value, the type comes first in its square brackets, followed by the value in parentheses.

```
val greetStrings = new Array[String](3)

greetStrings(0) = "Hello"
greetStrings(1) = ", "
greetStrings(2) = "world!\n"

for (i <- 0 to 2)
  print(greetStrings(i))
```

Listing 3.1 · Parameterizing an array with a type.

**Note**

Although the code in Listing 3.1 demonstrates important concepts, it does not show the recommended way to create and initialize an array in Scala. You'll see a better way in Listing 3.2 on page 41.

Had you been in a more explicit mood, you could have specified the type of greetStrings explicitly like this:

```
val greetStrings: Array[String] = new Array[String](3)
```

Given Scala's type inference, this line of code is semantically equivalent to the actual first line of Listing 3.1. But this form demonstrates that while the type parameterization portion (the type names in square brackets) forms part of the type of the instance, the value parameterization part (the values in parentheses) does not. The type of greetStrings is Array[String], not Array[String](3).

The next three lines of code in Listing 3.1 initialize each element of the greetStrings array:

```
greetStrings(0) = "Hello"
greetStrings(1) = ", "
greetStrings(2) = "world!\n"
```

As mentioned previously, arrays in Scala are accessed by placing the index inside parentheses, not square brackets as in Java. Thus the zeroth element of the array is greetStrings(0), not greetStrings[0].

These three lines of code illustrate an important concept to understand about Scala concerning the meaning of val. When you define a variable with val, the variable can't be reassigned, but the object to which it refers could potentially still be changed. So in this case, you couldn't reassign greetStrings to a different array; greetStrings will always point to the same Array[String] instance with which it was initialized. But you *can* change the elements of that Array[String] over time, so the array itself is mutable.

The final two lines in Listing 3.1 contain a for expression that prints out each greetStrings array element in turn:

```
for (i <- 0 to 2)
  print(greetStrings(i))
```

The first line of code in this for expression illustrates another general rule of Scala: if a method takes only one parameter, you can call it without a dot or parentheses. The to in this example is actually a method that takes one Int argument. The code 0 to 2 is transformed into the method call (0).to(2).[1] Note that this syntax only works if you explicitly specify the receiver of the method call. You cannot write "println 10", but you can write "Console println 10".

Scala doesn't technically have operator overloading, because it doesn't actually have operators in the traditional sense. Instead, characters such as +, -, *, and / can be used in method names. Thus, when you typed 1 + 2 into the Scala interpreter in Step 1, you were actually invoking a method named + on the Int object 1, passing in 2 as a parameter. As illustrated in Figure 3.1, you could alternatively have written 1 + 2 using traditional method invocation syntax, (1).+(2).

Another important idea illustrated by this example will give you insight into why arrays are accessed with parentheses in Scala. Scala has fewer special cases than Java. Arrays are simply instances of classes like any other class in Scala. When you apply parentheses surrounding one or more values to a variable, Scala will transform the code into an invocation of a method named apply on that variable. So greetStrings(i) gets transformed into greetStrings.apply(i). Thus accessing an element of an array in Scala is simply a method call like any other. This principle is not restricted to arrays: any application of an object to some arguments in parentheses will be transformed to an apply method call. Of course this will compile only

---

[1]This to method actually returns not an array but a different kind of sequence, containing the values 0, 1, and 2, which the for expression iterates over. Sequences and other collections will be described in Chapter 17.

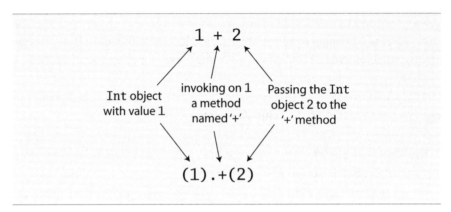

Figure 3.1 · All operations are method calls in Scala.

if that type of object actually defines an `apply` method. So it's not a special case; it's a general rule.

Similarly, when an assignment is made to a variable to which parentheses and one or more arguments have been applied, the compiler will transform that into an invocation of an `update` method that takes the arguments in parentheses as well as the object to the right of the equals sign. For example:

```
greetStrings(0) = "Hello"
```

will be transformed into:

```
greetStrings.update(0, "Hello")
```

Thus, the following is semantically equivalent to the code in Listing 3.1:

```
val greetStrings = new Array[String](3)
greetStrings.update(0, "Hello")
greetStrings.update(1, ", ")
greetStrings.update(2, "world!\n")
for (i <- 0.to(2))
  print(greetStrings.apply(i))
```

Scala achieves a conceptual simplicity by treating everything, from arrays to expressions, as objects with methods. You don't have to remember special cases, such as the differences in Java between primitive and their corresponding wrapper types, or between arrays and regular objects. Moreover, this uniformity does not incur a significant performance cost. The Scala compiler uses Java arrays, primitive types, and native arithmetic where possible in the compiled code.

Although the examples you've seen so far in this step compile and run just fine, Scala provides a more concise way to create and initialize arrays that you would normally use (see Listing 3.2). This code creates a new array of length three, initialized to the passed strings, "zero", "one", and "two". The compiler infers the type of the array to be Array[String], because you passed strings to it.

```
val numNames = Array("zero", "one", "two")
```

Listing 3.2 · Creating and initializing an array.

What you're actually doing in Listing 3.2 is calling a factory method, named apply, which creates and returns the new array. This apply method takes a variable number of arguments[2] and is defined on the Array *companion object*. You'll learn more about companion objects in Section 4.3. If you're a Java programmer, you can think of this as calling a static method named apply on class Array. A more verbose way to call the same apply method is:

```
val numNames2 = Array.apply("zero", "one", "two")
```

## Step 8. Use lists

One of the big ideas of the functional style of programming is that methods should not have side effects. A method's only act should be to compute and return a value. Some benefits gained when you take this approach are that methods become less entangled, and therefore more reliable and reusable. Another benefit (in a statically typed language) is that everything that goes into and out of a method is checked by a type checker, so logic errors are more likely to manifest themselves as type errors. Applying this functional philosophy to the world of objects means making objects immutable.

As you've seen, a Scala array is a mutable sequence of objects that all share the same type. An Array[String] contains only strings, for example.

---

[2]Variable-length argument lists, or *repeated parameters*, are described in Section 8.8.

Although you can't change the length of an array after it is instantiated, you can change its element values. Thus, arrays are mutable objects.

For an immutable sequence of objects that share the same type you can use Scala's List class. As with arrays, a List[String] contains only strings. Scala's List, scala.List, differs from Java's java.util.List type in that Scala Lists are always immutable (whereas Java Lists can be mutable). More generally, Scala's List is designed to enable a functional style of programming. Creating a list is easy, and Listing 3.3 shows how:

```
val oneTwoThree = List(1, 2, 3)
```

Listing 3.3 · Creating and initializing a list.

The code in Listing 3.3 establishes a new val named oneTwoThree, initialized with a new List[Int] with the integer elements 1, 2, and 3.[3] Because Lists are immutable, they behave a bit like Java strings: when you call a method on a list that might seem by its name to imply the list will mutate, it instead creates and returns a new list with the new value. For example, List has a method named ':::' for list concatenation. Here's how you use it:

```
val oneTwo = List(1, 2)
val threeFour = List(3, 4)
val oneTwoThreeFour = oneTwo ::: threeFour
println(oneTwo + " and " + threeFour + " were not mutated.")
println("Thus, " + oneTwoThreeFour + " is a new list.")
```

If you run this script, you'll see:

```
List(1, 2) and List(3, 4) were not mutated.
Thus, List(1, 2, 3, 4) is a new list.
```

Perhaps the most common operator you'll use with lists is '::', which is pronounced "cons." Cons prepends a new element to the beginning of an existing list and returns the resulting list. For example, if you run this script:

```
val twoThree = List(2, 3)
val oneTwoThree = 1 :: twoThree
println(oneTwoThree)
```

---

[3]You don't need to say new List because "List.apply()" is defined as a factory method on the scala.List *companion object*. You'll read more on companion objects in Section 4.3.

You'll see:

```
List(1, 2, 3)
```

> **Note**
>
> In the expression "1 :: twoThree", :: is a method of its *right* operand,
> the list, twoThree. You might suspect there's something amiss with the
> associativity of the :: method, but it is actually a simple rule to
> remember: If a method is used in operator notation, such as a * b, the
> method is invoked on the left operand, as in a.*(b)—unless the method
> name ends in a colon. If the method name ends in a colon, the method is
> invoked on the *right* operand. Therefore, in 1 :: twoThree, the :: method
> is invoked on twoThree, passing in 1, like this: twoThree.::(1).
> Operator associativity will be described in more detail in Section 5.9.

Given that a shorthand way to specify an empty list is Nil, one way to
initialize new lists is to string together elements with the cons operator, with
Nil as the last element.[4] For example, the following script will produce the
same output as the previous one, "List(1, 2, 3)":

```
val oneTwoThree = 1 :: 2 :: 3 :: Nil
println(oneTwoThree)
```

Scala's List is packed with useful methods, many of which are shown in
Table 3.1. The full power of lists will be revealed in Chapter 16.

---

**Why not append to lists?**

Class List does offer an "append" operation—it's written :+ and is
explained in Chapter 24—but this operation is rarely used, because
the time it takes to append to a list grows linearly with the size of the
list, whereas prepending with :: takes constant time. If you want to
build a list efficiently by appending elements, you can prepend them
and when you're done call reverse. Or you can use a ListBuffer, a
mutable list that does offer an append operation, and when you're done
call toList. ListBuffer will be described in Section 22.2.

---

[4]The reason you need Nil at the end is that :: is defined on class List. If you try to just
say 1 :: 2 :: 3, it won't compile because 3 is an Int, which doesn't have a :: method.

Table 3.1 · Some `List` methods and usages

| What it is | What it does |
|---|---|
| `List()` or `Nil` | The empty List |
| `List("Cool", "tools", "rule")` | Creates a new `List[String]` with the three values `"Cool"`, `"tools"`, and `"rule"` |
| `val thrill = "Will" :: "fill" ::`<br>`"until" :: Nil` | Creates a new `List[String]` with the three values `"Will"`, `"fill"`, and `"until"` |
| `List("a", "b") ::: List("c", "d")` | Concatenates two lists (returns a new `List[String]` with values `"a"`, `"b"`, `"c"`, and `"d"`) |
| `thrill(2)` | Returns the element at index 2 (zero based) of the `thrill` list (returns `"until"`) |
| `thrill.count(s => s.length == 4)` | Counts the number of string elements in `thrill` that have length 4 (returns 2) |
| `thrill.drop(2)` | Returns the `thrill` list without its first 2 elements (returns `List("until")`) |
| `thrill.dropRight(2)` | Returns the `thrill` list without its rightmost 2 elements (returns `List("Will")`) |
| `thrill.exists(s => s == "until")` | Determines whether a string element exists in `thrill` that has the value `"until"` (returns `true`) |
| `thrill.filter(s => s.length == 4)` | Returns a list of all elements, in order, of the `thrill` list that have length 4 (returns `List("Will", "fill")`) |
| `thrill.forall(s =>`<br>`s.endsWith("l"))` | Indicates whether all elements in the `thrill` list end with the letter `"l"` (returns `true`) |
| `thrill.foreach(s => print(s))` | Executes the `print` statement on each of the strings in the `thrill` list (prints `"Willfilluntil"`) |

44

## Table 3.1 · continued

| | |
|---|---|
| thrill.foreach(print) | Same as the previous, but more concise (also prints "Willfilluntil") |
| thrill.head | Returns the first element in the thrill list (returns "Will") |
| thrill.init | Returns a list of all but the last element in the thrill list (returns List("Will", "fill")) |
| thrill.isEmpty | Indicates whether the thrill list is empty (returns false) |
| thrill.last | Returns the last element in the thrill list (returns "until") |
| thrill.length | Returns the number of elements in the thrill list (returns 3) |
| thrill.map(s => s + "y") | Returns a list resulting from adding a "y" to each string element in the thrill list (returns List("Willy", "filly", "untily")) |
| thrill.mkString(", ") | Makes a string with the elements of the list (returns "Will, fill, until") |
| thrill.filterNot(s => s.length == 4) | Returns a list of all elements, in order, of the thrill list *except those* that have length 4 (returns List("until")) |
| thrill.reverse | Returns a list containing all elements of the thrill list in reverse order (returns List("until", "fill", "Will")) |
| thrill.sort((s, t) => s.charAt(0).toLower < t.charAt(0).toLower) | Returns a list containing all elements of the thrill list in alphabetical order of the first character lowercased (returns List("fill", "until", "Will")) |
| thrill.tail | Returns the thrill list minus its first element (returns List("fill", "until")) |

45

# Step 9. Use tuples

Another useful container object is the *tuple*. Like lists, tuples are immutable, but unlike lists, tuples can contain different types of elements. Whereas a list might be a List[Int] or a List[String], a tuple could contain both an integer and a string at the same time. Tuples are very useful, for example, if you need to return multiple objects from a method. Whereas in Java you would often create a JavaBean-like class to hold the multiple return values, in Scala you can simply return a tuple. And it is simple: To instantiate a new tuple that holds some objects, just place the objects in parentheses, separated by commas. Once you have a tuple instantiated, you can access its elements individually with a dot, underscore, and the one-based index of the element. An example is shown in Listing 3.4:

```
val pair = (99, "Luftballons")
println(pair._1)
println(pair._2)
```

Listing 3.4 · Creating and using a tuple.

In the first line of Listing 3.4, you create a new tuple that contains the integer 99, as its first element, and the string, "Luftballons", as its second element. Scala infers the type of the tuple to be Tuple2[Int, String], and gives that type to the variable pair as well. In the second line, you access the _1 field, which will produce the first element, 99. The "." in the second line is the same dot you'd use to access a field or invoke a method. In this case you are accessing a field named _1. If you run this script, you'll see:

```
99
Luftballons
```

The actual type of a tuple depends on the number of elements it contains and the types of those elements. Thus, the type of (99, "Luftballons") is Tuple2[Int, String]. The type of ('u', 'r', "the", 1, 4, "me") is Tuple6[Char, Char, String, Int, Int, String].[5]

---

[5]Although conceptually you could create tuples of any length, currently the Scala library only defines them up to Tuple22.

### Accessing the elements of a tuple

You may be wondering why you can't access the elements of a tuple
like the elements of a list, for example, with "pair(0)". The reason
is that a list's apply method always returns the same type, but each
element of a tuple may be a different type: _1 can have one result type,
_2 another, and so on. These _N numbers are one-based, instead of
zero-based, because starting with 1 is a tradition set by other languages
with statically typed tuples, such as Haskell and ML.

## Step 10. Use sets and maps

Because Scala aims to help you take advantage of both functional and imperative styles, its collections libraries make a point to differentiate between
mutable and immutable collections. For example, arrays are always mutable; lists are always immutable. Scala also provides mutable and immutable
alternatives for sets and maps, but uses the same simple names for both versions. For sets and maps, Scala models mutability in the class hierarchy.

For example, the Scala API contains a base *trait* for sets, where a trait
is similar to a Java interface. (You'll find out more about traits in Chapter 12.) Scala then provides two subtraits, one for mutable sets and another
for immutable sets.

As you can see in Figure 3.2, these three traits all share the same simple name, Set. Their fully qualified names differ, however, because each
resides in a different package. Concrete set classes in the Scala API, such
as the HashSet classes shown in Figure 3.2, extend either the mutable or
immutable Set trait. (Although in Java you "implement" interfaces, in Scala
you "extend" or "mix in" traits.) Thus, if you want to use a HashSet, you
can choose between mutable and immutable varieties depending upon your
needs. The default way to create a set is shown in Listing 3.5:

```
var jetSet = Set("Boeing", "Airbus")
jetSet += "Lear"
println(jetSet.contains("Cessna"))
```

Listing 3.5 · Creating, initializing, and using an immutable set.

47

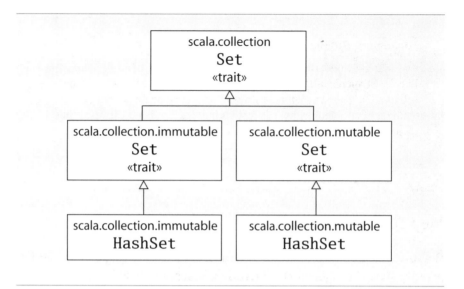

Figure 3.2 · Class hierarchy for Scala sets.

In the first line of code in Listing 3.5, you define a new var named jetSet and initialize it with an immutable set containing the two strings, "Boeing" and "Airbus". As this example shows, you can create sets in Scala similarly to how you create lists and arrays: by invoking a factory method named apply on a Set companion object. In Listing 3.5, you invoke apply on the companion object for scala.collection.immutable.Set, which returns an instance of a default, immutable Set. The Scala compiler infers jetSet's type to be the immutable Set[String].

To add a new element to a set, you call + on the set, passing in the new element. On both mutable and immutable sets, the + method will create and return a new set with the element added. In Listing 3.5, you're working with an immutable set. Although mutable sets offer an actual += method, immutable sets do not.

In this case, the second line of code, "jetSet += "Lear"", is essentially a shorthand for:

```
jetSet = jetSet + "Lear"
```

Thus, in the second line of Listing 3.5, you reassign the jetSet var with a new set containing "Boeing", "Airbus", and "Lear". Finally, the last line of Listing 3.5 prints out whether or not the set contains the string "Cessna". (As you'd expect, it prints false.)

If you want a mutable set, you'll need to use an *import*, as shown in Listing 3.6:

```
import scala.collection.mutable

val movieSet = mutable.Set("Hitch", "Poltergeist")
movieSet += "Shrek"
println(movieSet)
```

Listing 3.6 · Creating, initializing, and using a mutable set.

In the first line of Listing 3.6 you import the mutable Set. As with Java, an import statement allows you to use a simple name, such as Set, instead of the longer, fully qualified name. As a result, when you say Set on the third line, the compiler knows you mean scala.collection.mutable.Set. On that line, you initialize movieSet with a new mutable set that contains the strings "Hitch" and "Poltergeist". The subsequent line adds "Shrek" to the mutable set by calling the += method on the set, passing in the string "Shrek". As mentioned previously, += is an actual method defined on mutable sets. Had you wanted to, instead of writing movieSet += "Shrek", you could have written movieSet.+=("Shrek").[6]

Although the default set implementations produced by the mutable and immutable Set factory methods shown thus far will likely be sufficient for most situations, occasionally you may want an explicit set class. Fortunately, the syntax is similar. Simply import that class you need, and use the factory method on its companion object. For example, if you need an immutable HashSet, you could do this:

```
import scala.collection.immutable.HashSet

val hashSet = HashSet("Tomatoes", "Chilies")
println(hashSet + "Coriander")
```

Another useful collection class in Scala is Map. As with sets, Scala provides mutable and immutable versions of Map, using a class hierarchy. As

---

[6]Because the set in Listing 3.6 is mutable, there is no need to reassign movieSet, which is why it can be a val. By contrast, using += with the immutable set in Listing 3.5 required reassigning jetSet, which is why it must be a var.

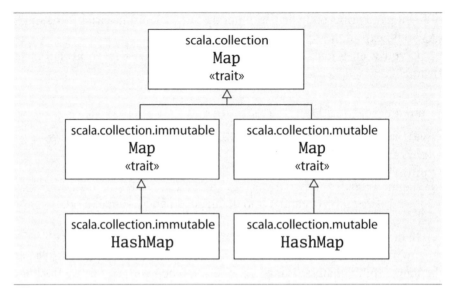

Figure 3.3 · Class hierarchy for Scala maps.

you can see in Figure 3.3, the class hierarchy for maps looks a lot like the one for sets. There's a base Map trait in package `scala.collection`, and two subtrait Maps: a mutable Map in `scala.collection.mutable` and an immutable one in `scala.collection.immutable`.

Implementations of Map, such as the HashMaps shown in the class hierarchy in Figure 3.3, extend either the mutable or immutable trait. You can create and initialize maps using factory methods similar to those used for arrays, lists, and sets.

```
import scala.collection.mutable

val treasureMap = mutable.Map[Int, String]()
treasureMap += (1 -> "Go to island.")
treasureMap += (2 -> "Find big X on ground.")
treasureMap += (3 -> "Dig.")
println(treasureMap(2))
```

Listing 3.7 · Creating, initializing, and using a mutable map.

For example, Listing 3.7 shows a mutable map in action. On the first line of Listing 3.7, you import the mutable `Map`. You then define a `val` named `treasureMap`, and initialize it with an empty mutable `Map` that has integer keys and string values. The map is empty because you pass nothing to the factory method (the parentheses in "`Map[Int, String]()`" are empty).[7] On the next three lines you add key/value pairs to the map using the `->` and `+=` methods. As illustrated previously, the Scala compiler transforms a binary operation expression like `1 -> "Go to island."` into `(1).->("Go to island.")`. Thus, when you say `1 -> "Go to island."`, you are actually calling a method named `->` on an integer with the value 1, passing in a string with the value `"Go to island."` This `->` method, which you can invoke on any object in a Scala program, returns a two-element tuple containing the key and value.[8] You then pass this tuple to the `+=` method of the map object to which `treasureMap` refers. Finally, the last line prints the value that corresponds to the key 2 in the `treasureMap`.

If you run this code, it will print:

```
Find big X on ground.
```

If you prefer an immutable map, no import is necessary, as immutable is the default map. An example is shown in Listing 3.8:

```
val romanNumeral = Map(
  1 -> "I", 2 -> "II", 3 -> "III", 4 -> "IV", 5 -> "V"
)
println(romanNumeral(4))
```

Listing 3.8 · Creating, initializing, and using an immutable map.

Given there are no imports, when you say `Map` in the first line of Listing 3.8, you'll get the default: a `scala.collection.immutable.Map`. You pass five key/value tuples to the map's factory method, which returns an immutable `Map` containing the passed key/value pairs. If you run the code in Listing 3.8 it will print "`IV`".

---

[7]The explicit type parameterization, "`[Int, String]`", is required in Listing 3.7 because without any values passed to the factory method, the compiler is unable to infer the map's type parameters. By contrast, the compiler can infer the type parameters from the values passed to the map factory shown in Listing 3.8, thus no explicit type parameters are needed.

[8]The Scala mechanism that allows you to invoke `->` on any object, *implicit conversion*, will be covered in Chapter 21.

## Step 11. Learn to recognize the functional style

As mentioned in Chapter 1, Scala allows you to program in an imperative style, but encourages you to adopt a more functional style. If you are coming to Scala from an imperative background—for example, if you are a Java programmer—one of the main challenges you may face when learning Scala is figuring out how to program in the functional style. We realize this style might be unfamiliar at first, and in this book we try hard to guide you through the transition. It will require some work on your part, and we encourage you to make the effort. If you come from an imperative background, we believe that learning to program in a functional style will not only make you a better Scala programmer, it will expand your horizons and make you a better programmer in general.

The first step is to recognize the difference between the two styles in code. One telltale sign is that if code contains any vars, it is probably in an imperative style. If the code contains no vars at all—*i.e.*, it contains *only* vals—it is probably in a functional style. One way to move towards a functional style, therefore, is to try to program without vars.

If you're coming from an imperative background, such as Java, C++, or C#, you may think of var as a regular variable and val as a special kind of variable. On the other hand, if you're coming from a functional background, such as Haskell, OCaml, or Erlang, you might think of val as a regular variable and var as akin to blasphemy. The Scala perspective, however, is that val and var are just two different tools in your toolbox, both useful, neither inherently evil. Scala encourages you to lean towards vals, but ultimately reach for the best tool given the job at hand. Even if you agree with this balanced philosophy, however, you may still find it challenging at first to figure out how to get rid of vars in your code.

Consider the following while loop example, adapted from Chapter 2, which uses a var and is therefore in the imperative style:

```scala
def printArgs(args: Array[String]): Unit = {
  var i = 0
  while (i < args.length) {
    println(args(i))
    i += 1
  }
}
```

You can transform this bit of code into a more functional style by getting rid of the var, for example, like this:

```
def printArgs(args: Array[String]): Unit = {
  for (arg <- args)
    println(arg)
}
```

or this:

```
def printArgs(args: Array[String]): Unit = {
  args.foreach(println)
}
```

This example illustrates one benefit of programming with fewer vars. The refactored (more functional) code is clearer, more concise, and less error-prone than the original (more imperative) code. The reason Scala encourages a functional style is that it can help you write more understandable, less error-prone code.

But you can go even further. The refactored printArgs method is not *purely* functional because it has side effects—in this case, its side effect is printing to the standard output stream. The telltale sign of a function with side effects is that its result type is Unit. If a function isn't returning any interesting value, which is what a result type of Unit means, the only way that function can make a difference in the world is through some kind of side effect. A more functional approach would be to define a method that formats the passed args for printing, but just returns the formatted string, as shown in Listing 3.9:

```
def formatArgs(args: Array[String]) = args.mkString("\n")
```

Listing 3.9 · A function without side effects or vars.

Now you're really functional: no side effects or vars in sight. The mkString method, which you can call on any iterable collection (including arrays, lists, sets, and maps), returns a string consisting of the result of calling toString on each element, separated by the passed string. Thus if args contains three elements "zero", "one", and "two", formatArgs will return "zero\none\ntwo". Of course, this function doesn't actually print anything out like the printArgs methods did, but you can easily pass its result to println to accomplish that:

```
println(formatArgs(args))
```

Every useful program is likely to have side effects of some form; otherwise, it wouldn't be able to provide value to the outside world. Preferring methods without side effects encourages you to design programs where side-effecting code is minimized. One benefit of this approach is that it can help make your programs easier to test.

For example, to test any of the three printArgs methods shown earlier in this section, you'd need to redefine println, capture the output passed to it, and make sure it is what you expect. By contrast, you could test the formatArgs function simply by checking its result:

```
val res = formatArgs(Array("zero", "one", "two"))
assert(res == "zero\none\ntwo")
```

Scala's assert method checks the passed Boolean and if it is false, throws AssertionError. If the passed Boolean is true, assert just returns quietly. You'll learn more about assertions and tests in Chapter 14.

That said, bear in mind that neither vars nor side effects are inherently evil. Scala is not a pure functional language that forces you to program everything in the functional style. Scala is a hybrid imperative/functional language. You may find that in some situations an imperative style is a better fit for the problem at hand, and in such cases you should not hesitate to use it. To help you learn how to program without vars, however, we'll show you many specific examples of code with vars and how to transform those vars to vals in Chapter 7.

---

### A balanced attitude for Scala programmers

Prefer vals, immutable objects, and methods without side effects. Reach for them first. Use vars, mutable objects, and methods with side effects when you have a specific need and justification for them.

---

# Step 12. Read lines from a file

Scripts that perform small, everyday tasks often need to process lines in files. In this section, you'll build a script that reads lines from a file and prints them out prepended with the number of characters in each line. The first version is shown in Listing 3.10:

```
import scala.io.Source

if (args.length > 0) {

  for (line <- Source.fromFile(args(0)).getLines())
    println(line.length + " " + line)
}
else
  Console.err.println("Please enter filename")
```

Listing 3.10 · Reading lines from a file.

This script starts with an import of a class named Source from package scala.io. It then checks to see if at least one argument was specified on the command line. If so, the first argument is interpreted as a filename to open and process. The expression Source.fromFile(args(0)) attempts to open the specified file and returns a Source object, on which you call getLines. The getLines method returns an Iterator[String], which provides one line on each iteration, excluding the end-of-line character. The for expression iterates through these lines and prints for each the length of the line, a space, and the line itself. If there were no arguments supplied on the command line, the final else clause will print a message to the standard error stream. If you place this code in a file named countchars1.scala, and run it on itself with:

```
$ scala countchars1.scala countchars1.scala
```

You should see:

```
22 import scala.io.Source
0
22 if (args.length > 0) {
0
51   for (line <- Source.fromFile(args(0)).getLines())
```

55

```
37     println(line.length + " " + line)
1 }
4 else
46   Console.err.println("Please enter filename")
```

Although the script in its current form prints out the needed information, you may wish to line up the numbers, right adjusted, and add a pipe character, so that the output looks instead like:

```
22 | import scala.io.Source
 0 |
22 | if (args.length > 0) {
 0 |
51 |   for (line <- Source.fromFile(args(0)).getLines())
37 |     println(line.length + " " + line)
 1 | }
 4 | else
46 |   Console.err.println("Please enter filename")
```

To accomplish this, you can iterate through the lines twice. The first time through you'll determine the maximum width required by any line's character count. The second time through you'll print the output, using the maximum width calculated previously. Because you'll be iterating through the lines twice, you may as well assign them to a variable:

```
val lines = Source.fromFile(args(0)).getLines().toList
```

The final toList is required because the getLines method returns an iterator. Once you've iterated through an iterator, it is spent. By transforming it into a list via the toList call, you gain the ability to iterate as many times as you wish, at the cost of storing all lines from the file in memory at once. The lines variable, therefore, references a list of strings that contains the contents of the file specified on the command line. Next, because you'll be calculating the width of each line's character count twice, once per iteration, you might factor that expression out into a small function, which calculates the character width of the passed string's length:

```
def widthOfLength(s: String) = s.length.toString.length
```

With this function, you could calculate the maximum width like this:

```
var maxWidth = 0
for (line <- lines)
  maxWidth = maxWidth.max(widthOfLength(line))
```

Here you iterate through each line with a for expression, calculate the character width of that line's length, and, if it is larger than the current maximum, assign it to maxWidth, a var that was initialized to 0. (The max method, which you can invoke on any Int, returns the greater of the value on which it was invoked and the value passed to it.) Alternatively, if you prefer to find the maximum without vars, you could first find the longest line like this:

```
val longestLine = lines.reduceLeft(
  (a, b) => if (a.length > b.length) a else b
)
```

The reduceLeft method applies the passed function to the first two elements in lines, then applies it to the result of the first application and the next element in lines, and so on, all the way through the list. On each such application, the result will be the longest line encountered so far because the passed function, (a, b) => if (a.length > b.length) a else b, returns the longest of the two passed strings. "reduceLeft" will return the result of the last application of the function, which in this case will be the longest string element contained in lines.

Given this result, you can calculate the maximum width by passing the longest line to widthOfLength:

```
val maxWidth = widthOfLength(longestLine)
```

All that remains is to print out the lines with proper formatting. You can do that like this:

```
for (line <- lines) {
  val numSpaces = maxWidth - widthOfLength(line)
  val padding = " " * numSpaces
  println(padding + line.length + " | " + line)
}
```

In this for expression, you once again iterate through the lines. For each line, you first calculate the number of spaces required before the line length and assign it to numSpaces. Then you create a string containing numSpaces

spaces with the expression " " * numSpaces. Finally, you print out the information with the desired formatting. The entire script looks as shown in Listing 3.11:

```scala
import scala.io.Source

def widthOfLength(s: String) = s.length.toString.length

if (args.length > 0) {

  val lines = Source.fromFile(args(0)).getLines().toList

  val longestLine = lines.reduceLeft(
    (a, b) => if (a.length > b.length) a else b
  )

  val maxWidth = widthOfLength(longestLine)

  for (line <- lines) {
    val numSpaces = maxWidth - widthOfLength(line)
    val padding = " " * numSpaces
    println(padding + line.length + " | " + line)
  }
}
else
  Console.err.println("Please enter filename")
```

Listing 3.11 · Printing formatted character counts for the lines of a file.

## Conclusion

With the knowledge you've gained in this chapter, you should be able to start using Scala for small tasks, especially scripts. In later chapters, we will dive further into these topics and introduce other topics that weren't even hinted at here.

# Chapter 4

# Classes and Objects

You've now seen the basics of classes and objects in Scala from the previous two chapters. In this chapter, we'll take you a bit deeper. You'll learn more about classes, fields, and methods, and get an overview of semicolon inference. We'll discuss singleton objects, including how to use them to write and run a Scala application. If you are familiar with Java, you'll find that the concepts in Scala are similar, but not exactly the same. So even if you're a Java guru, it will pay to read on.

## 4.1 Classes, fields, and methods

A class is a blueprint for objects. Once you define a class, you can create objects from the class blueprint with the keyword new. For example, given the class definition:

```
class ChecksumAccumulator {
  // class definition goes here
}
```

You can create ChecksumAccumulator objects with:

```
new ChecksumAccumulator
```

Inside a class definition, you place fields and methods, which are collectively called *members*. Fields, which you define with either val or var, are variables that refer to objects. Methods, which you define with def, contain executable code. The fields hold the state, or data, of an object, whereas the methods use that data to do the computational work of the object. When you instantiate a class, the runtime sets aside some memory to hold the image of that object's state (*i.e.*, the content of its variables). For example, if you defined a ChecksumAccumulator class and gave it a var field named sum:

```
class ChecksumAccumulator {
  var sum = 0
}
```

and you instantiated it twice with:

```
val acc = new ChecksumAccumulator
val csa = new ChecksumAccumulator
```

The image of the objects in memory might look like this:

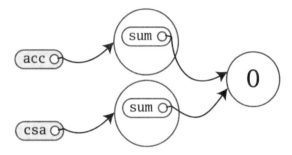

Since sum, a field declared inside class ChecksumAccumulator, is a var, not a val, you can later reassign to sum a different Int value, like this:

```
acc.sum = 3
```

Now the picture would look like this:

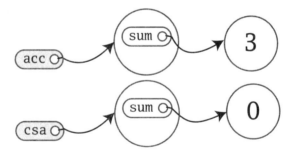

One thing to notice about this picture is that there are two sum variables, one in the object referenced by acc and the other in the object referenced by csa. Fields are also known as *instance variables*, because every instance gets its own set of the variables. Collectively, an object's instance variables make up the memory image of the object. You see this illustrated here not only in that you see two sum variables, but also that when you changed one, the other was unaffected.

Another thing to note in this example is that you were able to mutate the object acc referred to, even though acc is a val. What you can't do with acc (or csa), given that they are vals, not vars, is reassign a different object to them. For example, the following attempt would fail:

```
// Won't compile, because acc is a val
acc = new ChecksumAccumulator
```

What you can count on, therefore, is that acc will always refer to the same ChecksumAccumulator object with which you initialize it, but the fields contained inside that object might change over time.

One important way to pursue robustness of an object is to ensure that the object's state—the values of its instance variables—remains valid during its entire lifetime. The first step is to prevent outsiders from accessing the fields directly by making the fields *private*. Because private fields can only be accessed by methods defined in the same class, all the code that can update the state will be localized to the class. To declare a field private, you place a private access modifier in front of the field, like this:

```
class ChecksumAccumulator {
  private var sum = 0
}
```

Given this definition of ChecksumAccumulator, any attempt to access sum from the outside of the class would fail:

```
val acc = new ChecksumAccumulator
acc.sum = 5 // Won't compile, because sum is private
```

**Note**
The way you make members public in Scala is by not explicitly specifying any access modifier. Put another way, where you'd say "public" in Java, you simply say nothing in Scala. Public is Scala's default access level.

Now that sum is private, the only code that can access sum is code defined inside the body of the class itself. Thus, ChecksumAccumulator won't be of much use to anyone unless we define some methods in it:

```
class ChecksumAccumulator {
  private var sum = 0
  def add(b: Byte): Unit = {
    sum += b
  }
  def checksum(): Int = {
    return ~(sum & 0xFF) + 1
  }
}
```

The ChecksumAccumulator now has two methods, add and checksum, both of which exhibit the basic form of a function definition, shown in Figure 2.1 on page 28.

Any parameters to a method can be used inside the method. One important characteristic of method parameters in Scala is that they are vals, not vars.[1] If you attempt to reassign a parameter inside a method in Scala, therefore, it won't compile:

```
def add(b: Byte): Unit = {
  b = 1      // This won't compile, because b is a val
  sum += b
}
```

Although add and checksum in this version of ChecksumAccumulator correctly implement the desired functionality, you can express them using a more concise style. First, the return at the end of the checksum method is superfluous and can be dropped. In the absence of any explicit return statement, a Scala method returns the last value computed by the method.

The recommended style for methods is in fact to avoid having explicit, and especially multiple, return statements. Instead, think of each method as an expression that yields one value, which is returned. This philosophy will encourage you to make methods quite small, to factor larger methods

---

[1]The reason parameters are vals is that vals are easier to reason about. You needn't look further to determine if a val is reassigned, as you must do with a var.

into multiple smaller ones. On the other hand, design choices depend on the design context, and Scala makes it easy to write methods that have multiple, explicit returns if that's what you desire.

Because all checksum does is calculate a value, it does not need an explicit return. Another shorthand for methods is that you can leave off the curly braces if a method computes only a single result expression. If the result expression is short, it can even be placed on the same line as the def itself. For the utmost in conciseness, you can leave off the result type and Scala will infer it. With these changes, class ChecksumAccumulator looks like this:

```scala
class ChecksumAccumulator {
  private var sum = 0
  def add(b: Byte) = sum += b
  def checksum() = ~(sum & 0xFF) + 1
}
```

Although the Scala compiler will correctly infer the result types of the add and checksum methods shown in the previous example, readers of the code will also need to *mentally infer* the result types by studying the bodies of the methods. As a result it is often better to explicitly provide the result types of public methods declared in a class even when the compiler would infer it for you. Listing 4.1 shows this style.

---

```scala
// In file ChecksumAccumulator.scala
class ChecksumAccumulator {
  private var sum = 0
  def add(b: Byte): Unit = { sum += b }
  def checksum(): Int = ~(sum & 0xFF) + 1
}
```

---

Listing 4.1 · Final version of class ChecksumAccumulator.

Methods with a result type of Unit, such as ChecksumAccumulator's add method, are executed for their side effects. A side effect is generally defined as mutating state somewhere external to the method or performing an I/O action. In add's case, the side effect is that sum is reassigned. A method that is executed only for its side effects is known as a *procedure*.

63

## 4.2 Semicolon inference

In a Scala program, a semicolon at the end of a statement is usually optional. You can type one if you want but you don't have to if the statement appears by itself on a single line. On the other hand, a semicolon is required if you write multiple statements on a single line:

```
val s = "hello"; println(s)
```

If you want to enter a statement that spans multiple lines, most of the time you can simply enter it and Scala will separate the statements in the correct place. For example, the following is treated as one four-line statement:

```
if (x < 2)
  println("too small")
else
  println("ok")
```

Occasionally, however, Scala will split a statement into two parts against your wishes:

```
x
+ y
```

This parses as two statements x and +y. If you intend it to parse as one statement x + y, you can always wrap it in parentheses:

```
(x
+ y)
```

Alternatively, you can put the + at the end of a line. For just this reason, whenever you are chaining an infix operation such as +, it is a common Scala style to put the operators at the end of the line instead of the beginning:

```
x +
y +
z
```

---

### The rules of semicolon inference

The precise rules for statement separation are surprisingly simple for how well they work. In short, a line ending is treated as a semicolon unless one of the following conditions is true:

1. The line in question ends in a word that would not be legal as the end of a statement, such as a period or an infix operator.

2. The next line begins with a word that cannot start a statement.

3. The line ends while inside parentheses (...) or brackets [...], because these cannot contain multiple statements anyway.

---

## 4.3 Singleton objects

As mentioned in Chapter 1, one way in which Scala is more object-oriented than Java is that classes in Scala cannot have static members. Instead, Scala has *singleton objects*. A singleton object definition looks like a class definition, except instead of the keyword `class` you use the keyword `object`. Listing 4.2 shows an example.

The singleton object in this figure is named `ChecksumAccumulator`, the same name as the class in the previous example. When a singleton object shares the same name with a class, it is called that class's *companion object*. You must define both the class and its companion object in the same source file. The class is called the *companion class* of the singleton object. A class and its companion object can access each other's private members.

The `ChecksumAccumulator` singleton object has one method, named `calculate`, which takes a `String` and calculates a checksum for the characters in the `String`. It also has one private field, `cache`, a mutable map in which previously calculated checksums are cached.[2] The first line of the method, "`if (cache.contains(s))`", checks the cache to see if the passed string is already contained as a key in the map. If so, it just returns the mapped value, `cache(s)`. Otherwise, it executes the else clause, which cal-

---

[2]We used a cache here to show a singleton object with a field. A cache such as this is a performance optimization that trades off memory for computation time. In general, you would likely use such a cache only if you encountered a performance problem that the cache solves, and might use a weak map, such as `WeakHashMap` in `scala.collection.jcl`, so that entries in the cache could be garbage collected if memory becomes scarce.

```
// In file ChecksumAccumulator.scala
import scala.collection.mutable

object ChecksumAccumulator {

  private val cache = mutable.Map.empty[String, Int]

  def calculate(s: String): Int =
    if (cache.contains(s))
      cache(s)
    else {
      val acc = new ChecksumAccumulator
      for (c <- s)
        acc.add(c.toByte)
      val cs = acc.checksum()
      cache += (s -> cs)
      cs
    }
}
```

Listing 4.2 · Companion object for class ChecksumAccumulator.

culates the checksum. The first line of the else clause defines a val named acc and initializes it with a new ChecksumAccumulator instance.[3] The next line is a for expression, which cycles through each character in the passed string, converts the character to a Byte by invoking toByte on it, and passes that to the add method of the ChecksumAccumulator instances to which acc refers. After the for expression completes, the next line of the method invokes checksum on acc, which gets the checksum for the passed String, and stores it into a val named cs. In the next line, cache += (s -> cs), the passed string key is mapped to the integer checksum value, and this key-value pair is added to the cache map. The last expression of the method, cs, ensures the checksum is the result of the method.

If you are a Java programmer, one way to think of singleton objects is as the home for any static methods you might have written in Java. You can invoke methods on singleton objects using a similar syntax: the name of the singleton object, a dot, and the name of the method. For example, you can

---

[3]Because the keyword new is only used to instantiate classes, the new object created here is an instance of the ChecksumAccumulator class, not the singleton object of the same name.

invoke the `calculate` method of singleton object `ChecksumAccumulator` like this:

```
ChecksumAccumulator.calculate("Every value is an object.")
```

A singleton object is more than a holder of static methods, however. It is a first-class object. You can think of a singleton object's name, therefore, as a "name tag" attached to the object:

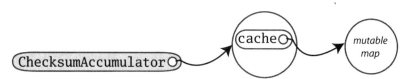

Defining a singleton object doesn't define a type (at the Scala level of abstraction). Given just a definition of object `ChecksumAccumulator`, you can't make a variable of type `ChecksumAccumulator`. Rather, the type named `ChecksumAccumulator` is defined by the singleton object's companion class. However, singleton objects extend a superclass and can mix in traits. Given each singleton object is an instance of its superclasses and mixed-in traits, you can invoke its methods via these types, refer to it from variables of these types, and pass it to methods expecting these types. We'll show some examples of singleton objects inheriting from classes and traits in Chapter 13.

One difference between classes and singleton objects is that singleton objects cannot take parameters, whereas classes can. Because you can't instantiate a singleton object with the new keyword, you have no way to pass parameters to it. Each singleton object is implemented as an instance of a *synthetic class* referenced from a static variable, so they have the same initialization semantics as Java statics.[4] In particular, a singleton object is initialized the first time some code accesses it.

A singleton object that does not share the same name with a companion class is called a *standalone object*. You can use standalone objects for many purposes, including collecting related utility methods together or defining an entry point to a Scala application. This use case is shown in the next section.

---

[4]The name of the synthetic class is the object name plus a dollar sign. Thus the synthetic class for the singleton object named `ChecksumAccumulator` is `ChecksumAccumulator$`.

## 4.4   A Scala application

To run a Scala program, you must supply the name of a standalone singleton object with a `main` method that takes one parameter, an `Array[String]`, and has a result type of `Unit`. Any standalone object with a `main` method of the proper signature can be used as the entry point into an application. An example is shown in Listing 4.3:

```
// In file Summer.scala
import ChecksumAccumulator.calculate

object Summer {
  def main(args: Array[String]) = {
    for (arg <- args)
      println(arg + ": " + calculate(arg))
  }
}
```

Listing 4.3 · The Summer application.

The name of the singleton object in Listing 4.3 is Summer. Its `main` method has the proper signature, so you can use it as an application. The first statement in the file is an import of the `calculate` method defined in the `ChecksumAccumulator` object in the previous example. This import statement allows you to use the method's simple name in the rest of the file.[5] The body of the `main` method simply prints out each argument and the checksum for the argument, separated by a colon.

**Note**
Scala implicitly imports members of packages `java.lang` and `scala`, as well as the members of a singleton object named `Predef`, into every Scala source file. `Predef`, which resides in package `scala`, contains many useful methods. For example, when you say `println` in a Scala source file, you're actually invoking `println` on `Predef`. (`Predef.println` turns around and invokes `Console.println`, which does the real work.) When you say `assert`, you're invoking `Predef.assert`.

---

[5]If you're a Java programmer, you can think of this import as similar to the static import feature introduced in Java 5. One difference in Scala, however, is that you can import members from any object, not just singleton objects.

To run the Summer application, place the code from Listing 4.3 into a file named Summer.scala. Because Summer uses ChecksumAccumulator, place the code for ChecksumAccumulator, both the class shown in Listing 4.1 and its companion object shown in Listing 4.2, into a file named ChecksumAccumulator.scala.

One difference between Scala and Java is that whereas Java requires you to put a public class in a file named after the class—for example, you'd put class SpeedRacer in file SpeedRacer.java—in Scala, you can name .scala files anything you want, no matter what Scala classes or code you put in them. In general in the case of non-scripts, however, it is recommended style to name files after the classes they contain as is done in Java, so that programmers can more easily locate classes by looking at file names. This is the approach we've taken with the two files in this example, Summer.scala and ChecksumAccumulator.scala.

Neither ChecksumAccumulator.scala nor Summer.scala are scripts, because they end in a definition. A script, by contrast, must end in a result expression. Thus if you try to run Summer.scala as a script, the Scala interpreter will complain that Summer.scala does not end in a result expression (assuming of course you didn't add any expression of your own after the Summer object definition). Instead, you'll need to actually compile these files with the Scala compiler, then run the resulting class files. One way to do this is to use scalac, which is the basic Scala compiler, like this:

```
$ scalac ChecksumAccumulator.scala Summer.scala
```

This compiles your source files, but there may be a perceptible delay before the compilation finishes. The reason is that every time the compiler starts up, it spends time scanning the contents of jar files and doing other initial work before it even looks at the fresh source files you submit to it. For this reason, the Scala distribution also includes a Scala compiler *daemon* called fsc (for fast Scala compiler). You use it like this:

```
$ fsc ChecksumAccumulator.scala Summer.scala
```

The first time you run fsc, it will create a local server daemon attached to a port on your computer. It will then send the list of files to compile to the daemon via the port, and the daemon will compile the files. The next time you run fsc, the daemon will already be running, so fsc will simply send the file list to the daemon, which will immediately compile the files. Using fsc, you only need to wait for the Java runtime to startup the first time. If you ever want to stop the fsc daemon, you can do so with fsc -shutdown.

Running either of these scalac or fsc commands will produce Java class files that you can then run via the scala command, the same command you used to invoke the interpreter in previous examples. However, instead of giving it a filename with a .scala extension containing Scala code to interpret as you did in every previous example,[6] in this case you'll give it the name of a standalone object containing a main method of the proper signature. You can run the Summer application, therefore, by typing:

```
$ scala Summer of love
```

You will see checksums printed for the two command line arguments:

```
of: -213
love: -182
```

## 4.5   The App trait

Scala provides a trait, scala.App, that can save you some finger typing. Although we haven't yet covered everything you'll need to understand exactly how this trait works, we figured you'd want to know about it now anyway. Listing 4.4 shows an example:

```
import ChecksumAccumulator.calculate

object FallWinterSpringSummer extends App {

  for (season <- List("fall", "winter", "spring"))
    println(season + ": " + calculate(season))
}
```

Listing 4.4 · Using the App trait.

To use the trait, you first write "extends App" after the name of your singleton object. Then instead of writing a main method, you place the code

---

[6]The actual mechanism that the scala program uses to "interpret" a Scala source file is that it compiles the Scala source code to Java bytecodes, loads them immediately via a class loader, and executes them.

you would have put in the `main` method directly between the curly braces of the singleton object. You can access command-line arguments via an array of strings named `args`. That's it. You can compile and run this application just like any other.

## 4.6 Conclusion

This chapter has given you the basics of classes and objects in Scala, and shown you how to compile and run applications. In the next chapter, you'll learn about Scala's basic types and how to use them.

# Chapter 5

# Basic Types and Operations

Now that you've seen classes and objects in action, it's a good time to look at Scala's basic types and operations in more depth. If you're familiar with Java, you'll be glad to find that Java's basic types and operators have the same meaning in Scala. However, there are some interesting differences that will make this chapter worth reading even if you're an experienced Java developer. Because some aspects of Scala covered in this chapter are essentially the same as in Java, we've inserted notes indicating what sections Java developers can safely skip.

In this chapter, you'll get an overview of Scala's basic types, including `Strings` and the value types `Int`, `Long`, `Short`, `Byte`, `Float`, `Double`, `Char`, and `Boolean`. You'll learn the operations you can perform on these types, including how operator precedence works in Scala expressions. You'll also learn how implicit conversions can "enrich" variants of these basic types, giving you additional operations beyond those supported by Java.

## 5.1  Some basic types

Several fundamental types of Scala, along with the ranges of values instances of these types may have, are shown in Table 5.1. Collectively, types `Byte`, `Short`, `Int`, `Long`, and `Char` are called *integral types*. The integral types plus `Float` and `Double` are called *numeric types*.

Other than `String`, which resides in package `java.lang`, all of the types shown in Table 5.1 are members of package `scala`.[1] For example, the full

---

[1]Packages, which were briefly described in Step 1 in Chapter 2, will be covered in depth in Chapter 13.

Table 5.1 · Some basic types

| Basic type | Range |
|---|---|
| Byte | 8-bit signed two's complement integer ($-2^7$ to $2^7$ - 1, inclusive) |
| Short | 16-bit signed two's complement integer ($-2^{15}$ to $2^{15}$ - 1, inclusive) |
| Int | 32-bit signed two's complement integer ($-2^{31}$ to $2^{31}$ - 1, inclusive) |
| Long | 64-bit signed two's complement integer ($-2^{63}$ to $2^{63}$ - 1, inclusive) |
| Char | 16-bit unsigned Unicode character (0 to $2^{16}$ - 1, inclusive) |
| String | a sequence of Chars |
| Float | 32-bit IEEE 754 single-precision float |
| Double | 64-bit IEEE 754 double-precision float |
| Boolean | true or false |

name of Int is scala.Int. However, given that all the members of package scala and java.lang are automatically imported into every Scala source file, you can just use the simple names (*i.e.*, names like Boolean, Char, or String) everywhere.

Savvy Java developers will note that Scala's basic types have the exact same ranges as the corresponding types in Java. This enables the Scala compiler to transform instances of Scala *value types*, such as Int or Double, down to Java primitive types in the bytecodes it produces.

## 5.2   Literals

All of the basic types listed in Table 5.1 can be written with *literals*. A literal is a way to write a constant value directly in code.

### Fast track for Java programmers

The syntax of most literals shown in this section are exactly the same as in Java, so if you're a Java master, you can safely skip much of this section. Some differences you should read about are Scala's literals for raw strings and symbols, described starting on page 78, and string interpolation, described starting on page 80. Also, Scala does not support octal literals; integer literals that start with a 0, such as 031, do not compile.

**Integer literals**

Integer literals for the types Int, Long, Short, and Byte come in two forms: decimal and hexadecimal. The way an integer literal begins indicates the base of the number. If the number begins with a 0x or 0X, it is hexadecimal (base 16), and may contain 0 through 9 as well as upper or lowercase digits A through F. Some examples are:

```
scala> val hex = 0x5
hex: Int = 5

scala> val hex2 = 0x00FF
hex2: Int = 255

scala> val magic = 0xcafebabe
magic: Int = -889275714
```

Note that the Scala shell always prints integer values in base 10, no matter what literal form you may have used to initialize it. Thus the interpreter displays the value of the hex2 variable you initialized with literal 0x00FF as decimal 255. (Of course, you don't need to take our word for it. A good way to start getting a feel for the language is to try these statements out in the interpreter as you read this chapter.) If the number begins with a non-zero digit, and is otherwise undecorated, it is decimal (base 10). For example:

```
scala> val dec1 = 31
dec1: Int = 31

scala> val dec2 = 255
dec2: Int = 255

scala> val dec3 = 20
dec3: Int = 20
```

If an integer literal ends in an L or l, it is a Long; otherwise it is an Int. Some examples of Long integer literals are:

```
scala> val prog = 0XCAFEBABEL
prog: Long = 3405691582

scala> val tower = 35L
tower: Long = 35
```

```
scala> val of = 31l
of: Long = 31
```

If an Int literal is assigned to a variable of type Short or Byte, the literal is treated as if it were a Short or Byte type so long as the literal value is within the valid range for that type. For example:

```
scala> val little: Short = 367
little: Short = 367

scala> val littler: Byte = 38
littler: Byte = 38
```

## Floating point literals

Floating point literals are made up of decimal digits, optionally containing a decimal point, and optionally followed by an E or e and an exponent. Some examples of floating-point literals are:

```
scala> val big = 1.2345
big: Double = 1.2345

scala> val bigger = 1.2345e1
bigger: Double = 12.345

scala> val biggerStill = 123E45
biggerStill: Double = 1.23E47
```

Note that the exponent portion means the power of 10 by which the other portion is multiplied. Thus, 1.2345e1 is 1.2345 *times* $10^1$, which is 12.345. If a floating-point literal ends in an F or f, it is a Float; otherwise it is a Double. Optionally, a Double floating-point literal can end in D or d. Some examples of Float literals are:

```
scala> val little = 1.2345F
little: Float = 1.2345

scala> val littleBigger = 3e5f
littleBigger: Float = 300000.0
```

That last value expressed as a Double could take these (and other) forms:

```
scala> val anotherDouble = 3e5
anotherDouble: Double = 300000.0

scala> val yetAnother = 3e5D
yetAnother: Double = 300000.0
```

**Character literals**

Character literals are composed of any Unicode character between single quotes, such as:

```
scala> val a = 'A'
a: Char = A
```

In addition to providing an explicit character between the single quotes, you can identify a character using its Unicode code point. To do so, write \u followed by four hex digits with the code point, like this:

```
scala> val d = '\u0041'
d: Char = A

scala> val f = '\u0044'
f: Char = D
```

In fact, such Unicode characters can appear anywhere in a Scala program. For instance you could also write an identifier like this:

```
scala> val B\u0041\u0044 = 1
BAD: Int = 1
```

This identifier is treated as identical to BAD, the result of expanding the two Unicode characters in the code above. In general, it is a bad idea to name identifiers like this because it is hard to read. Rather, this syntax is intended to allow Scala source files that include non-ASCII Unicode characters to be represented in ASCII.

Finally, there are also a few character literals represented by special escape sequences, shown in Table 5.2. For example:

```
scala> val backslash = '\\'
backslash: Char = \
```

77

Table 5.2 · Special character literal escape sequences

| Literal | Meaning |
| --- | --- |
| \n | line feed (\u000A) |
| \b | backspace (\u0008) |
| \t | tab (\u0009) |
| \f | form feed (\u000C) |
| \r | carriage return (\u000D) |
| \" | double quote (\u0022) |
| \' | single quote (\u0027) |
| \\ | backslash (\u005C) |

**String literals**

A string literal is composed of characters surrounded by double quotes:

```
scala> val hello = "hello"
hello: String = hello
```

The syntax of the characters within the quotes is the same as with character literals. For example:

```
scala> val escapes = "\\\"\'"
escapes: String = \"'
```

Because this syntax is awkward for strings that contain a lot of escape sequences or strings that span multiple lines, Scala includes a special syntax for *raw strings*. You start and end a raw string with three double quotation marks in a row ("""). The interior of a raw string may contain any characters whatsoever, including newlines, quotation marks, and special characters, except of course three quotes in a row. For example, the following program prints out a message using a raw string:

```
println("""Welcome to Ultamix 3000.
           Type "HELP" for help.""")
```

However, running this code does not produce quite what is desired:

```
Welcome to Ultamix 3000.
           Type "HELP" for help.
```

78

The issue is that the leading spaces before the second line are included in the string! To help with this common situation, you can call `stripMargin` on strings. To use this method, put a pipe character (|) at the front of each line, and then call `stripMargin` on the whole string:

```
println("""|Welcome to Ultamix 3000.
           |Type "HELP" for help.""".stripMargin)
```

Now the code behaves as desired:

```
Welcome to Ultamix 3000.
Type "HELP" for help.
```

**Symbol literals**

A symbol literal is written '*ident*, where *ident* can be any alphanumeric identifier. Such literals are mapped to instances of the predefined class `scala.Symbol`. Specifically, the literal `'cymbal` will be expanded by the compiler to a factory method invocation: `Symbol("cymbal")`. Symbol literals are typically used in situations where you would use just an identifier in a dynamically typed language. For instance, you might want to define a method that updates a record in a database:

```
scala> def updateRecordByName(r: Symbol, value: Any) = {
         // code goes here
       }
updateRecordByName: (Symbol,Any)Unit
```

The method takes as parameters a symbol indicating the name of a record field and a value with which the field should be updated in the record. In a dynamically typed language, you could invoke this operation passing an undeclared field identifier to the method, but in Scala this would not compile:

```
scala> updateRecordByName(favoriteAlbum, "OK Computer")
<console>:6: error: not found: value favoriteAlbum
       updateRecordByName(favoriteAlbum, "OK Computer")
                          ^
```

Instead, and almost as concisely, you can pass a symbol literal:

```
scala> updateRecordByName('favoriteAlbum, "OK Computer")
```

79

There is not much you can do with a symbol, except find out its name:

```
scala> val s = 'aSymbol
s: Symbol = 'aSymbol

scala> val nm = s.name
nm: String = aSymbol
```

Another thing that's noteworthy is that symbols are *interned*. If you write the same symbol literal twice, both expressions will refer to the exact same Symbol object.

### Boolean literals

The Boolean type has two literals, true and false:

```
scala> val bool = true
bool: Boolean = true

scala> val fool = false
fool: Boolean = false
```

That's all there is to it. You are now literally[2] an expert in Scala.

## 5.3   String interpolation

Scala includes a flexible mechanism for string interpolation, which allows you to embed expressions within string literals. Its most common use case is to provide a concise and readable alternative to string concatenation. Here's an example:

```
val name = "reader"
println(s"Hello, $name!")
```

The expression, s"Hello, $name!" is a *processed* string literal. Because the letter s immediately precedes the open quote, Scala will use the *s string interpolator* to process the literal. The s interpolator will evaluate each embedded expression, invoke toString on each result, and replace the embedded expressions in the literal with those results. Thus s"Hello, $name!" yields "Hello, reader!", the same result as "Hello, " + name + "!".

---

[2]figuratively speaking

You can place any expression after a dollar sign ($) in a processed string literal. For single-variable expressions, you can often just place the variable name after the dollar sign. Scala will interpret all characters up to the first non-identifier character as the expression. If the expression includes non-identifier characters, you must place it in curly braces, with the open curly brace immediately following the dollar sign. Here's an example:

```
scala> s"The answer is ${6 * 7}."
res0: String = The answer is 42.
```

Scala provides two other string interpolators by default: raw and f. The raw string interpolator behaves like s, except it does not recognize character literal escape sequences (such as those shown in Table 5.2). For example, the following statement prints four backslashes, not two:

```
println(raw"No\\\\escape!") // prints: No\\\\escape!
```

The f string interpolator allows you to attach printf-style formatting instructions to embedded expressions. You place the instructions after the expression, starting with a percent sign (%), using the syntax specified by java.util.Formatter. For example, here's how you might format pi:

```
scala> f"${math.Pi}%.5f"
res1: String = 3.14159
```

If you provide no formatting instructions for an embedded expression, the f string interpolator will default to %s, which means the toString value will be substituted, just like the s string interpolator. For example:

```
scala> val pi = "Pi"
pi: String = Pi

scala> f"$pi is approximately ${math.Pi}%.8f."
res2: String = Pi is approximately 3.14159265.
```

In Scala, string interpolation is implemented by rewriting code at compile time. The compiler will treat any expression consisting of an identifier followed immediately by the open double quote of a string literal as a string interpolator expression. The s, f, and raw string interpolators are implemented via this general mechanism. Libraries and users can define other string interpolators for other purposes.

## 5.4 Operators are methods

Scala provides a rich set of operators for its basic types. As mentioned in previous chapters, these operators are actually just a nice syntax for ordinary method calls. For example, 1 + 2 really means the same thing as 1.+(2). In other words, class Int contains a method named + that takes an Int and returns an Int result. This + method is invoked when you add two Ints:

```
scala> val sum = 1 + 2     // Scala invokes 1.+(2)
sum: Int = 3
```

To prove this to yourself, you can write the expression explicitly as a method invocation:

```
scala> val sumMore = 1.+(2)
sumMore: Int = 3
```

In fact, Int contains several *overloaded* + methods that take different parameter types.[3] For example, Int has another method, also named +, that takes and returns a Long. If you add a Long to an Int, this alternate + method will be invoked, as in:

```
scala> val longSum = 1 + 2L     // Scala invokes 1.+(2L)
longSum: Long = 3
```

The + symbol is an operator—an infix operator to be specific. Operator notation is not limited to methods like + that look like operators in other languages. You can use *any* method in operator notation. For example, class String has a method, indexOf, that takes one Char parameter. The indexOf method searches the string for the first occurrence of the specified character and returns its index or –1 if it doesn't find the character. You can use indexOf as an operator, like this:

```
scala> val s = "Hello, world!"
s: String = Hello, world!

scala> s indexOf 'o'     // Scala invokes s.indexOf('o')
res0: Int = 4
```

---

[3]*Overloaded* methods have the same name but different argument types. More on method overloading in Section 6.11.

In addition, `String` offers an overloaded `indexOf` method that takes two parameters, the character for which to search and an index at which to start. (The other `indexOf` method, shown previously, starts at index zero, the beginning of the `String`.) Even though this `indexOf` method takes two arguments, you can use it in operator notation. But whenever you call a method that takes multiple arguments using operator notation, you have to place those arguments in parentheses. For example, here's how you use this other `indexOf` form as an operator (continuing from the previous example):

```
scala> s indexOf ('o', 5) // Scala invokes s.indexOf('o', 5)
res1: Int = 8
```

---

### Any method can be an operator

In Scala operators are not special language syntax; any method can be an operator. What makes a method an operator is how you *use* it. When you write "s.indexOf('o')", indexOf is not an operator. But when you write "s indexOf 'o'", indexOf *is* an operator, because you're using it in operator notation.

---

So far, you've seen examples of *infix* operator notation, which means the method to invoke sits between the object and the parameter (or parameters) you wish to pass to the method, as in "7 + 2". Scala also has two other operator notations: prefix and postfix. In prefix notation, you put the method name before the object on which you are invoking the method (for example, the '-' in -7). In postfix notation, you put the method after the object (for example, the "toLong" in "7 toLong").

In contrast to the infix operator notation—in which operators take two operands, one to the left and the other to the right—prefix and postfix operators are *unary*: they take just one operand. In prefix notation, the operand is to the right of the operator. Some examples of prefix operators are -2.0, !found, and ~0xFF. As with the infix operators, these prefix operators are a shorthand way of invoking methods. In this case, however, the name of the method has "unary_" prepended to the operator character. For instance, Scala will transform the expression -2.0 into the method invocation "(2.0).unary_-". You can demonstrate this to yourself by typing the method call both via operator notation and explicitly:

```
scala> -2.0                    // Scala invokes (2.0).unary_-
res2: Double = -2.0

scala> (2.0).unary_-
res3: Double = -2.0
```

The only identifiers that can be used as prefix operators are +, -, !, and ~. Thus, if you define a method named unary_!, you could invoke that method on a value or variable of the appropriate type using prefix operator notation, such as !p. But if you define a method named unary_*, you wouldn't be able to use prefix operator notation because * isn't one of the four identifiers that can be used as prefix operators. You could invoke the method normally, as in p.unary_*, but if you attempted to invoke it via *p, Scala will parse it as if you'd written *.p, which is probably not what you had in mind![4]

Postfix operators are methods that take no arguments, when they are invoked without a dot or parentheses. In Scala, you can leave off empty parentheses on method calls. The convention is that you include parentheses if the method has side effects, such as println(), but you can leave them off if the method has no side effects, such as toLowerCase invoked on a String:

```
scala> val s = "Hello, world!"
s: String = Hello, world!

scala> s.toLowerCase
res4: String = hello, world!
```

In this latter case of a method that requires no arguments, you can alternatively leave off the dot and use postfix operator notation:

```
scala> s toLowerCase
res5: String = hello, world!
```

In this case, toLowerCase is used as a postfix operator on the operand s.

Therefore, to see what operators you can use with Scala's basic types, all you really need to do is look at the methods declared in the type's classes in the Scala API documentation. Given that this is a Scala tutorial, however, we'll give you a quick tour of most of these methods in the next few sections.

---

[4]All is not necessarily lost, however. There is an extremely slight chance your program with the *p might compile as C++.

> **Fast track for Java programmers**
> Many aspects of Scala described in the remainder of this chapter are the
> same as in Java. If you're a Java guru in a rush, you can safely skip to
> Section 5.8 on page 89, which describes how Scala differs from Java in the
> area of object equality.

## 5.5 Arithmetic operations

You can invoke arithmetic methods via infix operator notation for addition
(+), subtraction (–), multiplication (∗), division (/), and remainder (%) on any
numeric type. Here are some examples:

```
scala> 1.2 + 2.3
res6: Double = 3.5

scala> 3 - 1
res7: Int = 2

scala> 'b' - 'a'
res8: Int = 1

scala> 2L * 3L
res9: Long = 6

scala> 11 / 4
res10: Int = 2

scala> 11 % 4
res11: Int = 3

scala> 11.0f / 4.0f
res12: Float = 2.75

scala> 11.0 % 4.0
res13: Double = 3.0
```

When both the left and right operands are integral types (`Int`, `Long`,
`Byte`, `Short`, or `Char`), the / operator will tell you the whole number portion of the quotient, excluding any remainder. The % operator indicates the
remainder of an implied integer division.

The floating-point remainder you get with % is not the one defined by the
IEEE 754 standard. The IEEE 754 remainder uses rounding division, not
truncating division, in calculating the remainder, so it is quite different from

the integer remainder operation. If you really want an IEEE 754 remainder, you can call IEEEremainder on scala.math, as in:

```
scala> math.IEEEremainder(11.0, 4.0)
res14: Double = -1.0
```

The numeric types also offer unary prefix operators + (method unary_+) and – (method unary_-), which allow you to indicate whether a literal number is positive or negative, as in –3 or +4.0. If you don't specify a unary + or –, a literal number is interpreted as positive. Unary + exists solely for symmetry with unary –, but has no effect. The unary – can also be used to negate a variable. Here are some examples:

```
scala> val neg = 1 + -3
neg: Int = -2

scala> val y = +3
y: Int = 3

scala> -neg
res15: Int = 2
```

## 5.6   Relational and logical operations

You can compare numeric types with relational methods greater than (>), less than (<), greater than or equal to (>=), and less than or equal to (<=), which yield a Boolean result. In addition, you can use the unary '!' operator (the unary_! method) to invert a Boolean value. Here are a few examples:

```
scala> 1 > 2
res16: Boolean = false

scala> 1 < 2
res17: Boolean = true

scala> 1.0 <= 1.0
res18: Boolean = true

scala> 3.5f >= 3.6f
res19: Boolean = false
```

```
scala> 'a' >= 'A'
res20: Boolean = true

scala> val untrue = !true
untrue: Boolean = false
```

Logical methods, logical-and (&& and &) and logical-or (|| and |), take Boolean operands in infix notation and yield a Boolean result. For example:

```
scala> val toBe = true
toBe: Boolean = true

scala> val question = toBe || !toBe
question: Boolean = true

scala> val paradox = toBe && !toBe
paradox: Boolean = false
```

The && and || operations *short-circuit* as in Java: expressions built from these operators are only evaluated as far as needed to determine the result. In other words, the right-hand side of && and || expressions won't be evaluated if the left-hand side determines the result. For example, if the left-hand side of a && expression evaluates to false, the result of the expression will definitely be false, so the right-hand side is not evaluated. Likewise, if the left-hand side of a || expression evaluates to true, the result of the expression will definitely be true, so the right-hand side is not evaluated.

```
scala> def salt() = { println("salt"); false }
salt: ()Boolean

scala> def pepper() = { println("pepper"); true }
pepper: ()Boolean

scala> pepper() && salt()
pepper
salt
res21: Boolean = false

scala> salt() && pepper()
salt
res22: Boolean = false
```

In the first expression, pepper and salt are invoked, but in the second, only salt is invoked. Given salt returns false, there's no need to call pepper.

If you want to evaluate the right hand side no matter what, use & and |
instead. The & method performs a logical-and operation, and | a logical-or,
but don't short-ciruit like && and ||. Here's an example:

```
scala> salt() & pepper()
salt
pepper
res23: Boolean = false
```

> **Note**
>
> You may be wondering how short-circuiting can work given operators are
> just methods. Normally, all arguments are evaluated before entering a
> method, so how can a method avoid evaluating its second argument? The
> answer is that all Scala methods have a facility for delaying the evaluation
> of their arguments, or even declining to evaluate them at all. The facility is
> called *by-name parameters* and is discussed in Section 9.5.

## 5.7  Bitwise operations

Scala enables you to perform operations on individual bits of integer types
with several bitwise methods. The bitwise methods are: bitwise-and (&),
bitwise-or (|), and bitwise-xor (^).[5] The unary bitwise complement operator
(~, the method unary_~) inverts each bit in its operand. For example:

```
scala> 1 & 2
res24: Int = 0

scala> 1 | 2
res25: Int = 3

scala> 1 ^ 3
res26: Int = 2

scala> ~1
res27: Int = -2
```

The first expression, 1 & 2, bitwise-ands each bit in 1 (0001) and 2 (0010),
which yields 0 (0000). The second expression, 1 | 2, bitwise-ors each bit in

---

[5]The bitwise-xor method performs an *exclusive or* on its operands. Identical bits yield a
0. Different bits yield a 1. Thus 0011 ^ 0101 yields 0110.

the same operands, yielding 3 (0011). The third expression, 1 ^ 3, bitwise-xors each bit in 1 (0001) and 3 (0011), yielding 2 (0010). The final expression, ~1, inverts each bit in 1 (0001), yielding -2, which in binary looks like 11111111111111111111111111111110.

Scala integer types also offer three shift methods: shift left (<<), shift right (>>), and unsigned shift right (>>>). The shift methods, when used in infix operator notation, shift the integer value on the left of the operator by the amount specified by the integer value on the right. Shift left and unsigned shift right fill with zeroes as they shift. Shift right fills with the highest bit (the sign bit) of the left-hand value as it shifts. Here are some examples:

```
scala> -1 >> 31
res28: Int = -1

scala> -1 >>> 31
res29: Int = 1

scala> 1 << 2
res30: Int = 4
```

-1 in binary is 11111111111111111111111111111111. In the first example, −1 >> 31, -1 is shifted to the right 31 bit positions. Since an Int consists of 32 bits, this operation effectively moves the leftmost bit over until it becomes the rightmost bit.[6] Since the >> method fills with ones as it shifts right, because the leftmost bit of -1 is 1, the result is identical to the original left operand, 32 one bits, or -1. In the second example, −1 >>> 31, the leftmost bit is again shifted right until it is in the rightmost position, but this time filling with zeroes along the way. Thus the result this time is binary 00000000000000000000000000000001, or 1. In the final example, 1 << 2, the left operand, 1, is shifted left two positions (filling in with zeroes), resulting in binary 00000000000000000000000000000100, or 4.

## 5.8 Object equality

If you want to compare two objects for equality, you can use either == or its inverse !=. Here are a few simple examples:

---

[6]The leftmost bit in an integer type is the sign bit. If the leftmost bit is 1, the number is negative. If 0, the number is positive.

```
scala> 1 == 2
res31: Boolean = false

scala> 1 != 2
res32: Boolean = true

scala> 2 == 2
res33: Boolean = true
```

These operations actually apply to all objects, not just basic types. For example, you can use == to compare lists:

```
scala> List(1, 2, 3) == List(1, 2, 3)
res34: Boolean = true

scala> List(1, 2, 3) == List(4, 5, 6)
res35: Boolean = false
```

Going further, you can compare two objects that have different types:

```
scala> 1 == 1.0
res36: Boolean = true

scala> List(1, 2, 3) == "hello"
res37: Boolean = false
```

You can even compare against null, or against things that might be null. No exception will be thrown:

```
scala> List(1, 2, 3) == null
res38: Boolean = false

scala> null == List(1, 2, 3)
res39: Boolean = false
```

As you see, == has been carefully crafted so that you get just the equality comparison you want in most cases. This is accomplished with a very simple rule: First check the left side for null. If it is not null, call the equals method. Since equals is a method, the precise comparison you get depends on the type of the left-hand argument. Since there is an automatic null check, you do not have to do the check yourself.[7]

---

[7]The automatic check does not look at the right-hand side, but any reasonable equals method should return false if its argument is null.

This kind of comparison will yield `true` on different objects, so long as their contents are the same and their `equals` method is written to be based on contents. For example, here is a comparison between two strings that happen to have the same five letters in them:

```
scala> ("he" + "llo") == "hello"
res40: Boolean = true
```

---

### How Scala's == differs from Java's

In Java, you can use == to compare both primitive and reference types. On primitive types, Java's == compares value equality, as in Scala. On reference types, however, Java's == compares *reference equality*, which means the two variables point to the same object on the JVM's heap. Scala provides a facility for comparing reference equality, as well, under the name eq. However, eq and its opposite, ne, only apply to objects that directly map to Java objects. The full details about eq and ne are given in Sections 11.1 and 11.2. Also, see Chapter 30 on how to write a good equals method.

---

## 5.9 Operator precedence and associativity

Operator precedence determines which parts of an expression are evaluated before the other parts. For example, the expression 2 + 2 * 7 evaluates to 16, not 28, because the * operator has a higher precedence than the + operator. Thus the multiplication part of the expression is evaluated before the addition part. You can of course use parentheses in expressions to clarify evaluation order or to override precedence. For example, if you really wanted the result of the expression above to be 28, you could write the expression like this:

```
(2 + 2) * 7
```

Given that Scala doesn't have operators, per se, just a way to use methods in operator notation, you may be wondering how operator precedence works. Scala decides precedence based on the first character of the methods used in operator notation (there's one exception to this rule, which will be discussed in the following pages). If the method name starts with a *, for example,

it will have a higher precedence than a method that starts with a +. Thus
2 + 2 * 7 will be evaluated as 2 + (2 * 7). Similarly, a +++ b *** c (in which
a, b, and c are variables, and +++ and *** are methods) will be evaluated
a +++ (b *** c), because the *** method has a higher precedence than the
+++ method.

Table 5.3 · Operator precedence

| |
| --- |
| (all other special characters) |
| * / % |
| + − |
| : |
| = ! |
| < > |
| & |
| ^ |
| \| |
| (all letters) |
| (all assignment operators) |

Table 5.3 shows the precedence given to the first character of a method
in decreasing order of precedence, with characters on the same line having
the same precedence. The higher a character is in this table, the higher the
precedence of methods that start with that character. Here's an example that
illustrates the influence of precedence:

```
scala> 2 << 2 + 2
res41: Int = 32
```

The << method starts with the character <, which appears lower in Table 5.3 than the character +, which is the first and only character of the +
method. Thus << will have lower precedence than +, and the expression
will be evaluated by first invoking the + method, then the << method, as in
2 << (2 + 2). 2 + 2 is 4, by our math, and 2 << 4 yields 32. If you swap the
operators, you'll get a different result:

```
scala> 2 + 2 << 2
res42: Int = 16
```

Since the first characters are the same as in the previous example, the methods will be invoked in the same order. First the + method will be invoked, then the << method. So 2 + 2 will again yield 4, and 4 << 2 is 16.

The one exception to the precedence rule, alluded to earlier, concerns *assignment operators*, which end in an equals character. If an operator ends in an equals character (=), and the operator is not one of the comparison operators <=, >=, ==, or !=, then the precedence of the operator is the same as that of simple assignment (=). That is, it is lower than the precedence of any other operator. For instance:

```
x *= y + 1
```

means the same as:

```
x *= (y + 1)
```

because *= is classified as an assignment operator whose precedence is lower than +, even though the operator's first character is *, which would suggest a precedence higher than +.

When multiple operators of the same precedence appear side by side in an expression, the *associativity* of the operators determines the way operators are grouped. The associativity of an operator in Scala is determined by its *last* character. As mentioned on page 43 of Chapter 3, any method that ends in a ':' character is invoked on its right operand, passing in the left operand. Methods that end in any other character are the other way around: They are invoked on their left operand, passing in the right operand. So a * b yields a.*(b), but a ::: b yields b.:::(a).

No matter what associativity an operator has, however, its operands are always evaluated left to right. So if a is an expression that is not just a simple reference to an immutable value, then a ::: b is more precisely treated as the following block:

```
{ val x = a; b.:::(x) }
```

In this block a is still evaluated before b, and then the result of this evaluation is passed as an operand to b's ::: method.

This associativity rule also plays a role when multiple operators of the same precedence appear side by side. If the methods end in ':', they are grouped right to left; otherwise, they are grouped left to right. For example, a ::: b ::: c is treated as a ::: (b ::: c). But a * b * c, by contrast, is treated as (a * b) * c.

Operator precedence is part of the Scala language. You needn't be afraid to use it. Nevertheless, it is good style to use parentheses to clarify what operators are operating upon what expressions. Perhaps the only precedence you can truly count on other programmers knowing without looking up is that multiplicative operators, *, /, and %, have a higher precedence than the additive ones + and −. Thus even if a + b << c yields the result you want without parentheses, the extra clarity you get by writing (a + b) << c may reduce the frequency with which your peers utter your name in operator notation, for example, by shouting in disgust, "bills !*&^%~ code!".[8]

## 5.10   Rich wrappers

You can invoke many more methods on Scala's basic types than were described in the previous sections. A few examples are shown in Table 5.4. These methods are available via *implicit conversions*, a technique that will be described in detail in Chapter 21. All you need to know for now is that for each basic type described in this chapter, there is also a "rich wrapper" that provides several additional methods. To see all the available methods on the basic types, therefore, you should look at the API documentation on the rich wrapper for each basic type. Those classes are listed in Table 5.5.

## 5.11   Conclusion

The main take-aways from this chapter are that operators in Scala are method calls, and that implicit conversions to rich variants exist for Scala's basic types that add even more useful methods. In the next chapter, we'll show you what it means to design objects in a functional style that gives new implementations of some of the operators that you have seen in this chapter.

---

[8]By now you should be able to figure out that given this code, the Scala compiler would invoke (bills.!*&^%~(code)).!().

94

Table 5.4 · Some rich operations

| Code | Result |
|------|--------|
| 0 max 5 | 5 |
| 0 min 5 | 0 |
| -2.7 abs | 2.7 |
| -2.7 round | -3L |
| 1.5 isInfinity | false |
| (1.0 / 0) isInfinity | true |
| 4 to 6 | Range(4, 5, 6) |
| "bob" capitalize | "Bob" |
| "robert" drop 2 | "bert" |

Table 5.5 · Rich wrapper classes

| Basic type | Rich wrapper |
|------------|--------------|
| Byte | scala.runtime.RichByte |
| Short | scala.runtime.RichShort |
| Int | scala.runtime.RichInt |
| Long | scala.runtime.RichLong |
| Char | scala.runtime.RichChar |
| Float | scala.runtime.RichFloat |
| Double | scala.runtime.RichDouble |
| Boolean | scala.runtime.RichBoolean |
| String | scala.collection.immutable.StringOps |

# Chapter 6

# Functional Objects

With the understanding of Scala basics you've gained from previous chapters, you're ready to design more full-featured classes in Scala. In this chapter, the emphasis is on classes that define functional objects, or objects that do not have any mutable state. As a running example, we'll create several variants of a class that models rational numbers as immutable objects. Along the way, we'll show you more aspects of object-oriented programming in Scala: class parameters and constructors, methods and operators, private members, overriding, checking preconditions, overloading, and self references.

## 6.1 A specification for class `Rational`

A *rational number* is a number that can be expressed as a ratio $\frac{n}{d}$, where $n$ and $d$ are integers, except that $d$ cannot be zero. $n$ is called the *numerator* and $d$ the *denominator*. Examples of rational numbers are $\frac{1}{2}$, $\frac{2}{3}$, $\frac{112}{239}$, and $\frac{2}{1}$. Compared to floating-point numbers, rational numbers have the advantage that fractions are represented exactly, without rounding or approximation.

   The class we'll design in this chapter must model the behavior of rational numbers, including allowing them to be added, subtracted, multiplied, and divided. To add two rationals, you must first obtain a common denominator, then add the two numerators. For example, to add $\frac{1}{2} + \frac{2}{3}$, you multiply both parts of the left operand by 3 and both parts of the right operand by 2, which gives you $\frac{3}{6} + \frac{4}{6}$. Adding the two numerators yields the result, $\frac{7}{6}$. To multiply two rational numbers, you can simply multiply their numerators and multiply their denominators. Thus, $\frac{1}{2} * \frac{2}{5}$ gives $\frac{2}{10}$, which can be represented more compactly in its "normalized" form as $\frac{1}{5}$. You divide by swapping the

97

numerator and denominator of the right operand and then multiplying. For instance $\frac{1}{2}/\frac{3}{5}$ is the same as $\frac{1}{2} * \frac{5}{3}$, or $\frac{5}{6}$.

One, maybe rather trivial, observation is that in mathematics, rational numbers do not have mutable state. You can add one rational number to another, but the result will be a new rational number. The original numbers will not have "changed." The immutable Rational class we'll design in this chapter will have the same property. Each rational number will be represented by one Rational object. When you add two Rational objects, you'll create a new Rational object to hold the sum.

This chapter will give you a glimpse of some of the ways Scala enables you to write libraries that feel like native language support. For example, at the end of this chapter you'll be able to do this with class Rational:

```
scala> val oneHalf = new Rational(1, 2)
oneHalf: Rational = 1/2

scala> val twoThirds = new Rational(2, 3)
twoThirds: Rational = 2/3

scala> (oneHalf / 7) + (1 - twoThirds)
res0: Rational = 17/42
```

## 6.2 Constructing a Rational

A good place to start designing class Rational is to consider how client programmers will create a new Rational object. Given we've decided to make Rational objects immutable, we'll require that clients provide all data needed by an instance (in this case, a numerator and a denominator) when they construct the instance. Thus, we will start the design with this:

```
class Rational(n: Int, d: Int)
```

One of the first things to note about this line of code is that if a class doesn't have a body, you don't need to specify empty curly braces (though you could, of course, if you wanted to). The identifiers n and d in the parentheses after the class name, Rational, are called *class parameters*. The Scala compiler will gather up these two class parameters and create a *primary constructor* that takes the same two parameters.

98

### Immutable object trade-offs

Immutable objects offer several advantages over mutable objects, and
one potential disadvantage. First, immutable objects are often easier to
reason about than mutable ones, because they do not have complex state
spaces that change over time. Second, you can pass immutable objects
around quite freely, whereas you may need to make defensive copies
of mutable objects before passing them to other code. Third, there is
no way for two threads concurrently accessing an immutable to corrupt
its state once it has been properly constructed, because no thread can
change the state of an immutable. Fourth, immutable objects make safe
hash table keys. If a mutable object is mutated after it is placed into a
HashSet, for example, that object may not be found the next time you
look into the HashSet.

The main disadvantage of immutable objects is that they sometimes
require that a large object graph be copied, whereas an update could
be done in its place. In some cases this can be awkward to express
and might also cause a performance bottleneck. As a result, it is not
uncommon for libraries to provide mutable alternatives to immutable
classes. For example, class StringBuilder is a mutable alternative to
the immutable String. We'll give you more information on designing
mutable objects in Scala in Chapter 18.

**Note**

This initial Rational example highlights a difference between Java and
Scala. In Java, classes have constructors, which can take parameters;
whereas in Scala, classes can take parameters directly. The Scala notation
is more concise—class parameters can be used directly in the body of the
class; there's no need to define fields and write assignments that copy
constructor parameters into fields. This can yield substantial savings in
boilerplate code, especially for small classes.

The Scala compiler will compile any code you place in the class body,
which isn't part of a field or a method definition, into the primary constructor.
For example, you could print a debug message like this:

```scala
class Rational(n: Int, d: Int) {
  println("Created " + n + "/" + d)
}
```

Given this code, the Scala compiler would place the call to `println` into Rational's primary constructor. The `println` call will, therefore, print its debug message whenever you create a new `Rational` instance:

```
scala> new Rational(1, 2)
Created 1/2
res0: Rational = Rational@2591e0c9
```

## 6.3   Reimplementing the `toString` method

When we created an instance of `Rational` in the previous example, the interpreter printed "Rational@90110a". The interpreter obtained this somewhat funny looking string by calling `toString` on the `Rational` object. By default, class `Rational` inherits the implementation of `toString` defined in class `java.lang.Object`, which just prints the class name, an @ sign, and a hexadecimal number. The result of `toString` is primarily intended to help programmers by providing information that can be used in debug print statements, log messages, test failure reports, and interpreter and debugger output. The result currently provided by `toString` is not especially helpful because it doesn't give any clue about the rational number's value. A more useful implementation of `toString` would print out the values of the Rational's numerator and denominator. You can *override* the default implementation by adding a method `toString` to class `Rational`, like this:

```
class Rational(n: Int, d: Int) {
  override def toString = n + "/" + d
}
```

The override modifier in front of a method definition signals that a previous method definition is overridden (more on this in Chapter 10). Since Rational numbers will display nicely now, we removed the debug `println` statement we put into the body of previous version of class `Rational`. You can test the new behavior of `Rational` in the interpreter:

```
scala> val x = new Rational(1, 3)
x: Rational = 1/3

scala> val y = new Rational(5, 7)
y: Rational = 5/7
```

## 6.4   Checking preconditions

As a next step, we will turn our attention to a problem with the current behavior of the primary constructor. As mentioned at the beginning of this chapter, rational numbers may not have a zero in the denominator. Currently, however, the primary constructor accepts a zero passed as d:

```
scala> new Rational(5, 0)
res1: Rational = 5/0
```

One of the benefits of object-oriented programming is that it allows you to encapsulate data inside objects so that you can ensure the data is valid throughout its lifetime. In the case of an immutable object such as Rational, this means that you should ensure the data is valid when the object is constructed. Given that a zero denominator is an invalid state for a Rational number, you should not let a Rational be constructed if a zero is passed in the d parameter.

The best way to approach this problem is to define as a *precondition* of the primary constructor that d must be non-zero. A precondition is a constraint on values passed into a method or constructor, a requirement which callers must fulfill. One way to do that is to use require,[1] like this:

```
class Rational(n: Int, d: Int) {
  require(d != 0)
  override def toString = n + "/" + d
}
```

The require method takes one boolean parameter. If the passed value is true, require will return normally. Otherwise, require will prevent the object from being constructed by throwing an IllegalArgumentException.

## 6.5   Adding fields

Now that the primary constructor is properly enforcing its precondition, we will turn our attention to supporting addition. To do so, we'll define a public add method on class Rational that takes another Rational as a parameter. To keep Rational immutable, the add method must not add the passed

---

[1]The require method is defined in standalone object, Predef. As mentioned in Section 4.4, Predef's members are imported automatically into every Scala source file.

rational number to itself. Rather, it must create and return a new `Rational` that holds the sum. You might think you could write `add` this way:

```
class Rational(n: Int, d: Int) { // This won't compile
  require(d != 0)
  override def toString = n + "/" + d
  def add(that: Rational): Rational =
    new Rational(n * that.d + that.n * d, d * that.d)
}
```

However, given this code the compiler will complain:

```
<console>:11: error: value d is not a member of Rational
        new Rational(n * that.d + that.n * d, d * that.d)
                            ^
<console>:11: error: value d is not a member of Rational
        new Rational(n * that.d + that.n * d, d * that.d)
                                                     ^
```

Although class parameters n and d are in scope in the code of your add method, you can only access their value on the object on which add was invoked. Thus, when you say n or d in add's implementation, the compiler is happy to provide you with the values for these class parameters. But it won't let you say `that.n` or `that.d` because that does not refer to the `Rational` object on which add was invoked.[2] To access the numerator and denominator on that, you'll need to make them into fields. Listing 6.1 shows how you could add these fields to class `Rational`.[3]

In the version of `Rational` shown in Listing 6.1, we added two fields named numer and denom, and initialized them with the values of class parameters n and d.[4] We also changed the implementation of `toString` and add so that they use the fields, not the class parameters. This version of class `Rational` compiles. You can test it by adding some rational numbers:

---

[2] Actually, you could add a `Rational` to itself, in which case that would refer to the object on which add was invoked. But because you can pass any `Rational` object to add, the compiler still won't let you say `that.n`.

[3] In Section 10.6 you'll find out about *parametric fields*, which provide a shorthand for writing the same code.

[4] Even though n and d are used in the body of the class, given they are only used inside constructors, the Scala compiler will not emit fields for them. Thus, given this code the Scala compiler will generate a class with two Int fields, one for numer and one for denom.

```
class Rational(n: Int, d: Int) {
  require(d != 0)
  val numer: Int = n
  val denom: Int = d
  override def toString = numer + "/" + denom
  def add(that: Rational): Rational =
    new Rational(
      numer * that.denom + that.numer * denom,
      denom * that.denom
    )
}
```

Listing 6.1 · Rational with fields.

```
scala> val oneHalf = new Rational(1, 2)
oneHalf: Rational = 1/2

scala> val twoThirds = new Rational(2, 3)
twoThirds: Rational = 2/3

scala> oneHalf add twoThirds
res2: Rational = 7/6
```

One other thing you can do now that you couldn't do before is access the numerator and denominator values from outside the object. Simply access the public numer and denom fields, like this:

```
scala> val r = new Rational(1, 2)
r: Rational = 1/2

scala> r.numer
res3: Int = 1

scala> r.denom
res4: Int = 2
```

## 6.6   Self references

The keyword this refers to the object instance on which the currently exe-cuting method was invoked, or if used in a constructor, the object instance being constructed. As an example, consider adding a method, lessThan, which tests whether the given Rational is smaller than a parameter:

```
def lessThan(that: Rational) =
  this.numer * that.denom < that.numer * this.denom
```

Here, this.numer refers to the numerator of the object on which lessThan was invoked. You can also leave off the this prefix and write just numer; the two notations are equivalent.

As an example of where you can't do without this, consider adding a max method to class Rational that returns the greater of the given rational number and an argument:

```
def max(that: Rational) =
  if (this.lessThan(that)) that else this
```

Here, the first this is redundant. You could have left it off and written: lessThan(that). But the second this represents the result of the method in the case where the test returns false; were you to omit it, there would be nothing left to return!

## 6.7 Auxiliary constructors

Sometimes you need multiple constructors in a class. In Scala, constructors other than the primary constructor are called *auxiliary constructors*. For example, a rational number with a denominator of 1 can be written more succinctly as simply the numerator. Instead of $\frac{5}{1}$, for example, you can just write 5. It might be nice, therefore, if instead of writing new Rational(5, 1), client programmers could simply write new Rational(5). This would require adding an auxiliary constructor to Rational that takes only one argument, the numerator, with the denominator predefined to be 1. Listing 6.2 shows what that would look like.

Auxiliary constructors in Scala start with def this(...). The body of Rational's auxiliary constructor merely invokes the primary constructor, passing along its lone argument, n, as the numerator and 1 as the denominator. You can see the auxiliary constructor in action by typing the following into the interpreter:

```
scala> val y = new Rational(3)
y: Rational = 3/1
```

```scala
class Rational(n: Int, d: Int) {

  require(d != 0)

  val numer: Int = n
  val denom: Int = d

  def this(n: Int) = this(n, 1) // auxiliary constructor

  override def toString = numer + "/" + denom

  def add(that: Rational): Rational =
    new Rational(
      numer * that.denom + that.numer * denom,
      denom * that.denom
    )
}
```

Listing 6.2 · `Rational` with an auxiliary constructor.

In Scala, every auxiliary constructor must invoke another constructor of the same class as its first action. In other words, the first statement in every auxiliary constructor in every Scala class will have the form "this(...)". The invoked constructor is either the primary constructor (as in the `Rational` example), or another auxiliary constructor that comes textually before the calling constructor. The net effect of this rule is that every constructor invocation in Scala will end up eventually calling the primary constructor of the class. The primary constructor is thus the single point of entry of a class.

**Note**

If you're familiar with Java, you may wonder why Scala's rules for constructors are a bit more restrictive than Java's. In Java, a constructor must either invoke another constructor of the same class, or directly invoke a constructor of the superclass, as its first action. In a Scala class, only the primary constructor can invoke a superclass constructor. The increased restriction in Scala is really a design trade-off that needed to be paid in exchange for the greater conciseness and simplicity of Scala's constructors compared to Java's. Superclasses and the details of how constructor invocation and inheritance interact will be explained in Chapter 10.

## 6.8 Private fields and methods

In the previous version of Rational, we simply initialized numer with n and denom with d. As a result, the numerator and denominator of a Rational can be larger than needed. For example, the fraction $\frac{66}{42}$ could be normalized to an equivalent reduced form, $\frac{11}{7}$, but Rational's primary constructor doesn't currently do this:

```
scala> new Rational(66, 42)
res5: Rational = 66/42
```

To normalize in this way, you need to divide the numerator and denominator by their *greatest common divisor*. For example, the greatest common divisor of 66 and 42 is 6. (In other words, 6 is the largest integer that divides evenly into both 66 and 42.) Dividing both the numerator and denominator of $\frac{66}{42}$ by 6 yields its reduced form, $\frac{11}{7}$. Listing 6.3 shows one way to do this:

```
class Rational(n: Int, d: Int) {

  require(d != 0)

  private val g = gcd(n.abs, d.abs)
  val numer = n / g
  val denom = d / g

  def this(n: Int) = this(n, 1)

  def add(that: Rational): Rational =
    new Rational(
      numer * that.denom + that.numer * denom,
      denom * that.denom
    )

  override def toString = numer + "/" + denom

  private def gcd(a: Int, b: Int): Int =
    if (b == 0) a else gcd(b, a % b)
}
```

Listing 6.3 · Rational with a private field and method.

In this version of Rational, we added a private field, g, and modified the initializers for numer and denom. (An *initializer* is the code that initializes a variable; for example, the "n / g" that initializes numer.) Because g is private, it can be accessed inside the body of the class, but not outside. We also added a private method, gcd, which calculates the greatest common divisor of two passed Ints. For example, gcd(12, 8) is 4. As you saw in Section 4.1, to make a field or method private you simply place the private keyword in front of its definition. The purpose of the private "helper method" gcd is to factor out code needed by some other part of the class, in this case, the primary constructor. To ensure g is always positive, we pass the absolute value of n and d, which we obtain by invoking abs on them, a method you can invoke on any Int to get its absolute value.

The Scala compiler will place the code for the initializers of Rational's three fields into the primary constructor in the order in which they appear in the source code. Thus, g's initializer, gcd(n.abs, d.abs), will execute before the other two, because it appears first in the source. Field g will be initialized with the result, the greatest common divisor of the absolute value of the class parameters, n and d. Field g is then used in the initializers of numer and denom. By dividing n and d by their greatest common divisor, g, every Rational will be constructed in its normalized form:

```
scala> new Rational(66, 42)
res6: Rational = 11/7
```

## 6.9  Defining operators

The current implementation of Rational addition is OK, but could be made more convenient to use. You might ask yourself why you can write:

```
x + y
```

if x and y are integers or floating-point numbers, but you have to write:

```
x.add(y)
```

or at least:

```
x add y
```

if they are rational numbers. There's no convincing reason why this should be so. Rational numbers are numbers just like other numbers. In a mathematical sense they are even more natural than, say, floating-point numbers.

Why should you not use the natural arithmetic operators on them? In Scala you can do this. In the rest of this chapter, we'll show you how.

The first step is to replace add by the usual mathematical symbol. This is straightforward, as + is a legal identifier in Scala. We can simply define a method with + as its name. While we're at it, we may as well implement a method named * that performs multiplication. The result is shown in Listing 6.4:

```scala
class Rational(n: Int, d: Int) {

  require(d != 0)

  private val g = gcd(n.abs, d.abs)
  val numer = n / g
  val denom = d / g

  def this(n: Int) = this(n, 1)

  def + (that: Rational): Rational =
    new Rational(
      numer * that.denom + that.numer * denom,
      denom * that.denom
    )

  def * (that: Rational): Rational =
    new Rational(numer * that.numer, denom * that.denom)

  override def toString = numer + "/" + denom

  private def gcd(a: Int, b: Int): Int =
    if (b == 0) a else gcd(b, a % b)
}
```

Listing 6.4 · Rational with operator methods.

With class Rational defined in this manner, you can now write:

```scala
scala> val x = new Rational(1, 2)
x: Rational = 1/2

scala> val y = new Rational(2, 3)
y: Rational = 2/3

scala> x + y
res7: Rational = 7/6
```

As always, the operator syntax on the last input line is equivalent to a method call. You could also write:

```
scala> x.+(y)
res8: Rational = 7/6
```

but this is not as readable.

Another thing to note is that given Scala's rules for operator precedence, which were described in Section 5.9, the $*$ method will bind more tightly than the $+$ method for `Rational`s. In other words, expressions involving $+$ and $*$ operations on `Rational`s will behave as expected. For example, $x + x * y$ will execute as $x + (x * y)$, not $(x + x) * y$:

```
scala> x + x * y
res9: Rational = 5/6

scala> (x + x) * y
res10: Rational = 2/3

scala> x + (x * y)
res11: Rational = 5/6
```

## 6.10   Identifiers in Scala

You have now seen the two most important ways to form an identifier in Scala: alphanumeric and operator. Scala has very flexible rules for forming identifiers. Besides the two forms you have seen there are also two others. All four forms of identifier formation are described in this section.

An *alphanumeric identifier* starts with a letter or underscore, which can be followed by further letters, digits, or underscores. The '$' character also counts as a letter; however, it is reserved for identifiers generated by the Scala compiler. Identifiers in user programs should not contain '$' characters, even though it will compile; if they do, this might lead to name clashes with identifiers generated by the Scala compiler.

Scala follows Java's convention of using camel-case[5] identifiers, such as `toString` and `HashSet`. Although underscores are legal in identifiers, they are not used that often in Scala programs, in part to be consistent with Java,

---

[5]This style of naming identifiers is called *camel case* because the identifiersHaveHumps consisting of the embedded capital letters.

but also because underscores have many other non-identifier uses in Scala code. As a result, it is best to avoid identifiers like `to_string`, `__init__`, or `name_`. Camel-case names of fields, method parameters, local variables, and functions should start with a lower case letter, for example: `length`, `flatMap`, and `s`. Camel-case names of classes and traits should start with an upper case letter, for example: `BigInt`, `List`, and `UnbalancedTreeMap`.[6]

**Note**
One consequence of using a trailing underscore in an identifier is that if you attempt, for example, to write a declaration like this, "`val name_: Int = 1`", you'll get a compiler error. The compiler will think you are trying to declare a `val` named "`name_:`". To get this to compile, you would need to insert an extra space before the colon, as in: "`val name_ : Int = 1`".

One way in which Scala's conventions depart from Java's involves constant names. In Scala, the word *constant* does not just mean `val`. Even though a `val` does remain constant after it is initialized, it is still a variable. For example, method parameters are `val`s, but each time the method is called those `val`s can hold different values. A constant is more permanent. For example, `scala.math.Pi` is defined to be the double value closest to the real value of $\pi$, the ratio of a circle's circumference to its diameter. This value is unlikely to change ever; thus, `Pi` is clearly a constant. You can also use constants to give names to values that would otherwise be *magic numbers* in your code: literal values with no explanation, which in the worst case appear in multiple places. You may also want to define constants for use in pattern matching, a use case that will be described in Section 15.2. In Java, the convention is to give constants names that are all upper case, with underscores separating the words, such as `MAX_VALUE` or `PI`. In Scala, the convention is merely that the first character should be upper case. Thus, constants named in the Java style, such as `X_OFFSET`, will work as Scala constants, but the Scala convention is to use camel case for constants, such as `XOffset`.

An *operator identifier* consists of one or more operator characters. Operator characters are printable ASCII characters such as +, :, ?, ~ or #.[7] Here

---

[6]In Section 16.5, you'll see that sometimes you may want to give a special kind of class known as a *case class* a name consisting solely of operator characters. For example, the Scala API contains a class named ::, which facilitates pattern matching on `List`s.

[7]More precisely, an operator character belongs to the Unicode set of mathematical symbols(Sm) or other symbols(So), or to the 7-bit ASCII characters that are not letters, digits,

are some examples of operator identifiers:

+   ++   :::   <?>   :->

The Scala compiler will internally "mangle" operator identifiers to turn them into legal Java identifiers with embedded $ characters. For instance, the identifier :-> would be represented internally as `$colon$minus$greater`. If you ever wanted to access this identifier from Java code, you'd need to use this internal representation.

Because operator identifiers in Scala can become arbitrarily long, there is a small difference between Java and Scala. In Java, the input x<-y would be parsed as four lexical symbols, so it would be equivalent to x < - y. In Scala, <- would be parsed as a single identifier, giving x <- y. If you want the first interpretation, you need to separate the < and the - characters by a space. This is unlikely to be a problem in practice, as very few people would write x<-y in Java without inserting spaces or parentheses between the operators.

A *mixed identifier* consists of an alphanumeric identifier, which is followed by an underscore and an operator identifier. For example, `unary_+` used as a method name defines a unary + operator. Or, `myvar_=` used as method name defines an assignment operator. In addition, the mixed identifier form `myvar_=` is generated by the Scala compiler to support *properties* (more on that in Chapter 18).

A *literal identifier* is an arbitrary string enclosed in back ticks (` ... `). Some examples of literal identifiers are:

`` `x`   `<clinit>`   `yield` ``

The idea is that you can put any string that's accepted by the runtime as an identifier between back ticks. The result is always a Scala identifier. This works even if the name contained in the back ticks would be a Scala reserved word. A typical use case is accessing the static `yield` method in Java's Thread class. You cannot write `Thread.yield()` because `yield` is a reserved word in Scala. However, you can still name the method in back ticks, *e.g.*, `` Thread.`yield`() ``.

---

parentheses, square brackets, curly braces, single or double quote, or an underscore, period, semi-colon, comma, or back tick character.

## 6.11 Method overloading

Back to class Rational. With the latest changes, you can now do addition and multiplication operations in a natural style on rational numbers. But one thing still missing is mixed arithmetic. For instance, you cannot multiply a rational number by an integer because the operands of * always have to be Rationals. So for a rational number r you can't write r * 2. You must write r * new Rational(2), which is not as nice.

To make Rational even more convenient, we'll add new methods to the class that perform mixed addition and multiplication on rational numbers and integers. While we're at it, we'll add methods for subtraction and division too. The result is shown in Listing 6.5.

There are now two versions each of the arithmetic methods: one that takes a rational as its argument and another that takes an integer. In other words, each of these method names is *overloaded* because each name is now being used by multiple methods. For example, the name + is used by one method that takes a Rational and another that takes an Int. In a method call, the compiler picks the version of an overloaded method that correctly matches the types of the arguments. For instance, if the argument y in x.+(y) is a Rational, the compiler will pick the method + that takes a Rational parameter. But if the argument is an integer, the compiler will pick the method + that takes an Int parameter instead. If you try this:

```
scala> val x = new Rational(2, 3)
x: Rational = 2/3

scala> x * x
res12: Rational = 4/9

scala> x * 2
res13: Rational = 4/3
```

You'll see that the * method invoked is determined in each case by the type of the right operand.

> **Note**
> Scala's process of overloaded method resolution is very similar to Java's. In every case, the chosen overloaded version is the one that best matches the static types of the arguments. Sometimes there is no unique best matching version; in that case the compiler will give you an "ambiguous reference" error.

```scala
class Rational(n: Int, d: Int) {

  require(d != 0)

  private val g = gcd(n.abs, d.abs)
  val numer = n / g
  val denom = d / g

  def this(n: Int) = this(n, 1)

  def + (that: Rational): Rational =
    new Rational(
      numer * that.denom + that.numer * denom,
      denom * that.denom
    )

  def + (i: Int): Rational =
    new Rational(numer + i * denom, denom)

  def - (that: Rational): Rational =
    new Rational(
      numer * that.denom - that.numer * denom,
      denom * that.denom
    )

  def - (i: Int): Rational =
    new Rational(numer - i * denom, denom)

  def * (that: Rational): Rational =
    new Rational(numer * that.numer, denom * that.denom)

  def * (i: Int): Rational =
    new Rational(numer * i, denom)

  def / (that: Rational): Rational =
    new Rational(numer * that.denom, denom * that.numer)

  def / (i: Int): Rational =
    new Rational(numer, denom * i)

  override def toString = numer + "/" + denom

  private def gcd(a: Int, b: Int): Int =
    if (b == 0) a else gcd(b, a % b)
}
```

Listing 6.5 · Rational with overloaded methods.

## 6.12   Implicit conversions

Now that you can write r * 2, you might also want to swap the operands, as
in 2 * r. Unfortunately this does not work yet:

```
scala> 2 * r
<console>:10: error: overloaded method value * with
alternatives:
  (x: Double)Double <and>
  (x: Float)Float <and>
  (x: Long)Long <and>
  (x: Int)Int <and>
  (x: Char)Int <and>
  (x: Short)Int <and>
  (x: Byte)Int
 cannot be applied to (Rational)
              2 * r
                ^
```

The problem here is that 2 * r is equivalent to 2.*(r), so it is a method
call on the number 2, which is an integer. But the Int class contains no
multiplication method that takes a Rational argument—it couldn't because
class Rational is not a standard class in the Scala library.

However, there is another way to solve this problem in Scala: You can
create an implicit conversion that automatically converts integers to rational
numbers when needed. Try adding this line in the interpreter:

```
scala> implicit def intToRational(x: Int) = new Rational(x)
```

This defines a conversion method from Int to Rational. The implicit
modifier in front of the method tells the compiler to apply it automatically in
a number of situations. With the conversion defined, you can now retry the
example that failed before:

```
scala> val r = new Rational(2,3)
r: Rational = 2/3

scala> 2 * r
res15: Rational = 4/3
```

For an implicit conversion to work, it needs to be in scope. If you place the implicit method definition inside class `Rational`, it won't be in scope in the interpreter. For now, you'll need to define it directly in the interpreter.

As you can glimpse from this example, implicit conversions are a very powerful technique for making libraries more flexible and more convenient to use. Because they are so powerful, they can also be easily misused. You'll find out more on implicit conversions, including ways to bring them into scope where they are needed, in Chapter 21.

## 6.13   A word of caution

As this chapter has demonstrated, creating methods with operator names and defining implicit conversions can help you design libraries for which client code is concise and easy to understand. Scala gives you a great deal of power to design such easy-to-use libraries. But please bear in mind that with power comes responsibility.

If used unartfully, both operator methods and implicit conversions can give rise to client code that is hard to read and understand. Because implicit conversions are applied implicitly by the compiler, not explicitly written down in the source code, it can be non-obvious to client programmers what implicit conversions are being applied. And although operator methods will usually make client code more concise, they will only make it more readable to the extent client programmers will be able to recognize and remember the meaning of each operator.

The goal you should keep in mind as you design libraries is not merely enabling concise client code, but readable, understandable client code. Conciseness will often be a big part of that readability, but you can take conciseness too far. By designing libraries that enable tastefully concise and at the same time understandable client code, you can help those client programmers work productively.

## 6.14   Conclusion

In this chapter, you saw more aspects of classes in Scala. You saw how to add parameters to a class, define several constructors, define operators as methods, and customize classes so that they are natural to use. Maybe most importantly, you saw that defining and using immutable objects is a quite natural way to code in Scala.

Although the final version of `Rational` shown in this chapter fulfills the requirements set forth at the beginning of the chapter, it could still be improved. We will in fact return to this example later in the book. For example, in Chapter 30, you'll learn how to override `equals` and `hashcode` to allow `Rational`s to behave better when compared with == or placed into hash tables. In Chapter 21, you'll learn how to place implicit method definitions in a companion object for `Rational`, so they can be more easily placed into scope when client programmers are working with `Rational`s.

# Chapter 7

# Built-in Control Structures

Scala has only a handful of built-in control structures. The only control structures are if, while, for, try, match, and function calls. The reason Scala has so few is that it has included function literals since its inception. Instead of accumulating one higher-level control structure after another in the base syntax, Scala accumulates them in libraries. (Chapter 9 will show precisely how that is done.) This chapter will show those few control structures that are built in.

One thing you will notice is that almost all of Scala's control structures result in some value. This is the approach taken by functional languages, where programs are viewed as computing a value, thus the components of a program should also compute values. You can also view this approach as the logical conclusion of a trend already present in imperative languages. In imperative languages, function calls can return a value, even though having the called function update an output variable passed as an argument would work just as well. In addition, imperative languages often have a ternary operator (such as the ?: operator of C, C++, and Java), which behaves exactly like if, but results in a value. Scala adopts this ternary operator model, but calls it if. In other words, Scala's if can result in a value. Scala then continues this trend by having for, try, and match also result in values.

Programmers can use these result values to simplify their code, just as they use return values of functions. Without this facility, the programmer must create temporary variables just to hold results that are calculated inside a control structure. Removing these temporary variables makes the code a little simpler, and it also prevents many bugs where you set the variable in one branch but forget to set it in another.

Overall, Scala's basic control structures, minimal as they are, provide all of the essentials from imperative languages. Further, they allow you to shorten your code by consistently having result values. To show you how this works, we'll take a closer look at each of Scala's basic control structures.

## 7.1  If expressions

Scala's if works just like in many other languages. It tests a condition and then executes one of two code branches depending on whether the condition holds true. Here is a common example, written in an imperative style:

```
var filename = "default.txt"
if (!args.isEmpty)
  filename = args(0)
```

This code declares a variable, filename, and initializes it to a default value. It then uses an if expression to check whether any arguments were supplied to the program. If so, it changes the variable to hold the value specified in the argument list. If no arguments were supplied, it leaves the variable set to the default value.

This code can be written more nicely because, as mentioned in Step 3 in Chapter 2, Scala's if is an expression that results in a value. Listing 7.1 shows how you can accomplish the same effect as the previous example, without using any vars:

```
val filename =
  if (!args.isEmpty) args(0)
  else "default.txt"
```

Listing 7.1 · Scala's idiom for conditional initialization.

This time, the if has two branches. If args is not empty, the initial element, args(0), is chosen; otherwise, the default value is chosen. The if expression results in the chosen value, and the filename variable is initialized with that value. This code is slightly shorter, but its real advantage is that it uses a val instead of a var. Using a val is the functional style, and it helps you in much the same way as a final variable in Java. It tells readers of the code that the variable will never change, saving them from scanning all code in the variable's scope to see if it ever changes.

A second advantage to using a val instead of a var is that it better supports *equational reasoning*. The introduced variable is *equal* to the expression that computes it, assuming that expression has no side effects. Thus, any time you are about to write the variable name, you could instead write the expression. Instead of println(filename), for example, you could just write this:

```
println(if (!args.isEmpty) args(0) else "default.txt")
```

The choice is yours. You can write it either way. Using vals helps you safely make this kind of refactoring as your code evolves over time.

> Look for opportunities to use vals. They can make your code both easier to read and easier to refactor.

## 7.2  While loops

Scala's while loop behaves as in other languages. It has a condition and a body, and the body is executed over and over as long as the condition holds true. Listing 7.2 shows an example:

```
def gcdLoop(x: Long, y: Long): Long = {
  var a = x
  var b = y
  while (a != 0) {
    val temp = a
    a = b % a
    b = temp
  }
  b
}
```

Listing 7.2 · Calculating greatest common divisor with a while loop.

Scala also has a do-while loop. This works like the while loop except that it tests the condition after the loop body instead of before. Listing 7.3

119

shows a Scala script that uses a do-while to echo lines read from the standard input, until an empty line is entered:

```
var line = ""
do {
  line = readLine()
  println("Read: " + line)
} while (line != "")
```

Listing 7.3 · Reading from the standard input with do-while.

The while and do-while constructs are called "loops," not expressions, because they don't result in an interesting value. The type of the result is Unit. It turns out that a value (and in fact, only one value) exists whose type is Unit. It is called the *unit value* and is written (). The existence of () is how Scala's Unit differs from Java's void. Try this in the interpreter:

```
scala> def greet() = { println("hi") }
greet: ()Unit

scala> () == greet()
hi
res0: Boolean = true
```

Because no equals sign precedes its body, greet is defined to be a procedure with a result type of Unit. Therefore, greet returns the unit value, (). This is confirmed in the next line: comparing the greet's result for equality with the unit value, (), yields true.

One other construct that results in the unit value, which is relevant here, is reassignment to vars. For example, were you to attempt to read lines in Scala using the following while loop idiom from Java (and C and C++), you'll run into trouble:

```
var line = ""
while ((line = readLine()) != "") // This doesn't work!
  println("Read: " + line)
```

When you compile this code, Scala will give you a warning that comparing values of type Unit and String using != will always yield true. Whereas in Java, assignment results in the value assigned (in this case a line from the standard input), in Scala assignment always results in the unit value, (). Thus, the value of the assignment "line = readLine()" will always be () and never be "". As a result, this while loop's condition will never be false, and the loop will, therefore, never terminate.

Because the while loop results in no value, it is often left out of pure functional languages. Such languages have expressions, not loops. Scala includes the while loop nonetheless because sometimes an imperative solution can be more readable, especially to programmers with a predominantly imperative background. For example, if you want to code an algorithm that repeats a process until some condition changes, a while loop can express it directly while the functional alternative, which likely uses recursion, may be less obvious to some readers of the code.

For example, Listing 7.4 shows an alternate way to determine a greatest common divisor of two numbers.[1] Given the same two values for x and y, the gcd function shown in Listing 7.4 will return the same result as the gcdLoop function, shown in Listing 7.2. The difference between these two approaches is that gcdLoop is written in an imperative style, using vars and and a while loop, whereas gcd is written in a more functional style that involves recursion (gcd calls itself) and requires no vars.

```
def gcd(x: Long, y: Long): Long =
  if (y == 0) x else gcd(y, x % y)
```

Listing 7.4 · Calculating greatest common divisor with recursion.

In general, we recommend you challenge while loops in your code in the same way you challenge vars. In fact, while loops and vars often go hand in hand. Because while loops don't result in a value, to make any kind of difference to your program, a while loop will usually either need to update vars or perform I/O. You can see this in action in the gcdLoop example shown previously. As that while loop does its business, it updates vars a and b. Thus, we suggest you be a bit suspicious of while loops in your code. If there isn't a good justification for a particular while or do-while loop, try to find a way to do the same thing without it.

---

[1] The gcd function shown in Listing 7.4 uses the same approach used by the like-named function, first shown in Listing 6.3, to calculate greatest common divisors for class Rational. The main difference is that instead of Ints the gcd of Listing 7.4 works with Longs.

## 7.3 For expressions

Scala's for expression is a Swiss army knife of iteration. It lets you combine a few simple ingredients in different ways to express a wide variety of iterations. Simple uses enable common tasks such as iterating through a sequence of integers. More advanced expressions can iterate over multiple collections of different kinds, filter out elements based on arbitrary conditions, and produce new collections.

### Iteration through collections

The simplest thing you can do with for is to iterate through all the elements of a collection. For example, Listing 7.5 shows some code that prints out all files in the current directory. The I/O is performed using the Java API. First, we create a java.io.File on the current directory, ".", and call its listFiles method. This method returns an array of File objects, one per directory and file contained in the current directory. We store the resulting array in the filesHere variable.

```
val filesHere = (new java.io.File(".")).listFiles

for (file <- filesHere)
  println(file)
```

Listing 7.5 · Listing files in a directory with a for expression.

With the "file <- filesHere" syntax, which is called a *generator*, we iterate through the elements of filesHere. In each iteration, a new val named file is initialized with an element value. The compiler infers the type of file to be File, because filesHere is an Array[File]. For each iteration, the body of the for expression, println(file), will be executed. Because File's toString method yields the name of the file or directory, the names of all the files and directories in the current directory will be printed.

The for expression syntax works for any kind of collection, not just arrays.[2] One convenient special case is the Range type, which you briefly saw

---

[2]To be precise, the expression to the right of the <- symbol in a for expression can be any type that has certain methods (in this case foreach) with appropriate signatures. Details on how the Scala compiler processes for expressions are described in Chapter 23.

in Table 5.4 on page 95. You can create Ranges using syntax like "1 to 5" and can iterate through them with a for. Here is a simple example:

```
scala> for (i <- 1 to 4)
         println("Iteration " + i)
Iteration 1
Iteration 2
Iteration 3
Iteration 4
```

If you don't want to include the upper bound of the range in the values that are iterated over, use until instead of to:

```
scala> for (i <- 1 until 4)
         println("Iteration " + i)
Iteration 1
Iteration 2
Iteration 3
```

Iterating through integers like this is common in Scala, but not nearly as much as in other languages. In other languages, you might use this facility to iterate through an array, like this:

```
// Not common in Scala...
for (i <- 0 to filesHere.length - 1)
  println(filesHere(i))
```

This for expression introduces a variable i, sets it in turn to each integer between 0 and filesHere.length – 1, and executes the body of the for expression for each setting of i. For each setting of i, the i'th element of filesHere is extracted and processed.

The reason this kind of iteration is less common in Scala is that you can just iterate over the collection directly. When you do, your code becomes shorter and you sidestep many of the off-by-one errors that can arise when iterating through arrays. Should you start at 0 or 1? Should you add -1, +1, or nothing to the final index? Such questions are easily answered, but also easily answered wrong. It is safer to avoid such questions entirely.

## Filtering

Sometimes you don't want to iterate through a collection in its entirety; you want to filter it down to some subset. You can do this with a for expression by adding a *filter*, an if clause inside the for's parentheses. For example, the code shown in Listing 7.6 lists only those files in the current directory whose names end with ".scala":

```
val filesHere = (new java.io.File(".")).listFiles

for (file <- filesHere if file.getName.endsWith(".scala"))
  println(file)
```

Listing 7.6 · Finding .scala files using a for with a filter.

You could alternatively accomplish the same goal with this code:

```
for (file <- filesHere)
  if (file.getName.endsWith(".scala"))
    println(file)
```

This code yields the same output as the previous code, and likely looks more familiar to programmers with an imperative background. The imperative form, however, is only an option because this particular for expression is executed for its printing side-effects and results in the unit value (). As demonstrated later in this section, the for expression is called an "expression" because it can result in an interesting value, a collection whose type is determined by the for expression's <- clauses.

You can include more filters if you want. Just keep adding if clauses. For example, to be extra defensive, the code in Listing 7.7 prints only files and not directories. It does so by adding a filter that checks the file's isFile method.

```
for (
  file <- filesHere
  if file.isFile
  if file.getName.endsWith(".scala")
) println(file)
```

Listing 7.7 · Using multiple filters in a for expression.

## Nested iteration

If you add multiple <- clauses, you will get nested "loops." For example, the for expression shown in Listing 7.8 has two nested loops. The outer loop iterates through filesHere, and the inner loop iterates through fileLines(file) for any file that ends with .scala.

```
def fileLines(file: java.io.File) =
  scala.io.Source.fromFile(file).getLines().toList

def grep(pattern: String) =
  for (
    file <- filesHere
    if file.getName.endsWith(".scala");
    line <- fileLines(file)
    if line.trim.matches(pattern)
  ) println(file + ": " + line.trim)

grep(".*gcd.*")
```

Listing 7.8 · Using multiple generators in a for expression.

If you prefer, you can use curly braces instead of parentheses to surround the generators and filters. One advantage to using curly braces is that you can leave off some of the semicolons that are needed when you use parentheses because, as explained in Section 4.2, the Scala compiler will not infer semicolons while inside parentheses.

## Mid-stream variable bindings

Note that the previous code repeats the expression line.trim. This is a non-trivial computation, so you might want to only compute it once. You can do this by binding the result to a new variable using an equals sign (=). The bound variable is introduced and used just like a val, only with the val keyword left out. Listing 7.9 shows an example.

In Listing 7.9, a variable named trimmed is introduced halfway through the for expression. That variable is initialized to the result of line.trim. The rest of the for expression then uses the new variable in two places, once in an if and once in println.

```scala
def grep(pattern: String) =
  for {
    file <- filesHere
    if file.getName.endsWith(".scala")
    line <- fileLines(file)
    trimmed = line.trim
    if trimmed.matches(pattern)
  } println(file + ": " + trimmed)
grep(".*gcd.*")
```

Listing 7.9 · Mid-stream assignment in a for expression.

**Producing a new collection**

While all of the examples so far have operated on the iterated values and then forgotten them, you can also generate a value to remember for each iteration. To do so, you prefix the body of the for expression by the keyword yield. For example, here is a function that identifies the .scala files and stores them in an array:

```scala
def scalaFiles =
  for {
    file <- filesHere
    if file.getName.endsWith(".scala")
  } yield file
```

Each time the body of the for expression executes, it produces one value, in this case simply file. When the for expression completes, the result will include all of the yielded values contained in a single collection. The type of the resulting collection is based on the kind of collections processed in the iteration clauses. In this case the result is an Array[File], because filesHere is an array and the type of the yielded expression is File.

Be careful, by the way, where you place the yield keyword. The syntax of a for-yield expression is like this:

**for** *clauses* **yield** *body*

The yield goes before the entire body. Even if the body is a block surrounded by curly braces, put the yield before the first curly brace, not before the last expression of the block. Avoid the temptation to write things like this:

```
for (file <- filesHere if file.getName.endsWith(".scala")) {
  yield file   // Syntax error!
}
```

For example, the for expression shown in Listing 7.10 first transforms the Array[File] named filesHere, which contains all files in the current directory, to one that contains only .scala files. For each of these it generates an Iterator[String], the result of the fileLines method, whose definition is shown in Listing 7.8. An Iterator offers methods next and hasNext that allow you to iterate over a collection of elements. This initial iterator is transformed into another Iterator[String] containing only trimmed lines that include the substring "for". Finally, for each of these, an integer length is yielded. The result of this for expression is an Array[Int] containing those lengths.

```
val forLineLengths =
  for {
    file <- filesHere
    if file.getName.endsWith(".scala")
    line <- fileLines(file)
    trimmed = line.trim
    if trimmed.matches(".*for.*")
  } yield trimmed.length
```

Listing 7.10 · Transforming an Array[File] to Array[Int] with a for.

At this point, you have seen all the major features of Scala's for expression, but we went through them rather quickly. A more thorough coverage of for expressions is given in Chapter 23.

## 7.4 Exception handling with try expressions

Scala's exceptions behave just like in many other languages. Instead of returning a value in the normal way, a method can terminate by throwing an exception. The method's caller can either catch and handle that exception, or it can itself simply terminate, in which case the exception propagates to the caller's caller. The exception propagates in this way, unwinding the call stack, until a method handles it or there are no more methods left.

## Throwing exceptions

Throwing an exception in Scala looks the same as in Java. You create an exception object and then throw it with the `throw` keyword:

```
throw new IllegalArgumentException
```

Although it may seem somewhat paradoxical, in Scala, `throw` is an expression that has a result type. Here's an example where result type matters:

```
val half =
  if (n % 2 == 0)
    n / 2
  else
    throw new RuntimeException("n must be even")
```

What happens here is that if n is even, `half` will be initialized to half of n. If n is not even, an exception will be thrown before `half` can be initialized to anything at all. Because of this, it is safe to treat a thrown exception as any kind of value whatsoever. Any context that tries to use the return from a `throw` will never get to do so, and thus no harm will come.

Technically, an exception throw has type `Nothing`. You can use a `throw` as an expression even though it will never actually evaluate to anything. This little bit of technical gymnastics might sound weird, but is frequently useful in cases like the previous example. One branch of an `if` computes a value, while the other throws an exception and computes `Nothing`. The type of the whole `if` expression is then the type of that branch which does compute something. Type `Nothing` is discussed further in Section 11.3.

## Catching exceptions

You catch exceptions using the syntax shown in Listing 7.11 The syntax for `catch` clauses was chosen for its consistency with an important part of Scala: *pattern matching*. Pattern matching, a powerful feature, is described briefly in this chapter and in more detail in Chapter 15.

128

```scala
import java.io.FileReader
import java.io.FileNotFoundException
import java.io.IOException

try {
  val f = new FileReader("input.txt")
  // Use and close file
} catch {
  case ex: FileNotFoundException => // Handle missing file
  case ex: IOException => // Handle other I/O error
}
```

Listing 7.11 · A try-catch clause in Scala.

The behavior of this try-catch expression is the same as in other languages with exceptions. The body is executed, and if it throws an exception, each catch clause is tried in turn. In this example, if the exception is of type FileNotFoundException, the first clause will execute. If it is of type IOException, the second clause will execute. If the exception is of neither type, the try-catch will terminate and the exception will propagate further.

**Note**

One difference you'll quickly notice in Scala is that, unlike Java, Scala does not require you to catch checked exceptions or declare them in a throws clause. You can declare a throws clause if you wish with the @throws annotation, but it is not required. See Section 31.2 for more information on @throws.

**The finally clause**

You can wrap an expression with a finally clause if you want to cause some code to execute no matter how the expression terminates. For example, you might want to be sure an open file gets closed even if a method exits by throwing an exception. Listing 7.12 shows an example.[3]

---

[3]Although you must always surround the case statements of a catch clause in parentheses, try and finally do not require parentheses if they contain only one expression. For example, you could write: try t() catch { case e: Exception => ... } finally f().

```
import java.io.FileReader

val file = new FileReader("input.txt")
try {
  // Use the file
} finally {
  file.close()  // Be sure to close the file
}
```

Listing 7.12 · A try-finally clause in Scala.

**Note**
Listing 7.12 shows the idiomatic way to ensure a non-memory resource, such as a file, socket, or database connection, is closed. First you acquire the resource. Then you start a try block in which you use the resource. Lastly, you close the resource in a finally block. This idiom is the same in Scala as in Java; alternatively, in Scala you can employ a technique called the *loan pattern* to achieve the same goal more concisely. The loan pattern will be described in Section 9.4.

**Yielding a value**

As with most other Scala control structures, try-catch-finally results in a value. For example, Listing 7.13 shows how you can try to parse a URL but use a default value if the URL is badly formed. The result is that of the try clause if no exception is thrown, or the relevant catch clause if an exception is thrown and caught. If an exception is thrown but not caught, the expression has no result at all. The value computed in the finally clause, if there is one, is dropped. Usually finally clauses do some kind of clean up, such as closing a file. Normally, they should not change the value computed in the main body or a catch clause of the try.

If you're familiar with Java, it's worth noting that Scala's behavior differs from Java only because Java's try-finally does not result in a value. As in Java, if a finally clause includes an explicit return statement, or throws an exception, that return value or exception will "overrule" any previous one that originated in the try block or one of its catch clauses. For example, given this, rather contrived, function definition:

```
def f(): Int = try return 1 finally return 2
```

```
import java.net.URL
import java.net.MalformedURLException

def urlFor(path: String) =
  try {
    new URL(path)
  } catch {
    case e: MalformedURLException =>
      new URL("http://www.scala-lang.org")
  }
```

Listing 7.13 · A catch clause that yields a value.

calling f() results in 2. By contrast, given:

```
def g(): Int = try 1 finally 2
```

calling g() results in 1. Both of these functions exhibit behavior that could surprise most programmers, so it's usually best to avoid returning values from finally clauses. The best way to think of finally clauses is as a way to ensure some side effect happens, such as closing an open file.

## 7.5 Match expressions

Scala's match expression lets you select from a number of *alternatives*, just like switch statements in other languages. In general a match expression lets you select using arbitrary *patterns*, which will be described in Chapter 15. The general form can wait. For now, just consider using match to select among a number of alternatives.

As an example, the script in Listing 7.14 reads a food name from the argument list and prints a companion to that food. This match expression examines firstArg, which has been set to the first argument out of the argument list. If it is the string "salt", it prints "pepper", while if it is the string "chips", it prints "salsa", and so on. The default case is specified with an underscore (_), a wildcard symbol frequently used in Scala as a placeholder for a completely unknown value.

There are a few important differences from Java's switch statement. One is that any kind of constant, as well as other things, can be used in cases

```scala
val firstArg = if (args.length > 0) args(0) else ""

firstArg match {
  case "salt" => println("pepper")
  case "chips" => println("salsa")
  case "eggs" => println("bacon")
  case _ => println("huh?")
}
```

Listing 7.14 · A match expression with side effects.

in Scala, not just the integer-type, enum, and string constants of Java's case statements. In Listing 7.14, the alternatives are strings. Another difference is that there are no breaks at the end of each alternative. Instead the break is implicit, and there is no fall through from one alternative to the next. The common case—not falling through—becomes shorter, and a source of errors is avoided because programmers can no longer fall through by accident.

The most significant difference from Java's switch, however, may be that match expressions result in a value. In the previous example, each alternative in the match expression prints out a value. It would work just as well to yield the value rather than print it, as shown in Listing 7.15. The value that results from this match expression is stored in the friend variable. Aside from the code getting shorter (in number of tokens anyway), the code now disentangles two separate concerns: first it chooses a food and then prints it.

```scala
val firstArg = if (!args.isEmpty) args(0) else ""
val friend =
  firstArg match {
    case "salt" => "pepper"
    case "chips" => "salsa"
    case "eggs" => "bacon"
    case _ => "huh?"
  }
println(friend)
```

Listing 7.15 · A match expression that yields a value.

## 7.6 Living without break and continue

You may have noticed that there has been no mention of break or continue.
Scala leaves out these commands because they do not mesh well with function literals, a feature described in the next chapter. It is clear what continue
means inside a while loop, but what would it mean inside a function literal?
While Scala supports both imperative and functional styles of programming,
in this case it leans slightly towards functional programming in exchange
for simplifying the language. Do not worry, though. There are many ways to
program without break and continue, and if you take advantage of function
literals, those alternatives can often be shorter than the original code.

The simplest approach is to replace every continue by an if and every break by a boolean variable. The boolean variable indicates whether
the enclosing while loop should continue. For example, suppose you are
searching through an argument list for a string that ends with ".scala" but
does not start with a hyphen. In Java you could—if you were quite fond of
while loops, break, and continue—write the following:

```
int i = 0;                  // This is Java
boolean foundIt = false;
while (i < args.length) {
  if (args[i].startsWith("-")) {
    i = i + 1;
    continue;
  }
  if (args[i].endsWith(".scala")) {
    foundIt = true;
    break;
  }
  i = i + 1;
}
```

To transliterate this Java code directly to Scala, instead of doing an if
and then a continue, you could write an if that surrounds the entire remainder of the while loop. To get rid of the break, you would normally
add a boolean variable indicating whether to keep going, but in this case you
can reuse foundIt. Using both of these tricks, the code ends up looking as
shown in Listing 7.16.

133

```
var i = 0
var foundIt = false

while (i < args.length && !foundIt) {
  if (!args(i).startsWith("-")) {
    if (args(i).endsWith(".scala"))
      foundIt = true
  }
  i = i + 1
}
```

Listing 7.16 · Looping without **break** or continue.

This Scala code in Listing 7.16 is quite similar to the original Java code. All the basic pieces are still there and in the same order. There are two reassignable variables and a while loop. Inside the loop, there is a test that i is less than args.length, a check for "-", and a check for ".scala".

If you wanted to get rid of the vars in Listing 7.16, one approach you could try is to rewrite the loop as a recursive function. You could, for example, define a searchFrom function that takes an integer as an input, searches forward from there, and then returns the index of the desired argument. Using this technique the code would look as shown in Listing 7.17:

```
def searchFrom(i: Int): Int =
  if (i >= args.length) -1
  else if (args(i).startsWith("-")) searchFrom(i + 1)
  else if (args(i).endsWith(".scala")) i
  else searchFrom(i + 1)
val i = searchFrom(0)
```

Listing 7.17 · A recursive alternative to looping with vars.

The version in Listing 7.17 gives a human-meaningful name to what the function does, and it uses recursion to substitute for looping. Each continue is replaced by a recursive call with i + 1 as the argument, effectively skipping to the next integer. Many people find this style of programming easier to understand, once they get used to the recursion.

134

**Note**

The Scala compiler will not actually emit a recursive function for the code shown in Listing 7.17. Because all of the recursive calls are in *tail-call* position, the compiler will generate code similar to a while loop. Each recursive call will be implemented as a jump back to the beginning of the function. Tail-call optimization is discussed in Section 8.9.

If after all this discussion you still feel the need to use break, there's help in Scala's standard library. Class Breaks in package scala.util.control offers a break method, which can be used to exit an enclosing block that's marked with breakable. Here is an example how this library-supplied break method could be applied:

```
import scala.util.control.Breaks._
import java.io._

val in = new BufferedReader(new InputStreamReader(System.in))

breakable {
  while (true) {
    println("? ")
    if (in.readLine() == "") break
  }
}
```

This will repeatedly read non-empty lines from the standard input. Once the user enters an empty line, control flow exits from the enclosing breakable, and with it the while loop.

The Breaks class implements break by throwing an exception that is caught by an enclosing application of the breakable method. Therefore, the call to break does not need to be in the same method as the call to breakable.

## 7.7  Variable scope

Now that you've seen Scala's built-in control structures, we'll use them in this section to explain how scoping works in Scala.

135

**Fast track for Java programmers**

If you're a Java programmer, you'll find that Scala's scoping rules are almost identical to Java's. One difference between Java and Scala is that Scala allows you to define variables of the same name in nested scopes. So if you're a Java programmer, you may wish to at least skim this section.

Variable declarations in Scala programs have a *scope* that defines where you can use the name. The most common example of scoping is that curly braces generally introduce a new scope, so anything defined inside curly braces leaves scope after the final closing brace.[4] As an illustration, consider the function shown in Listing 7.18.

The `printMultiTable` function shown in Listing 7.18 prints out a multiplication table.[5] The first statement of this function introduces a variable named i and initializes it to the integer 1. You can then use the name i for the remainder of the function.

The next statement in `printMultiTable` is a while loop:

```
while (i <= 10) {
  var j = 1
  ...
}
```

You can use i here because it is still in scope. In the first statement inside that while loop, you introduce another variable, this time named j, and again initialize it to 1. Because the variable j was defined inside the open curly brace of the while loop, it can be used only within that while loop. If you were to attempt to do something with j after the closing curly brace of this while loop, after the comment that says j, prod, and k are out of scope, your program would not compile.

All variables defined in this example—i, j, prod, and k—are *local variables*. Such variables are "local" to the function in which they are defined. Each time a function is invoked, a new set of its local variables is used.

Once a variable is defined, you can't define a new variable with the same name in the same scope. For example, the following script with two variables named a in the same scope would not compile:

---

[4]There are a few exceptions to this rule because in Scala you can sometimes use curly braces in place of parentheses. One example of this kind of curly-brace use is the alternative for expression syntax described in Section 7.3.

[5]The `printMultiTable` function shown in Listing 7.18 is written in an imperative style. We'll refactor it into a functional style in the next section.

```
def printMultiTable() = {
  var i = 1
  // only i in scope here
  while (i <= 10) {
    var j = 1
    // both i and j in scope here
    while (j <= 10) {
      val prod = (i * j).toString
      // i, j, and prod in scope here
      var k = prod.length
      // i, j, prod, and k in scope here
      while (k < 4) {
        print(" ")
        k += 1
      }
      print(prod)
      j += 1
    }
    // i and j still in scope; prod and k out of scope
    println()
    i += 1
  }
  // i still in scope; j, prod, and k out of scope
}
```

Listing 7.18 · Variable scoping when printing a multiplication table.

```
val a = 1
val a = 2 // Does not compile
println(a)
```

You can, on the other hand, define a variable in an inner scope that has the same name as a variable in an outer scope. The following script would compile and run:

```
val a = 1;
{
  val a = 2 // Compiles just fine
  println(a)
}
println(a)
```

When executed, the script shown previously would print 2 then 1, because the a defined inside the curly braces is a different variable, which is in scope only until the closing curly brace.[6] One difference to note between Scala and Java is that Java will not let you create a variable in an inner scope that has the same name as a variable in an outer scope. In a Scala program, an inner variable is said to *shadow* a like-named outer variable, because the outer variable becomes invisible in the inner scope.

You might have already noticed something that looks like shadowing in the interpreter:

```
scala> val a = 1
a: Int = 1

scala> val a = 2
a: Int = 2

scala> println(a)
2
```

In the interpreter, you can reuse variable names to your heart's content. Among other things, this allows you to change your mind if you made a mistake when you defined a variable the first time in the interpreter. You can do this because conceptually the interpreter creates a new nested scope

---

[6]By the way, the semicolon is required in this case after the first definition of a because Scala's semicolon inference mechanism will not place one there.

for each new statement you type in. Thus, you could visualize the previous interpreted code like this:

```
val a = 1;
{
  val a = 2;
  {
    println(a)
  }
}
```

This code will compile and run as a Scala script, and like the code typed into the interpreter, will print 2. Keep in mind that such code can be very confusing to readers, because variable names adopt new meanings in nested scopes. It is usually better to choose a new, meaningful variable name rather than to shadow an outer variable.

## 7.8  Refactoring imperative-style code

To help you gain insight into the functional style, in this section we'll refactor the imperative approach to printing a multiplication table shown in Listing 7.18. Our functional alternative is shown in Listing 7.19.

The imperative style reveals itself in Listing 7.18 in two ways. First, invoking printMultiTable has a side effect: printing a multiplication table to the standard output. In Listing 7.19, we refactored the function so that it returns the multiplication table as a string. Since the function no longer prints, we renamed it multiTable. As mentioned previously, one advantage of side-effect-free functions is they are easier to unit test. To test printMultiTable, you would need to somehow redefine print and println so you could check the output for correctness. You could test multiTable more easily by checking its string result.

The other telltale sign of the imperative style in printMultiTable is its while loop and vars. By contrast, the multiTable function uses vals, for expressions, *helper functions*, and calls to mkString.

We factored out the two helper functions, makeRow and makeRowSeq, to make the code easier to read. Function makeRowSeq uses a for expression whose generator iterates through column numbers 1 through 10. The body of this for calculates the product of row and column, determines the padding

139

```scala
// Returns a row as a sequence
def makeRowSeq(row: Int) =
  for (col <- 1 to 10) yield {
    val prod = (row * col).toString
    val padding = " " * (4 - prod.length)
    padding + prod
  }

// Returns a row as a string
def makeRow(row: Int) = makeRowSeq(row).mkString

// Returns table as a string with one row per line
def multiTable() = {

  val tableSeq = // a sequence of row strings
    for (row <- 1 to 10)
    yield makeRow(row)

  tableSeq.mkString("\n")
}
```

Listing 7.19 · A functional way to create a multiplication table.

needed for the product, and yields the result of concatenating the padding and product strings. The result of the for expression will be a sequence (some subclass of scala.Seq) containing these yielded strings as elements. The other helper function, makeRow, simply invokes mkString on the result returned by makeRowSeq. mkString will concatenate the strings in the sequence and return them as one string.

The multiTable method first initializes tableSeq with the result of a for expression whose generator iterates through row numbers 1 to 10, and for each calls makeRow to get the string for that row. This string is yielded; thus the result of this for expression will be a sequence of row strings. The only remaining task is to convert the sequence of strings into a single string. The call to mkString accomplishes this, and because we pass "\n", we get an end of line character inserted between each string. If you pass the string returned by multiTable to println, you'll see the same output that's produced by calling printMultiTable.

| 1  | 2  | 3  | 4  | 5  | 6  | 7  | 8  | 9  | 10  |
|----|----|----|----|----|----|----|----|----|-----|
| 2  | 4  | 6  | 8  | 10 | 12 | 14 | 16 | 18 | 20  |
| 3  | 6  | 9  | 12 | 15 | 18 | 21 | 24 | 27 | 30  |
| 4  | 8  | 12 | 16 | 20 | 24 | 28 | 32 | 36 | 40  |
| 5  | 10 | 15 | 20 | 25 | 30 | 35 | 40 | 45 | 50  |
| 6  | 12 | 18 | 24 | 30 | 36 | 42 | 48 | 54 | 60  |
| 7  | 14 | 21 | 28 | 35 | 42 | 49 | 56 | 63 | 70  |
| 8  | 16 | 24 | 32 | 40 | 48 | 56 | 64 | 72 | 80  |
| 9  | 18 | 27 | 36 | 45 | 54 | 63 | 72 | 81 | 90  |
| 10 | 20 | 30 | 40 | 50 | 60 | 70 | 80 | 90 | 100 |

## 7.9 Conclusion

Scala's built-in control structures are minimal, but they do the job. They act much like their imperative equivalents, but because they tend to result in a value, they support a functional style, too. Just as important, they are careful in what they omit, thus leaving room for one of Scala's most powerful features, the function literal, which will be described in the next chapter.

# Chapter 8

# Functions and Closures

When programs get larger, you need some way to divide them into smaller, more manageable pieces. For dividing up control flow, Scala offers an approach familiar to all experienced programmers: divide the code into functions. In fact, Scala offers several ways to define functions that are not present in Java. Besides methods, which are functions that are members of some object, there are also functions nested within functions, function literals, and function values. This chapter takes you on a tour through all of these flavors of functions in Scala.

## 8.1 Methods

The most common way to define a function is as a member of some object; such a function is called a *method*. As an example, Listing 8.1 shows two methods that together read a file with a given name and print out all lines whose length exceeds a given width. Every printed line is prefixed with the name of the file it appears in.

The processFile method takes a filename and width as parameters. It creates a Source object from the file name and, in the generator of the for expression, calls getLines on the source. As mentioned in Step 12 of Chapter 3, getLines returns an iterator that provides one line from the file on each iteration, excluding the end-of-line character. The for expression processes each of these lines by calling the helper method, processLine. The processLine method takes three parameters: a filename, a width, and a line. It tests whether the length of the line is greater than the given width, and, if so, it prints the filename, a colon, and the line.

```
import scala.io.Source

object LongLines {

  def processFile(filename: String, width: Int) = {
    val source = Source.fromFile(filename)
    for (line <- source.getLines())
      processLine(filename, width, line)
  }

  private def processLine(filename: String,
      width: Int, line: String) = {

    if (line.length > width)
      println(filename + ": " + line.trim)
  }
}
```

Listing 8.1 · LongLines with a private processLine method.

To use LongLines from the command line, we'll create an application
that expects the line width as the first command-line argument, and interprets
subsequent arguments as filenames:[1]

```
object FindLongLines {
  def main(args: Array[String]) = {
    val width = args(0).toInt
    for (arg <- args.drop(1))
      LongLines.processFile(arg, width)
  }
}
```

Here's how you'd use this application to find the lines in LongLines.scala
that are over 45 characters in length (there's just one):

```
$ scala FindLongLines 45 LongLines.scala
LongLines.scala: def processFile(filename: String, width: Int) = {
```

---

[1]In this book, we usually won't check command-line arguments for validity in example
applications, both to save trees and reduce boilerplate code that can obscure the example's
important code. The trade-off is that instead of producing a helpful error message when given
bad input, our example applications will throw an exception.

So far, this is very similar to what you would do in any object-oriented language. However, the concept of a function in Scala is more general than a method. Scala's other ways to express functions will be explained in the following sections.

## 8.2 Local functions

The construction of the processFile method in the previous section demonstrated an important design principle of the functional programming style: programs should be decomposed into many small functions that each do a well-defined task. Individual functions are often quite small. The advantage of this style is that it gives a programmer many building blocks that can be flexibly composed to do more difficult things. Each building block should be simple enough to be understood individually.

One problem with this approach is that all the helper function names can pollute the program namespace. In the interpreter this is not so much of a problem, but once functions are packaged in reusable classes and objects, it's desirable to hide the helper functions from clients of a class. They often do not make sense individually, and you often want to keep enough flexibility to delete the helper functions if you later rewrite the class a different way.

In Java, your main tool for this purpose is the private method. This private-method approach works in Scala as well, as demonstrated in Listing 8.1, but Scala offers an additional approach: you can define functions inside other functions. Just like local variables, such local functions are visible only in their enclosing block. Here's an example:

```
def processFile(filename: String, width: Int) = {
  def processLine(filename: String,
      width: Int, line: String) = {
    if (line.length > width)
      println(filename + ": " + line.trim)
  }
  val source = Source.fromFile(filename)
  for (line <- source.getLines()) {
    processLine(filename, width, line)
  }
}
```

In this example, we refactored the original LongLines version, shown in Listing 8.1, by transforming private method, processLine, into a local function of processFile. To do so we removed the private modifier, which can only be applied (and is only needed) for members, and placed the definition of processLine inside the definition of processFile. As a local function, processLine is in scope inside processFile, but inaccessible outside.

Now that processLine is defined inside processFile, however, another improvement becomes possible. Notice how filename and width are passed unchanged into the helper function? This is not necessary because local functions can access the parameters of their enclosing function. You can just use the parameters of the outer processLine function, as shown in Listing 8.2.

```scala
import scala.io.Source
object LongLines {
  def processFile(filename: String, width: Int) = {
    def processLine(line: String) = {
      if (line.length > width)
        println(filename + ": " + line.trim)
    }
    val source = Source.fromFile(filename)
    for (line <- source.getLines())
      processLine(line)
  }
}
```

Listing 8.2 · LongLines with a local processLine function.

Simpler, isn't it? This use of an enclosing function's parameters is a common and useful example of the general nesting Scala provides. The nesting and scoping described in Section 7.7 applies to all Scala constructs, including functions. It's a simple principle, but very powerful, especially in a language with first-class functions.

## 8.3 First-class functions

Scala has *first-class functions*. Not only can you define functions and call them, but you can write down functions as unnamed *literals* and then pass them around as *values*. We introduced function literals in Chapter 2 and showed the basic syntax in Figure 2.2 on page 34.

A function literal is compiled into a class that when instantiated at runtime is a *function value*.[2] Thus the distinction between function literals and values is that function literals exist in the source code, whereas function values exist as objects at runtime. The distinction is much like that between classes (source code) and objects (runtime).

Here is a simple example of a function literal that adds one to a number:

```
(x: Int) => x + 1
```

The => designates that this function converts the thing on the left (any integer x) to the thing on the right (x + 1). So, this is a function mapping any integer x to x + 1.

Function values are objects, so you can store them in variables if you like. They are functions, too, so you can invoke them using the usual parentheses function-call notation. Here is an example of both activities:

```
scala> var increase = (x: Int) => x + 1
increase: Int => Int = <function1>

scala> increase(10)
res0: Int = 11
```

Because increase, in this example, is a var, you can assign a different function value to it later on.

```
scala> increase = (x: Int) => x + 9999
increase: Int => Int = <function1>

scala> increase(10)
res1: Int = 10009
```

---

[2]Every function value is an instance of some class that extends one of several FunctionN traits in package scala, such as Function0 for functions with no parameters, Function1 for functions with one parameter, and so on. Each FunctionN trait has an apply method used to invoke the function.

147

If you want to have more than one statement in the function literal, surround its body by curly braces and put one statement per line, thus forming a block. Just like a method, when the function value is invoked, all of the statements will be executed, and the value returned from the function is whatever results from evaluating the last expression.

```
scala> increase = (x: Int) => {
         println("We")
         println("are")
         println("here!")
         x + 1
       }
increase: Int => Int = <function1>

scala> increase(10)
We
are
here!
res2: Int = 11
```

So now you have seen the nuts and bolts of function literals and function values. Many Scala libraries give you opportunities to use them. For example, a foreach method is available for all collections.[3] It takes a function as an argument and invokes that function on each of its elements. Here is how it can be used to print out all of the elements of a list:

```
scala> val someNumbers = List(-11, -10, -5, 0, 5, 10)
someNumbers: List[Int] = List(-11, -10, -5, 0, 5, 10)

scala> someNumbers.foreach((x: Int) => println(x))
-11
-10
-5
0
5
10
```

---

[3]A foreach method is defined in trait Traversable, a common supertrait of List, Set, Array, and Map. See Chapter 17 for the details.

148

As another example, collection types also have a `filter` method. This method selects those elements of a collection that pass a test the user supplies. That test is supplied using a function. For example, the function `(x: Int) => x > 0` could be used for filtering. This function maps positive integers to true and all others to false. Here is how to use it with `filter`:

```
scala> someNumbers.filter((x: Int) => x > 0)
res4: List[Int] = List(5, 10)
```

Methods like `foreach` and `filter` are described further later in the book. Chapter 16 talks about their use in class `List`. Chapter 17 discusses their use with other collection types.

## 8.4   Short forms of function literals

Scala provides a number of ways to leave out redundant information and write function literals more briefly. Keep your eyes open for these opportunities, because they allow you to remove clutter from your code.

One way to make a function literal more brief is to leave off the parameter types. Thus, the previous example with filter could be written like this:

```
scala> someNumbers.filter((x) => x > 0)
res5: List[Int] = List(5, 10)
```

The Scala compiler knows that x must be an integer, because it sees that you are immediately using the function to filter a list of integers (referred to by someNumbers). This is called *target typing* because the targeted usage of an expression (in this case, an argument to `someNumbers.filter()`) is allowed to influence the typing of that expression (in this case to determine the type of the x parameter). The precise details of target typing are not important. You can simply start by writing a function literal without the argument type, and if the compiler gets confused, add in the type. Over time you'll get a feel for which situations the compiler can and cannot puzzle out.

A second way to remove useless characters is to leave out parentheses around a parameter whose type is inferred. In the previous example, the parentheses around x are unnecessary:

```
scala> someNumbers.filter(x => x > 0)
res6: List[Int] = List(5, 10)
```

## 8.5 Placeholder syntax

To make a function literal even more concise, you can use underscores as placeholders for one or more parameters, so long as each parameter appears only one time within the function literal. For example, _ > 0 is very short notation for a function that checks whether a value is greater than zero:

```
scala> someNumbers.filter(_ > 0)
res7: List[Int] = List(5, 10)
```

You can think of the underscore as a "blank" in the expression that needs to be "filled in." This blank will be filled in with an argument to the function each time the function is invoked. For example, given that someNumbers was initialized on page 148 to the value List(-11, -10, -5, 0, 5, 10), the filter method will replace the blank in _ > 0 first with -11, as in -11 > 0, then with -10, as in -10 > 0, then with -5, as in -5 > 0, and so on to the end of the List. The function literal _ > 0, therefore, is equivalent to the slightly more verbose x => x > 0, as demonstrated here:

```
scala> someNumbers.filter(x => x > 0)
res8: List[Int] = List(5, 10)
```

Sometimes when you use underscores as placeholders for parameters, the compiler might not have enough information to infer missing parameter types. For example, suppose you write _ + _ by itself:

```
scala> val f = _ + _
<console>:7: error: missing parameter type for expanded
function ((x$1, x$2) => x$1.$plus(x$2))
       val f = _ + _
               ^
```

In such cases, you can specify the types using a colon, like this:

```
scala> val f = (_: Int) + (_: Int)
f: (Int, Int) => Int = <function2>

scala> f(5, 10)
res9: Int = 15
```

Note that _ + _ expands into a literal for a function that takes two parameters. This is why you can use this short form only if each parameter appears in the function literal exactly once. Multiple underscores mean multiple parameters, not reuse of a single parameter repeatedly. The first underscore represents the first parameter, the second underscore the second parameter, the third underscore the third parameter, and so on.

## 8.6 Partially applied functions

Although the previous examples substitute underscores in place of individual parameters, you can also replace an entire parameter list with an underscore. For example, rather than writing println(_), you could write println _. Here's an example:

```
someNumbers.foreach(println _)
```

Scala treats this short form exactly as if you had written the following:

```
someNumbers.foreach(x => println(x))
```

Thus, the underscore in this case is not a placeholder for a single parameter. It is a placeholder for an entire parameter list. Remember that you need to leave a space between the function name and the underscore; otherwise, the compiler will think you are referring to a different symbol, such as, for example, a method named println_, which likely does not exist.

When you use an underscore in this way, you are writing a *partially applied function*. In Scala, when you invoke a function, passing in any needed arguments, you *apply* that function *to* the arguments. For example, given the following function:

```
scala> def sum(a: Int, b: Int, c: Int) = a + b + c
sum: (a: Int, b: Int, c: Int)Int
```

You could apply the function sum to the arguments 1, 2, and 3 like this:

```
scala> sum(1, 2, 3)
res10: Int = 6
```

A partially applied function is an expression in which you don't supply all of the arguments needed by the function. Instead, you supply some, or none, of the needed arguments. For example, to create a partially applied function expression involving sum, in which you supply none of the three required arguments, you just place an underscore after "sum". The resulting function can then be stored in a variable. Here's an example:

```
scala> val a = sum _
a: (Int, Int, Int) => Int = <function3>
```

Given this code, the Scala compiler instantiates a function value that takes the three integer parameters missing from the partially applied function expression, sum _, and assigns a reference to that new function value to the variable a. When you apply three arguments to this new function value, it will turn around and invoke sum, passing in those same three arguments:

```
scala> a(1, 2, 3)
res11: Int = 6
```

Here's what just happened: The variable named a refers to a function value object. This function value is an instance of a class generated automatically by the Scala compiler from sum _, the partially applied function expression. The class generated by the compiler has an apply method that takes three arguments.[4] The generated class's apply method takes three arguments because three is the number of arguments missing in the sum _ expression. The Scala compiler translates the expression a(1, 2, 3) into an invocation of the function value's apply method, passing in the three arguments 1, 2, and 3. Thus, a(1, 2, 3) is a short form for:

```
scala> a.apply(1, 2, 3)
res12: Int = 6
```

This apply method, defined in the class generated automatically by the Scala compiler from the expression sum _, simply forwards those three missing parameters to sum, and returns the result. In this case apply invokes sum(1, 2, 3), and returns what sum returns, which is 6.

Another way to think about this kind of expression, in which an underscore is used to represent an entire parameter list, is as a way to transform a def into a function value. For example, if you have a local function, such as sum(a: Int, b: Int, c: Int): Int, you can "wrap" it in a function value whose apply method has the same parameter list and result types. When you apply this function value to some arguments, it in turn applies sum to

---

[4]The generated class extends trait Function3, which declares a three-arg apply method.

those same arguments and returns the result. Although you can't assign a method or nested function to a variable, or pass it as an argument to another function, you can do these things if you wrap the method or nested function in a function value by placing an underscore after its name.

Now, although sum _ is indeed a partially applied function, it may not be obvious to you why it is called this. It has this name because you are not applying that function to all of its arguments. In the case of sum _, you are applying it to *none* of its arguments. But you can also express a partially applied function by supplying only *some* of the required arguments. Here's an example:

```scala
scala> val b = sum(1, _: Int, 3)
b: Int => Int = <function1>
```

In this case, you've supplied the first and last argument to sum, but not the middle argument. Since only one argument is missing, the Scala compiler generates a new function class whose apply method takes one argument. When invoked with that one argument, this generated function's apply method invokes sum, passing in 1, the argument passed to the function, and 3. Here are some examples:

```scala
scala> b(2)
res13: Int = 6
```

In this case, b.apply invoked sum(1, 2, 3).

```scala
scala> b(5)
res14: Int = 9
```

And in this case, b.apply invoked sum(1, 5, 3).

If you are writing a partially applied function expression in which you leave off all parameters, such as println _ or sum _, you can express it more concisely by leaving off the underscore if a function is required at that point in the code. For example, instead of printing out each of the numbers in someNumbers (defined on page 148) like this:

```scala
someNumbers.foreach(println _)
```

You could just write:

```scala
someNumbers.foreach(println)
```

This last form is allowed only in places where a function is required, such as the invocation of foreach in this example. The compiler knows a function is required in this case, because foreach requires that a function be passed as an argument. In situations where a function is not required, attempting to use this form will cause a compilation error. Here's an example:

```
scala> val c = sum
<console>:8: error: missing arguments for method sum;
follow this method with `_' if you want to treat it as a
partially applied function
        val c = sum
              ^

scala> val d = sum _
d: (Int, Int, Int) => Int = <function3>

scala> d(10, 20, 30)
res14: Int = 60
```

## 8.7 Closures

So far in this chapter, all the examples of function literals have referred only to passed parameters. For example, in (x: Int) => x > 0, the only variable used in the function body, x > 0, is x, which is defined as a parameter to the function. You can, however, refer to variables defined elsewhere:

```
(x: Int) => x + more   // how much more?
```

This function adds "more" to its argument, but what is more? From the point of view of this function, more is a *free variable* because the function literal does not itself give a meaning to it. The x variable, by contrast, is a *bound variable* because it does have a meaning in the context of the function: it is defined as the function's lone parameter, an Int. If you try using this function literal by itself, without any more defined in its scope, the compiler will complain:

```
scala> (x: Int) => x + more
<console>:8: error: not found: value more
            (x: Int) => x + more
                          ^
```

---

## Why the trailing underscore?

Scala's syntax for partially applied functions highlights a difference
in the design trade-offs of Scala and classical functional languages,
such as Haskell or ML. In these languages, partially applied functions
are considered the normal case. Furthermore, these languages have
a fairly strict static type system that will usually highlight every error
with partial applications that you can make. Scala bears a much closer
relation to imperative languages, such as Java, where a method that's
not applied to all its arguments is considered an error. Furthermore,
the object-oriented tradition of subtyping and a universal root type
accepts some programs that would be considered erroneous in classical
functional languages.

For instance, say you mistook the drop(n: Int) method of List
for tail(), and therefore forgot you need to pass a number to drop.
You might write, "println(drop)". Had Scala adopted the classical
functional tradition that partially applied functions are OK everywhere,
this code would type check. However, you might be surprised to find
out that the output printed by this println statement would always be
<function>! What would have happened is that the expression drop
would have been treated as a function object. Because println takes
objects of any type, this would have compiled OK, but it would have
given an unexpected result.

To avoid situations like this, Scala normally requires you to specify
function arguments that are left out explicitly, even if the indication is
as simple as a '_'. Scala allows you to leave off even the _ only when a
function type is expected.

---

On the other hand, the same function literal will work fine so long as
there is something available named more:

```
scala> var more = 1
more: Int = 1

scala> val addMore = (x: Int) => x + more
addMore: Int => Int = <function1>

scala> addMore(10)
res16: Int = 11
```

The function value (the object) that's created at runtime from this function literal is called a *closure*. The name arises from the act of "closing" the function literal by "capturing" the bindings of its free variables. A function literal with no free variables, such as (x: Int) => x + 1, is called a *closed term*, where a *term* is a bit of source code. Thus a function value created at runtime from this function literal is not a closure in the strictest sense, because (x: Int) => x + 1 is already closed as written. But any function literal with free variables, such as (x: Int) => x + more, is an *open term*. Therefore, any function value created at runtime from (x: Int) => x + more will, by definition, require that a binding for its free variable, more, be captured. The resulting function value, which will contain a reference to the captured more variable, is called a closure because the function value is the end product of the act of closing the open term, (x: Int) => x + more.

This example brings up a question: What happens if more changes after the closure is created? In Scala, the answer is that the closure sees the change. For example:

```
scala> more = 9999
more: Int = 9999

scala> addMore(10)
res17: Int = 10009
```

Intuitively, Scala's closures capture variables themselves, not the value to which variables refer.[5] As the previous example shows, the closure created for (x: Int) => x + more sees the change to more made outside the closure. The same is true in the opposite direction. Changes made by a closure to a captured variable are visible outside the closure. Here's an example:

```
scala> val someNumbers = List(-11, -10, -5, 0, 5, 10)
someNumbers: List[Int] = List(-11, -10, -5, 0, 5, 10)

scala> var sum = 0
sum: Int = 0

scala> someNumbers.foreach(sum +=  _)
```

---

[5]By contrast, Java's inner classes do not allow you to access modifiable variables in surrounding scopes at all, so there is no difference between capturing a variable and capturing its currently held value.

156

```
scala> sum
res19: Int = -11
```

This example uses a roundabout way to sum the numbers in a List. Variable sum is in a surrounding scope from the function literal sum += _, which adds numbers to sum. Even though it is the closure modifying sum at runtime, the resulting total, -11, is still visible outside the closure.

What if a closure accesses some variable that has several different copies as the program runs? For example, what if a closure uses a local variable of some function, and the function is invoked many times? Which instance of that variable gets used at each access?

Only one answer is consistent with the rest of the language: the instance used is the one that was active at the time the closure was created. For example, here is a function that creates and returns "increase" closures:

```
def makeIncreaser(more: Int) = (x: Int) => x + more
```

Each time this function is called it will create a new closure. Each closure will access the more variable that was active when the closure was created.

```
scala> val inc1 = makeIncreaser(1)
inc1: Int => Int = <function1>

scala> val inc9999 = makeIncreaser(9999)
inc9999: Int => Int = <function1>
```

When you call makeIncreaser(1), a closure is created and returned that captures the value 1 as the binding for more. Similarly, when you call makeIncreaser(9999), a closure that captures the value 9999 for more is returned. When you apply these closures to arguments (in this case, there's just one argument, x, which must be passed in), the result that comes back depends on how more was defined when the closure was created:

```
scala> inc1(10)
res20: Int = 11

scala> inc9999(10)
res21: Int = 10009
```

It makes no difference that the more in this case is a parameter to a method call that has already returned. The Scala compiler rearranges things in cases

157

like these so that the captured parameter lives out on the heap, instead of the stack, and thus can outlive the method call that created it. This rearrangement is all taken care of automatically, so you don't have to worry about it. Capture any variable you like: val, var, or parameter.

## 8.8   Special function call forms

Most functions and function calls you encounter will be as you have seen so far in this chapter. The function will have a fixed number of parameters, the call will have an equal number of arguments, and the arguments will be specified in the same order and number as the parameters.

Since function calls are so central to programming in Scala, however, a few special forms of function definitions and function calls have been added to the language to address some special needs. Scala supports repeated parameters, named arguments, and default arguments.

### Repeated parameters

Scala allows you to indicate that the last parameter to a function may be repeated. This allows clients to pass variable length argument lists to the function. To denote a repeated parameter, place an asterisk after the type of the parameter. For example:

```
scala> def echo(args: String*) =
         for (arg <- args) println(arg)
echo: (args: String*)Unit
```

Defined this way, echo can be called with zero to many String arguments:

```
scala> echo()

scala> echo("one")
one

scala> echo("hello", "world!")
hello
world!
```

Inside the function, the type of the repeated parameter is an Array of the declared type of the parameter. Thus, the type of args inside the echo function, which is declared as type "String*" is actually Array[String]. Nevertheless, if you have an array of the appropriate type, and you attempt to pass it as a repeated parameter, you'll get a compiler error:

```
scala> val arr = Array("What's", "up", "doc?")
arr: Array[String] = Array(What's, up, doc?)

scala> echo(arr)
<console>:10: error: type mismatch;
 found   : Array[String]
 required: String
              echo(arr)
                   ^
```

To accomplish this, you'll need to append the array argument with a colon and an _* symbol, like this:

```
scala> echo(arr: _*)
What's
up
doc?
```

This notation tells the compiler to pass each element of arr as its own argument to echo, rather than all of it as a single argument.

### Named arguments

In a normal function call, the arguments in the call are matched one by one in the order of the parameters of the called function:

```
scala> def speed(distance: Float, time: Float): Float =
          distance / time
speed: (distance: Float, time: Float)Float

scala> speed(100, 10)
res27: Float = 10.0
```

In this call, the 100 is matched to distance and the 10 to time. The 100 and 10 are matched in the same order as the formal parameters are listed.

Named arguments allow you to pass arguments to a function in a different order. The syntax is simply that each argument is preceded by a parameter name and an equals sign. For example, the following call to speed is equivalent to speed(100,10):

```
scala> speed(distance = 100, time = 10)
res28: Float = 10.0
```

Called with named arguments, the arguments can be reversed without changing the meaning:

```
scala> speed(time = 10, distance = 100)
res29: Float = 10.0
```

It is also possible to mix positional and named arguments. In that case, the positional arguments come first. Named arguments are most frequently used in combination with default parameter values.

### Default parameter values

Scala lets you specify default values for function parameters. The argument for such a parameter can optionally be omitted from a function call, in which case the corresponding argument will be filled in with the default.

An example is shown in Listing 8.3. Function printTime has one parameter, out, and it has a default value of Console.out.

```
def printTime(out: java.io.PrintStream = Console.out) =
  out.println("time = " + System.currentTimeMillis())
```

Listing 8.3 · A parameter with a default value.

If you call the function as printTime(), thus specifying no argument to be used for out, then out will be set to its default value of Console.out. You could also call the function with an explicit output stream. For example, you could send logging to the standard error output by calling the function as printTime(Console.err).

Default parameters are especially helpful when used in combination with named parameters. In Listing 8.4, function printTime2 has two optional parameters. The out parameter has a default of Console.out, and the divisor parameter has a default value of 1.

Function printTime2 can be called as printTime2() to have both parameters filled in with their default values. Using named arguments, however, either one of the parameters can be specified while leaving the other as the default. To specify the output stream, call it like this:

```
def printTime2(out: java.io.PrintStream = Console.out,
               divisor: Int = 1) =
  out.println("time = " + System.currentTimeMillis()/divisor)
```

Listing 8.4 · A function with two parameters that have defaults.

```
printTime2(out = Console.err)
```

To specify the time divisor, call it like this:

```
printTime2(divisor = 1000)
```

## 8.9   Tail recursion

In Section 7.2, we mentioned that to transform a while loop that updates vars into a more functional style that uses only vals, you may sometimes need to use recursion. Here's an example of a recursive function that approximates a value by repeatedly improving a guess until it is good enough:

```
def approximate(guess: Double): Double =
  if (isGoodEnough(guess)) guess
  else approximate(improve(guess))
```

A function like this is often used in search problems, with appropriate implementations for isGoodEnough and improve. If you want the approximate function to run faster, you might be tempted to write it with a while loop to try and speed it up, like this:

```
def approximateLoop(initialGuess: Double): Double = {
  var guess = initialGuess
  while (!isGoodEnough(guess))
    guess = improve(guess)
  guess
}
```

Which of the two versions of approximate is preferable? In terms of brevity and var avoidance, the first, functional one wins. But is the imperative approach perhaps more efficient? In fact, if we measure execution times, it turns out that they are almost exactly the same!

161

This might seem surprising because a recursive call looks much more "expansive" than a simple jump from the end of a loop to its beginning. However, in the case of approximate above, the Scala compiler is able to apply an important optimization. Note that the recursive call is the last thing that happens in the evaluation of function approximate's body. Functions like approximate, which call themselves as their last action, are called *tail recursive*. The Scala compiler detects tail recursion and replaces it with a jump back to the beginning of the function, after updating the function parameters with the new values.

The moral is that you should not shy away from using recursive algorithms to solve your problem. Often, a recursive solution is more elegant and concise than a loop-based one. If the solution is tail recursive, there won't be any runtime overhead to be paid.

### Tracing tail-recursive functions

A tail-recursive function will not build a new stack frame for each call; all calls will execute in a single frame. This may surprise a programmer inspecting a stack trace of a program that failed. For example, this function calls itself some number of times then throws an exception:

```
def boom(x: Int): Int =
  if (x == 0) throw new Exception("boom!")
  else boom(x - 1) + 1
```

This function is *not* tail recursive, because it performs an increment operation after the recursive call. You'll get what you expect when you run it:

```
scala>  boom(3)
java.lang.Exception: boom!
      at .boom(<console>:5)
      at .boom(<console>:6)
      at .boom(<console>:6)
      at .boom(<console>:6)
      at .<init>(<console>:6)
...
```

If you now modify boom so that it does become tail recursive:

## Tail call optimization

The compiled code for `approximate` is essentially the same as the compiled code for `approximateLoop`. Both functions compile down to the same thirteen instructions of Java bytecodes. If you look through the bytecodes generated by the Scala compiler for the tail recursive method, `approximate`, you'll see that although both `isGoodEnough` and `improve` are invoked in the body of the method, `approximate` is not. The Scala compiler optimized away the recursive call:

```
public double approximate(double);
  Code:
    0:   aload_0
    1:   astore_3
    2:   aload_0
    3:   dload_1
    4:   invokevirtual    #24; //Method isGoodEnough:(D)Z
    7:   ifeq    12
    10:  dload_1
    11:  dreturn
    12:  aload_0
    13:  dload_1
    14:  invokevirtual    #27; //Method improve:(D)D
    17:  dstore_1
    18:  goto    2
```

```
def bang(x: Int): Int =
  if (x == 0) throw new Exception("bang!")
  else bang(x - 1)
```

You'll get:

```
scala> bang(5)
java.lang.Exception: bang!
    at .bang(<console>:5)
    at .<init>(<console>:6) ...
```

This time, you see only a single stack frame for bang. You might think that bang crashed before it called itself, but this is not the case. If you think you might be confused by tail-call optimizations when looking at a stack trace, you can turn them off by giving the following argument to the scala shell or to the scalac compiler:

```
-g:notailcalls
```

With that option specified, you will get a longer stack trace:

```
scala> bang(5)
java.lang.Exception: bang!
        at .bang(<console>:5)
        at .bang(<console>:5)
        at .bang(<console>:5)
        at .bang(<console>:5)
        at .bang(<console>:5)
        at .bang(<console>:5)
        at .<init>(<console>:6) ...
```

**Limits of tail recursion**

The use of tail recursion in Scala is fairly limited because the JVM instruction set makes implementing more advanced forms of tail recursion very difficult. Scala only optimizes directly recursive calls back to the same function making the call. If the recursion is indirect, as in the following example of two mutually recursive functions, no optimization is possible:

```
def isEven(x: Int): Boolean =
  if (x == 0) true else isOdd(x - 1)
def isOdd(x: Int): Boolean =
  if (x == 0) false else isEven(x - 1)
```

You also won't get a tail-call optimization if the final call goes to a function value. Consider for instance the following recursive code:

```
val funValue = nestedFun _
def nestedFun(x: Int) : Unit = {
  if (x != 0) { println(x); funValue(x - 1) }
}
```

The funValue variable refers to a function value that essentially wraps a call to nestedFun. When you apply the function value to an argument, it turns around and applies nestedFun to that same argument, and returns the result. Therefore, you might hope the Scala compiler would perform a tail-call optimization, but in this case it would not. Tail-call optimization is limited to situations where a method or nested function calls itself directly as its last operation, without going through a function value or some other intermediary. (If you don't fully understand tail recursion yet, see Section 8.9).

## 8.10 Conclusion

This chapter has given you a grand tour of functions in Scala. In addition to methods, Scala provides local functions, function literals, and function values. In addition to normal function calls, Scala provides partially applied functions and functions with repeated parameters. When possible, function calls are implemented as optimized tail calls, and thus many nice-looking recursive functions run just as quickly as hand-optimized versions that use while loops. The next chapter will build on these foundations and show how Scala's rich support for functions helps you abstract over control.

# Chapter 9

# Control Abstraction

In Chapter 7, we pointed out that Scala doesn't have many built-in control abstractions because it gives you the ability to create your own. In the previous chapter, you learned about function values. In this chapter, we'll show you how to apply function values to create new control abstractions. Along the way, you'll also learn about currying and by-name parameters.

## 9.1 Reducing code duplication

All functions are separated into common parts, which are the same in every invocation of the function, and non-common parts, which may vary from one function invocation to the next. The common parts are in the body of the function, while the non-common parts must be supplied via arguments. When you use a function value as an argument, the non-common part of the algorithm is itself some other algorithm! At each invocation of such a function, you can pass in a different function value as an argument, and the invoked function will, at times of its choosing, invoke the passed function value. These *higher-order functions*—functions that take functions as parameters—give you extra opportunities to condense and simplify code.

One benefit of higher-order functions is they enable you to create control abstractions that allow you to reduce code duplication. For example, suppose you are writing a file browser, and you want to provide an API that allows users to search for files matching some criterion. First, you add a facility to search for files whose names end in a particular string. This would enable your users to find, for example, all files with a ".scala" extension. You could provide such an API by defining a public `filesEnding` method inside

167

a singleton object like this:

```
object FileMatcher {
  private def filesHere = (new java.io.File(".")).listFiles

  def filesEnding(query: String) =
    for (file <- filesHere; if file.getName.endsWith(query))
      yield file
}
```

The `filesEnding` method obtains the list of all files in the current directory using the private helper method `filesHere`, then filters them based on whether each file name ends with the user-specified query. Given `filesHere` is private, the `filesEnding` method is the only accessible method defined in `FileMatcher`, the API you provide to your users.

So far so good, and there is no repeated code yet. Later on, though, you decide to let people search based on any part of the file name. This is good for when your users cannot remember if they named a file phb-important.doc, stupid-phb-report.doc, may2003salesdoc.phb, or something entirely different; they just know that "phb" appears in the name somewhere. You go back to work and add this function to your `FileMatcher` API:

```
def filesContaining(query: String) =
  for (file <- filesHere; if file.getName.contains(query))
    yield file
```

This function works just like `filesEnding`. It searches `filesHere`, checks the name, and returns the file if the name matches. The only difference is that this function uses `contains` instead of `endsWith`.

The months go by, and the program becomes more successful. Eventually, you give in to the requests of a few power users who want to search based on regular expressions. These sloppy guys have immense directories with thousands of files, and they would like to do things like find all "pdf" files that have "oopsla" in the title somewhere. To support them, you write this function:

```
def filesRegex(query: String) =
  for (file <- filesHere; if file.getName.matches(query))
    yield file
```

Experienced programmers will notice all of this repetition and wonder if it can be factored into a common helper function. Doing it the obvious way does not work, however. You would like to be able to do the following:

```
def filesMatching(query: String, method) =
  for (file <- filesHere; if file.getName.method(query))
    yield file
```

This approach would work in some dynamic languages, but Scala does not allow pasting together code at runtime like this. So what do you do?

Function values provide an answer. While you cannot pass around a method name as a value, you can get the same effect by passing around a function value that calls the method for you. In this case, you could add a matcher parameter to the method whose sole purpose is to check a file name against a query:

```
def filesMatching(query: String,
    matcher: (String, String) => Boolean) = {

  for (file <- filesHere; if matcher(file.getName, query))
    yield file
}
```

In this version of the method, the if clause now uses matcher to check the file name against the query. Precisely what this check does depends on what is specified as the matcher. Take a look, now, at the type of matcher itself. It is a function, and thus has a => in the type. This function takes two string arguments—the file name and the query—and returns a boolean, so the type of this function is (String, String) => Boolean.

Given this new filesMatching helper method, you can simplify the three searching methods by having them call the helper method, passing in an appropriate function:

```
def filesEnding(query: String) =
  filesMatching(query, _.endsWith(_))

def filesContaining(query: String) =
  filesMatching(query, _.contains(_))

def filesRegex(query: String) =
  filesMatching(query, _.matches(_))
```

The function literals shown in this example use the placeholder syntax, introduced in the previous chapter, which may not as yet feel very natural to you. So here's a clarification of how placeholders are used: The function literal _.endsWith(_), used in the filesEnding method, means the same thing as:

```
(fileName: String, query: String) => fileName.endsWith(query)
```

Because filesMatching takes a function that requires two String arguments, you need not specify the types of the arguments; you could just write (fileName, query) => fileName.endsWith(query). Since the parameters are each used only once in the body of the function (*i.e.*, the first parameter, fileName, is used first in the body, and the second parameter, query, is used second), you can use the placeholder syntax: _.endsWith(_). The first underscore is a placeholder for the first parameter, the file name, and the second underscore a placeholder for the second parameter, the query string.

This code is already simplified, but it can actually be even shorter. Notice that the query gets passed to filesMatching, but filesMatching does nothing with the query except to pass it back to the passed matcher function. This passing back and forth is unnecessary because the caller already knew the query to begin with! You might as well remove the query parameter from filesMatching and matcher, thus simplifying the code as shown in Listing 9.1.

This example demonstrates the way in which first-class functions can help you eliminate code duplication where it would be very difficult to do so without them. In Java, for example, you could create an interface containing a method that takes one String and returns a Boolean, then create and pass anonymous inner class instances that implement this interface to filesMatching. Although this approach would remove the code duplication you are trying to eliminate, it would, at the same time, add as much or more new code. Thus the benefit is not worth the cost, and you may as well live with the duplication.

Moreover, this example demonstrates how closures can help you reduce code duplication. The function literals used in the previous example, such as _.endsWith(_) and _.contains(_), are instantiated at runtime into function values that are *not* closures because they don't capture any free variables. Both variables used in the expression, _.endsWith(_), for example, are represented by underscores, which means they are taken from arguments

```
object FileMatcher {
  private def filesHere = (new java.io.File(".")).listFiles

  private def filesMatching(matcher: String => Boolean) =
    for (file <- filesHere; if matcher(file.getName))
      yield file

  def filesEnding(query: String) =
    filesMatching(_.endsWith(query))

  def filesContaining(query: String) =
    filesMatching(_.contains(query))

  def filesRegex(query: String) =
    filesMatching(_.matches(query))
}
```

Listing 9.1 · Using closures to reduce code duplication.

to the function. Thus, _.endsWith(_) uses two bound variables, and no free variables. By contrast, the function literal _.endsWith(query), used in the most recent example, contains one bound variable, the argument represented by the underscore, and one free variable named query. It is only because Scala supports closures that you were able to remove the query parameter from filesMatching in the most recent example, thereby simplifying the code even further.

## 9.2 Simplifying client code

The previous example demonstrated that higher-order functions can help reduce code duplication as you implement an API. Another important use of higher-order functions is to put them in an API itself to make client code more concise. A good example is provided by the special-purpose looping methods of Scala's collection types.[1] Many of these are listed in Table 3.1 in Chapter 3, but take a look at just one example for now to see why these methods are so useful.

---

[1]These special-purpose looping methods are defined in trait Traversable, which is extended by List, Set, and Map. See Chapter 17 for a discussion.

Consider `exists`, a method that determines whether a passed value is contained in a collection. You could, of course, search for an element by having a var initialized to false, looping through the collection checking each item, and setting the var to true if you find what you are looking for. Here's a method that uses this approach to determine whether a passed `List` contains a negative number:

```
def containsNeg(nums: List[Int]): Boolean = {
  var exists = false
  for (num <- nums)
    if (num < 0)
      exists = true
  exists
}
```

If you define this method in the interpreter, you can call it like this:

```
scala> containsNeg(List(1, 2, 3, 4))
res0: Boolean = false

scala> containsNeg(List(1, 2, -3, 4))
res1: Boolean = true
```

A more concise way to define the method, though, is by calling the higher-order function `exists` on the passed `List`, like this:

```
def containsNeg(nums: List[Int]) = nums.exists(_ < 0)
```

This version of `containsNeg` yields the same results as the previous:

```
scala> containsNeg(Nil)
res2: Boolean = false

scala> containsNeg(List(0, -1, -2))
res3: Boolean = true
```

The `exists` method represents a control abstraction. It is a special-purpose looping construct provided by the Scala library, rather than built into the Scala language like `while` or `for`. In the previous section, the higher-order function, `filesMatching`, reduces code duplication in the implementation of the object `FileMatcher`. The `exists` method provides a similar benefit, but because `exists` is public in Scala's collections API, the code duplication it reduces is client code of that API. If `exists` didn't exist, and you wanted to write a `containsOdd` method to test whether a list contains odd numbers, you might write it like this:

```
def containsOdd(nums: List[Int]): Boolean = {
  var exists = false
  for (num <- nums)
    if (num % 2 == 1)
      exists = true
  exists
}
```

If you compare the body of `containsNeg` with that of `containsOdd`, you'll find that everything is repeated except the test condition of an `if` expression. Using `exists`, you could write this instead:

```
def containsOdd(nums: List[Int]) = nums.exists(_ % 2 == 1)
```

The body of the code in this version is again identical to the body of the corresponding `containsNeg` method (the version that uses `exists`), except the condition for which to search is different. Yet the amount of code duplication is much smaller because all of the looping infrastructure is factored out into the `exists` method itself.

There are many other looping methods in Scala's standard library. As with `exists`, they can often shorten your code if you recognize opportunities to use them.

## 9.3  Currying

In Chapter 1, we said that Scala allows you to create new control abstractions that "feel like native language support." Although the examples you've seen so far are indeed control abstractions, it is unlikely anyone would mistake them for native language support. To understand how to make control abstractions that feel more like language extensions, you first need to understand the functional programming technique called *currying*.

A curried function is applied to multiple argument lists, instead of just one. Listing 9.2 shows a regular, non-curried function, which adds two `Int` parameters, x and y.

```
scala> def plainOldSum(x: Int, y: Int) = x + y
plainOldSum: (x: Int, y: Int)Int

scala> plainOldSum(1, 2)
res4: Int = 3
```

Listing 9.2 · Defining and invoking a "plain old" function.

By contrast, Listing 9.3 shows a similar function that's curried. Instead of one list of two Int parameters, you apply this function to two lists of one Int parameter each.

```
scala> def curriedSum(x: Int)(y: Int) = x + y
curriedSum: (x: Int)(y: Int)Int

scala> curriedSum(1)(2)
res5: Int = 3
```

Listing 9.3 · Defining and invoking a curried function.

What's happening here is that when you invoke curriedSum, you actually get two traditional function invocations back to back. The first function invocation takes a single Int parameter named x, and returns a function value for the second function. This second function takes the Int parameter y. Here's a function named first that does in spirit what the first traditional function invocation of curriedSum would do:

```
scala> def first(x: Int) = (y: Int) => x + y
first: (x: Int)Int => Int
```

Applying the first function to 1—in other words, invoking the first function and passing in 1—yields the second function:

```
scala> val second = first(1)
second: Int => Int = <function1>
```

Applying the second function to 2 yields the result:

```
scala> second(2)
res6: Int = 3
```

These `first` and `second` functions are just an illustration of the currying process. They are not directly connected to the `curriedSum` function. Nevertheless, there is a way to get an actual reference to `curriedSum`'s "second" function. You can use the placeholder notation to use `curriedSum` in a partially applied function expression, like this:

```
scala> val onePlus = curriedSum(1)_
onePlus: Int => Int = <function1>
```

The underscore in `curriedSum(1)_` is a placeholder for the second parameter list.[2] The result is a reference to a function that, when invoked, adds one to its sole `Int` argument and returns the result:

```
scala> onePlus(2)
res7: Int = 3
```

And here's how you'd get a function that adds two to its sole `Int` argument:

```
scala> val twoPlus = curriedSum(2)_
twoPlus: Int => Int = <function1>

scala> twoPlus(2)
res8: Int = 4
```

## 9.4  Writing new control structures

In languages with first-class functions, you can effectively make new control structures even though the syntax of the language is fixed. All you need to do is create methods that take functions as arguments.

For example, here is the "twice" control structure, which repeats an operation two times and returns the result:

```
scala> def twice(op: Double => Double, x: Double) = op(op(x))
twice: (op: Double => Double, x: Double)Double

scala> twice(_ + 1, 5)
res9: Double = 7.0
```

---

[2]In the previous chapter, when the placeholder notation was used on traditional methods, like `println _`, you had to leave a space between the name and the underscore. In this case you don't, because whereas `println_` is a legal identifier in Scala, `curriedSum(1)_` is not.

The type of op in this example is Double => Double, which means it is a function that takes one Double as an argument and returns another Double.

Any time you find a control pattern repeated in multiple parts of your code, you should think about implementing it as a new control structure. Earlier in the chapter you saw filesMatching, a very specialized control pattern. Consider now a more widely used coding pattern: open a resource, operate on it, and then close the resource. You can capture this in a control abstraction using a method like the following:

```
def withPrintWriter(file: File, op: PrintWriter => Unit) = {
  val writer = new PrintWriter(file)
  try {
    op(writer)
  } finally {
    writer.close()
  }
}
```

Given such a method, you can use it like this:

```
withPrintWriter(
  new File("date.txt"),
  writer => writer.println(new java.util.Date)
)
```

The advantage of using this method is that it's withPrintWriter, not user code, that assures the file is closed at the end. So it's impossible to forget to close the file. This technique is called the *loan pattern*, because a control-abstraction function, such as withPrintWriter, opens a resource and "loans" it to a function. For instance, withPrintWriter in the previous example loans a PrintWriter to the function, op. When the function completes, it signals that it no longer needs the "borrowed" resource. The resource is then closed in a finally block, to ensure it is indeed closed, regardless of whether the function completes by returning normally or throwing an exception.

One way in which you can make the client code look a bit more like a built-in control structure is to use curly braces instead of parentheses to surround the argument list. In any method invocation in Scala in which you're passing in exactly one argument, you can opt to use curly braces to surround the argument instead of parentheses.

For example, instead of:

```
scala> println("Hello, world!")
Hello, world!
```

You could write:

```
scala> println { "Hello, world!" }
Hello, world!
```

In the second example, you used curly braces instead of parentheses to surround the arguments to `println`. This curly braces technique will work, however, only if you're passing in one argument. Here's an attempt at violating that rule:

```
scala> val g = "Hello, world!"
g: String = Hello, world!

scala> g.substring { 7, 9 }
<console>:1: error: ';' expected but ',' found.
       g.substring { 7, 9 }
                      ^
```

Because you are attempting to pass in two arguments to `substring`, you get an error when you try to surround those arguments with curly braces. Instead, you'll need to use parentheses:

```
scala> g.substring(7, 9)
res12: String = wo
```

The purpose of this ability to substitute curly braces for parentheses for passing in one argument is to enable client programmers to write function literals between curly braces. This can make a method call feel more like a control abstraction. Take the `withPrintWriter` method defined previously as an example. In its most recent form, `withPrintWriter` takes two arguments, so you can't use curly braces. Nevertheless, because the function passed to `withPrintWriter` is the last argument in the list, you can use currying to pull the first argument, the `File`, into a separate argument list. This will leave the function as the lone parameter of the second argument list. Listing 9.4 shows how you'd need to redefine `withPrintWriter`.

```
def withPrintWriter(file: File)(op: PrintWriter => Unit) = {
  val writer = new PrintWriter(file)
  try {
    op(writer)
  } finally {
    writer.close()
  }
}
```

Listing 9.4 · Using the loan pattern to write to a file.

The new version differs from the old one only in that there are now two parameter lists with one parameter each instead of one parameter list with two parameters. Look between the two parameters. In the previous version of withPrintWriter, shown on page 176, you see ...File, op.... But in this version, you see ...File)(op.... Given the above definition, you can call the method with a more pleasing syntax:

```
val file = new File("date.txt")

withPrintWriter(file) { writer =>
  writer.println(new java.util.Date)
}
```

In this example, the first argument list, which contains one File argument, is written surrounded by parentheses. The second argument list, which contains one function argument, is surrounded by curly braces.

## 9.5 By-name parameters

The withPrintWriter method shown in the previous section differs from built-in control structures of the language, such as if and while, in that the code between the curly braces takes an argument. The function passed to withPrintWriter requires one argument of type PrintWriter. This argument shows up as the "writer =>" in:

```
withPrintWriter(file) { writer =>
  writer.println(new java.util.Date)
}
```

But what if you want to implement something more like if or while, where there is no value to pass into the code between the curly braces? To help with such situations, Scala provides by-name parameters.

As a concrete example, suppose you want to implement an assertion construct called myAssert.[3] The myAssert function will take a function value as input and consult a flag to decide what to do. If the flag is set, myAssert will invoke the passed function and verify that it returns true. If the flag is turned off, myAssert will quietly do nothing at all.

Without using by-name parameters, you could write myAssert like this:

```
var assertionsEnabled = true

def myAssert(predicate: () => Boolean) =
  if (assertionsEnabled && !predicate())
    throw new AssertionError
```

The definition is fine, but using it is a little bit awkward:

```
myAssert(() => 5 > 3)
```

You would really prefer to leave out the empty parameter list and => symbol in the function literal and write the code like this:

```
myAssert(5 > 3) // Won't work, because missing () =>
```

By-name parameters exist precisely so that you can do this. To make a by-name parameter, you give the parameter a type starting with => instead of () =>. For example, you could change myAssert's predicate parameter into a by-name parameter by changing its type, "() => Boolean", into "=> Boolean". Listing 9.5 shows how that would look:

```
def byNameAssert(predicate: => Boolean) =
  if (assertionsEnabled && !predicate)
    throw new AssertionError
```

Listing 9.5 · Using a by-name parameter.

Now you can leave out the empty parameter in the property you want to assert. The result is that using byNameAssert looks exactly like using a built-in control structure:

---

[3]You'll call this myAssert, not assert, because Scala provides an assert of its own, which will be described in Section 14.1.

179

```
byNameAssert(5 > 3)
```

A by-name type, in which the empty parameter list, (), is left out, is only allowed for parameters. There is no such thing as a by-name variable or a by-name field.

Now, you may be wondering why you couldn't simply write myAssert using a plain old Boolean for the type of its parameter, like this:

```
def boolAssert(predicate: Boolean) =
  if (assertionsEnabled && !predicate)
    throw new AssertionError
```

This formulation is also legal, of course, and the code using this version of boolAssert would still look exactly as before:

```
boolAssert(5 > 3)
```

Nevertheless, one difference exists between these two approaches that is important to note. Because the type of boolAssert's parameter is Boolean, the expression inside the parentheses in boolAssert(5 > 3) is evaluated *before* the call to boolAssert. The expression 5 > 3 yields true, which is passed to boolAssert. By contrast, because the type of byNameAssert's predicate parameter is => Boolean, the expression inside the parentheses in byNameAssert(5 > 3) is *not* evaluated before the call to byNameAssert. Instead a function value will be created whose apply method will evaluate 5 > 3, and this function value will be passed to byNameAssert.

The difference between the two approaches, therefore, is that if assertions are disabled, you'll see any side effects that the expression inside the parentheses may have in boolAssert, but not in byNameAssert. For example, if assertions are disabled, attempting to assert on "x / 0 == 0" will yield an exception in boolAssert's case:

```
scala> val x = 5
x: Int = 5

scala> var assertionsEnabled = false
assertionsEnabled: Boolean = false

scala> boolAssert(x / 0 == 0)
java.lang.ArithmeticException: / by zero
  ... 33 elided
```

But attempting to assert on the same code in byNameAssert's case will *not* yield an exception:

```scala
scala> byNameAssert(x / 0 == 0)
```

## 9.6 Conclusion

This chapter has shown you how to build on Scala's rich function support to build control abstractions. You can use functions within your code to factor out common control patterns, and you can take advantage of higher-order functions in the Scala library to reuse control patterns that are common across all programmers' code. We also discussed how to use currying and by-name parameters so that your own higher-order functions can be used with a concise syntax.

In the previous chapter and this one, you have seen quite a lot of information about functions. The next few chapters will go back to discussing more object-oriented features of the language.

# Chapter 10

# Composition and Inheritance

Chapter 6 introduced some basic object-oriented aspects of Scala. This chapter picks up where Chapter 6 left off and dives into Scala's support for object-oriented programming in much greater detail.

We'll compare two fundamental relationships between classes: composition and inheritance. Composition means one class holds a reference to another, using the referenced class to help it fulfill its mission. Inheritance is the superclass/subclass relationship.

In addition to these topics, we'll discuss abstract classes, parameterless methods, extending classes, overriding methods and fields, parametric fields, invoking superclass constructors, polymorphism and dynamic binding, final members and classes, and factory objects and methods.

## 10.1 A two-dimensional layout library

As a running example in this chapter, we'll create a library for building and rendering two-dimensional layout elements. Each element will represent a rectangle filled with text. For convenience, the library will provide factory methods named "elem" that construct new elements from passed data. For example, you'll be able to create a layout element containing a string using a factory method with the following signature:

```
elem(s: String): Element
```

As you can see, elements will be modeled with a type named Element. You'll be able to call above or beside on an element, passing in a second element, to get a new element that combines the two. For example, the following expression would construct a larger element consisting of two columns, each with a height of two:

```
val column1 = elem("hello") above elem("***")
val column2 = elem("***") above elem("world")
column1 beside column2
```

Printing the result of this expression would give you:

```
hello ***
*** world
```

Layout elements are a good example of a system in which objects can be constructed from simple parts with the aid of composing operators. In this chapter, we'll define classes that enable element objects to be constructed from arrays, lines, and rectangles. These basic element objects will be the simple parts. We'll also define composing operators above and beside. Such composing operators are also often called *combinators* because they combine elements of some domain into new elements.

Thinking in terms of combinators is generally a good way to approach library design: it pays to think about the fundamental ways to construct objects in an application domain. What are the simple objects? In what ways can more interesting objects be constructed out of simpler ones? How do combinators hang together? What are the most general combinations? Do they satisfy any interesting laws? If you have good answers to these questions, your library design is on track.

## 10.2   Abstract classes

Our first task is to define type `Element`, which represents layout elements. Since elements are two dimensional rectangles of characters, it makes sense to include a member, `contents`, that refers to the contents of a layout element. The contents can be represented as an array of strings, where each string represents a line. Hence, the type of the result returned by `contents` will be `Array[String]`. Listing 10.1 shows what it will look like.

In this class, `contents` is declared as a method that has no implementation. In other words, the method is an *abstract* member of class `Element`. A class with abstract members must itself be declared abstract, which is done by writing an `abstract` modifier in front of the `class` keyword:

184

```
abstract class Element {
  def contents: Array[String]
}
```

Listing 10.1 · Defining an abstract method and class.

**abstract class** Element ...

The abstract modifier signifies that the class may have abstract members that do not have an implementation. As a result, you cannot instantiate an abstract class. If you try to do so, you'll get a compiler error:

```
scala> new Element
<console>:5: error: class Element is abstract;
    cannot be instantiated
        new Element
          ^
```

Later in this chapter, you'll see how to create subclasses of class Element, which you'll be able to instantiate because they fill in the missing definition for contents.

Note that the contents method in class Element does not carry an abstract modifier. A method is abstract if it does not have an implementation (*i.e.*, no equals sign or body). Unlike Java, no abstract modifier is necessary (or allowed) on method declarations. Methods that have an implementation are called *concrete*.

Another bit of terminology distinguishes between *declarations* and *definitions*. Class Element *declares* the abstract method contents, but currently *defines* no concrete methods. In the next section, however, we'll enhance Element by defining some concrete methods.

## 10.3 Defining parameterless methods

As a next step, we'll add methods to Element that reveal its width and height, as shown in Listing 10.2. The height method returns the number of lines in contents. The width method returns the length of the first line, or if there are no lines in the element, returns zero. (This means you cannot define an element with a height of zero and a non-zero width.)

```
abstract class Element {
  def contents: Array[String]
  def height: Int = contents.length
  def width: Int = if (height == 0) 0 else contents(0).length
}
```

Listing 10.2 · Defining parameterless methods width and height.

Note that none of Element's three methods has a parameter list, not even an empty one. For example, instead of:

```
def width(): Int
```

the method is defined without parentheses:

```
def width: Int
```

Such *parameterless methods* are quite common in Scala. By contrast, methods defined with empty parentheses, such as def height(): Int, are called *empty-paren methods*. The recommended convention is to use a parameterless method whenever there are no parameters *and* the method accesses mutable state only by reading fields of the containing object (in particular, it does not change mutable state). This convention supports the *uniform access principle*,[1] which says that client code should not be affected by a decision to implement an attribute as a field or method.

For instance, we could implement width and height as fields, instead of methods, simply by changing the def in each definition to a val:

```
abstract class Element {
  def contents: Array[String]
  val height = contents.length
  val width =
    if (height == 0) 0 else contents(0).length
}
```

The two pairs of definitions are completely equivalent from a client's point of view. The only difference is that field accesses might be slightly faster than method invocations because the field values are pre-computed when the

---

[1]Meyer, *Object-Oriented Software Construction* [Mey00]

class is initialized, instead of being computed on each method call. On the other hand, the fields require extra memory space in each `Element` object. So it depends on the usage profile of a class whether an attribute is better represented as a field or method, and that usage profile might change over time. The point is that clients of the `Element` class should not be affected when its internal implementation changes.

In particular, a client of class `Element` should not need to be rewritten if a field of that class gets changed into an access function, so long as the access function is *pure* (*i.e.*, it does not have any side effects and does not depend on mutable state). The client should not need to care either way.

So far so good. But there's still a slight complication that has to do with the way Java handles things. The problem is that Java does not implement the uniform access principle. So it's `string.length()` in Java, not `string.length`, even though it's `array.length`, not `array.length()`. Needless to say, this is very confusing.

To bridge that gap, Scala is very liberal when it comes to mixing parameterless and empty-paren methods. In particular, you can override a parameterless method with an empty-paren method, and *vice versa*. You can also leave off the empty parentheses on an invocation of any function that takes no arguments. For instance, the following two lines are both legal in Scala:

```
Array(1, 2, 3).toString
"abc".length
```

In principle it's possible to leave out all empty parentheses in Scala function calls. However, it's still recommended to write the empty parentheses when the invoked method represents more than a property of its receiver object. For instance, empty parentheses are appropriate if the method performs I/O, writes reassignable variables (`vars`), or reads `vars` other than the receiver's fields, either directly or indirectly by using mutable objects. That way, the parameter list acts as a visual clue that some interesting computation is triggered by the call. For instance:

```
"hello".length  // no () because no side-effect
println()       // better to not drop the ()
```

To summarize, it is encouraged in Scala to define methods that take no parameters and have no side effects as parameterless methods (*i.e.*, leaving off the empty parentheses). On the other hand, you should never define a method that has side-effects without parentheses, because invocations of that method would then look like a field selection. So your clients might be surprised to see the side effects.

Similarly, whenever you invoke a function that has side effects, be sure to include the empty parentheses when you write the invocation. Another way to think about this is if the function you're calling performs an operation, use the parentheses. But if it merely provides access to a property, leave the parentheses off.

## 10.4  Extending classes

We still need to be able to create new element objects. You have already seen that "new Element" cannot be used for this because class Element is abstract. To instantiate an element, therefore, we will need to create a subclass that extends Element and implements the abstract contents method. Listing 10.3 shows one possible way to do that:

```
class ArrayElement(conts: Array[String]) extends Element {
  def contents: Array[String] = conts
}
```

Listing 10.3 · Defining ArrayElement as a subclass of Element.

Class ArrayElement is defined to *extend* class Element. Just like in Java, you use an extends clause after the class name to express this:

```
... extends Element ...
```

Such an extends clause has two effects: It makes class ArrayElement *inherit* all non-private members from class Element, and it makes the type ArrayElement a *subtype* of the type Element. Given ArrayElement extends Element, class ArrayElement is called a *subclass* of class Element. Conversely, Element is a *superclass* of ArrayElement. If you leave out an extends clause, the Scala compiler implicitly assumes your class extends from scala.AnyRef, which on the Java platform is the same as class java.lang.Object. Thus, class Element implicitly extends class AnyRef. You can see these inheritance relationships in Figure 10.1.

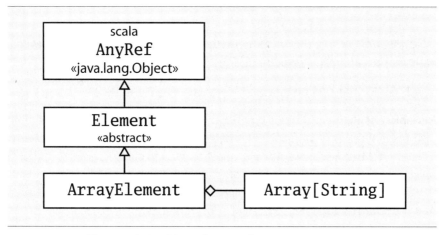

Figure 10.1 · Class diagram for `ArrayElement`.

*Inheritance* means that all members of the superclass are also members of the subclass, with two exceptions. First, private members of the superclass are not inherited in a subclass. Second, a member of a superclass is not inherited if a member with the same name and parameters is already implemented in the subclass. In that case we say the member of the subclass *overrides* the member of the superclass. If the member in the subclass is concrete and the member of the superclass is abstract, we also say that the concrete member *implements* the abstract one.

For example, the `contents` method in `ArrayElement` overrides (or alternatively: implements) abstract method `contents` in class `Element`.[2] By contrast, class `ArrayElement` inherits the `width` and `height` methods from class `Element`. For example, given an `ArrayElement ae`, you can query its width using `ae.width`, as if `width` were defined in class `ArrayElement`:

```
scala> val ae = new ArrayElement(Array("hello", "world"))
ae: ArrayElement = ArrayElement@39274bf7

scala> ae.width
```

---

[2]One flaw with this design is that because the returned array is mutable, clients could change it. For the book we'll keep things simple, but were `ArrayElement` part of a real project, you might consider returning a *defensive copy* of the array instead. Another problem is we aren't currently ensuring that every `String` element of the `contents` array has the same length. This could be solved by checking the precondition in the primary constructor and throwing an exception if it is violated.

```
res0: Int = 5
```

*Subtyping* means that a value of the subclass can be used wherever a value of the superclass is required. For example:

```
val e: Element = new ArrayElement(Array("hello"))
```

Variable e is defined to be of type Element, so its initializing value should also be an Element. In fact, the initializing value's type is ArrayElement. This is OK, because class ArrayElement extends class Element, and as a result, the type ArrayElement is compatible with the type Element.[3]

Figure 10.1 also shows the *composition* relationship that exists between ArrayElement and Array[String]. This relationship is called composition because class ArrayElement is "composed" out of class Array[String], in that the Scala compiler will place into the binary class it generates for ArrayElement a field that holds a reference to the passed conts array. We'll discuss some design considerations concerning composition and inheritance later in this chapter, in Section 10.11.

## 10.5   Overriding methods and fields

The uniform access principle is just one aspect where Scala treats fields and methods more uniformly than Java. Another difference is that in Scala, fields and methods belong to the same namespace. This makes it possible for a field to override a parameterless method. For instance, you could change the implementation of contents in class ArrayElement from a method to a field without having to modify the abstract method definition of contents in class Element, as shown in Listing 10.4:

```
class ArrayElement(conts: Array[String]) extends Element {
  val contents: Array[String] = conts
}
```

Listing 10.4 · Overriding a parameterless method with a field.

Field contents (defined with a val) in this version of ArrayElement is a perfectly good implementation of the parameterless method contents

---

[3]For more perspective on the difference between subclass and subtype, see the glossary entry for *subtype*.

(declared with a def) in class Element. On the other hand, in Scala it is forbidden to define a field and method with the same name in the same class, whereas this is allowed in Java.

For example, this Java class would compile just fine:

```
// This is Java
class CompilesFine {
  private int f = 0;
  public int f() {
    return 1;
  }
}
```

But the corresponding Scala class would not compile:

```
class WontCompile {
  private var f = 0 // Won't compile, because a field
  def f = 1         // and method have the same name
}
```

Generally, Scala has just two namespaces for definitions in place of Java's four. Java's four namespaces are fields, methods, types, and packages. By contrast, Scala's two namespaces are:

- values (fields, methods, packages, and singleton objects)

- types (class and trait names)

The reason Scala places fields and methods into the same namespace is precisely so you can override a parameterless method with a val, something you can't do with Java.[4]

## 10.6 Defining parametric fields

Consider again the definition of class ArrayElement shown in the previous section. It has a parameter conts whose sole purpose is to be copied into the

---

[4]The reason that packages share the same namespace as fields and methods in Scala is to enable you to import packages (in addition to just the names of types) and the fields and methods of singleton objects. This is also something you can't do in Java. It will be described in Section 13.3.

contents field. The name `conts` of the parameter was chosen just so that it would look similar to the field name `contents` without actually clashing with it. This is a "code smell," a sign that there may be some unnecessary redundancy and repetition in your code.

You can avoid the code smell by combining the parameter and the field in a single *parametric field* definition, as shown in Listing 10.5:

```
class ArrayElement(
  val contents: Array[String]
) extends Element
```

Listing 10.5 · Defining `contents` as a parametric field.

Note that now the `contents` parameter is prefixed by `val`. This is a shorthand that defines at the same time a parameter and field with the same name. Specifically, class `ArrayElement` now has an (unreassignable) field `contents`, which can be accessed from outside the class. The field is initialized with the value of the parameter. It's as if the class had been written as follows, where x123 is an arbitrary fresh name for the parameter:

```
class ArrayElement(x123: Array[String]) extends Element {
  val contents: Array[String] = x123
}
```

You can also prefix a class parameter with `var`, in which case the corresponding field would be reassignable. Finally, it is possible to add modifiers, such as `private`, `protected`,[5] or override to these parametric fields, just as you can for any other class member. Consider, for instance, the following class definitions:

```
class Cat {
  val dangerous = false
}
class Tiger(
  override val dangerous: Boolean,
  private var age: Int
) extends Cat
```

---

[5]The `protected` modifier, which grants access to subclasses, will be covered in detail in Chapter 13.

Tiger's definition is a shorthand for the following alternate class definition with an overriding member dangerous and a private member age:

```
class Tiger(param1: Boolean, param2: Int) extends Cat {
  override val dangerous = param1
  private var age = param2
}
```

Both members are initialized from the corresponding parameters. We chose the names of those parameters, param1 and param2, arbitrarily. The important thing was that they not clash with any other name in scope.

## 10.7   Invoking superclass constructors

You now have a complete system consisting of two classes: an abstract class Element, which is extended by a concrete class ArrayElement. You might also envision other ways to express an element. For example, clients might want to create a layout element consisting of a single line given by a string. Object-oriented programming makes it easy to extend a system with new data-variants. You can simply add subclasses. For example, Listing 10.6 shows a LineElement class that extends ArrayElement:

```
class LineElement(s: String) extends ArrayElement(Array(s)) {
  override def width = s.length
  override def height = 1
}
```

Listing 10.6 · Invoking a superclass constructor.

Since LineElement extends ArrayElement, and ArrayElement's constructor takes a parameter (an Array[String]), LineElement needs to pass an argument to the primary constructor of its superclass. To invoke a superclass constructor, you simply place the argument or arguments you want to pass in parentheses following the name of the superclass. For example, class LineElement passes Array(s) to ArrayElement's primary constructor by placing it in parentheses after the superclass ArrayElement's name:

```
... extends ArrayElement(Array(s)) ...
```

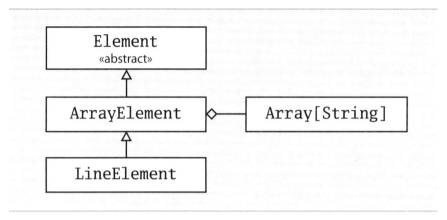

Figure 10.2 · Class diagram for LineElement.

With the new subclass, the inheritance hierarchy for layout elements now looks as shown in Figure 10.2.

## 10.8 Using override modifiers

Note that the definitions of width and height in LineElement carry an override modifier. In Section 6.3, you saw this modifier in the definition of a toString method. Scala requires such a modifier for all members that override a concrete member in a parent class. The modifier is optional if a member implements an abstract member with the same name. The modifier is forbidden if a member does not override or implement some other member in a base class. Since height and width in class LineElement override concrete definitions in class Element, override is required.

This rule provides useful information for the compiler that helps avoid some hard-to-catch errors and makes system evolution safer. For instance, if you happen to misspell the method or accidentally give it a different parameter list, the compiler will respond with an error message:

```
$ scalac LineElement.scala
.../LineElement.scala:50:
error: method hight overrides nothing
  override def hight = 1
            ^
```

The override convention is even more important when it comes to system evolution. Say you defined a library of 2D drawing methods. You made it publicly available, and it is widely used. In the next version of the library you want to add to your base class Shape a new method with this signature:

```
def hidden(): Boolean
```

Your new method will be used by various drawing methods to determine whether a shape needs to be drawn. This could lead to a significant speedup, but you cannot do this without the risk of breaking client code. After all, a client could have defined a subclass of Shape with a different implementation of hidden. Perhaps the client's method actually makes the receiver object disappear instead of testing whether the object is hidden. Because the two versions of hidden override each other, your drawing methods would end up making objects disappear, which is certainly not what you want!

These "accidental overrides" are the most common manifestation of what is called the "fragile base class" problem. The problem is that if you add new members to base classes (which we usually call superclasses) in a class hierarchy, you risk breaking client code. Scala cannot completely solve the fragile base class problem, but it improves on the situation compared to Java.[6] If the drawing library and its clients were written in Scala, then the client's original implementation of hidden could not have had an override modifier, because at the time there was no other method with that name.

Once you add the hidden method to the second version of your shape class, a recompile of the client would give an error like the following:

```
.../Shapes.scala:6: error: error overriding method
    hidden in class Shape of type ()Boolean;
 method hidden needs `override' modifier
 def hidden(): Boolean =
   ^
```

That is, instead of wrong behavior your client would get a compile-time error, which is usually much preferable.

---

[6]In Java 1.5, an @Override annotation was introduced that works similarly to Scala's override modifier, but unlike Scala's override, is not required.

## 10.9 Polymorphism and dynamic binding

You saw in Section 10.4 that a variable of type Element could refer to an object of type ArrayElement. The name for this phenomenon is *polymorphism*, which means "many shapes" or "many forms." In this case, Element objects can have many forms.[7]

So far, you've seen two such forms: ArrayElement and LineElement. You can create more forms of Element by defining new Element subclasses. For example, you could define a new form of Element that has a given width and height, and is filled everywhere with a given character:

```scala
class UniformElement(
  ch: Char,
  override val width: Int,
  override val height: Int
) extends Element {
  private val line = ch.toString * width
  def contents = Array.fill(height)(line)
}
```

The inheritance hierarchy for class Element now looks as shown in Figure 10.3. As a result, Scala will accept all of the following assignments, because the type of the assigning expression conforms to the type of the defined variable:

```scala
val e1: Element = new ArrayElement(Array("hello", "world"))
val ae: ArrayElement = new LineElement("hello")
val e2: Element = ae
val e3: Element = new UniformElement('x', 2, 3)
```

If you check the inheritance hierarchy, you'll find that in each of these four val definitions, the type of the expression to the right of the equals sign is below the type of the val being initialized to the left of the equals sign.

The other half of the story, however, is that method invocations on variables and expressions are *dynamically bound*. This means that the actual method implementation invoked is determined at run time based on the class of the object, not the type of the variable or expression. To demonstrate this

---

[7]This kind of polymorphism is called *subtyping polymorphism*. Another kind of polymorphism in Scala called *universal polymorphism* is discussed in Chapter 19.

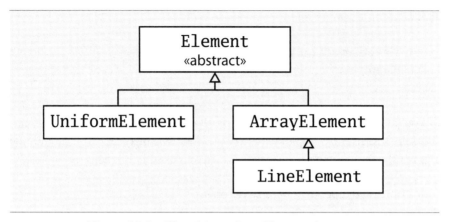

Figure 10.3 · Class hierarchy of layout elements.

behavior, we'll temporarily remove all existing members from our Element classes and add a method named demo to Element. We'll override demo in ArrayElement and LineElement, but not in UniformElement:

```scala
abstract class Element {
  def demo() = {
    println("Element's implementation invoked")
  }
}
class ArrayElement extends Element {
  override def demo() = {
    println("ArrayElement's implementation invoked")
  }
}
class LineElement extends ArrayElement {
  override def demo() = {
    println("LineElement's implementation invoked")
  }
}
// UniformElement inherits Element's demo
class UniformElement extends Element
```

If you enter this code into the interpreter, you can then define this method

197

that takes an Element and invokes demo on it:

```
def invokeDemo(e: Element) = {
  e.demo()
}
```

If you pass an ArrayElement to invokeDemo, you'll see a message indicating ArrayElement's implementation of demo was invoked, even though the type of the variable, e, on which demo was invoked is Element:

```
scala> invokeDemo(new ArrayElement)
ArrayElement's implementation invoked
```

Similarly, if you pass a LineElement to invokeDemo, you'll see a message that indicates LineElement's demo implementation was invoked:

```
scala> invokeDemo(new LineElement)
LineElement's implementation invoked
```

The behavior when passing a UniformElement may at first glance look suspicious, but it is correct:

```
scala> invokeDemo(new UniformElement)
Element's implementation invoked
```

Because UniformElement does not override demo, it inherits the implementation of demo from its superclass, Element. Thus, Element's implementation is the correct implementation of demo to invoke when the class of the object is UniformElement.

## 10.10   Declaring final members

Sometimes when designing an inheritance hierarchy, you want to ensure that a member cannot be overridden by subclasses. In Scala, as in Java, you do this by adding a final modifier to the member. As shown in Listing 10.7, you could place a final modifier on ArrayElement's demo method.

Given this version of ArrayElement, an attempt to override demo in its subclass, LineElement, would not compile:

```
class ArrayElement extends Element {
  final override def demo() = {
    println("ArrayElement's implementation invoked")
  }
}
```

Listing 10.7 · Declaring a final method.

```
elem.scala:18: error: error overriding method demo
    in class ArrayElement of type ()Unit;
method demo cannot override final member
    override def demo() = {
                 ^
```

You may also at times want to ensure that an entire class not be sub-classed. To do this you simply declare the entire class final by adding a final modifier to the class declaration. For example, Listing 10.8 shows how you would declare ArrayElement final:

```
final class ArrayElement extends Element {
  override def demo() = {
    println("ArrayElement's implementation invoked")
  }
}
```

Listing 10.8 · Declaring a final class.

With this version of ArrayElement, any attempt at defining a subclass would fail to compile:

```
elem.scala: 18: error: illegal inheritance from final class
    ArrayElement
  class LineElement extends ArrayElement {
                ^
```

We'll now remove the final modifiers and demo methods, and go back to the earlier implementation of the Element family. We'll focus our attention in the remainder of this chapter to completing a working version of the layout library.

## 10.11   Using composition and inheritance

Composition and inheritance are two ways to define a new class in terms of another existing class. If what you're after is primarily code reuse, you should in general prefer composition to inheritance. Only inheritance suffers from the fragile base class problem, in which you can inadvertently break subclasses by changing a superclass.

One question you can ask yourself about an inheritance relationship is whether it models an *is-a* relationship.[8] For example, it would be reasonable to say that ArrayElement *is-an* Element. Another question you can ask is whether clients will want to use the subclass type as a superclass type.[9] In the case of ArrayElement, we do indeed expect clients will want to use an ArrayElement as an Element.

If you ask these questions about the inheritance relationships shown in Figure 10.3, do any of the relationships seem suspicious? In particular, does it seem obvious to you that a LineElement *is-an* ArrayElement? Do you think clients would ever need to use a LineElement as an ArrayElement?

In fact, we defined LineElement as a subclass of ArrayElement primarily to reuse ArrayElement's definition of contents. Perhaps it would be better, therefore, to define LineElement as a direct subclass of Element, like this:

```
class LineElement(s: String) extends Element {
  val contents = Array(s)
  override def width = s.length
  override def height = 1
}
```

In the previous version, LineElement had an inheritance relationship with ArrayElement, from which it inherited contents. It now has a composition relationship with Array: it holds a reference to an array of strings from its own contents field.[10] Given this implementation of LineElement, the inheritance hierarchy for Element now looks as shown in Figure 10.4.

---

[8]Meyers, *Effective C++* [Mey91]

[9]Eckel, *Thinking in Java* [Eck98]

[10]Class ArrayElement also has a composition relationship with Array, because its parametric contents field holds a reference to an array of strings. The code for ArrayElement is shown in Listing 10.5 on page 192. Its composition relationship is represented in class diagrams by a diamond, as shown, for example, in Figure 10.1 on page 189.

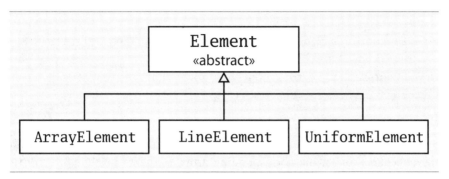

Figure 10.4 · Class hierarchy with revised LineElement.

## 10.12   Implementing above, beside, and toString

As a next step, we'll implement method above in class Element. Putting one element above another means concatenating the two contents values of the elements. So a first draft of method above could look like this:

```
def above(that: Element): Element =
  new ArrayElement(this.contents ++ that.contents)
```

The ++ operation concatenates two arrays. Arrays in Scala are represented as Java arrays, but support many more methods. Specifically, arrays in Scala can be converted to instances of a class scala.Seq, which represents sequence-like structures and contains a number of methods for accessing and transforming sequences. Some other array methods will be explained in this chapter and a more comprehensive discussion will be given in Chapter 17.

In fact, the code shown previously is not quite sufficient because it does not let you put elements of different widths on top of each other. To keep things simple in this section, however, we'll leave this as is and only pass elements of the same length to above. In Section 10.14, we'll make an enhancement to above so that clients can use it to combine elements of different widths.

The next method to implement is beside. To put two elements beside each other, we'll create a new element in which every line results from concatenating corresponding lines of the two elements. As before, to keep things simple, we'll start by assuming the two elements have the same height. This leads to the following design of method beside:

201

```
def beside(that: Element): Element = {
  val contents = new Array[String](this.contents.length)
  for (i <- 0 until this.contents.length)
    contents(i) = this.contents(i) + that.contents(i)
  new ArrayElement(contents)
}
```

The beside method first allocates a new array, contents, and fills it with the concatenation of the corresponding array elements in this.contents and that.contents. It finally produces a new ArrayElement containing the new contents.

Although this implementation of beside works, it is in an imperative style, the telltale sign of which is the loop in which we index through arrays. Alternatively, the method could be abbreviated to one expression:

```
new ArrayElement(
  for (
    (line1, line2) <- this.contents zip that.contents
  ) yield line1 + line2
)
```

Here, the two arrays, this.contents and that.contents, are transformed into an array of pairs (as Tuple2s are called) using the zip operator. The zip operator picks corresponding elements in its two operands and forms an array of pairs. For instance, this expression:

```
Array(1, 2, 3) zip Array("a", "b")
```

will evaluate to:

```
Array((1, "a"), (2, "b"))
```

If one of the two operand arrays is longer than the other, zip will drop the remaining elements. In the expression above, the third element of the left operand, 3, does not form part of the result, because it does not have a corresponding element in the right operand.

The zipped array is then iterated over by a for expression. Here, the syntax "for ((line1, line2) <- ... )" allows you to name both elements of a pair in one *pattern* (*i.e.*, line1 stands now for the first element of the pair, and line2 stands for the second). Scala's pattern-matching system will be described in detail in Chapter 15. For now, you can just think of this as a way to define two `vals`, `line1` and `line2`, for each step of the iteration.

The `for` expression has a `yield` part and therefore yields a result. The result is of the same kind as the expression iterated over (*i.e.*, it is an array). Each element of the array is the result of concatenating the corresponding lines, `line1` and `line2`. So the end result of this code is the same as in the first version of `beside`, but because it avoids explicit array indexing, the result is obtained in a less error-prone way.

You still need a way to display elements. As usual, this is done by defining a `toString` method that returns an element formatted as a string. Here is its definition:

```
override def toString = contents mkString "\n"
```

The implementation of `toString` makes use of `mkString`, which is defined for all sequences, including arrays. As you saw in Section 7.8, an expression like "`arr mkString sep`" returns a string consisting of all elements of the array `arr`. Each element is mapped to a string by calling its `toString` method. A separator string `sep` is inserted between consecutive element strings. So the expression, "`contents mkString "\n"`" formats the `contents` array as a string, where each array element appears on a line by itself.

Note that `toString` does not carry an empty parameter list. This follows the recommendations for the uniform access principle, because `toString` is a pure method that does not take any parameters. With the addition of these three methods, class `Element` now looks as shown in Listing 10.9.

## 10.13   Defining a factory object

You now have a hierarchy of classes for layout elements. This hierarchy could be presented to your clients "as is," but you might also choose to hide the hierarchy behind a factory object.

A factory object contains methods that construct other objects. Clients would then use these factory methods to construct objects, rather than constructing the objects directly with new. An advantage of this approach is that object creation can be centralized and the details of how objects are represented with classes can be hidden. This hiding will both make your library simpler for clients to understand, because less detail is exposed, and provide

```
abstract class Element {

  def contents: Array[String]

  def width: Int =
    if (height == 0) 0 else contents(0).length

  def height: Int = contents.length

  def above(that: Element): Element =
    new ArrayElement(this.contents ++ that.contents)

  def beside(that: Element): Element =
    new ArrayElement(
      for (
        (line1, line2) <- this.contents zip that.contents
      ) yield line1 + line2
    )

  override def toString = contents mkString "\n"
}
```

Listing 10.9 · Class `Element` with `above`, `beside`, and `toString`.

you with more opportunities to change your library's implementation later without breaking client code.

The first task in constructing a factory for layout elements is to choose where the factory methods should be located. Should they be members of a singleton object or of a class? What should the containing object or class be called? There are many possibilities. A straightforward solution is to create a companion object of class Element and make this the factory object for layout elements. That way, you need to expose only the class/object combo of Element to your clients, and you can hide the three implementation classes ArrayElement, LineElement, and UniformElement.

Listing 10.10 is a design of the Element object that follows this scheme. The Element object contains three overloaded variants of an elem method and each constructs a different kind of layout object.

With the advent of these factory methods, it makes sense to change the implementation of class Element so that it goes through the elem factory methods rather than creating new ArrayElement instances explicitly. To call the factory methods without qualifying them with Element, the name of the

```
object Element {

  def elem(contents: Array[String]): Element =
    new ArrayElement(contents)

  def elem(chr: Char, width: Int, height: Int): Element =
    new UniformElement(chr, width, height)

  def elem(line: String): Element =
    new LineElement(line)
}
```

Listing 10.10 · A factory object with factory methods.

singleton object, we will import `Element.elem` at the top of the source file.
In other words, instead of invoking the factory methods with `Element.elem`
inside class `Element`, we'll import `Element.elem` so we can just call the
factory methods by their simple name, `elem`. Listing 10.11 shows what class
`Element` will look like after these changes.

In addition, given the factory methods, the subclasses, `ArrayElement`,
`LineElement`, and `UniformElement`, could now be private because they
no longer need to be accessed directly by clients. In Scala, you can define classes and singleton objects inside other classes and singleton objects.
One way to make the `Element` subclasses private is to place them inside the
`Element` singleton object and declare them private there. The classes will
still be accessible to the three `elem` factory methods, where they are needed.
Listing 10.12 shows how that will look.

## 10.14 Heighten and widen

We need one last enhancement. The version of `Element` shown in Listing 10.11 is not quite sufficient because it does not allow clients to place
elements of different widths on top of each other, or place elements of different heights beside each other.

For example, evaluating the following expression won't work correctly,
because the second line in the combined element is longer than the first:

```
new ArrayElement(Array("hello")) above
new ArrayElement(Array("world!"))
```

```
import Element.elem

abstract class Element {

  def contents: Array[String]

  def width: Int =
    if (height == 0) 0 else contents(0).length

  def height: Int = contents.length

  def above(that: Element): Element =
    elem(this.contents ++ that.contents)

  def beside(that: Element): Element =
    elem(
      for (
        (line1, line2) <- this.contents zip that.contents
      ) yield line1 + line2
    )

  override def toString = contents mkString "\n"
}
```

Listing 10.11 · Class Element refactored to use factory methods.

Similarly, evaluating the following expression would not work properly, because the first ArrayElement has a height of two and the second a height of only one:

```
new ArrayElement(Array("one", "two")) beside
new ArrayElement(Array("one"))
```

Listing 10.13 shows a private helper method, widen, which takes a width and returns an Element of that width. The result contains the contents of this Element, centered, padded to the left and right by any spaces needed to achieve the required width. Listing 10.13 also shows a similar method, heighten, which performs the same function in the vertical direction. The widen method is invoked by above to ensure that Elements placed above each other have the same width. Similarly, the heighten method is invoked by beside to ensure that elements placed beside each other have the same height. With these changes, the layout library is ready for use.

```
object Element {
  private class ArrayElement(
    val contents: Array[String]
  ) extends Element

  private class LineElement(s: String) extends Element {
    val contents = Array(s)
    override def width = s.length
    override def height = 1
  }

  private class UniformElement(
    ch: Char,
    override val width: Int,
    override val height: Int
  ) extends Element {
    private val line = ch.toString * width
    def contents = Array.fill(height)(line)
  }

  def elem(contents:  Array[String]): Element =
    new ArrayElement(contents)

  def elem(chr: Char, width: Int, height: Int): Element =
    new UniformElement(chr, width, height)

  def elem(line: String): Element =
    new LineElement(line)
}
```

Listing 10.12 · Hiding implementation with private classes.

207

```
import Element.elem

abstract class Element {
  def contents:  Array[String]

  def width: Int = contents(0).length
  def height: Int = contents.length

  def above(that: Element): Element = {
    val this1 = this widen that.width
    val that1 = that widen this.width
    elem(this1.contents ++ that1.contents)
  }

  def beside(that: Element): Element = {
    val this1 = this heighten that.height
    val that1 = that heighten this.height
    elem(
      for ((line1, line2) <- this1.contents zip that1.contents)
      yield line1 + line2)
  }

  def widen(w: Int): Element =
    if (w <= width) this
    else {
      val left = elem(' ', (w - width) / 2, height)
      val right = elem(' ', w - width - left.width, height)
      left beside this beside right
    }

  def heighten(h: Int): Element =
    if (h <= height) this
    else {
      val top = elem(' ', width, (h - height) / 2)
      val bot = elem(' ', width, h - height - top.height)
      top above this above bot
    }

  override def toString = contents mkString "\n"
}
```

Listing 10.13 · Element with widen and heighten methods.

## 10.15   Putting it all together

A fun way to exercise almost all elements of the layout library is to write a program that draws a spiral with a given number of edges. This `Spiral` program, shown in Listing 10.14, will do just that.

```
import Element.elem

object Spiral {

  val space = elem(" ")
  val corner = elem("+")

  def spiral(nEdges: Int, direction: Int): Element = {
    if (nEdges == 1)
      elem("+")
    else {
      val sp = spiral(nEdges - 1, (direction + 3) % 4)
      def verticalBar = elem('|', 1, sp.height)
      def horizontalBar = elem('-', sp.width, 1)
      if (direction == 0)
        (corner beside horizontalBar) above (sp beside space)
      else if (direction == 1)
        (sp above space) beside (corner above verticalBar)
      else if (direction == 2)
        (space beside sp) above (horizontalBar beside corner)
      else
        (verticalBar above corner) beside (space above sp)
    }
  }

  def main(args: Array[String]) = {
    val nSides = args(0).toInt
    println(spiral(nSides, 0))
  }
}
```

Listing 10.14 · The `Spiral` application.

Because `Spiral` is a standalone object with a `main` method with the proper signature, it is a Scala application. `Spiral` takes one command-line argument, an integer, and draws a spiral with the specified number of edges. For example, you could draw a six-edge spiral, as shown on the left, and larger spirals, as shown on the right.

```
$ scala Spiral 6      $ scala Spiral 11     $ scala Spiral 17
+-----                 +----------            +----------------
|                      |                      |
| +-+                  | +------+             | +------------+
| + |                  | |      |             | |            |
| | |                  | | +--+ |             | | +--------+ |
+---+                  | | |  | |             | | |        | |
                       | | ++ | |             | | | +----+ | |
                       | |    | |             | | | |    | | |
                       | +----+ |             | | | | ++ | | |
                       |        |             | | | | || | | |
                       +--------+             | | | +--+ | | |
                                             | | |      | | |
                                             | | +------+ | |
                                             | |          | |
                                             | +----------+ |
                                             |              |
                                             +--------------+
```

## 10.16 Conclusion

In this section, you saw more concepts related to object-oriented programming in Scala. Among others, you encountered abstract classes, inheritance and subtyping, class hierarchies, parametric fields, and method overriding. You should have developed a feel for constructing a non-trivial class hierarchy in Scala. We'll work with the layout library again in Chapter 14.

# Chapter 11

# Scala's Hierarchy

Now that you've seen the details of class inheritance in the previous chapter, it is a good time to take a step back and look at Scala's class hierarchy as a whole. In Scala, every class inherits from a common superclass named Any. Because every class is a subclass of Any, the methods defined in Any are "universal" methods: they may be invoked on any object. Scala also defines some interesting classes at the bottom of the hierarchy, Null and Nothing, which essentially act as common *sub*classes. For example, just as Any is a superclass of every other class, Nothing is a subclass of every other class. In this chapter, we'll give you a tour of Scala's class hierarchy.

## 11.1 Scala's class hierarchy

Figure 11.1 shows an outline of Scala's class hierarchy. At the top of the hierarchy is class Any, which defines methods that include the following:

```
final def ==(that: Any): Boolean
final def !=(that: Any): Boolean
def equals(that: Any): Boolean
def ##: Int
def hashCode: Int
def toString: String
```

Because every class inherits from Any, every object in a Scala program can be compared using ==, !=, or equals; hashed using ## or hashCode; and formatted using toString. The equality and inequality methods, == and !=, are declared final in class Any, so they cannot be overridden in subclasses.

The == method is essentially the same as equals and != is always the negation of equals.[1] So individual classes can tailor what == or != means by overriding the equals method. We'll show an example later in this chapter.

The root class Any has two subclasses: AnyVal and AnyRef. AnyVal is the parent class of *value classes* in Scala. While you can define your own value classes (see Section 11.4), there are nine value classes built into Scala: Byte, Short, Char, Int, Long, Float, Double, Boolean, and Unit. The first eight of these correspond to Java's primitive types, and their values are represented at run time as Java's primitive values. The instances of these classes are all written as literals in Scala. For example, 42 is an instance of Int, 'x' is an instance of Char, and false an instance of Boolean. You cannot create instances of these classes using new. This is enforced by the "trick" that value classes are all defined to be both abstract and final.

So if you were to write:

```
scala> new Int
```

you would get:

```
<console>:5: error: class Int is abstract; cannot be
instantiated
      new Int
      ^
```

The other value class, Unit, corresponds roughly to Java's void type; it is used as the result type of a method that does not otherwise return an interesting result. Unit has a single instance value, which is written (), as discussed in Section 7.2.

As explained in Chapter 5, the value classes support the usual arithmetic and boolean operators as methods. For instance, Int has methods named + and *, and Boolean has methods named || and &&. Value classes also inherit all methods from class Any. You can test this in the interpreter:

---

[1]The only case where == does not directly call equals is for Java's boxed numeric classes, such as Integer or Long. In Java, a new Integer(1) does not equal a new Long(1) even though for primitive values 1 == 1L. Since Scala is a more regular language than Java, it was necessary to correct this discrepancy by special-casing the == method for these classes. Likewise, the ## method provides a Scala version of hashing that is the same as Java's hashCode, except for boxed numeric types, where it works consistently with ==. For instance new Integer(1) and new Long(1) hash the same with ## even though their Java hashCodes are different.

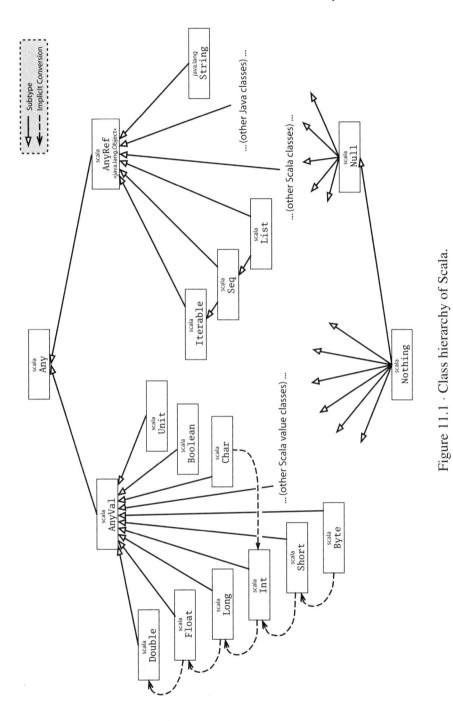

Figure 11.1 · Class hierarchy of Scala.

```
scala> 42.toString
res1: String = 42

scala> 42.hashCode
res2: Int = 42

scala> 42 equals 42
res3: Boolean = true
```

Note that the value class space is flat; all value classes are subtypes of scala.AnyVal, but they do not subclass each other. Instead there are implicit conversions between different value class types. For example, an instance of class scala.Int is automatically widened (by an implicit conversion) to an instance of class scala.Long when required.

As mentioned in Section 5.10, implicit conversions are also used to add more functionality to value types. For instance, the type Int supports all of the operations below:

```
scala> 42 max 43
res4: Int = 43

scala> 42 min 43
res5: Int = 42

scala> 1 until 5
res6: scala.collection.immutable.Range = Range(1, 2, 3, 4)

scala> 1 to 5
res7: scala.collection.immutable.Range.Inclusive
  = Range(1, 2, 3, 4, 5)

scala> 3.abs
res8: Int = 3

scala> (-3).abs
res9: Int = 3
```

Here's how this works: The methods min, max, until, to, and abs are all defined in a class scala.runtime.RichInt, and there is an implicit conversion from class Int to RichInt. The conversion is applied whenever a method is invoked on an Int that is undefined in Int but defined in RichInt. Similar "booster classes" and implicit conversions exist for the other value classes. Implicit conversions will be discussed in detail in Chapter 21.

The other subclass of the root class Any is class AnyRef. This is the base class of all *reference classes* in Scala. As mentioned previously, on the Java platform AnyRef is in fact just an alias for class java.lang.Object. So classes written in Java, as well as classes written in Scala, all inherit from AnyRef.[2] One way to think of java.lang.Object, therefore, is as the way AnyRef is implemented on the Java platform. Thus, although you can use Object and AnyRef interchangeably in Scala programs on the Java platform, the recommended style is to use AnyRef everywhere.

## 11.2 How primitives are implemented

How is all this implemented? In fact, Scala stores integers in the same way as Java—as 32-bit words. This is important for efficiency on the JVM and also for interoperability with Java libraries. Standard operations like addition or multiplication are implemented as primitive operations. However, Scala uses the "backup" class java.lang.Integer whenever an integer needs to be seen as a (Java) object. This happens for instance when invoking the toString method on an integer number or when assigning an integer to a variable of type Any. Integers of type Int are converted transparently to "boxed integers" of type java.lang.Integer whenever necessary.

All this sounds a lot like auto-boxing in Java 5 and it is indeed quite similar. There's one crucial difference though: Boxing in Scala is much less visible than boxing in Java. Try the following in Java:

```
// This is Java
boolean isEqual(int x, int y) {
  return x == y;
}
System.out.println(isEqual(421, 421));
```

You will surely get true. Now, change the argument types of isEqual to java.lang.Integer (or Object, the result will be the same):

---

[2]The reason AnyRef alias exists, instead of just using the name java.lang.Object, is because Scala was originally designed to work on both the Java and .NET platforms. On .NET, AnyRef was an alias for System.Object.

```
// This is Java
boolean isEqual(Integer x, Integer y) {
  return x == y;
}
System.out.println(isEqual(421, 421));
```

You will find that you get false! What happens is that the number 421 gets boxed twice, so that the arguments for x and y are two different objects. Because == means reference equality on reference types, and Integer is a reference type, the result is false. This is one aspect where it shows that Java is not a pure object-oriented language. There is a difference between primitive types and reference types that can be clearly observed.

Now try the same experiment in Scala:

```
scala> def isEqual(x: Int, y: Int) = x == y
isEqual: (x: Int, y: Int)Boolean

scala> isEqual(421, 421)
res10: Boolean = true

scala> def isEqual(x: Any, y: Any) = x == y
isEqual: (x: Any, y: Any)Boolean

scala> isEqual(421, 421)
res11: Boolean = true
```

The equality operation == in Scala is designed to be transparent with respect to the type's representation. For value types, it is the natural (numeric or boolean) equality. For reference types other than Java's boxed numeric types, == is treated as an alias of the equals method inherited from Object. That method is originally defined as reference equality, but is overridden by many subclasses to implement their natural notion of equality. This also means that in Scala you never fall into Java's well-known trap concerning string comparisons. In Scala, string comparison works as it should:

```
scala> val x = "abcd".substring(2)
x: String = cd

scala> val y = "abcd".substring(2)
y: String = cd

scala> x == y
res12: Boolean = true
```

In Java, the result of comparing x with y would be false. The programmer should have used `equals` in this case, but it is easy to forget.

However, there are situations where you need reference equality instead of user-defined equality. For example, in some situations where efficiency is paramount, you would like to *hash cons* with some classes and compare their instances with reference equality.[3] For these cases, class `AnyRef` defines an additional `eq` method, which cannot be overridden and is implemented as reference equality (*i.e.*, it behaves like `==` in Java for reference types). There's also the negation of `eq`, which is called `ne`. For example:

```
scala> val x = new String("abc")
x: String = abc

scala> val y = new String("abc")
y: String = abc

scala> x == y
res13: Boolean = true

scala> x eq y
res14: Boolean = false

scala> x ne y
res15: Boolean = true
```

Equality in Scala is discussed further in Chapter 30.

## 11.3   Bottom types

At the bottom of the type hierarchy in Figure 11.1 you see the two classes `scala.Null` and `scala.Nothing`. These are special types that handle some "corner cases" of Scala's object-oriented type system in a uniform way.

Class `Null` is the type of the `null` reference; it is a subclass of every reference class (*i.e.*, every class that itself inherits from `AnyRef`). `Null` is not compatible with value types. You cannot, for example, assign a `null` value to an integer variable:

---

[3] You hash cons instances of a class by caching all instances you have created in a weak collection. Then, any time you want a new instance of the class, you first check the cache. If the cache already has an element equal to the one you are about to create, you can reuse the existing instance. As a result of this arrangement, any two instances that are equal with `equals()` are also equal with reference equality.

```
scala> val i: Int = null
<console>:7: error: an expression of type Null is ineligible
for implicit conversion
         val i: Int = null
                  ^
```

Type Nothing is at the very bottom of Scala's class hierarchy; it is a sub-type of every other type. However, there exist no values of this type whatso-ever. Why does it make sense to have a type without values? As discussed in Section 7.4, one use of Nothing is that it signals abnormal termination.

For instance there's the error method in the Predef object of Scala's standard library, which is defined like this:

```
def error(message: String): Nothing =
  throw new RuntimeException(message)
```

The return type of error is Nothing, which tells users that the method will not return normally (it throws an exception instead). Because Nothing is a subtype of every other type, you can use methods like error in very flexible ways. For instance:

```
def divide(x: Int, y: Int): Int =
  if (y != 0) x / y
  else error("can't divide by zero")
```

The "then" branch of the conditional, x / y, has type Int, whereas the else branch, the call to error, has type Nothing. Because Nothing is a subtype of Int, the type of the whole conditional is Int, as required.

## 11.4 Defining your own value classes

As mentioned in Section 11.1, you can define your own value classes to augment the ones that are built in. Like the built-in value classes, an instance of your value class will usually compile to Java bytecode that does not use the wrapper class. In contexts where a wrapper is needed, such as with generic code, the value will get boxed and unboxed automatically.

Only certain classes can be made into value classes. For a class to be a value class, it must have exactly one parameter and it must have nothing inside it except defs. Furthermore, no other class can extend a value class, and a value class cannot redefine equals or hashCode.

To define a value class, make it a subclass of `AnyVal`, and put `val` before the one parameter. Here is an example value class:

```
class Dollars(val amount: Int) extends AnyVal {
  override def toString() = "$" + amount
}
```

As described in Section 10.6, the `val` prefix allows the `amount` parameter to be accessed as a field. For example, the following code creates an instance of the value class, then retrieves the amount from it:

```
scala> val money = new Dollars(1000000)
money: Dollars = $1000000
scala> money.amount
res16: Int = 1000000
```

In this example, `money` refers to an instance of the value class. It is of type `Dollars` in Scala source code, but the compiled Java bytecode will use type `Int` directly.

This example defines a `toString` method, and the compiler figures out when to use it. That's why printing `money` gives $1000000, with a dollar sign, but printing `money.amount` gives 1000000. You can even define multiple value types that are all backed by the same `Int` value. For example:

```
class SwissFrancs(val amount: Int) extends AnyVal {
  override def toString() = amount + " CHF"
}
```

Even though both `Dollars` and `SwissFrancs` are represented as integers, it works fine to use them in the same scope:

```
scala> val dollars = new Dollars(1000)
dollars: Dollars = $1000
scala> val francs = new SwissFrancs(1000)
francs: SwissFrancs = 1000 CHF
```

**Avoiding a types monoculture**

To get the most benefit from the Scala class hierarchy, try to define a new class for each domain concept, even when it would be possible to reuse the same class for different purposes. Even if such a class is a so-called *tiny type* with no methods or fields, defining the additional class is a way to help the compiler be helpful to you.

For example, suppose you are writing some code to generate HTML. In HTML, a style name is represented as a string. So are anchor identifiers. HTML itself is also a string, so if you wanted, you could define helper code using strings to represent all of these things, like this:

```
def title(text: String, anchor: String, style: String): String =
  s"<a id='$anchor'><h1 class='$style'>$text</h1></a>"
```

That type signature has four strings in it! Such *stringly typed* code is technically strongly typed, but since everything in sight is of type String, the compiler cannot help you detect the use of one when you meant to write the other. For example, it won't stop you from this travesty:

```
scala> title("chap:vcls", "bold", "Value Classes")
res17: String = <a id='bold'><h1 class='Value
    Classes'>chap:vcls</h1></a>
```

This HTML is mangled. The intended display text "Value Classes" is being used as a style class, and the text being displayed is "chap.vcls," which was supposed to be an anchor. To top it off, the actual anchor identifier is "bold," which is supposed to be a style class. Despite this comedy of errors, the compiler utters not a peep.

The compiler can be more helpful if you define a tiny type for each domain concept. For example, you could define a small class for styles, anchor identifiers, display text, and HTML. Since these classes have one parameter and no members, they can be defined as value classes:

```
class Anchor(val value: String) extends AnyVal
class Style(val value: String) extends AnyVal
class Text(val value: String) extends AnyVal
class Html(val value: String) extends AnyVal
```

Given these classes, it is possible to write a version of title that has a less trivial type signature:

```
def title(text: Text, anchor: Anchor, style: Style): Html =
  new Html(
    s"<a id='${anchor.value}'>" +
        s"<h1 class='${style.value}'>" +
        text.value +
        "</h1></a>"
  )
```

If you try to use this version with the arguments in the wrong order, the compiler can now detect the error. For example:

```
scala> title(new Anchor("chap:vcls"), new Style("bold"),
            new Text("Value Classes"))
<console>:18: error: type mismatch;
 found    : Anchor
 required: Text
                new Anchor("chap:vcls"),
                ^
<console>:19: error: type mismatch;
 found    : Style
 required: Anchor
                new Style("bold"),
                ^
<console>:20: error: type mismatch;
 found    : Text
 required: Style
                new Text("Value Classes"))
                ^
```

## 11.5   Conclusion

In this chapter we showed you the classes at the top and bottom of Scala's class hierarchy. Now that you've gotten a good foundation on class inheritance in Scala, you're ready to understand mixin composition. In the next chapter, you'll learn about traits.

# Chapter 12

# Traits

Traits are a fundamental unit of code reuse in Scala. A trait encapsulates method and field definitions, which can then be reused by mixing them into classes. Unlike class inheritance, in which each class must inherit from just one superclass, a class can mix in any number of traits. This chapter shows you how traits work and shows two of the most common ways they are useful: widening thin interfaces to rich ones, and defining stackable modifications. It also shows how to use the Ordered trait and compares traits to the multiple inheritance of other languages.

## 12.1 How traits work

A trait definition looks just like a class definition except that it uses the keyword trait. An example is shown in Listing 12.1:

```
trait Philosophical {
  def philosophize() = {
    println("I consume memory, therefore I am!")
  }
}
```

Listing 12.1 · The definition of trait Philosophical.

This trait is named Philosophical. It does not declare a superclass, so like a class, it has the default superclass of AnyRef. It defines one method, named philosophize, which is concrete. It's a simple trait, just enough to show how traits work.

223

Once a trait is defined, it can be *mixed in* to a class using either the extends or with keywords. Scala programmers "mix in" traits rather than inherit from them, because mixing in a trait has important differences from the multiple inheritance found in many other languages. This issue is discussed in Section 12.6. For example, Listing 12.2 shows a class that mixes in the Philosophical trait using extends:

```
class Frog extends Philosophical {
  override def toString = "green"
}
```

Listing 12.2 · Mixing in a trait using extends.

You can use the extends keyword to mix in a trait; in that case you implicitly inherit the trait's superclass. For instance, in Listing 12.2, class Frog subclasses AnyRef (the superclass of Philosophical) and mixes in Philosophical. Methods inherited from a trait can be used just like methods inherited from a superclass. Here's an example:

```
scala> val frog = new Frog
frog: Frog = green

scala> frog.philosophize()
I consume memory, therefore I am!
```

A trait also defines a type. Here's an example in which Philosophical is used as a type:

```
scala> val phil: Philosophical = frog
phil: Philosophical = green

scala> phil.philosophize()
I consume memory, therefore I am!
```

The type of phil is Philosophical, a trait. Thus, variable phil could have been initialized with any object whose class mixes in Philosophical.

If you wish to mix a trait into a class that explicitly extends a superclass, you use extends to indicate the superclass and with to mix in the trait. Listing 12.3 shows an example. If you want to mix in multiple traits, you add more with clauses. For example, given a trait HasLegs, you could mix both Philosophical and HasLegs into Frog as shown in Listing 12.4.

```
class Animal
class Frog extends Animal with Philosophical {
  override def toString = "green"
}
```

Listing 12.3 · Mixing in a trait using with.

```
class Animal
trait HasLegs
class Frog extends Animal with Philosophical with HasLegs {
  override def toString = "green"
}
```

Listing 12.4 · Mixing in multiple traits.

In the examples you've seen so far, class Frog has inherited an implementation of philosophize from trait Philosophical. Alternatively, Frog could override philosophize. The syntax looks the same as overriding a method declared in a superclass. Here's an example:

```
class Animal
class Frog extends Animal with Philosophical {
  override def toString = "green"
  override def philosophize() = {
    println("It ain't easy being " + toString + "!")
  }
}
```

Because this new definition of Frog still mixes in trait Philosophical, you can still use it from a variable of that type. But because Frog overrides Philosophical's implementation of philosophize, you'll get a new behavior when you call it:

```
scala> val phrog: Philosophical = new Frog
phrog: Philosophical = green

scala> phrog.philosophize()
It ain't easy being green!
```

At this point you might philosophize that traits are like Java interfaces with concrete methods, but they can actually do much more. Traits can, for example, declare fields and maintain state. In fact, you can do anything in a trait definition that you can do in a class definition, and the syntax looks exactly the same, with only two exceptions.

First, a trait cannot have any "class" parameters (*i.e.*, parameters passed to the primary constructor of a class). In other words, although you could define a class like this:

```
class Point(x: Int, y: Int)
```

The following attempt to define a trait would not compile:

```
trait NoPoint(x: Int, y: Int) // Does not compile
```

You'll find out in Section 20.5 how to work around this restriction.

The other difference between classes and traits is that whereas in classes, super calls are statically bound, in traits, they are dynamically bound. If you write "super.toString" in a class, you know exactly which method implementation will be invoked. When you write the same thing in a trait, however, the method implementation to invoke for the super call is undefined when you define the trait. Rather, the implementation to invoke will be determined anew each time the trait is mixed into a concrete class. This curious behavior of super is key to allowing traits to work as *stackable modifications*, which will be described in Section 12.5. The rules for resolving super calls will be given in Section 12.6.

## 12.2   Thin versus rich interfaces

One major use of traits is to automatically add methods to a class in terms of methods the class already has. That is, traits can enrich a *thin* interface, making it into a *rich* interface.

Thin versus rich interfaces represents a commonly faced trade-off in object-oriented design. The trade-off is between the implementers and the clients of an interface. A rich interface has many methods, which make it convenient for the caller. Clients can pick a method that exactly matches the functionality they need. A thin interface, on the other hand, has fewer methods, and thus is easier on the implementers. Clients calling into a thin interface, however, have to write more code. Given the smaller selection of methods to call, they may have to choose a less than perfect match for their needs and write extra code to use it.

Java's interfaces are more often thin than rich. For example, interface CharSequence, which was introduced in Java 1.4, is a thin interface common to all string-like classes that hold a sequence of characters. Here's its definition when seen as a Scala trait:

```scala
trait CharSequence {
  def charAt(index: Int): Char
  def length: Int
  def subSequence(start: Int, end: Int): CharSequence
  def toString(): String
}
```

Although most of the dozens of methods in class String would apply to any CharSequence, Java's CharSequence interface declares only four methods. Had CharSequence instead included the full String interface, it would have placed a large burden on implementers of CharSequence. Every programmer that implemented CharSequence in Java would have had to define dozens more methods. Because Scala traits can contain concrete methods, they make rich interfaces far more convenient.

Adding a concrete method to a trait tilts the thin-rich trade-off heavily towards rich interfaces. Unlike in Java, adding a concrete method to a Scala trait is a one-time effort. You only need to implement the method once, in the trait itself, instead of needing to reimplement it for every class that mixes in the trait. Thus, rich interfaces are less work to provide in Scala than in a language without traits.

To enrich an interface using traits, simply define a trait with a small number of abstract methods—the thin part of the trait's interface—and a potentially large number of concrete methods, all implemented in terms of the abstract methods. Then you can mix the enrichment trait into a class, implement the thin portion of the interface, and end up with a class that has all of the rich interface available.

## 12.3  Example: Rectangular objects

Graphics libraries often have many different classes that represent something rectangular. Some examples are windows, bitmap images, and regions selected with a mouse. To make these rectangular objects convenient to use, it is nice if the library provides geometric queries, such as width, height, left, right, topLeft, and so on. However, many such methods exist that would be nice to have, so it can be a large burden on library writers to provide all of them for all rectangular objects in a Java library. If such a library were written in Scala, by contrast, the library writer could use traits to easily supply all of these convenience methods on all the classes they like.

To see how, first imagine what the code would look like without traits. There would be some basic geometric classes like Point and Rectangle:

```scala
class Point(val x: Int, val y: Int)

class Rectangle(val topLeft: Point, val bottomRight: Point) {
  def left = topLeft.x
  def right = bottomRight.x
  def width = right - left
  // and many more geometric methods...
}
```

This Rectangle class takes two points in its primary constructor: the coordinates of the top-left and bottom-right corners. It then implements many convenience methods, such as left, right, and width, by performing simple calculations on these two points.

Another class a graphics library might have is a 2-D graphical widget:

```scala
abstract class Component {
  def topLeft: Point
  def bottomRight: Point

  def left = topLeft.x
  def right = bottomRight.x
  def width = right - left
  // and many more geometric methods...
}
```

Notice that the definitions of left, right, and width are exactly the same in the two classes. They will also be the same, aside from minor variations, in any other classes for rectangular objects.

This repetition can be eliminated with an enrichment trait. The trait will have two abstract methods: one that returns the top-left coordinate of the object, and another that returns the bottom-right coordinate. It can then supply concrete implementations of all the other geometric queries. Listing 12.5 shows what it will look like:

```scala
trait Rectangular {
  def topLeft: Point
  def bottomRight: Point

  def left = topLeft.x
  def right = bottomRight.x
  def width = right - left
  // and many more geometric methods...
}
```

Listing 12.5 · Defining an enrichment trait.

Class Component can mix in this trait to get all the geometric methods provided by Rectangular:

```scala
abstract class Component extends Rectangular {
  // other methods...
}
```

Similarly, Rectangle itself can mix in the trait:

```scala
class Rectangle(val topLeft: Point, val bottomRight: Point)
    extends Rectangular {
  // other methods...
}
```

Given these definitions, you can create a Rectangle and call geometric methods such as width and left on it:

```scala
scala> val rect = new Rectangle(new Point(1, 1),
           new Point(10, 10))
rect: Rectangle = Rectangle@5f5da68c

scala> rect.left
res2: Int = 1

scala> rect.right
res3: Int = 10
```

229

```
scala> rect.width
res4: Int = 9
```

## 12.4   The Ordered trait

Comparison is another domain where a rich interface is convenient. Whenever you compare two objects that are ordered, it is convenient if you use a single method call to ask about the precise comparison you want. If you want "is less than," you would like to call <, and if you want "is less than or equal," you would like to call <=. With a thin comparison interface, you might just have the < method, and you would sometimes have to write things like "(x < y) || (x == y)". A rich interface would provide you with all of the usual comparison operators, thus allowing you to directly write things like "x <= y".

Before looking at Ordered, imagine what you might do without it. Suppose you took the Rational class from Chapter 6 and added comparison operations to it. You would end up with something like this:[1]

```
class Rational(n: Int, d: Int) {
  // ...
  def < (that: Rational) =
    this.numer * that.denom < that.numer * this.denom
  def > (that: Rational) = that < this
  def <= (that: Rational) = (this < that) || (this == that)
  def >= (that: Rational) = (this > that) || (this == that)
}
```

This class defines four comparison operators (<, >, <=, and >=), and it's a classic demonstration of the costs of defining a rich interface. First, notice that three of the comparison operators are defined in terms of the first one. For example, > is defined as the reverse of <, and <= is defined as literally "less than or equal." Next, notice that all three of these methods would be the same for any other class that is comparable. There is nothing special about rational numbers regarding <=. In a comparison context, <= is *always* used to mean "less than or equals." Overall, there is quite a lot of boilerplate code

---

[1]This example is based on the Rational class shown in Listing 6.5 on page 113, with equals, hashCode, and modifications to ensure a positive denom added.

in this class which would be the same in any other class that implements comparison operations.

This problem is so common that Scala provides a trait to help with it. The trait is called Ordered. To use it, you replace all of the individual comparison methods with a single compare method. The Ordered trait then defines <, >, <=, and >= for you in terms of this one method. Thus, trait Ordered allows you to enrich a class with comparison methods by implementing only one method, compare.

Here is how it looks if you define comparison operations on Rational by using the Ordered trait:

```
class Rational(n: Int, d: Int) extends Ordered[Rational] {
  // ...
  def compare(that: Rational) =
    (this.numer * that.denom) - (that.numer * this.denom)
}
```

There are just two things to do. First, this version of Rational mixes in the Ordered trait. Unlike the traits you have seen so far, Ordered requires you to specify a *type parameter* when you mix it in. Type parameters are not discussed in detail until Chapter 19, but for now all you need to know is that when you mix in Ordered, you must actually mix in Ordered[C], where C is the class whose elements you compare. In this case, Rational mixes in Ordered[Rational].

The second thing you need to do is define a compare method for comparing two objects. This method should compare the receiver, this, with the object passed as an argument to the method. It should return an integer that is zero if the objects are the same, negative if receiver is less than the argument, and positive if the receiver is greater than the argument.

In this case, the comparison method of Rational uses a formula based on converting the fractions to a common denominator and then subtracting the resulting numerators. Given this mixin and the definition of compare, class Rational now has all four comparison methods:

```
scala> val half = new Rational(1, 2)
half: Rational = 1/2

scala> val third = new Rational(1, 3)
third: Rational = 1/3
```

231

```
scala> half < third
res5: Boolean = false

scala> half > third
res6: Boolean = true
```

Any time you implement a class that is ordered by some comparison, you should consider mixing in the `Ordered` trait. If you do, you will provide the class's users with a rich set of comparison methods.

Beware that the `Ordered` trait does not define `equals` for you, because it is unable to do so. The problem is that implementing `equals` in terms of `compare` requires checking the type of the passed object, and because of type erasure, `Ordered` itself cannot do this test. Thus, you need to define `equals` yourself, even if you inherit `Ordered`. You'll find out how to go about this in Chapter 30.

## 12.5 Traits as stackable modifications

You have now seen one major use of traits: turning a thin interface into a rich one. Now we'll turn to a second major use: providing stackable modifications to classes. Traits let you *modify* the methods of a class, and they do so in a way that allows you to *stack* those modifications with each other.

As an example, consider stacking modifications to a queue of integers. The queue will have two operations: `put`, which places integers in the queue, and `get`, which takes them back out. Queues are first-in, first-out, so `get` should return the integers in the same order they were put in the queue.

Given a class that implements such a queue, you could define traits to perform modifications such as these:

- `Doubling`: double all integers that are put in the queue

- `Incrementing`: increment all integers that are put in the queue

- `Filtering`: filter out negative integers from a queue

These three traits represent *modifications*, because they modify the behavior of an underlying queue class rather than defining a full queue class themselves. The three are also *stackable*. You can select any of the three you like, mix them into a class, and obtain a new class that has all of the modifications you chose.

An abstract `IntQueue` class is shown in Listing 12.6. `IntQueue` has a put method that adds new integers to the queue, and a get method that removes and returns them. A basic implementation of `IntQueue` that uses an `ArrayBuffer` is shown in Listing 12.7.

```
abstract class IntQueue {
  def get(): Int
  def put(x: Int)
}
```

Listing 12.6 · Abstract class `IntQueue`.

```
import scala.collection.mutable.ArrayBuffer

class BasicIntQueue extends IntQueue {
  private val buf = new ArrayBuffer[Int]
  def get() = buf.remove(0)
  def put(x: Int) = { buf += x }
}
```

Listing 12.7 · A `BasicIntQueue` implemented with an `ArrayBuffer`.

Class `BasicIntQueue` has a private field holding an array buffer. The get method removes an entry from one end of the buffer, while the put method adds elements to the other end. Here's how this implementation looks when you use it:

```
scala> val queue = new BasicIntQueue
queue: BasicIntQueue = BasicIntQueue@23164256

scala> queue.put(10)

scala> queue.put(20)

scala> queue.get()
res9: Int = 10

scala> queue.get()
res10: Int = 20
```

So far so good. Now take a look at using traits to modify this behavior. Listing 12.8 shows a trait that doubles integers as they are put in the queue.

233

The Doubling trait has two funny things going on. The first is that it declares a superclass, IntQueue. This declaration means that the trait can only be mixed into a class that also extends IntQueue. Thus, you can mix Doubling into BasicIntQueue, but not into Rational.

```
trait Doubling extends IntQueue {
  abstract override def put(x: Int) = { super.put(2 * x) }
}
```

Listing 12.8 · The Doubling stackable modification trait.

The second funny thing is that the trait has a super call on a method declared abstract. Such calls are illegal for normal classes because they will certainly fail at run time. For a trait, however, such a call can actually succeed. Since super calls in a trait are dynamically bound, the super call in trait Doubling will work so long as the trait is mixed in *after* another trait or class that gives a concrete definition to the method.

This arrangement is frequently needed with traits that implement stackable modifications. To tell the compiler you are doing this on purpose, you must mark such methods as abstract override. This combination of modifiers is only allowed for members of traits, not classes, and it means that the trait must be mixed into some class that has a concrete definition of the method in question.

There is a lot going on with such a simple trait, isn't there! Here's how it looks to use the trait:

```
scala> class MyQueue extends BasicIntQueue with Doubling
defined class MyQueue

scala> val queue = new MyQueue
queue: MyQueue = MyQueue@44bbf788

scala> queue.put(10)

scala> queue.get()
res12: Int = 20
```

In the first line in this interpreter session, we define class MyQueue, which extends BasicIntQueue and mixes in Doubling. We then put a 10 in the queue, but because Doubling has been mixed in, the 10 is doubled. When we get an integer from the queue, it is a 20.

Note that MyQueue defines no new code. It simply identifies a class and mixes in a trait. In this situation, you could supply "BasicIntQueue with Doubling" directly to new instead of defining a named class. It would look as shown in Listing 12.9:

```
scala> val queue = new BasicIntQueue with Doubling
queue: BasicIntQueue with Doubling = $anon$1@141f05bf

scala> queue.put(10)

scala> queue.get()
res14: Int = 20
```

Listing 12.9 · Mixing in a trait when instantiating with new.

To see how to stack modifications, we need to define the other two modification traits, Incrementing and Filtering. Implementations of these traits are shown in Listing 12.10:

```
trait Incrementing extends IntQueue {
  abstract override def put(x: Int) = { super.put(x + 1) }
}
trait Filtering extends IntQueue {
  abstract override def put(x: Int) = {
    if (x >= 0) super.put(x)
  }
}
```

Listing 12.10: Stackable modification traits Incrementing and Filtering.

Given these modifications, you can now pick and choose which ones you want for a particular queue. For example, here is a queue that both filters negative numbers and adds one to all numbers that it keeps:

```
scala> val queue = (new BasicIntQueue
          with Incrementing with Filtering)
queue: BasicIntQueue with Incrementing with Filtering...

scala> queue.put(-1); queue.put(0); queue.put(1)

scala> queue.get()
res16: Int = 1
```

```
scala> queue.get()
res17: Int = 2
```

The order of mixins is significant.[2] The precise rules are given in the following section, but, roughly speaking, traits further to the right take effect first. When you call a method on a class with mixins, the method in the trait furthest to the right is called first. If that method calls super, it invokes the method in the next trait to its left, and so on. In the previous example, Filtering's put is invoked first, so it removes integers that were negative to begin with. Incrementing's put is invoked second, so it adds one to those integers that remain.

If you reverse the order, first integers will be incremented, and *then* the integers that are still negative will be discarded:

```
scala> val queue = (new BasicIntQueue
            with Filtering with Incrementing)
queue: BasicIntQueue with Filtering with Incrementing...

scala> queue.put(-1); queue.put(0); queue.put(1)

scala> queue.get()
res19: Int = 0

scala> queue.get()
res20: Int = 1

scala> queue.get()
res21: Int = 2
```

Overall, code written in this style gives you a great deal of flexibility. You can define sixteen different classes by mixing in these three traits in different combinations and orders. That's a lot of flexibility for a small amount of code, so you should keep your eyes open for opportunities to arrange code as stackable modifications.

## 12.6 Why not multiple inheritance?

Traits are a way to inherit from multiple class-like constructs, but they differ in important ways from the multiple inheritance present in many languages.

---

[2]Once a trait is mixed into a class, you can alternatively call it a *mixin*.

One difference is especially important: the interpretation of super. With multiple inheritance, the method called by a super call can be determined right where the call appears. With traits, the method called is determined by a *linearization* of the classes and traits that are mixed into a class. This is the difference that enables the stacking of modifications described in the previous section.

Before looking at linearization, take a moment to consider how to stack modifications in a language with traditional multiple inheritance. Imagine the following code, but this time interpreted as multiple inheritance instead of trait mixin:

```
// Multiple inheritance thought experiment
val q = new BasicIntQueue with Incrementing with Doubling
q.put(42)  // which put would be called?
```

The first question is: Which put method would get invoked by this call? Perhaps the rule would be that the last superclass wins, in which case Doubling would get called. Doubling would double its argument and call super.put, and that would be it. No incrementing would happen! Likewise, if the rule were that the first superclass wins, the resulting queue would increment integers but not double them. Thus neither ordering would work.

You might also entertain the possibility of allowing programmers to identify exactly which superclass method they want when they say super. For example, imagine the following Scala-like code, in which super appears to be explicitly invoked on both Incrementing and Doubling:

```
// Multiple inheritance thought experiment
trait MyQueue extends BasicIntQueue
    with Incrementing with Doubling {
  def put(x: Int) = {
    Incrementing.super.put(x) // (Not real Scala)
    Doubling.super.put(x)
  }
}
```

This approach would give us new problems (with the verbosity of this attempt being the least of its problems). What would happen is that the base class's put method would get called *twice*—once with an incremented

237

value and once with a doubled value, but neither time with an incremented, doubled value.

There is simply no good solution to this problem using multiple inheritance. You would have to back up in your design and factor the code differently. By contrast, the traits solution in Scala is straightforward. You simply mix in `Incrementing` and `Doubling`, and Scala's special treatment of super in traits makes it all work out. Something is clearly different here from traditional multiple inheritance, but what?

As hinted previously, the answer is linearization. When you instantiate a class with new, Scala takes the class, and all of its inherited classes and traits, and puts them in a single, *linear* order. Then, whenever you call super inside one of those classes, the invoked method is the next one up the chain. If all of the methods but the last call super, the net result is stackable behavior.

The precise order of the linearization is described in the language specification. It is a little bit complicated, but the main thing you need to know is that, in any linearization, a class is always linearized in front of *all* its superclasses and mixed in traits. Thus, when you write a method that calls super, that method is definitely modifying the behavior of the superclasses and mixed in traits, not the other way around.

> **Note**
> The remainder of this section describes the details of linearization. You can safely skip the rest of this section if you are not interested in understanding those details right now.

The main properties of Scala's linearization are illustrated by the following example: Say you have a class `Cat`, which inherits from a superclass `Animal` and two supertraits `Furry` and `FourLegged`. `FourLegged` extends in turn another trait `HasLegs`:

```
class Animal
trait Furry extends Animal
trait HasLegs extends Animal
trait FourLegged extends HasLegs
class Cat extends Animal with Furry with FourLegged
```

Class `Cat`'s inheritance hierarchy and linearization are shown in Figure 12.1. Inheritance is indicated using traditional UML notation:[3] arrows

---

[3]Rumbaugh, *et al.*, *The Unified Modeling Language Reference Manual*. [Rum04]

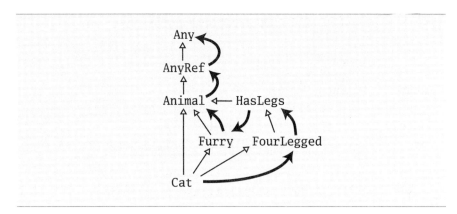

Figure 12.1 · Inheritance hierarchy and linearization of class Cat.

with white, triangular arrowheads indicate inheritance, with the arrowhead pointing to the supertype. The arrows with darkened, non-triangular arrowheads depict linearization. The darkened arrowheads point in the direction in which super calls will be resolved.

The linearization of Cat is computed from back to front as follows. The *last* part of the linearization of Cat is the linearization of its superclass, Animal. This linearization is copied over without any changes. (The linearization of each of these types is shown in Table 12.1 on page 240.) Because Animal doesn't explicitly extend a superclass or mix in any supertraits, it by default extends AnyRef, which extends Any. Animal's linearization, therefore, looks like:

$$\text{Animal} \twoheadrightarrow \text{AnyRef} \twoheadrightarrow \text{Any}$$

The second to last part is the linearization of the first mixin, trait Furry, but all classes that are already in the linearization of Animal are left out now, so that each class appears only once in Cat's linearization. The result is:

$$\text{Furry} \twoheadrightarrow \text{Animal} \twoheadrightarrow \text{AnyRef} \twoheadrightarrow \text{Any}$$

This is preceded by the linearization of FourLegged, where again any classes that have already been copied in the linearizations of the superclass or the first mixin are left out:

$$\text{FourLegged} \twoheadrightarrow \text{HasLegs} \twoheadrightarrow \text{Furry} \twoheadrightarrow \text{Animal} \twoheadrightarrow \text{AnyRef} \twoheadrightarrow \text{Any}$$

239

Table 12.1 · Linearization of types in Cat's hierarchy

| Type | Linearization |
|------|---------------|
| Animal | Animal, AnyRef, Any |
| Furry | Furry, Animal, AnyRef, Any |
| FourLegged | FourLegged, HasLegs, Animal, AnyRef, Any |
| HasLegs | HasLegs, Animal, AnyRef, Any |
| Cat | Cat, FourLegged, HasLegs, Furry, Animal, AnyRef, Any |

Finally, the first class in the linearization of Cat is Cat itself:

Cat → FourLegged → HasLegs → Furry → Animal → AnyRef → Any

When any of these classes and traits invokes a method via super, the implementation invoked will be the first implementation to its right in the linearization.

## 12.7   To trait or not to trait?

Whenever you implement a reusable collection of behavior, you will have to decide whether you want to use a trait or an abstract class. There is no firm rule, but this section contains a few guidelines to consider.

*If the behavior will not be reused*, then make it a concrete class. It is not reusable behavior after all.

*If it might be reused in multiple, unrelated classes*, make it a trait. Only traits can be mixed into different parts of the class hierarchy.

*If you want to inherit from it in Java code*, use an abstract class. Since traits with code do not have a close Java analog, it tends to be awkward to inherit from a trait in a Java class. Inheriting from a Scala class, meanwhile, is exactly like inheriting from a Java class. As one exception, a Scala trait with only abstract members translates directly to a Java interface, so you should feel free to define such traits even if you expect Java code to inherit from it. See Chapter 31 for more information on working with Java and Scala together.

*If you plan to distribute it in compiled form*, and you expect outside groups to write classes inheriting from it, you might lean towards using an abstract class. The issue is that when a trait gains or loses a member, any classes that inherit from it must be recompiled, even if they have not changed.

If outside clients will only call into the behavior, instead of inheriting from it, then using a trait is fine.

*If you still do not know*, after considering the above, then start by making it as a trait. You can always change it later, and in general using a trait keeps more options open.

## 12.8 Conclusion

This chapter has shown you how traits work and how to use them in several common idioms. You saw that traits are similar to multiple inheritance. But because traits interpret super using linearization, they both avoid some of the difficulties of traditional multiple inheritance and allow you to stack behaviors. You also saw the Ordered trait and learned how to write your own enrichment traits.

Now that you have seen all of these facets, it is worth stepping back and taking another look at traits as a whole. Traits do not merely support the idioms described in this chapter; they are a fundamental unit of code that is reusable through inheritance. As such, many experienced Scala programmers start with traits when they are at the early stages of implementation. Each trait can hold less than an entire concept, a mere fragment of a concept. As the design solidifies, the fragments can be combined into more complete concepts through trait mixin.

# Chapter 13

# Packages and Imports

When working on a program, especially a large one, it is important to minimize *coupling*—the extent to which the various parts of the program rely on the other parts. Low coupling reduces the risk that a small, seemingly innocuous change in one part of the program will have devastating consequences in another part. One way to minimize coupling is to write in a modular style. You divide the program into a number of smaller modules, each of which has an inside and an outside. When working on the inside of a module—its *implementation*—you need only coordinate with other programmers working on that very same module. Only when you must change the outside of a module—its *interface*—is it necessary to coordinate with developers working on other modules.

This chapter shows several constructs that help you program in a modular style. It shows how to place things in packages, make names visible through imports, and control the visibility of definitions through access modifiers. The constructs are similar in spirit to constructs in Java, but there are some differences—usually ways that are more consistent—so it's worth reading this chapter even if you already know Java.

## 13.1   Putting code in packages

Scala code resides in the Java platform's global hierarchy of packages. The example code you've seen so far in this book has been in the *unnamed* package. You can place code into named packages in Scala in two ways. First, you can place the contents of an entire file into a package by putting a package clause at the top of the file, as shown in Listing 13.1.

```
package bobsrockets.navigation
class Navigator
```

Listing 13.1 · Placing the contents of an entire file into a package.

The `package` clause of Listing 13.1 places class `Navigator` into the package named `bobsrockets.navigation`. Presumably, this is the navigation software developed by Bob's Rockets, Inc.

> **Note**
> Because Scala code is part of the Java ecosystem, it is recommended to
> follow Java's reverse-domain-name convention for Scala packages that
> you release to the public. Thus, a better name for `Navigator`'s package
> might be `com.bobsrockets.navigation`. In this chapter, however, we'll
> leave off the "com." to make the examples easier to understand.

The other way you can place code into packages in Scala is more like C# namespaces. You follow a package clause by a section in curly braces that contains the definitions that go into the package. This syntax is called a *packaging*. The packaging shown in Listing 13.2 has the same effect as the code in Listing 13.1:

```
package bobsrockets.navigation {
  class Navigator
}
```

Listing 13.2 · Long form of a simple package declaration.

For such simple examples, you might as well use the syntactic sugar shown in Listing 13.1. However, one use of the more general notation is to have different parts of a file in different packages. For example, you might include a class's tests in the same file as the original code, but put the tests in a different package, as shown in Listing 13.3.

## 13.2   Concise access to related code

When code is divided into a package hierarchy, it doesn't just help people browse through the code. It also tells the compiler that code in the same

```
package bobsrockets {
  package navigation {

    // In package bobsrockets.navigation
    class Navigator

    package tests {

      // In package bobsrockets.navigation.tests
      class NavigatorSuite
    }
  }
}
```

Listing 13.3 · Multiple packages in the same file.

```
package bobsrockets {
  package navigation {
    class Navigator {
      // No need to say bobsrockets.navigation.StarMap
      val map = new StarMap
    }
    class StarMap
  }
  class Ship {
    // No need to say bobsrockets.navigation.Navigator
    val nav = new navigation.Navigator
  }
  package fleets {
    class Fleet {
      // No need to say bobsrockets.Ship
      def addShip() = { new Ship }
    }
  }
}
```

Listing 13.4 · Concise access to classes and packages.

```
package bobsrockets {
  class Ship
}

package bobsrockets.fleets {
  class Fleet {
    // Doesn't compile! Ship is not in scope.
    def addShip() = { new Ship }
  }
}
```

Listing 13.5 · Symbols in enclosing packages not automatically available.

```
// In file launch.scala
package launch {
  class Booster3
}
// In file bobsrockets.scala
package bobsrockets {
  package navigation {
    package launch {
      class Booster1
    }
    class MissionControl {
      val booster1 = new launch.Booster1
      val booster2 = new bobsrockets.launch.Booster2
      val booster3 = new _root_.launch.Booster3
    }
  }
  package launch {
    class Booster2
  }
}
```

Listing 13.6 · Accessing hidden package names.

package is related in some way to each other. Scala takes advantage of this relatedness by allowing short, unqualified names when accessing code that is in the same package.

Listing 13.4 gives three simple examples. First, as you would expect, a class can be accessed from within its own package without needing a prefix. That's why new `StarMap` compiles. Class `StarMap` is in the same package, `bobsrockets.navigation`, as the new expression that accesses it, so the package name doesn't need to be prefixed.

Second, a package itself can be accessed from its containing package without needing a prefix. In Listing 13.4, look at how class `Navigator` is instantiated. The new expression appears in package `bobsrockets`, which is the containing package of `bobsrockets.navigation`. Thus, it can access package `bobsrockets.navigation` as simply `navigation`.

Third, when using the curly-braces packaging syntax, all names accessible in scopes outside the packaging are also available inside it. An example in Listing 13.4 is the way `addShip()` creates a new `Ship`. The method is defined within two packagings: an outer one for `bobsrockets`, and an inner one for `bobsrockets.fleets`. Since `Ship` is accessible in the outer packaging, it can be referenced from within `addShip()`.

Note that this kind of access is only available if you explicitly nest the packagings. If you stick to one package per file, then—like in Java—the only names available will be the ones defined in the current package. In Listing 13.5, the packaging of `bobsrockets.fleets` has been moved to the top level. Since it is no longer enclosed in a packaging for `bobsrockets`, names from `bobsrockets` are not immediately in scope. As a result, new `Ship` gives a compile error. If nesting packages with braces shifts your code uncomfortably to the right, you can also use multiple package clauses without the braces.[1] For instance, the code below also defines class `Fleet` in two nested packages bobrockets and fleets, just like you saw it in Listing 13.4:

```
package bobsrockets
package fleets
class Fleet {
  // No need to say bobsrockets.Ship
  def addShip() = { new Ship }
}
```

---

[1]This style of multiple package clauses without braces is called *chained package clauses*.

One final trick is important to know. Sometimes, you end up coding in a heavily crowded scope where package names are hiding each other. In Listing 13.6, the scope of class `MissionControl` includes three separate packages named `launch`! There's one `launch` in `bobsrockets.navigation`, one in `bobsrockets`, and one at the top level. How would you reference each of `Booster1`, `Booster2`, and `Booster3`?

Accessing the first one is easiest. A reference to `launch` by itself will get you to package `bobsrockets.navigation.launch`, because that is the `launch` package defined in the closest enclosing scope. Thus, you can refer to the first booster class as simply `launch.Booster1`. Referring to the second one also is not tricky. You can write `bobrockets.launch.Booster2` and be clear about which one you are referencing. That leaves the question of the third booster class: How can you access `Booster3`, considering that a nested `launch` package shadows the top-level one?

To help in this situation, Scala provides a package named `_root_` that is outside any package a user can write. Put another way, every top-level package you can write is treated as a member of package `_root_`. For example, both `launch` and `bobsrockets` of Listing 13.6 are members of package `_root_`. As a result, `_root_.launch` gives you the top-level `launch` package, and `_root_.launch.Booster3` designates the outermost booster class.

## 13.3  Imports

In Scala, packages and their members can be imported using `import` clauses. Imported items can then be accessed by a simple name like `File`, as opposed to requiring a qualified name like `java.io.File`. For example, consider the code shown in Listing 13.7.

An `import` clause makes members of a package or object available by their names alone without needing to prefix them by the package or object name. Here are some simple examples:

```
// easy access to Fruit
import bobsdelights.Fruit

// easy access to all members of bobsdelights
import bobsdelights._

// easy access to all members of Fruits
import bobsdelights.Fruits._
```

```
package bobsdelights

abstract class Fruit(
  val name: String,
  val color: String
)

object Fruits {
  object Apple extends Fruit("apple", "red")
  object Orange extends Fruit("orange", "orange")
  object Pear extends Fruit("pear", "yellowish")
  val menu = List(Apple, Orange, Pear)
}
```

Listing 13.7 · Bob's delightful fruits, ready for import.

The first of these corresponds to Java's single type import and the second to Java's *on-demand* import. The only difference is that Scala's on-demand imports are written with a trailing underscore (_) instead of an asterisk (*). (After all, * is a valid identifier in Scala!) The third import clause above corresponds to Java's import of static class fields.

These three imports give you a taste of what imports can do, but Scala imports are actually much more general. For one, imports in Scala can appear anywhere, not just at the beginning of a compilation unit. Also, they can refer to arbitrary values. For instance, the import shown in Listing 13.8 is possible:

```
def showFruit(fruit: Fruit) = {
  import fruit._
  println(name + "s are " + color)
}
```

Listing 13.8 · Importing the members of a regular (not singleton) object.

Method showFruit imports all members of its parameter fruit, which is of type Fruit. The subsequent println statement can refer to name and color directly. These two references are equivalent to fruit.name and fruit.color. This syntax is particularly useful when you use objects as modules, which will be described in Chapter 29.

---

### Scala's flexible imports

Scala's import clauses are quite a bit more flexible than Java's. There are three principal differences. In Scala, imports:

- may appear anywhere

- may refer to objects (singleton or regular) in addition to packages

- let you rename and hide some of the imported members

---

Another way Scala's imports are flexible is that they can import packages themselves, not just their non-package members. This is only natural if you think of nested packages being contained in their surrounding package. For example, in Listing 13.9, the package java.util.regex is imported. This makes regex usable as a simple name. To access the Pattern singleton object from the java.util.regex package, you can just say, regex.Pattern, as shown in Listing 13.9:

```
import java.util.regex

class AStarB {
  // Accesses java.util.regex.Pattern
  val pat = regex.Pattern.compile("a*b")
}
```

Listing 13.9 · Importing a package name.

Imports in Scala can also rename or hide members. This is done with an *import selector clause* enclosed in braces, which follows the object from which members are imported. Here are some examples:

```
import Fruits.{Apple, Orange}
```

This imports just members Apple and Orange from object Fruits.

```
import Fruits.{Apple => McIntosh, Orange}
```

This imports the two members Apple and Orange from object Fruits. However, the Apple object is renamed to McIntosh, so this object can be accessed with either `Fruits.Apple` or `McIntosh`. A renaming clause is always of the form "`<original-name> => <new-name>`".

```
import java.sql.{Date => SDate}
```

This imports the SQL date class as `SDate`, so that you can simultaneously import the normal Java date class as simply `Date`.

```
import java.{sql => S}
```

This imports the `java.sql` package as `S`, so that you can write things like `S.Date`.

```
import Fruits.{_}
```

This imports all members from object `Fruits`. It means the same thing as `import Fruits._`.

```
import Fruits.{Apple => McIntosh, _}
```

This imports all members from object `Fruits` but renames `Apple` to `McIntosh`.

```
import Fruits.{Pear => _, _}
```

This imports all members of `Fruits` *except* `Pear`. A clause of the form "`<original-name> => _`" excludes `<original-name>` from the names that are imported. In a sense, renaming something to '`_`' means hiding it altogether. This is useful to avoid ambiguities. Say you have two packages, `Fruits` and `Notebooks`, which both define a class `Apple`. If you want to get just the notebook named `Apple`, and not the fruit, you could still use two imports on demand like this:

```
  import Notebooks._
  import Fruits.{Apple => _, _}
```

This would import all `Notebooks` and all `Fruits`, except for `Apple`.

These examples demonstrate the great flexibility Scala offers when it comes to importing members selectively and possibly under different names. In summary, an import selector can consist of the following:

• A simple name x. This includes x in the set of imported names.

- A renaming clause x => y. This makes the member named x visible under the name y.

- A hiding clause x => _. This excludes x from the set of imported names.

- A *catch-all* '_'. This imports all members except those members mentioned in a preceding clause. If a catch-all is given, it must come last in the list of import selectors.

The simpler import clauses shown at the beginning of this section can be seen as special abbreviations of import clauses with a selector clause. For example, "import p._" is equivalent to "import p.{_}" and "import p.n" is equivalent to "import p.{n}".

## 13.4   Implicit imports

Scala adds some imports implicitly to every program. In essence, it is as if the following three import clauses had been added to the top of every source file with extension ".scala":

```
import java.lang._ // everything in the java.lang package
import scala._      // everything in the scala package
import Predef._     // everything in the Predef object
```

The java.lang package contains standard Java classes. It is always implicitly imported in Scala source files.[2] Because java.lang is imported implicitly, you can write Thread instead of java.lang.Thread, for instance.

As you have no doubt realized by now, the scala package contains the standard Scala library, with many common classes and objects. Because scala is imported implicitly, you can write List instead of scala.List, for instance.

The Predef object contains many definitions of types, methods, and implicit conversions that are commonly used on Scala programs. For example, because Predef is imported implicitly, you can write assert instead of Predef.assert.

---

[2]Scala also originally had an implementation on .NET., where namespace System, the .NET analogue of package java.lang, was imported instead.

These three import clauses are treated a bit specially in that later imports overshadow earlier ones. For instance, the `StringBuilder` class is defined both in package `scala` and, from Java version 1.5 on, also in package `java.lang`. Because the `scala` import overshadows the `java.lang` import, the simple name `StringBuilder` will refer to `scala.StringBuilder`, not `java.lang.StringBuilder`.

## 13.5 Access modifiers

Members of packages, classes, or objects can be labeled with the access modifiers `private` and `protected`. These modifiers restrict access to the members to certain regions of code. Scala's treatment of access modifiers roughly follows Java's but there are some important differences which are explained in this section.

### Private members

Private members in Scala are treated similarly to Java. A member labeled `private` is visible only inside the class or object that contains the member definition. In Scala, this rule applies also for inner classes. This treatment is more consistent, but differs from Java. Consider the example shown in Listing 13.10.

```
class Outer {
  class Inner {
    private def f() = { println("f") }
    class InnerMost {
      f() // OK
    }
  }
  (new Inner).f() // error: f is not accessible
}
```

Listing 13.10 · How private access differs in Scala and Java.

In Scala, the access (new Inner).f() is illegal because f is declared private in Inner and the access is not from within class Inner. By contrast, the first access to f in class InnerMost is OK, because that access is contained in the body of class Inner. Java would permit both accesses because it lets an outer class access private members of its inner classes.

## Protected members

Access to protected members in Scala is also a bit more restrictive than in Java. In Scala, a protected member is only accessible from subclasses of the class in which the member is defined. In Java such accesses are also possible from other classes in the same package. In Scala, there is another way to achieve this effect[3] so protected is free to be left as is. The example shown in Listing 13.11 illustrates protected accesses.

```
package p {
  class Super {
    protected def f() = { println("f") }
  }
  class Sub extends Super {
    f()
  }
  class Other {
    (new Super).f()  // error: f is not accessible
  }
}
```

Listing 13.11 · How protected access differs in Scala and Java.

In Listing 13.11, the access to f in class Sub is OK because f is declared protected in Super and Sub is a subclass of Super. By contrast the access to f in Other is not permitted, because Other does not inherit from Super. In Java, the latter access would be still permitted because Other is in the same package as Sub.

## Public members

Scala has no explicit modifier for public members: Any member not labeled private or protected is public. Public members can be accessed from anywhere.

---

[3]Using *qualifiers*, described in "Scope of protection" on page 255.

```
package bobsrockets
package navigation {
  private[bobsrockets] class Navigator {
    protected[navigation] def useStarChart() = {}
    class LegOfJourney {
      private[Navigator] val distance = 100
    }
    private[this] var speed = 200
  }
}
package launch {
  import navigation._
  object Vehicle {
    private[launch] val guide = new Navigator
  }
}
```

Listing 13.12 · Flexible scope of protection with access qualifiers.

### Scope of protection

Access modifiers in Scala can be augmented with qualifiers. A modifier of the form private[X] or protected[X] means that access is private or protected "up to" X, where X designates some enclosing package, class or singleton object.

Qualified access modifiers give you very fine-grained control over visibility. In particular they enable you to express Java's accessibility notions, such as package private, package protected, or private up to outermost class, which are not directly expressible with simple modifiers in Scala. But they also let you express accessibility rules that cannot be expressed in Java.

Listing 13.12 presents an example with many access qualifiers being used. In this listing, class Navigator is labeled private[bobsrockets]. This means that this class is visible in all classes and objects that are contained in package bobsrockets. In particular, the access to Navigator in object Vehicle is permitted because Vehicle is contained in package launch, which is contained in bobsrockets. On the other hand, all code outside the package bobsrockets cannot access class Navigator.

This technique is quite useful in large projects that span several packages. It allows you to define things that are visible in several sub-packages of your project but that remain hidden from clients external to your project. The same technique is not possible in Java. There, once a definition escapes its immediate package boundary, it is visible to the world at large.

Of course, the qualifier of a `private` may also be the directly enclosing package. An example is the access modifier of `guide` in object `Vehicle` in Listing 13.12. Such an access modifier is equivalent to Java's package-private access.

Table 13.1 · Effects of private qualifiers on `LegOfJourney.distance`

| *no access modifier* | public access |
| --- | --- |
| `private[bobsrockets]` | access within outer package |
| `private[navigation]` | same as package visibility in Java |
| `private[Navigator]` | same as `private` in Java |
| `private[LegOfJourney]` | same as `private` in Scala |
| `private[this]` | access only from same object |

All qualifiers can also be applied to `protected`, with the same meaning as `private`. That is, a modifier `protected[X]` in a class C allows access to the labeled definition in all subclasses of C and also within the enclosing package, class, or object X. For instance, the `useStarChart` method in Listing 13.12 is accessible in all subclasses of `Navigator` and also in all code contained in the enclosing package `navigation`. It thus corresponds exactly to the meaning of `protected` in Java.

The qualifiers of `private` can also refer to an enclosing class or object. For instance the `distance` variable in class `LegOfJourney` in Listing 13.12 is labeled `private[Navigator]`, so it is visible from everywhere in class `Navigator`. This gives the same access capabilities as for private members of inner classes in Java. A `private[C]` where C is the outermost enclosing class is the same as just `private` in Java.

Finally, Scala also has an access modifier that is even more restrictive than `private`. A definition labeled `private[this]` is accessible only from within the same object that contains the definition. Such a definition is called *object-private*. For instance, the definition of `speed` in class `Navigator` in Listing 13.12 is object-private. This means that any access must not only be within class `Navigator`, it must also be made from the very same instance of

Navigator. Thus the accesses "speed" and "this.speed" would be legal from within Navigator.

The following access, though, would not be allowed, even if it appeared inside class Navigator:

```
val other = new Navigator
other.speed // this line would not compile
```

Marking a member private[this] is a guarantee that it will not be seen from other objects of the same class. This can be useful for documentation. It also sometimes lets you write more general variance annotations (see Section 19.7 for details).

To summarize, Table 13.1 on page 256 lists the effects of private qualifiers. Each line shows a qualified private modifier and what it would mean if such a modifier were attached to the distance variable declared in class LegOfJourney in Listing 13.12.

**Visibility and companion objects**

In Java, static members and instance members belong to the same class, so access modifiers apply uniformly to them. You have already seen that in Scala there are no static members; instead you can have a companion object that contains members that exist only once. For instance, in Listing 13.13 object Rocket is a companion of class Rocket.

Scala's access rules privilege companion objects and classes when it comes to private or protected accesses. A class shares all its access rights with its companion object and *vice versa*. In particular, an object can access all private members of its companion class, just as a class can access all private members of its companion object.

For instance, the Rocket class in Listing 13.13 can access method fuel, which is declared private in object Rocket. Analogously, the Rocket object can access the private method canGoHomeAgain in class Rocket.

One exception where the similarity between Scala and Java breaks down concerns protected static members. A protected static member of a Java class C can be accessed in all subclasses of C. By contrast, a protected member in a companion object makes no sense, as singleton objects don't have any subclasses.

```
class Rocket {
  import Rocket.fuel
  private def canGoHomeAgain = fuel > 20
}

object Rocket {
  private def fuel = 10
  def chooseStrategy(rocket: Rocket) = {
    if (rocket.canGoHomeAgain)
      goHome()
    else
      pickAStar()
  }
  def goHome() = {}
  def pickAStar() = {}
}
```

Listing 13.13: Accessing private members of companion classes and objects.

## 13.6   Package objects

So far, the only code you have seen added to packages are classes, traits, and standalone objects. These are by far the most common definitions that are placed at the top level of a package. But Scala doesn't limit you to just those—Any kind of definition that you can put inside a class can also be at the top level of a package. If you have some helper method you'd like to be in scope for an entire package, go ahead and put it right at the top level of the package.

To do so, put the definitions in a *package object*. Each package is allowed to have one package object. Any definitions placed in a package object are considered members of the package itself.

An example is shown in Listing 13.14. File package.scala holds a package object for package bobsdelights. Syntactically, a package object looks much like one of the curly-braces packagings shown earlier in the chapter. The only difference is that it includes the object keyword. It's a package *object*, not a *package*. The contents of the curly braces can include any definitions you like. In this case, the package object includes the showFruit utility method from Listing 13.8.

Given that definition, any other code in any package can import the method just like it would import a class. For example, Listing 13.14 also shows the standalone object PrintMenu, which is located in a different package. PrintMenu can import the utility method showFruit in the same way it would import the class Fruit.

```scala
// In file bobsdelights/package.scala
package object bobsdelights {
  def showFruit(fruit: Fruit) = {
    import fruit._
    println(name + "s are " + color)
  }
}

// In file PrintMenu.scala
package printmenu
import bobsdelights.Fruits
import bobsdelights.showFruit

object PrintMenu {
  def main(args: Array[String]) = {
    for (fruit <- Fruits.menu) {
      showFruit(fruit)
    }
  }
}
```

Listing 13.14 · A package object.

Looking ahead, there are other uses of package objects for kinds of definitions you haven't seen yet. Package objects are frequently used to hold package-wide type aliases (Chapter 20) and implicit conversions (Chapter 21). The top-level scala package has a package object, and its definitions are available to all Scala code.

Package objects are compiled to class files named package.class that are the located in the directory of the package that they augment. It's useful to keep the same convention for source files. So you would typically put the source file of the package object bobsdelights of Listing 13.14 into a file named package.scala that resides in the bobsdelights directory.

259

## 13.7 Conclusion

In this chapter, you saw the basic constructs for dividing a program into packages. This gives you a simple and useful kind of modularity, so that you can work with very large bodies of code without different parts of the code trampling on each other. Scala's system is the same in spirit as Java's packages, but there are some differences where Scala chooses to be more consistent or more general.

Looking ahead, Chapter 29 describes a more flexible module system than division into packages. In addition to letting you separate code into several namespaces, that approach allows modules to be parameterized and inherit from each other. In the next chapter, we'll turn our attention to assertions and unit testing.

# Chapter 14

# Assertions and Tests

Assertions and tests are two important ways you can check that the software you write behaves as you expect. In this chapter, we'll show you several options you have in Scala to write and run them.

## 14.1 Assertions

Assertions in Scala are written as calls of a predefined method assert.[1] The expression assert(condition) throws an AssertionError if condition does not hold. There's also a two-argument version of assert: The expression assert(condition, explanation) tests condition and, if it does not hold, throws an AssertionError that contains the given explanation. The type of explanation is Any, so you can pass any object as the explanation. The assert method will call toString on it to get a string explanation to place inside the AssertionError. For example, in the method named "above" of class Element, shown in Listing 10.13 on page 208, you might place an assert after the calls to widen to make sure that the widened elements have equal widths. This is shown in Listing 14.1.

Another way you might choose to do this is to check the widths at the end of the widen method, right before you return the value. You can accomplish this by storing the result in a val, performing an assertion on the result, then mentioning the val last so the result is returned if the assertion succeeds. However, you can do this more concisely with a convenience method in Predef named ensuring, as shown in Listing 14.2.

---

[1]The assert method is defined in the Predef singleton object, whose members are automatically imported into every Scala source file.

261

```
def above(that: Element): Element = {
  val this1 = this widen that.width
  val that1 = that widen this.width
  assert(this1.width == that1.width)
  elem(this1.contents ++ that1.contents)
}
```

Listing 14.1 · Using an assertion.

The ensuring method can be used with any result type because of an implicit conversion. Although it looks in this code as if we're invoking ensuring on widen's result, which is type Element, we're actually invoking ensuring on a type to which Element is implicitly converted. The ensuring method takes one argument, a predicate function that takes a result type and returns Boolean, and passes the result to the predicate. If the predicate returns true, ensuring will return the result; otherwise, ensuring will throw an AssertionError.

In this example, the predicate is "w <= _.width". The underscore is a placeholder for the one argument passed to the predicate, the Element result of the widen method. If the width passed as w to widen is less than or equal to the width of the result Element, the predicate will result in true, and ensuring will result in the Element on which it was invoked. Because this is the last expression of the widen method, widen itself will then result in the Element.

Assertions can be enabled and disabled using the JVM's –ea and –da

```
private def widen(w: Int): Element =
  if (w <= width)
    this
  else {
    val left = elem(' ', (w - width) / 2, height)
    var right = elem(' ', w - width - left.width, height)
    left beside this beside right
  } ensuring (w <= _.width)
```

Listing 14.2 · Using ensuring to assert a function's result.

262

command-line flags. When enabled, each assertion serves as a little test that uses the actual data encountered as the software runs. In the remainder of this chapter, we'll focus on the writing of external tests, which provide their own test data and run independently from the application.

## 14.2 Testing in Scala

You have many options for testing in Scala, from established Java tools, such as JUnit and TestNG, to tools written in Scala, such as ScalaTest, specs2, and ScalaCheck. For the remainder of this chapter, we'll give you a quick tour of these tools. We'll start with ScalaTest.

ScalaTest is the most flexible Scala test framework: it can be easily customized to solve different problems. ScalaTest's flexibility means teams can use whatever testing style fits their needs best. For example, for teams familiar with JUnit, the FunSuite style will feel comfortable and familiar. Listing 14.3 shows an example.

```scala
import org.scalatest.FunSuite
import Element.elem

class ElementSuite extends FunSuite {

  test("elem result should have passed width") {
    val ele = elem('x', 2, 3)
    assert(ele.width == 2)
  }
}
```

Listing 14.3 · Writing tests with FunSuite.

The central concept in ScalaTest is the *suite*, a collection of tests. A *test* can be anything with a name that can start and either succeed, fail, be pending, or canceled. Trait Suite is the central unit of composition in Scala-Test. Suite declares "lifecycle" methods defining a default way to run tests, which can be overridden to customize how tests are written and run.

ScalaTest offers *style traits* that extend Suite and override lifecycle methods to support different testing styles. It also provides *mixin traits* that override lifecycle methods to address particular testing needs. You define

test classes by composing `Suite` style and mixin traits, and define test suites
by composing `Suite` instances.

FunSuite, which is extended by the test class shown in Listing 14.3, is
an example of a testing style. The "Fun" in FunSuite stands for function;
"test" is a method defined in `FunSuite`, which is invoked by the primary
constructor of `ElementSuite`. You specify the name of the test as a string
between the parentheses and the test code itself between curly braces. The
test code is a function passed as a by-name parameter to `test`, which regis-
ters it for later execution.

ScalaTest is integrated into common build tools (such as sbt and Maven)
and IDEs (such as IntelliJ IDEA and Eclipse). You can also run a `Suite` directly via ScalaTest's Runner application or from the Scala interpreter simply by invoking `execute` on it. Here's an example:

```
scala> (new ElementSuite).execute()
ElementSuite:
- elem result should have passed width
```

All ScalaTest styles, including `FunSuite`, are designed to encourage the
writing of focused tests with descriptive names. In addition, all styles generate specification-like output that can facilitate communication among stakeholders. The style you choose dictates only how the declarations of your tests
will look. Everything else in ScalaTest works consistently the same way no
matter what style you choose.[2]

## 14.3   Informative failure reports

The test in Listing 14.3 attempts to create an element of width 2 and assert
that the width of the resulting element is indeed 2. Were this assertion to
fail, the failure report would include the filename and line number of the
offending assertion, and an informative error message:

```
scala> val width = 3
width: Int = 3

scala> assert(width == 2)
org.scalatest.exceptions.TestFailedException:
    3 did not equal 2
```

---

[2]More detail on ScalaTest is available from `http://www.scalatest.org/`.

To provide descriptive error messages when assertions fail, ScalaTest analyzes the expressions passed to each `assert` invocation at compile time. If you prefer to see even more detailed information about assertion failures, you can use ScalaTest's `DiagrammedAssertions`, whose error messages display a diagram of the expression passed to `assert`:

```
scala> assert(List(1, 2, 3).contains(4))
org.scalatest.exceptions.TestFailedException:

assert(List(1, 2, 3).contains(4))
       |   | | | |         |
       |   1 2 3 false     4
       List(1, 2, 3)
```

ScalaTest's `assert` methods do not differentiate between the actual and expected result in error messages. They just indicate that the left operand did not equal the right operand, or show the values in a diagram. If you wish to emphasize the distinction between actual and expected, you can alternatively use ScalaTest's `assertResult` method, like this:

```
assertResult(2) {
  ele.width
}
```

With this expression you indicate that you expect the code between the curly braces to result in 2. Were the code between the braces to result in 3, you'd see the message, "`Expected 2, but got 3`" in the test failure report.

If you want to check that a method throws an expected exception, you can use ScalaTest's `assertThrows` method, like this:

```
assertThrows[IllegalArgumentException] {
  elem('x', -2, 3)
}
```

If the code between the curly braces throws a different exception than expected, or throws no exception, `assertThrows` will complete abruptly with a `TestFailedException`. You'll get a helpful error message in the failure report, such as:

```
Expected IllegalArgumentException to be thrown,
  but NegativeArraySizeException was thrown.
```

265

On the other hand, if the code completes abruptly with an instance of the passed exception class, assertThrows will return normally. If you wish to inspect the expected exception further, you can use intercept instead of assertThrows. The intercept method works the same as assertThrows, except if the expected exception is thrown, intercept returns it:

```
val caught =
  intercept[ArithmeticException] {
    1 / 0
  }
assert(caught.getMessage == "/ by zero")
```

In short, ScalaTest's assertions work hard to provide useful failure messages that will help you diagnose and fix problems in your code.

## 14.4   Tests as specifications

In the *behavior-driven development* (BDD) testing style, the emphasis is on writing human-readable specifications of the expected behavior of code and accompanying tests that verify the code has the specified behavior. ScalaTest includes several traits that facilitate this style of testing. An example using one such trait, FlatSpec, is shown in Listing 14.4.

In a FlatSpec, you write tests as *specifier clauses*. You start by writing a name for the *subject* under test as a string ("A UniformElement" in Listing 14.4), then should (or must or can), then a string that specifies a bit of behavior required of the subject, then in. In the curly braces following in, you write code that tests the specified behavior. In subsequent clauses you can write it to refer to the most recently given subject. When a FlatSpec is executed, it will run each specifier clause as a ScalaTest test. FlatSpec (and ScalaTest's other specification traits) generate output that reads like a specification when run. For example, here's what the output will look like if you run ElementSpec from Listing 14.4 in the interpreter:

```
scala> (new ElementSpec).execute()
A UniformElement
- should have a width equal to the passed value
- should have a height equal to the passed value
- should throw an IAE if passed a negative width
```

266

```
import org.scalatest.FlatSpec
import org.scalatest.Matchers
import Element.elem

class ElementSpec extends FlatSpec with Matchers {

  "A UniformElement" should
      "have a width equal to the passed value" in {
    val ele = elem('x', 2, 3)
    ele.width should be (2)
  }

  it should "have a height equal to the passed value" in {
    val ele = elem('x', 2, 3)
    ele.height should be (3)
  }

  it should "throw an IAE if passed a negative width" in {
    an [IllegalArgumentException] should be thrownBy {
      elem('x', -2, 3)
    }
  }
}
```

Listing 14.4 · Specifying and testing behavior with a ScalaTest FlatSpec.

Listing 14.4 also illustrates ScalaTest's *matchers* domain-specific language (DSL). By mixing in trait Matchers, you can write assertions that read more like natural language. ScalaTest provides many matchers in its DSL, and also enables you to define new matchers with custom failure messages. The matchers shown in Listing 14.4 include the "should be" and "an [ ... ] should be thrownBy { ... } " syntax. You can alternatively mix in MustMatchers if you prefer must to should. For example, mixing in MustMatchers would allow you to write expressions such as:

```
result must be >= 0
map must contain key 'c'
```

If the last assertion failed, you'd see an error message similar to:

```
Map('a' -> 1, 'b' -> 2) did not contain key 'c'
```

The specs2 testing framework, an open source tool written in Scala by Eric Torreborre, also supports the BDD style of testing but with a different syntax. For example, you could use specs2 to write the test shown in Listing 14.5:

```
import org.specs2._
import Element.elem

object ElementSpecification extends Specification {
  "A UniformElement" should {
    "have a width equal to the passed value" in {
      val ele = elem('x', 2, 3)
      ele.width must be_==(2)
    }
    "have a height equal to the passed value" in {
      val ele = elem('x', 2, 3)
      ele.height must be_==(3)
    }
    "throw an IAE if passed a negative width" in {
      elem('x', -2, 3) must
        throwA[IllegalArgumentException]
    }
  }
}
```

Listing 14.5 · Specifying and testing behavior with the specs2 framework.

Like ScalaTest, specs2 provides a matchers DSL. You can see some examples of specs2 matchers in action in Listing 14.5 in the lines that contain "must be_==" and "must throwA".[3] You can use specs2 standalone, but it is also integrated with ScalaTest and JUnit, so you can run specs2 tests with those tools as well.

One of the big ideas of BDD is that tests can be used to facilitate communication between the people who decide what a software system should do, the people who implement the software, and the people who determine whether the software is finished and working. Although any of ScalaTest's

---

[3] You can download specs2 from http://specs2.org/.

or specs2's styles can be used in this manner, ScalaTest's `FeatureSpec` in particular is designed for it. Listing 14.6 shows an example:

```
import org.scalatest._

class TVSetSpec extends FeatureSpec with GivenWhenThen {
  feature("TV power button") {
    scenario("User presses power button when TV is off") {
      Given("a TV set that is switched off")
      When("the power button is pressed")
      Then("the TV should switch on")
      pending
    }
  }
}
```

Listing 14.6 · Using tests to facilitate communication among stakeholders.

`FeatureSpec` is designed to guide conversations about software requirements: You must identify specific *features*, then specify those features in terms of *scenarios*. The `Given`, `When`, and `Then` methods (provided by trait `GivenWhenThen`) can help focus the conversation on the specifics of individual scenarios. The `pending` call at the end indicates that neither the test nor the actual behavior has been implemented—just the specification. Once all the tests and specified behavior have been implemented, the tests will pass and the requirements can be deemed to have been met.

## 14.5 Property-based testing

Another useful testing tool for Scala is ScalaCheck, an open source framework written by Rickard Nilsson. ScalaCheck enables you to specify properties that the code under test must obey. For each property, ScalaCheck will generate data and execute assertions that check whether the property holds. Listing 14.7 shows an example of using ScalaCheck from a ScalaTest `WordSpec` that mixes in trait `PropertyChecks`.

`WordSpec` is a ScalaTest style class. The `PropertyChecks` trait provides several `forAll` methods that allow you to mix property-based tests with traditional assertion-based or matcher-based tests. In this example, we

```scala
import org.scalatest.WordSpec
import org.scalatest.prop.PropertyChecks
import org.scalatest.MustMatchers._
import Element.elem

class ElementSpec extends WordSpec with PropertyChecks {
  "elem result" must {
    "have passed width" in {
      forAll { (w: Int) =>
        whenever (w > 0) {
          elem('x', w, 3).width must equal (w)
        }
      }
    }
  }
}
```

Listing 14.7 · Writing property-based tests with ScalaCheck.

check a property that the elem factory should obey. ScalaCheck properties are expressed as function values that take as parameters the data needed by the property's assertions. This data will be generated by ScalaCheck. In the property shown in Listing 14.7, the data is an integer named w that represents a width. Inside the body of the function, you see this code:

```scala
whenever (w > 0) {
  elem('x', w, 3).width must equal (w)
}
```

The whenever clause indicates that whenever the left hand expression is true, the expression on the right must hold true. Thus in this case, the expression in the block must hold true whenever w is greater than 0. The right-hand expression in this case will yield true if the width passed to the elem factory is the same as the width of the Element returned by the factory.

With this small amount of code, ScalaCheck will generate possibly hundreds of values for w and test each one, looking for a value for which the property doesn't hold. If the property holds true for every value ScalaCheck tries, the test will pass. Otherwise, the test will complete abruptly with a

`TestFailedException` that contains information including the value that caused the failure.

## 14.6 Organizing and running tests

Each framework mentioned in this chapter provides some mechanism for organizing and running tests. In this section, we'll give a quick overview of ScalaTest's approach. To get the full story on any of these frameworks, however, you'll need to consult their documentation.

In ScalaTest, you organize large test suites by nesting `Suites` inside `Suites`. When a `Suite` is executed, it will execute its nested `Suites` as well as its tests. The nested `Suites` will in turn execute their nested `Suites`, and so on. A large test suite, therefore, is represented as a tree of `Suite` objects. When you execute the root `Suite` in the tree, all `Suites` in the tree will be executed.

You can nest suites manually or automatically. To nest manually, you either override the `nestedSuites` method on your `Suites` or pass the `Suites` you want to nest to the constructor of class `Suites`, which ScalaTest provides for this purpose. To nest automatically, you provide package names to ScalaTest's `Runner`, which will discover `Suites` automatically, nest them under a root `Suite`, and execute the root `Suite`.

You can invoke ScalaTest's `Runner` application from the command line or via a build tool, such as sbt, maven, or ant. The simplest way to invoke `Runner` on the command line is via the `org.scalatest.run` application. This application expects a fully qualified test class name. For example, to run the test class shown in Listing 14.6, you must compile it with:

```
$ scalac -cp scalatest.jar TVSetSpec.scala
```

Then you can run it with:

```
$ scala -cp scalatest.jar org.scalatest.run TVSetSpec
```

With –cp you place ScalaTest's JAR file on the class path. (When downloaded, the JAR file name will include embedded Scala and ScalaTest version numbers.) The next token, `org.scalatest.run`, is the fully qualified application name. Scala will run this application and pass the remaining tokens as command line arguments. The `TVSetSpec` argument specifies the suite to execute. The result is shown in Figure 14.1.

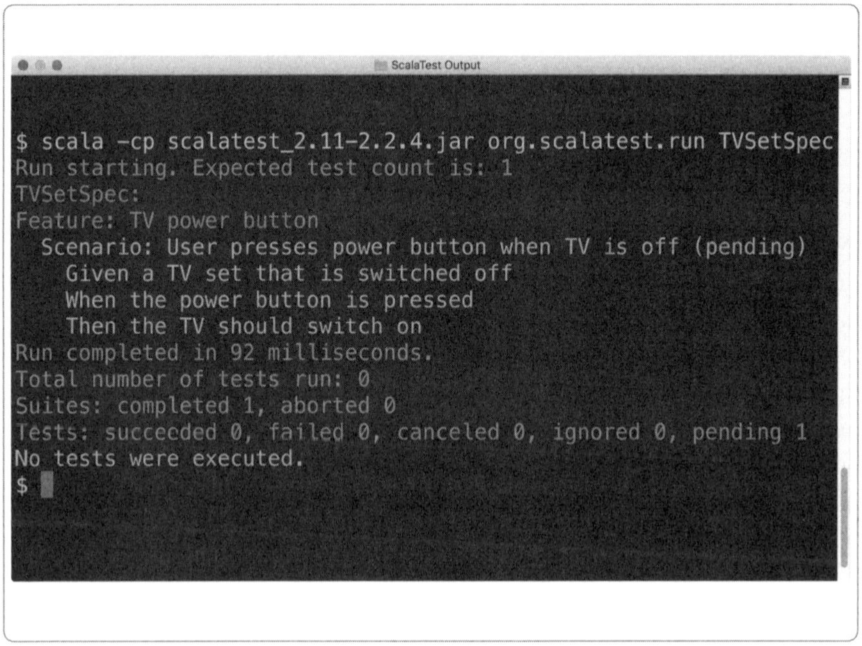

```
$ scala -cp scalatest_2.11-2.2.4.jar org.scalatest.run TVSetSpec
Run starting. Expected test count is: 1
TVSetSpec:
Feature: TV power button
  Scenario: User presses power button when TV is off (pending)
    Given a TV set that is switched off
    When the power button is pressed
    Then the TV should switch on
Run completed in 92 milliseconds.
Total number of tests run: 0
Suites: completed 1, aborted 0
Tests: succeeded 0, failed 0, canceled 0, ignored 0, pending 1
No tests were executed.
$
```

Figure 14.1 · The output of `org.scalatest.run`.

## 14.7 Conclusion

In this chapter you saw examples of mixing assertions directly in production code, as well as writing them externally in tests. You saw that as a Scala programmer, you can take advantage of popular testing tools from the Java community, such as JUnit and TestNG, as well as newer tools designed explicitly for Scala, such as ScalaTest, ScalaCheck, and specs2. Both in-code assertions and external tests can help you achieve your software quality goals. We felt that these techniques are important enough to justify the short detour from the Scala tutorial that this chapter represented. In the next chapter, however, we'll return to the language tutorial and cover a very useful aspect of Scala: pattern matching.

# Chapter 15

# Case Classes and Pattern Matching

This chapter introduces *case classes* and *pattern matching*, twin constructs that support you when writing regular, non-encapsulated data structures. These two constructs are particularly helpful for tree-like recursive data.

If you have programmed in a functional language before, then you will probably recognize pattern matching. But case classes will be new to you. Case classes are Scala's way to allow pattern matching on objects without requiring a large amount of boilerplate. Generally, all you need to do is add a single case keyword to each class that you want to be pattern matchable.

This chapter starts with a simple example of case classes and pattern matching. It then goes through all of the kinds of patterns that are supported, talks about the role of *sealed* classes, discusses the Option type, and shows some non-obvious places in the language where pattern matching is used. Finally, a larger, more realistic example of pattern matching is shown.

## 15.1 A simple example

Before delving into all the rules and nuances of pattern matching, it is worth looking at a simple example to get the general idea. Let's say you need to write a library that manipulates arithmetic expressions, perhaps as part of a domain-specific language you are designing.

A first step to tackling this problem is the definition of the input data. To keep things simple, we'll concentrate on arithmetic expressions consisting of variables, numbers, and unary and binary operations. This is expressed by the hierarchy of Scala classes shown in Listing 15.1.

```
abstract class Expr
case class Var(name: String) extends Expr
case class Number(num: Double) extends Expr
case class UnOp(operator: String, arg: Expr) extends Expr
case class BinOp(operator: String,
    left: Expr, right: Expr) extends Expr
```

Listing 15.1 · Defining case classes.

The hierarchy includes an abstract base class Expr with four subclasses, one for each kind of expression being considered.[1] The bodies of all five classes are empty. As mentioned previously, in Scala you can leave out the braces around an empty class body if you wish, so class C is the same as class C {}.

**Case classes**

The other noteworthy thing about the declarations of Listing 15.1 is that each subclass has a case modifier. Classes with such a modifier are called *case classes*. Using the modifier makes the Scala compiler add some syntactic conveniences to your class.

First, it adds a factory method with the name of the class. This means that, for instance, you can write Var("x") to construct a Var object, instead of the slightly longer new Var("x"):

```
scala> val v = Var("x")
v: Var = Var(x)
```

The factory methods are particularly nice when you nest them. Because there are no noisy new keywords sprinkled throughout the code, you can take in the expression's structure at a glance:

```
scala> val op = BinOp("+", Number(1), v)
op: BinOp = BinOp(+,Number(1.0),Var(x))
```

The second syntactic convenience is that all arguments in the parameter list of a case class implicitly get a val prefix, so they are maintained as fields:

---

[1]Instead of an abstract class, we could have also chosen to model the root of that class hierarchy as a trait. Modeling it as an abstract class may be slightly more efficient.

274

```
scala> v.name
res0: String = x

scala> op.left
res1: Expr = Number(1.0)
```

Third, the compiler adds "natural" implementations of methods `toString`, `hashCode`, and `equals` to your class. They will print, hash, and compare a whole tree consisting of the class and (recursively) all its arguments. Since `==` in Scala always delegates to `equals`, this means that elements of case classes are always compared structurally:

```
scala> println(op)
BinOp(+,Number(1.0),Var(x))

scala> op.right == Var("x")
res3: Boolean = true
```

Finally, the compiler adds a `copy` method to your class for making modified copies. This method is useful for making a new instance of the class that is the same as another one except that one or two attributes are different. The method works by using named and default parameters (see Section 8.8). You specify the changes you'd like to make by using named parameters. For any parameter you don't specify, the value from the old object is used. As an example, here is how you can make an operation just like op except that the operator has changed:

```
scala> op.copy(operator = "-")
res4: BinOp = BinOp(-,Number(1.0),Var(x))
```

All these conventions add a lot of convenience—at a small price. You have to write the case modifier, and your classes and objects become a bit larger. They are larger because additional methods are generated and an implicit field is added for each constructor parameter. However, the biggest advantage of case classes is that they support pattern matching.

**Pattern matching**

Say you want to simplify arithmetic expressions of the kinds just presented. There is a multitude of possible simplification rules. The following three rules just serve as an illustration:

```
UnOp("-", UnOp("-", e))  => e    // Double negation
BinOp("+", e, Number(0)) => e    // Adding zero
BinOp("*", e, Number(1)) => e    // Multiplying by one
```

Using pattern matching, these rules can be taken almost as they are to form the core of a simplification function in Scala, as shown in Listing 15.2. The function, `simplifyTop`, can be used like this:

```
scala> simplifyTop(UnOp("-", UnOp("-", Var("x"))))
res4: Expr = Var(x)
```

```
def simplifyTop(expr: Expr): Expr = expr match {
  case UnOp("-", UnOp("-", e))  => e    // Double negation
  case BinOp("+", e, Number(0)) => e    // Adding zero
  case BinOp("*", e, Number(1)) => e    // Multiplying by one
  case _ => expr
}
```

Listing 15.2 · The `simplifyTop` function, which does a pattern match.

The right-hand side of `simplifyTop` consists of a `match` expression. `match` corresponds to `switch` in Java, but it's written after the selector expression. In other words, it's:

*selector* **match** { *alternatives* }

instead of:

`switch` (*selector*) { *alternatives* }

A pattern match includes a sequence of *alternatives*, each starting with the keyword `case`. Each alternative includes a *pattern* and one or more expressions, which will be evaluated if the pattern matches. An arrow symbol => separates the pattern from the expressions.

A `match` expression is evaluated by trying each of the patterns in the order they are written. The first pattern that matches is selected, and the part following the arrow is selected and executed.

A *constant pattern* like "+" or 1 matches values that are equal to the constant with respect to ==. A *variable pattern* like e matches every value.

The variable then refers to that value in the right hand side of the case clause. In this example, note that the first three alternatives evaluate to e, a variable that is bound within the associated pattern. The *wildcard pattern* (_) also matches every value, but it does not introduce a variable name to refer to that value. In Listing 15.2, notice how the match ends with a default case that does nothing to the expression. Instead, it just results in expr, the expression matched upon.

A *constructor pattern* looks like UnOp("-", e). This pattern matches all values of type UnOp whose first argument matches "-" and whose second argument matches e. Note that the arguments to the constructor are themselves patterns. This allows you to write deep patterns using a concise notation. Here's an example:

```
UnOp("-", UnOp("-", e))
```

Imagine trying to implement this same functionality using the visitor design pattern![2] Almost as awkward, imagine implementing it as a long sequence of if statements, type tests, and type casts.

**match compared to switch**

Match expressions can be seen as a generalization of Java-style switches. A Java-style switch can be naturally expressed as a match expression, where each pattern is a constant and the last pattern may be a wildcard (which represents the default case of the switch).

However, there are three differences to keep in mind: First, match is an *expression* in Scala (*i.e.*, it always results in a value). Second, Scala's alternative expressions never "fall through" into the next case. Third, if none of the patterns match, an exception named MatchError is thrown. This means you always have to make sure that all cases are covered, even if it means adding a default case where there's nothing to do.

Listing 15.3 shows an example. The second case is necessary because without it, the match expression would throw a MatchError for every expr argument that is not a BinOp. In this example, no code is specified for that second case, so if that case runs it does nothing. The result of either case is the unit value '()', which is also the result of the entire match expression.

---

[2]Gamma, *et al.*, *Design Patterns* [Gam95]

```
expr match {
  case BinOp(op, left, right) =>
    println(expr + " is a binary operation")
  case _ =>
}
```

Listing 15.3 · A pattern match with an empty "default" case.

## 15.2   Kinds of patterns

The previous example showed several kinds of patterns in quick succession. Now take a minute to look at each pattern in detail.

The syntax of patterns is easy, so do not worry about that too much. All patterns look exactly like the corresponding expression. For instance, given the hierarchy of Listing 15.1, the pattern Var(x) matches any variable expression, binding x to the name of the variable. Used as an expression, Var(x)—exactly the same syntax—recreates an equivalent object, assuming x is already bound to the variable's name. Since the syntax of patterns is so transparent, the main thing to pay attention to is just what kinds of patterns are possible.

### Wildcard patterns

The wildcard pattern (_) matches any object whatsoever. You have already seen it used as a default, catch-all alternative, like this:

```
expr match {
  case BinOp(op, left, right) =>
    println(expr + " is a binary operation")
  case _ => // handle the default case
}
```

Wildcards can also be used to ignore parts of an object that you do not care about. For example, the previous example does not actually care what the elements of a binary operation are; it just checks whether or not it is a binary operation. Thus, the code can just as well use the wildcard pattern for the elements of the BinOp, as shown in Listing 15.4.

```
expr match {
  case BinOp(_, _, _) => println(expr + " is a binary operation")
  case _ => println("It's something else")
}
```

Listing 15.4 · A pattern match with wildcard patterns.

## Constant patterns

A constant pattern matches only itself. Any literal may be used as a constant. For example, 5, true, and "hello" are all constant patterns. Also, any val or singleton object can be used as a constant. For example, Nil, a singleton object, is a pattern that matches only the empty list. Listing 15.5 shows some examples of constant patterns:

```
def describe(x: Any) = x match {
  case 5 => "five"
  case true => "truth"
  case "hello" => "hi!"
  case Nil => "the empty list"
  case _ => "something else"
}
```

Listing 15.5 · A pattern match with constant patterns.

Here is how the pattern match shown in Listing 15.5 looks in action:

```
scala> describe(5)
res6: String = five

scala> describe(true)
res7: String = truth

scala> describe("hello")
res8: String = hi!

scala> describe(Nil)
res9: String = the empty list

scala> describe(List(1,2,3))
res10: String = something else
```

## Variable patterns

A variable pattern matches any object, just like a wildcard. But unlike a wildcard, Scala binds the variable to whatever the object is. You can then use this variable to act on the object further. For example, Listing 15.6 shows a pattern match that has a special case for zero, and a default case for all other values. The default case uses a variable pattern so that it has a name for the value, no matter what it is.

```
expr match {
  case 0 => "zero"
  case somethingElse => "not zero: " + somethingElse
}
```

Listing 15.6 · A pattern match with a variable pattern.

### Variable or constant?

Constant patterns can have symbolic names. You saw this already when we used Nil as a pattern. Here is a related example, where a pattern match involves the constants E (2.71828...) and Pi (3.14159...):

```
scala> import math.{E, Pi}
import math.{E, Pi}

scala> E match {
         case Pi => "strange math? Pi = " + Pi
         case _ => "OK"
       }
res11: String = OK
```

As expected, E does not match Pi, so the "strange math" case is not used.

How does the Scala compiler know that Pi is a constant imported from scala.math, and not a variable that stands for the selector value itself? Scala uses a simple lexical rule for disambiguation: a simple name starting with a lowercase letter is taken to be a pattern variable; all other references are taken to be constants. To see the difference, create a lowercase alias for pi and try with that:

```
scala> val pi = math.Pi
pi: Double = 3.141592653589793

scala> E match {
          case pi => "strange math? Pi = " + pi
       }
res12: String = strange math? Pi = 2.718281828459045
```

Here the compiler will not even let you add a default case at all. Since pi is a variable pattern, it will match all inputs, and so no cases following it can be reached:

```
scala> E match {
          case pi => "strange math? Pi = " + pi
          case _ => "OK"
       }
<console>:12: warning: unreachable code
              case _ => "OK"
                   ^
```

You can still use a lowercase name for a pattern constant, if you need to, by using one of two tricks. First, if the constant is a field of some object, you can prefix it with a qualifier. For instance, pi is a variable pattern, but this.pi or obj.pi are constants even though they start with lowercase letters. If that does not work (because pi is a local variable, say), you can alternatively enclose the variable name in back ticks. For instance, `pi` would again be interpreted as a constant, not as a variable:

```
scala> E match {
          case `pi` => "strange math? Pi = " + pi
          case _ => "OK"
       }
res14: String = OK
```

As you can see, the back-tick syntax for identifiers is used for two different purposes in Scala to help you code your way out of unusual circumstances. Here you see that it can be used to treat a lowercase identifier as a constant in a pattern match. Earlier on, in Section 6.10, you saw that it can also be used to treat a keyword as an ordinary identifier, *e.g.*, writing Thread.`yield`() treats yield as an identifier rather than a keyword.

## Constructor patterns

Constructors are where pattern matching becomes really powerful. A constructor pattern looks like "BinOp("+", e, Number(0))". It consists of a name (BinOp) and then a number of patterns within parentheses: "+", e, and Number(0). Assuming the name designates a case class, such a pattern means to first check that the object is a member of the named case class, and then to check that the constructor parameters of the object match the extra patterns supplied.

These extra patterns mean that Scala patterns support *deep matches*. Such patterns not only check the top-level object supplied, but also the contents of the object against further patterns. Since the extra patterns can themselves be constructor patterns, you can use them to check arbitrarily deep into an object. For example, the pattern shown in Listing 15.7 checks that the top-level object is a BinOp, that its third constructor parameter is a Number, and that the value field of that number is 0. This pattern is one line long yet checks three levels deep.

```
expr match {
  case BinOp("+", e, Number(0)) => println("a deep match")
  case _ =>
}
```

Listing 15.7 · A pattern match with a constructor pattern.

## Sequence patterns

You can match against sequence types, like List or Array, just like you match against case classes. Use the same syntax, but now you can specify any number of elements within the pattern. Listing 15.8 shows a pattern that checks for a three-element list starting with zero.

```
expr match {
  case List(0, _, _) => println("found it")
  case _ =>
}
```

Listing 15.8 · A sequence pattern with a fixed length.

If you want to match against a sequence without specifying how long it can be, you can specify _* as the last element of the pattern. This funny-looking pattern matches any number of elements within a sequence, including zero elements. Listing 15.9 shows an example that matches any list that starts with zero, regardless of how long the list is.

```
expr match {
  case List(0, _*) => println("found it")
  case _ =>
}
```

Listing 15.9 · A sequence pattern with an arbitrary length.

## Tuple patterns

You can match against tuples too. A pattern like (a, b, c) matches an arbitrary 3-tuple. An example is shown in Listing 15.10.

```
def tupleDemo(expr: Any) =
  expr match {
    case (a, b, c)  =>  println("matched " + a + b + c)
    case _ =>
  }
```

Listing 15.10 · A pattern match with a tuple pattern.

If you load the tupleDemo method shown in Listing 15.10 into the interpreter, and pass to it a tuple with three elements, you'll see:

```
scala> tupleDemo(("a ", 3, "-tuple"))
matched a 3-tuple
```

## Typed patterns

You can use a *typed pattern* as a convenient replacement for type tests and type casts. Listing 15.11 shows an example.

```
def generalSize(x: Any) = x match {
  case s: String => s.length
  case m: Map[_, _] => m.size
  case _ => -1
}
```

Listing 15.11 · A pattern match with typed patterns.

Here are a few examples of using generalSize in the Scala interpreter:

```
scala> generalSize("abc")
res16: Int = 3

scala> generalSize(Map(1 -> 'a', 2 -> 'b'))
res17: Int = 2

scala> generalSize(math.Pi)
res18: Int = -1
```

The generalSize method returns the size or length of objects of various types. Its argument is of type Any, so it could be any value. If the argument is a String, the method returns the string's length. The pattern "s: String" is a typed pattern; it matches every (non-null) instance of String. The pattern variable s then refers to that string.

Note that even though s and x refer to the same value, the type of x is Any, while the type of s is String. So you can write s.length in the alternative expression that corresponds to the pattern, but you could not write x.length, because the type Any does not have a length member.

An equivalent but more long-winded way that achieves the effect of a match against a typed pattern employs a type test followed by a type cast. Scala uses a different syntax than Java for these. To test whether an expression expr has type String, say, you write:

```
expr.isInstanceOf[String]
```

To cast the same expression to type String, you use:

```
expr.asInstanceOf[String]
```

Using a type test and cast, you could rewrite the first case of the previous match expression as shown in Listing 15.12.

```
if (x.isInstanceOf[String]) {
  val s = x.asInstanceOf[String]
  s.length
} else ...
```

Listing 15.12 · Using isInstanceOf and asInstanceOf (poor style).

The operators isInstanceOf and asInstanceOf are treated as predefined methods of class Any that take a type parameter in square brackets.
In fact, x.asInstanceOf[String] is a special case of a method invocation
with an explicit type parameter String.

As you will have noted by now, writing type tests and casts is rather
verbose in Scala. That's intentional because it is not encouraged practice.
You are usually better off using a pattern match with a typed pattern. That's
particularly true if you need to do both a type test and a type cast, because
both operations are then rolled into a single pattern match.

The second case of the match expression in Listing 15.11 contains the
typed pattern "m: Map[_, _]". This pattern matches any value that is a Map of
some arbitrary key and value types, and lets m refer to that value. Therefore,
m.size is well typed and returns the size of the map. The underscores in the
type pattern[3] are like wildcards in other patterns. You could have also used
(lowercase) type variables instead.

*Type erasure*

Can you also test for a map with specific element types? This would be
handy, say, for testing whether a given value is a map from type Int to type
Int. Let's try:

```
scala> def isIntIntMap(x: Any) = x match {
         case m: Map[Int, Int] => true
         case _ => false
       }
<console>:9: warning: non-variable type argument Int in type
pattern scala.collection.immutable.Map[Int,Int] (the
underlying of Map[Int,Int]) is unchecked since it is
```

---

[3]In the typed pattern, m: Map[_, _], the "Map[_, _]" portion is called a *type pattern*.

```
eliminated by erasure
        case m: Map[Int, Int] => true
                      ^
```

Scala uses the *erasure* model of generics, just like Java does. This means that no information about type arguments is maintained at runtime. Consequently, there is no way to determine at runtime whether a given Map object has been created with two Int arguments, rather than with arguments of different types. All the system can do is determine that a value is a Map of some arbitrary type parameters. You can verify this behavior by applying isIntIntMap to arguments of different instances of class Map:

```
scala> isIntIntMap(Map(1 -> 1))
res19: Boolean = true

scala> isIntIntMap(Map("abc" -> "abc"))
res20: Boolean = true
```

The first application returns true, which looks correct, but the second application also returns true, which might be a surprise. To alert you to the possibly non-intuitive runtime behavior, the compiler emits unchecked warnings like the one shown previously.

The only exception to the erasure rule is arrays, because they are handled specially in Java as well as in Scala. The element type of an array is stored with the array value, so you can pattern match on it. Here's an example:

```
scala> def isStringArray(x: Any) = x match {
        case a: Array[String] => "yes"
        case _ => "no"
      }
isStringArray: (x: Any)String

scala> val as = Array("abc")
as: Array[String] = Array(abc)

scala> isStringArray(as)
res21: String = yes

scala> val ai = Array(1, 2, 3)
ai: Array[Int] = Array(1, 2, 3)

scala> isStringArray(ai)
res22: String = no
```

### Variable binding

In addition to the standalone variable patterns, you can also add a variable to any other pattern. You simply write the variable name, an at sign (@), and then the pattern. This gives you a variable-binding pattern, which means the pattern is to perform the pattern match as normal, and if the pattern succeeds, set the variable to the matched object just as with a simple variable pattern.

As an example, Listing 15.13 shows a pattern match that looks for the absolute value operation being applied twice in a row. Such an expression can be simplified to only take the absolute value one time.

```
expr match {
  case UnOp("abs", e @ UnOp("abs", _)) => e
  case _ =>
}
```

Listing 15.13 · A pattern with a variable binding (via the @ sign).

The example shown in Listing 15.13 includes a variable-binding pattern with e as the variable and UnOp("abs", _) as the pattern. If the entire pattern match succeeds, then the portion that matched the UnOp("abs", _) part is made available as variable e. The result of the case is just e, because e has the same value as expr but with one less absolute value operation.

## 15.3   Pattern guards

Sometimes, syntactic pattern matching is not precise enough. For instance, say you are given the task of formulating a simplification rule that replaces sum expressions with two identical operands, such as e + e, by multiplications of two (*e.g.*, e * 2). In the language of Expr trees, an expression like:

```
BinOp("+", Var("x"), Var("x"))
```

would be transformed by this rule to:

```
BinOp("*", Var("x"), Number(2))
```

You might try to define this rule as follows:

287

```
scala> def simplifyAdd(e: Expr) = e match {
         case BinOp("+", x, x) => BinOp("*", x, Number(2))
         case _ => e
       }
<console>:14: error: x is already defined as value x
         case BinOp("+", x, x) => BinOp("*", x, Number(2))
                              ^
```

This fails because Scala restricts patterns to be *linear*: a pattern variable may only appear once in a pattern. However, you can re-formulate the match with a *pattern guard*, as shown in Listing 15.14:

```
scala> def simplifyAdd(e: Expr) = e match {
         case BinOp("+", x, y) if x == y =>
           BinOp("*", x, Number(2))
         case _ => e
       }
simplifyAdd: (e: Expr)Expr
```

Listing 15.14 · A match expression with a pattern guard.

A pattern guard comes after a pattern and starts with an `if`. The guard can be an arbitrary boolean expression, which typically refers to variables in the pattern. If a pattern guard is present, the match succeeds only if the guard evaluates to `true`. Hence, the first case above would only match binary operations with two equal operands.

Some other examples of guarded patterns are:

```
// match only positive integers
case n: Int if 0 < n => ...

// match only strings starting with the letter 'a'
case s: String if s(0) == 'a' => ...
```

## 15.4  Pattern overlaps

Patterns are tried in the order in which they are written. The version of `simplify` shown in Listing 15.15 presents an example where the order of the cases matters.

```
def simplifyAll(expr: Expr): Expr = expr match {
  case UnOp("-", UnOp("-", e)) =>
    simplifyAll(e)    // '-' is its own inverse
  case BinOp("+", e, Number(0)) =>
    simplifyAll(e)    // '0' is a neutral element for '+'
  case BinOp("*", e, Number(1)) =>
    simplifyAll(e)    // '1' is a neutral element for '*'
  case UnOp(op, e) =>
    UnOp(op, simplifyAll(e))
  case BinOp(op, l, r) =>
    BinOp(op, simplifyAll(l), simplifyAll(r))
  case _ => expr
}
```

Listing 15.15 · Match expression in which case order matters.

The version of simplify shown in Listing 15.15 will apply simplification rules everywhere in an expression, not just at the top, as simplifyTop did. It can be derived from simplifyTop by adding two more cases for general unary and binary expressions (cases four and five in Listing 15.15).

The fourth case has the pattern UnOp(op, e); *i.e.*, it matches every unary operation. The operator and operand of the unary operation can be arbitrary. They are bound to the pattern variables op and e, respectively. The alternative in this case applies simplifyAll recursively to the operand e and then rebuilds the same unary operation with the (possibly) simplified operand. The fifth case for BinOp is analogous: it is a "catch-all" case for arbitrary binary operations, which recursively applies the simplification method to its two operands.

In this example, it is important that the catch-all cases come *after* the more specific simplification rules. If you wrote them in the other order, then the catch-all case would be run in favor of the more specific rules. In many cases, the compiler will even complain if you try. For example, here's a match expression that won't compile because the first case will match anything that would be matched by the second case:

```
scala> def simplifyBad(expr: Expr): Expr = expr match {
         case UnOp(op, e) => UnOp(op, simplifyBad(e))
         case UnOp("-", UnOp("-", e)) => e
      }
<console>:21: warning: unreachable code
         case UnOp("-", UnOp("-", e)) => e
                                         ^
```

## 15.5   Sealed classes

Whenever you write a pattern match, you need to make sure you have covered all of the possible cases. Sometimes you can do this by adding a default case at the end of the match, but that only applies if there is a sensible default behavior. What do you do if there is no default? How can you ever feel safe that you covered all the cases?

You can enlist the help of the Scala compiler in detecting missing combinations of patterns in a match expression. To do this, the compiler needs to be able to tell which are the possible cases. In general, this is impossible in Scala because new case classes can be defined at any time and in arbitrary compilation units. For instance, nothing would prevent you from adding a fifth case class to the Expr class hierarchy in a different compilation unit from the one where the other four cases are defined.

The alternative is to make the superclass of your case classes *sealed*. A sealed class cannot have any new subclasses added except the ones in the same file. This is very useful for pattern matching because it means you only need to worry about the subclasses you already know about. What's more, you get better compiler support as well. If you match against case classes that inherit from a sealed class, the compiler will flag missing combinations of patterns with a warning message.

If you write a hierarchy of classes intended to be pattern matched, you should consider sealing them. Simply put the sealed keyword in front of the class at the top of the hierarchy. Programmers using your class hierarchy will then feel confident in pattern matching against it. The sealed keyword, therefore, is often a license to pattern match. Listing 15.16 shows an example in which Expr is turned into a sealed class.

```
sealed abstract class Expr
case class Var(name: String) extends Expr
case class Number(num: Double) extends Expr
case class UnOp(operator: String, arg: Expr) extends Expr
case class BinOp(operator: String,
    left: Expr, right: Expr) extends Expr
```

Listing 15.16 · A sealed hierarchy of case classes.

Now define a pattern match where some of the possible cases are left out:

```
def describe(e: Expr): String = e match {
  case Number(_) => "a number"
  case Var(_)    => "a variable"
}
```

You will get a compiler warning like the following:

```
warning: match is not exhaustive!
missing combination            UnOp
missing combination            BinOp
```

Such a warning tells you that there's a risk your code might produce a MatchError exception because some possible patterns (UnOp, BinOp) are not handled. The warning points to a potential source of runtime faults, so it is usually a welcome help in getting your program right.

However, at times you might encounter a situation where the compiler is too picky in emitting the warning. For instance, you might know from the context that you will only ever apply the describe method above to expressions that are either Numbers or Vars, so you know that no MatchError will be produced. To make the warning go away, you could add a third catch-all case to the method, like this:

```
def describe(e: Expr): String = e match {
  case Number(_) => "a number"
  case Var(_) => "a variable"
  case _ => throw new RuntimeException // Should not happen
}
```

291

That works, but it is not ideal. You will probably not be very happy that you were forced to add code that will never be executed (or so you think), just to make the compiler shut up.

A more lightweight alternative is to add an @unchecked annotation to the selector expression of the match. This is done as follows:

```
def describe(e: Expr): String = (e: @unchecked) match {
  case Number(_) => "a number"
  case Var(_)    => "a variable"
}
```

Annotations are described in Chapter 27. In general, you can add an annotation to an expression in the same way you add a type: follow the expression with a colon and the name of the annotation (preceded by an at sign). For example, in this case you add an @unchecked annotation to the variable e, with "e: @unchecked". The @unchecked annotation has a special meaning for pattern matching. If a match's selector expression carries this annotation, exhaustivity checking for the patterns that follow will be suppressed.

## 15.6   The Option type

Scala has a standard type named Option for optional values. Such a value can be of two forms: Some(x), where x is the actual value, or the None object, which represents a missing value.

Optional values are produced by some of the standard operations on Scala's collections. For instance, the get method of Scala's Map produces Some(value) if a value corresponding to a given key has been found, or None if the given key is not defined in the Map. Here's an example:

```
scala> val capitals =
         Map("France" -> "Paris", "Japan" -> "Tokyo")
capitals: scala.collection.immutable.Map[String,String] =
Map(France -> Paris, Japan -> Tokyo)

scala> capitals get "France"
res23: Option[String] = Some(Paris)

scala> capitals get "North Pole"
res24: Option[String] = None
```

The most common way to take optional values apart is through a pattern match. For instance:

```
scala> def show(x: Option[String]) = x match {
         case Some(s) => s
         case None => "?"
       }
show: (x: Option[String])String

scala> show(capitals get "Japan")
res25: String = Tokyo

scala> show(capitals get "France")
res26: String = Paris

scala> show(capitals get "North Pole")
res27: String = ?
```

The Option type is used frequently in Scala programs. Compare this to the dominant idiom in Java of using null to indicate no value. For example, the get method of java.util.HashMap returns either a value stored in the HashMap or null if no value was found. This approach works for Java but is error prone because it is difficult in practice to keep track of which variables in a program are allowed to be null.

If a variable is allowed to be null, then you must remember to check it for null every time you use it. When you forget to check, you open the possibility that a NullPointerException may result at runtime. Because such exceptions may not happen very often, it can be difficult to discover the bug during testing. For Scala, the approach would not work at all because it is possible to store value types in hash maps, and null is not a legal element for a value type. For instance, a HashMap[Int, Int] cannot return null to signify "no element."

By contrast, Scala encourages the use of Option to indicate an optional value. This approach to optional values has several advantages over Java's. First, it is far more obvious to readers of code that a variable whose type is Option[String] is an optional String than a variable of type String, which may sometimes be null. But most importantly, that programming error described earlier of using a variable that may be null without first checking it for null becomes a type error in Scala. If a variable is of type Option[String] and you try to use it as a String, your Scala program will not compile.

## 15.7 Patterns everywhere

Patterns are allowed in many parts of Scala, not just in standalone `match` expressions. Take a look at some other places you can use patterns.

### Patterns in variable definitions

Anytime you define a `val` or a `var`, you can use a pattern instead of a simple identifier. For example, you can take apart a tuple and assign each of its parts to its own variable, as shown in Listing 15.17:

```
scala> val myTuple = (123, "abc")
myTuple: (Int, String) = (123,abc)

scala> val (number, string) = myTuple
number: Int = 123
string: String = abc
```

Listing 15.17 · Defining multiple variables with one assignment.

This construct is quite useful when working with case classes. If you know the precise case class you are working with, then you can deconstruct it with a pattern. Here's an example:

```
scala> val exp = new BinOp("*", Number(5), Number(1))
exp: BinOp = BinOp(*,Number(5.0),Number(1.0))

scala> val BinOp(op, left, right) = exp
op: String = *
left: Expr = Number(5.0)
right: Expr = Number(1.0)
```

### Case sequences as partial functions

A sequence of cases (*i.e.*, alternatives) in curly braces can be used anywhere a function literal can be used. Essentially, a case sequence *is* a function literal, only more general. Instead of having a single entry point and list of parameters, a case sequence has multiple entry points, each with their own list of parameters. Each case is an entry point to the function, and the parameters are specified with the pattern. The body of each entry point is the right-hand side of the case.

Here is a simple example:

```
val withDefault: Option[Int] => Int = {
  case Some(x) => x
  case None => 0
}
```

The body of this function has two cases. The first case matches a Some, and returns the number inside the Some. The second case matches a None, and returns a default value of zero. Here is this function in use:

```
scala> withDefault(Some(10))
res28: Int = 10

scala> withDefault(None)
res29: Int = 0
```

This facility is quite useful for the Akka actors library, because it allows its `receive` method to be defined as a series of cases:

```
var sum = 0

def receive = {
  case Data(byte) =>
    sum += byte

  case GetChecksum(requester) =>
    val checksum = ~(sum & 0xFF) + 1
    requester ! checksum
}
```

One other generalization is worth noting: a sequence of cases gives you a *partial* function. If you apply such a function on a value it does not support, it will generate a run-time exception. For example, here is a partial function that returns the second element of a list of integers:

```
val second: List[Int] => Int = {
  case x :: y :: _ => y
}
```

When you compile this, the compiler will correctly warn that the match is not exhaustive:

```
<console>:17: warning: match is not exhaustive!
missing combination              Nil
```

This function will succeed if you pass it a three-element list, but not if you pass it an empty list:

```
scala> second(List(5, 6, 7))
res24: Int = 6

scala> second(List())
scala.MatchError: List()
      at $anonfun$1.apply(<console>:17)
      at $anonfun$1.apply(<console>:17)
```

If you want to check whether a partial function is defined, you must first tell the compiler that you know you are working with partial functions. The type List[Int] => Int includes all functions from lists of integers to integers, whether or not the functions are partial. The type that only includes *partial* functions from lists of integers to integers is written PartialFunction[List[Int],Int]. Here is the second function again, this time written with a partial function type:

```
val second: PartialFunction[List[Int],Int] = {
  case x :: y :: _ => y
}
```

Partial functions have a method isDefinedAt, which can be used to test whether the function is defined at a particular value. In this case, the function is defined for any list that has at least two elements:

```
scala> second.isDefinedAt(List(5,6,7))
res30: Boolean = true

scala> second.isDefinedAt(List())
res31: Boolean = false
```

The typical example of a partial function is a pattern matching function literal like the one in the previous example. In fact, such an expression gets translated by the Scala compiler to a partial function by translating the patterns twice—once for the implementation of the real function, and once to test whether the function is defined or not.

For instance, the function literal { case x :: y :: _ => y } gets translated to the following partial function value:

```
new PartialFunction[List[Int], Int] {
  def apply(xs: List[Int]) = xs match {
    case x :: y :: _ => y
  }
  def isDefinedAt(xs: List[Int]) = xs match {
    case x :: y :: _ => true
    case _ => false
  }
}
```

This translation takes effect whenever the declared type of a function literal is PartialFunction. If the declared type is just Function1, or is missing, the function literal is instead translated to a *complete function*.

In general, you should try to work with complete functions whenever possible, because using partial functions allows for runtime errors that the compiler cannot help you with. Sometimes partial functions are really helpful though. You might be sure that an unhandled value will never be supplied. Alternatively, you might be using a framework that expects partial functions and so will always check isDefinedAt before calling the function. An example of the latter is the react example given above, where the argument is a partially defined function, defined precisely for those messages that the caller wants to handle.

### Patterns in for expressions

You can also use a pattern in a for expression, as shown in Listing 15.18. This for expression retrieves all key/value pairs from the capitals map. Each pair is matched against the pattern (country, city), which defines the two variables country and city.

The pair pattern shown in Listing 15.18 was special because the match against it can never fail. Indeed, capitals yields a sequence of pairs, so you can be sure that every generated pair can be matched against a pair pattern.

```
scala> for ((country, city) <- capitals)
         println("The capital of " + country + " is " + city)
The capital of France is Paris
The capital of Japan is Tokyo
```

Listing 15.18 · A for expression with a tuple pattern.

But it is equally possible that a pattern might not match a generated value. Listing 15.19 shows an example where that is the case.

```
scala> val results = List(Some("apple"), None,
           Some("orange"))
results: List[Option[String]] = List(Some(apple), None,
Some(orange))

scala> for (Some(fruit) <- results) println(fruit)
apple
orange
```

Listing 15.19 · Picking elements of a list that match a pattern.

As you can see from this example, generated values that do not match the pattern are discarded. For instance, the second element None in the results list does not match the pattern Some(fruit); therefore it does not show up in the output.

## 15.8   A larger example

After having learned the different forms of patterns, you might be interested in seeing them applied in a larger example. The proposed task is to write an expression formatter class that displays an arithmetic expression in a two-dimensional layout. Divisions such as "x / (x + 1)" should be printed vertically, by placing the numerator on top of the denominator, like this:

```
      x
    -----
    x + 1
```

As another example, here's the expression ((a / (b * c) + 1 / n) / 3) in two dimensional layout:

```
        a       1
      ----- + -
      b * c     n
      ---------
          3
```

From these examples it looks like the class (we'll call it ExprFormatter) will have to do a fair bit of layout juggling, so it makes sense to use the layout library developed in Chapter 10. We'll also use the Expr family of case classes you saw previously in this chapter, and place both Chapter 10's layout library and this chapter's expression formatter into named packages. The full code for the example will be shown in Listings 15.20 and 15.21.

A useful first step is to concentrate on horizontal layout. A structured expression like:

```
BinOp("+",
      BinOp("*",
            BinOp("+", Var("x"), Var("y")),
            Var("z")),
      Number(1))
```

should print (x + y) * z + 1. Note that parentheses are mandatory around x + y, but would be optional around (x + y) * z. To keep the layout as legible as possible, your goal should be to omit parentheses wherever they are redundant, while ensuring that all necessary parentheses are present.

To know where to put parentheses, the code needs to know about the relative precedence of each operator, so it's a good idea to tackle this first. You could express the relative precedence directly as a map literal of the following form:

```
Map(
  "|" -> 0, "||" -> 0,
  "&" -> 1, "&&" -> 1, ...
)
```

However, this would involve some amount of pre-computation of precedences on your part. A more convenient approach is to just define groups of operators of increasing precedence and then calculate the precedence of each operator from that. Listing 15.20 shows the code.

The `precedence` variable is a map from operators to their precedences, which are integers starting with 0. It is calculated using a `for` expression with two generators. The first generator produces every index `i` of the `opGroups` array. The second generator produces every operator `op` in `opGroups(i)`. For each such operator the `for` expression yields an association from the operator `op` to its index `i`. Hence, the relative position of an operator in the array is taken to be its precedence.

Associations are written with an infix arrow, *e.g.*, `op -> i`. So far you have seen associations only as part of map constructions, but they are also values in their own right. In fact, the association `op -> i` is nothing else but the pair `(op, i)`.

Now that you have fixed the precedence of all binary operators except /, it makes sense to generalize this concept to also cover unary operators. The precedence of a unary operator is higher than the precedence of every binary operator. Thus we can set `unaryPrecedence` (shown in Listing 15.20) to the length of the `opGroups` array, which is one more than the precedence of the * and % operators. The precedence of a fraction is treated differently from the other operators because fractions use vertical layout. However, it will prove convenient to assign to the division operator the special precedence value –1, so we'll initialize `fractionPrecedence` to -1 (shown in Listing 15.20).

After these preparations, you are ready to write the main `format` method. This method takes two arguments: an expression e, of type `Expr`, and the precedence `enclPrec` of the operator directly enclosing the expression e. (If there's no enclosing operator, `enclPrec` should be zero.) The method yields a layout element that represents a two-dimensional array of characters.

Listing 15.21 shows the remainder of class `ExprFormatter`, which includes three methods. The first method, `stripDot`, is a helper method. The next method, the private `format` method, does most of the work to format expressions. The last method, also named `format`, is the lone public method in the library, which takes an expression to format. The private `format` method does its work by performing a pattern match on the kind of expression. The `match` expression has five cases. We'll discuss each case individually.

```scala
package org.stairwaybook.expr
import org.stairwaybook.layout.Element.elem

sealed abstract class Expr
case class Var(name: String) extends Expr
case class Number(num: Double) extends Expr
case class UnOp(operator: String, arg: Expr) extends Expr
case class BinOp(operator: String,
    left: Expr, right: Expr) extends Expr

class ExprFormatter {

  // Contains operators in groups of increasing precedence
  private val opGroups =
    Array(
      Set("|", "||"),
      Set("&", "&&"),
      Set("^"),
      Set("==", "!="),
      Set("<", "<=", ">", ">="),
      Set("+", "-"),
      Set("*", "%")
    )

  // A mapping from operators to their precedence
  private val precedence = {
    val assocs =
      for {
        i <- 0 until opGroups.length
        op <- opGroups(i)
      } yield op -> i
    assocs.toMap
  }

  private val unaryPrecedence = opGroups.length
  private val fractionPrecedence = -1

  // continued in Listing 15.21...
```

Listing 15.20 · The top half of the expression formatter.

```
// ...continued from Listing 15.20
import org.stairwaybook.layout.Element
private def format(e: Expr, enclPrec: Int): Element =
  e match {
    case Var(name) =>
      elem(name)
    case Number(num) =>
      def stripDot(s: String) =
        if (s endsWith ".0") s.substring(0, s.length - 2)
        else s
      elem(stripDot(num.toString))
    case UnOp(op, arg) =>
      elem(op) beside format(arg, unaryPrecedence)
    case BinOp("/", left, right) =>
      val top = format(left, fractionPrecedence)
      val bot = format(right, fractionPrecedence)
      val line = elem('-', top.width max bot.width, 1)
      val frac = top above line above bot
      if (enclPrec != fractionPrecedence) frac
      else elem(" ") beside frac beside elem(" ")
    case BinOp(op, left, right) =>
      val opPrec = precedence(op)
      val l = format(left, opPrec)
      val r = format(right, opPrec + 1)
      val oper = l beside elem(" " + op + " ") beside r
      if (enclPrec <= opPrec) oper
      else elem("(") beside oper beside elem(")")
  }
  def format(e: Expr): Element = format(e, 0)
}
```

Listing 15.21 · The bottom half of the expression formatter.

302

The first case is:

```
case Var(name) =>
  elem(name)
```

If the expression is a variable, the result is an element formed from the variable's name.

The second case is:

```
case Number(num) =>
  def stripDot(s: String) =
    if (s endsWith ".0") s.substring(0, s.length - 2)
    else s
  elem(stripDot(num.toString))
```

If the expression is a number, the result is an element formed from the number's value. The stripDot function cleans up the display of a floating-point number by stripping any ".0" suffix from a string.

The third case is:

```
case UnOp(op, arg) =>
  elem(op) beside format(arg, unaryPrecedence)
```

If the expression is a unary operation UnOp(op, arg) the result is formed from the operation op and the result of formatting the argument arg with the highest-possible environment precedence.[4] This means that if arg is a binary operation (but not a fraction) it will always be displayed in parentheses.

The fourth case is:

```
case BinOp("/", left, right) =>
  val top = format(left, fractionPrecedence)
  val bot = format(right, fractionPrecedence)
  val line = elem('-', top.width max bot.width, 1)
  val frac = top above line above bot
  if (enclPrec != fractionPrecedence) frac
  else elem(" ") beside frac beside elem(" ")
```

---

[4]The value of unaryPrecedence is the highest possible precedence, because it was initialized to one more than the precedence of the * and % operators.

303

If the expression is a fraction, an intermediate result `frac` is formed by placing the formatted operands `left` and `right` on top of each other, separated by an horizontal line element. The width of the horizontal line is the maximum of the widths of the formatted operands. This intermediate result is also the final result unless the fraction appears itself as an argument of another fraction. In the latter case, a space is added on each side of `frac`. To see the reason why, consider the expression "(a / b) / c".

Without the widening correction, formatting this expression would give:

```
a
-
b
-
c
```

The problem with this layout is evident—it's not clear where the top-level fractional bar is. The expression above could mean either "(a / b) / c" or "a / (b / c)". To disambiguate, a space should be added on each side to the layout of the nested fraction "a / b".

Then the layout becomes unambiguous:

```
 a
 -
 b
---
 c
```

The fifth and last case is:

```
case BinOp(op, left, right) =>
  val opPrec = precedence(op)
  val l = format(left, opPrec)
  val r = format(right, opPrec + 1)
  val oper = l beside elem(" " + op + " ") beside r
  if (enclPrec <= opPrec) oper
  else elem("(") beside oper beside elem(")")
```

This case applies for all other binary operations. Since it comes after the case starting with:

```
case BinOp("/", left, right) => ...
```

you know that the operator op in the pattern BinOp(op, left, right) cannot be a division. To format such a binary operation, one needs to format first its operands left and right. The precedence parameter for formatting the left operand is the precedence opPrec of the operator op, while for the right operand it is one more than that. This scheme ensures that parentheses also reflect the correct associativity.

For instance, the operation:

```
BinOp("-", Var("a"), BinOp("-", Var("b"), Var("c")))
```

would be correctly parenthesized as "a - (b - c)". The intermediate result oper is then formed by placing the formatted left and right operands side-by-side, separated by the operator. If the precedence of the current operator is smaller than the precedence of the enclosing operator, oper is placed between parentheses; otherwise, it is returned as is.

```
import org.stairwaybook.expr._

object Express extends App {
  val f = new ExprFormatter
  val e1 = BinOp("*", BinOp("/", Number(1), Number(2)),
                      BinOp("+", Var("x"), Number(1)))
  val e2 = BinOp("+", BinOp("/", Var("x"), Number(2)),
                      BinOp("/", Number(1.5), Var("x")))
  val e3 = BinOp("/", e1, e2)
  def show(e: Expr) = println(f.format(e)+ "\n\n")
  for (e <- Array(e1, e2, e3)) show(e)
}
```

Listing 15.22 · An application that prints formatted expressions.

This finishes the design of the private format function. The only remaining method is the public format method, which allows client programmers to format a top-level expression without passing a precedence argument. Listing 15.22 shows a demo program that exercises ExprFormatter.

Note that, even though this program does not define a `main` method, it is still a runnable application because it inherits from the App trait. You can run the Express program with the command:

```
scala Express
```

This will give the following output:

```
1
- * (x + 1)
2



x    1.5
- + ---
2    x



1
- * (x + 1)
2
-----------
  x    1.5
  - + ---
  2    x
```

## 15.9 Conclusion

In this chapter, you learned about Scala's case classes and pattern matching in detail. By using them, you can take advantage of several concise idioms not normally available in object-oriented languages. However, Scala's pattern matching goes further than this chapter describes. If you want to use pattern matching on one of your classes, but you do not want to open access to your classes the way case classes do, you can use the *extractors* described in Chapter 26. In the next chapter, we'll turn our attention to lists.

# Chapter 16

# Working with Lists

Lists are probably the most commonly used data structure in Scala programs. This chapter explains lists in detail. We will present many common operations that can be performed on lists. We'll also cover some important design principles for programs working on lists.

## 16.1   List literals

You saw lists already in the preceding chapters, so you know that a list containing the elements 'a', 'b', and 'c' is written List('a', 'b', 'c'). Here are some other examples:

```
val fruit = List("apples", "oranges", "pears")
val nums = List(1, 2, 3, 4)
val diag3 =
  List(
    List(1, 0, 0),
    List(0, 1, 0),
    List(0, 0, 1)
  )
val empty = List()
```

Lists are quite similar to arrays, but there are two important differences. First, lists are immutable. That is, elements of a list cannot be changed by assignment. Second, lists have a recursive structure (*i.e.*, a *linked list*),[1] whereas arrays are flat.

---

[1]For a graphical depiction of the structure of a List, see Figure 22.2 on page 476.

## 16.2 The List type

Like arrays, lists are *homogeneous*: the elements of a list all have the same type. The type of a list that has elements of type T is written List[T]. For instance, here are the same four lists with explicit types added:

```
val fruit: List[String] = List("apples", "oranges", "pears")
val nums: List[Int] = List(1, 2, 3, 4)
val diag3: List[List[Int]] =
  List(
    List(1, 0, 0),
    List(0, 1, 0),
    List(0, 0, 1)
  )
val empty: List[Nothing] = List()
```

The list type in Scala is *covariant*. This means that for each pair of types S and T, if S is a subtype of T, then List[S] is a subtype of List[T]. For instance, List[String] is a subtype of List[Object]. This is natural because every list of strings can also be seen as a list of objects.[2]

Note that the empty list has type List[Nothing]. You saw in Section 11.3 that Nothing is the bottom type in Scala's class hierarchy. It is a subtype of every other Scala type. Because lists are covariant, it follows that List[Nothing] is a subtype of List[T] for any type T. So the empty list object, which has type List[Nothing], can also be seen as an object of every other list type of the form List[T]. That's why it is permissible to write code like:

```
// List() is also of type List[String]!
val xs: List[String] = List()
```

## 16.3 Constructing lists

All lists are built from two fundamental building blocks, Nil and :: (pronounced "cons"). Nil represents the empty list. The infix operator, ::, expresses list extension at the front. That is, x :: xs represents a list whose

---

[2]Chapter 19 gives more details on covariance and other kinds of variance.

first element is x, followed by (the elements of) list xs. Hence, the previous list values could also have been defined as follows:

```
val fruit = "apples" :: ("oranges" :: ("pears" :: Nil))
val nums  = 1 :: (2 :: (3 :: (4 :: Nil)))
val diag3 = (1 :: (0 :: (0 :: Nil))) ::
            (0 :: (1 :: (0 :: Nil))) ::
            (0 :: (0 :: (1 :: Nil))) :: Nil
val empty = Nil
```

In fact the previous definitions of fruit, nums, diag3, and empty in terms of List(...) are just wrappers that expand to these definitions. For instance, List(1, 2, 3) creates the list 1 :: (2 :: (3 :: Nil)).

Because it ends in a colon, the :: operation associates to the right: A :: B :: C is interpreted as A :: (B :: C). Therefore, you can drop the parentheses in the previous definitions. For instance:

```
val nums = 1 :: 2 :: 3 :: 4 :: Nil
```

is equivalent to the previous definition of nums.

## 16.4   Basic operations on lists

All operations on lists can be expressed in terms of the following three:

| | |
|---|---|
| head | returns the first element of a list |
| tail | returns a list consisting of all elements except the first |
| isEmpty | returns true if the list is empty |

These operations are defined as methods of class List. Some examples are shown in Table 16.1. The head and tail methods are defined only for non-empty lists. When selected from an empty list, they throw an exception:

```
scala> Nil.head
java.util.NoSuchElementException: head of empty list
```

As an example of how lists can be processed, consider sorting the elements of a list of numbers into ascending order. One simple way to do so is *insertion sort*, which works as follows: To sort a non-empty list x :: xs, sort the remainder xs and insert the first element x at the right position in the result.

309

Table 16.1 · Basic list operations

| What it is | What it does |
|---|---|
| empty.isEmpty | returns true |
| fruit.isEmpty | returns false |
| fruit.head | returns "apples" |
| fruit.tail.head | returns "oranges" |
| diag3.head | returns List(1, 0, 0) |

Sorting an empty list yields the empty list. Expressed as Scala code, the insertion sort algorithm looks like:

```
def isort(xs: List[Int]): List[Int] =
  if (xs.isEmpty) Nil
  else insert(xs.head, isort(xs.tail))

def insert(x: Int, xs: List[Int]): List[Int] =
  if (xs.isEmpty || x <= xs.head) x :: xs
  else xs.head :: insert(x, xs.tail)
```

## 16.5   List patterns

Lists can also be taken apart using pattern matching. List patterns correspond one-by-one to list expressions. You can either match on all elements of a list using a pattern of the form List(...), or you take lists apart bit by bit using patterns composed from the :: operator and the Nil constant.

Here's an example of the first kind of pattern:

```
scala> val List(a, b, c) = fruit
a: String = apples
b: String = oranges
c: String = pears
```

The pattern List(a, b, c) matches lists of length 3, and binds the three elements to the pattern variables a, b, and c. If you don't know the number of list elements beforehand, it's better to match with :: instead. For instance, the pattern a :: b :: rest matches lists of length 2 or greater:

310

> ### About pattern matching on `Lists`
>
> If you review the possible forms of patterns explained in Chapter 15,
> you might find that neither `List(...)` nor `::` look like it fits the
> kinds of patterns defined there. In fact, `List(...)` is an instance of
> a library-defined *extractor* pattern. Such patterns will be discussed in
> Chapter 26. The "cons" pattern `x :: xs` is a special case of an infix
> operation pattern. As an expression, an infix operation is equivalent
> to a method call. For patterns, the rules are different: As a pattern, an
> infix operation such as `p op q` is equivalent to `op(p, q)`. That is, the
> infix operator `op` is treated as a constructor pattern. In particular, a cons
> pattern such as `x :: xs` is treated as `::(x, xs)`.
>
> This hints that there should be a class named `::` that corresponds
> to the pattern constructor. Indeed, there is such a class—it is named
> `scala.::` and is exactly the class that builds non-empty lists. So `::`
> exists twice in Scala, once as a name of a class in package `scala` and
> again as a method in class `List`. The effect of the method `::` is to
> produce an instance of the class `scala.::`. You'll find out more details
> about how the List class is implemented in Chapter 22.

```
scala> val a :: b :: rest = fruit
a: String = apples
b: String = oranges
rest: List[String] = List(pears)
```

Taking lists apart with patterns is an alternative to taking them apart with the
basic methods `head`, `tail`, and `isEmpty`. For instance, here's insertion sort
again, this time written with pattern matching:

```
def isort(xs: List[Int]): List[Int] = xs match {
  case List()   => List()
  case x :: xs1 => insert(x, isort(xs1))
}
def insert(x: Int, xs: List[Int]): List[Int] = xs match {
  case List()  => List(x)
  case y :: ys => if (x <= y) x :: xs
                  else y :: insert(x, ys)
}
```

Often, pattern matching over lists is clearer than decomposing them with methods, so pattern matching should be a part of your list processing toolbox.

This is all you need to know about lists in Scala to use them correctly. However, there are also a large number of methods that capture common patterns of operations over lists. These methods make list processing programs more concise and often clearer. The next two sections present the most important methods defined in the List class.

## 16.6  First-order methods on class List

This section explains most first-order methods defined in the List class. A method is *first-order* if it does not take any functions as arguments. We will also introduce some recommended techniques to structure programs that operate on lists by using two examples.

### Concatenating two lists

An operation similar to :: is list concatenation, written ':::'. Unlike ::, ::: takes two lists as operands. The result of xs ::: ys is a new list that contains all the elements of xs, followed by all the elements of ys.

Here are some examples:

```
scala> List(1, 2) ::: List(3, 4, 5)
res0: List[Int] = List(1, 2, 3, 4, 5)

scala> List() ::: List(1, 2, 3)
res1: List[Int] = List(1, 2, 3)

scala> List(1, 2, 3) ::: List(4)
res2: List[Int] = List(1, 2, 3, 4)
```

Like cons, list concatenation associates to the right. An expression like this:

```
xs ::: ys ::: zs
```

is interpreted like this:

```
xs ::: (ys ::: zs)
```

**The Divide and Conquer principle**

Concatenation (`:::`) is implemented as a method in class `List`. It would also be possible to implement concatenation "by hand," using pattern matching on lists. It's instructive to try to do that yourself, because it shows a common way to implement algorithms using lists. First, we'll settle on a signature for the concatenation method, which we'll call append. In order not to mix things up too much, assume that append is defined outside the `List` class, so it will take the two lists to be concatenated as parameters. These two lists must agree on their element type, but that element type can be arbitrary. This can be expressed by giving append a type parameter[3] that represents the element type of the two input lists:

```
def append[T](xs: List[T], ys: List[T]): List[T]
```

To design the implementation of append, it pays to remember the "divide and conquer" design principle for programs over recursive data structures such as lists. Many algorithms over lists first split an input list into simpler cases using a pattern match. That's the *divide* part of the principle. They then construct a result for each case. If the result is a non-empty list, some of its parts may be constructed by recursive invocations of the same algorithm. That's the *conquer* part of the principle.

To apply this principle to the implementation of the append method, the first question to ask is on which list to match. This is less trivial in the case of append than for many other methods because there are two choices. However, the subsequent "conquer" phase tells you that you need to construct a list consisting of all elements of both input lists. Since lists are constructed from the back towards the front, ys can remain intact, whereas xs will need to be taken apart and prepended to ys. Thus, it makes sense to concentrate on xs as a source for a pattern match. The most common pattern match over lists simply distinguishes an empty from a non-empty list. So this gives the following outline of an append method:

```
def append[T](xs: List[T], ys: List[T]): List[T] =
  xs match {
    case List() => ???
    case x :: xs1 => ???
  }
```

---

[3]Type parameters will be explained in more detail in Chapter 19.

All that remains is to fill in the two places marked with ???.[4] The first such place is the alternative where the input list xs is empty. In this case concatenation yields the second list:

```
case List() => ys
```

The second place left open is the alternative where the input list xs consists of some head x followed by a tail xs1. In this case the result is also a non-empty list. To construct a non-empty list you need to know what the head and the tail of that list should be. You know that the first element of the result list is x. As for the remaining elements, these can be computed by appending the second list, ys, to the rest of the first list, xs1.

This completes the design and gives:

```
def append[T](xs: List[T], ys: List[T]): List[T] =
  xs match {
    case List() => ys
    case x :: xs1 => x :: append(xs1, ys)
  }
```

The computation of the second alternative illustrated the "conquer" part of the divide and conquer principle: Think first what the shape of the desired output should be, then compute the individual parts of that shape, using recursive invocations of the algorithm where appropriate. Finally, construct the output from these parts.

### Taking the length of a list: length

The length method computes the length of a list.

```
scala> List(1, 2, 3).length
res3: Int = 3
```

On lists, unlike arrays, length is a relatively expensive operation. It needs to traverse the whole list to find its end, and therefore takes time proportional to the number of elements in the list. That's why it's not a good idea to replace a test such as xs.isEmpty by xs.length == 0. The result of the two tests is equivalent, but the second one is slower, in particular if the list xs is long.

---

[4]The ??? method, which throws scala.NotImplementedError and has result type Nothing, can be used as a temporary implementation during development.

314

**Accessing the end of a list: init and last**

You know already the basic operations head and tail, which respectively take the first element of a list, and the rest of the list except the first element. They each have a dual operation: last returns the last element of a (non-empty) list, whereas init returns a list consisting of all elements except the last one:

```
scala> val abcde = List('a', 'b', 'c', 'd', 'e')
abcde: List[Char] = List(a, b, c, d, e)

scala> abcde.last
res4: Char = e

scala> abcde.init
res5: List[Char] = List(a, b, c, d)
```

Like head and tail, these methods throw an exception when applied to an empty list:

```
scala> List().init
java.lang.UnsupportedOperationException: Nil.init
        at scala.List.init(List.scala:544)
        at ...

scala> List().last
java.util.NoSuchElementException: Nil.last
        at scala.List.last(List.scala:563)
        at ...
```

Unlike head and tail, which both run in constant time, init and last need to traverse the whole list to compute their result. As a result, they take time proportional to the length of the list.

It's a good idea to organize your data so that most accesses are at the head of a list, rather than the last element.

**Reversing lists: reverse**

If at some point in the computation an algorithm demands frequent accesses to the end of a list, it's sometimes better to reverse the list first and work with the result instead. Here's how to do the reversal:

```
scala> abcde.reverse
res6: List[Char] = List(e, d, c, b, a)
```

Like all other list operations, reverse creates a new list rather than changing the one it operates on. Since lists are immutable, such a change would not be possible anyway. To verify this, check that the original value of abcde is unchanged after the reverse operation:

```
scala> abcde
res7: List[Char] = List(a, b, c, d, e)
```

The reverse, init, and last operations satisfy some laws that can be used for reasoning about computations and for simplifying programs.

1. reverse is its own inverse:

    xs.reverse.reverse   *equals*   xs

2. reverse turns init to tail and last to head, except that the elements are reversed:

    xs.reverse.init   *equals*   xs.tail.reverse
    xs.reverse.tail   *equals*   xs.init.reverse
    xs.reverse.head   *equals*   xs.last
    xs.reverse.last   *equals*   xs.head

Reverse could be implemented using concatenation (:::), like in the following method, rev:

```
def rev[T](xs: List[T]): List[T] = xs match {
  case List() => xs
  case x :: xs1 => rev(xs1) ::: List(x)
}
```

However, this method is less efficient than one would hope for. To study the complexity of rev, assume that the list xs has length n. Notice that there are n recursive calls to rev. Each call except the last involves a list concatenation. List concatenation xs ::: ys takes time proportional to the length of its first argument xs. Hence, the total complexity of rev is:

$$n + (n-1) + ... + 1 = (1+n) * n/2$$

In other words, rev's complexity is quadratic in the length of its input argument. This is disappointing when compared to the standard reversal of a mutable, linked list, which has linear complexity. However, the current implementation of rev is not the best implementation possible. In the example starting on page 330, you will see how to speed it up.

**Prefixes and suffixes: drop, take, and splitAt**

The drop and take operations generalize tail and init in that they return arbitrary prefixes or suffixes of a list. The expression "xs take n" returns the first n elements of the list xs. If n is greater than xs.length, the whole list xs is returned. The operation "xs drop n" returns all elements of the list xs, except for the first n ones. If n is greater than xs.length, the empty list is returned.

The splitAt operation splits the list at a given index, returning a pair of two lists.[5] It is defined by the equality:

xs splitAt n   *equals*   (xs take n, xs drop n)

However, splitAt avoids traversing the list xs twice. Here are some examples of these three methods:

```
scala> abcde take 2
res8: List[Char] = List(a, b)

scala> abcde drop 2
res9: List[Char] = List(c, d, e)

scala> abcde splitAt 2
res10: (List[Char], List[Char]) = (List(a, b),List(c, d, e))
```

---

[5]As mentioned in Section 10.12, the term *pair* is an informal name for Tuple2.

### Element selection: `apply` and `indices`

Random element selection is supported through the `apply` method; however it is a less common operation for lists than it is for arrays.

```
scala> abcde apply 2 // rare in Scala
res11: Char = c
```

As for all other types, `apply` is implicitly inserted when an object appears in the function position in a method call. So the line above can be shortened to:

```
scala> abcde(2)        // rare in Scala
res12: Char = c
```

One reason why random element selection is less popular for lists than for arrays is that `xs(n)` takes time proportional to the index n. In fact, `apply` is simply defined by a combination of `drop` and `head`:

$$\text{xs apply n} \quad equals \quad \text{(xs drop n).head}$$

This definition also makes clear that list indices range from 0 up to the length of the list minus one, the same as for arrays. The `indices` method returns a list consisting of all valid indices of a given list:

```
scala> abcde.indices
res13: scala.collection.immutable.Range
  = Range(0, 1, 2, 3, 4)
```

### Flattening a list of lists: `flatten`

The `flatten` method takes a list of lists and flattens it out to a single list:

```
scala> List(List(1, 2), List(3), List(), List(4, 5)).flatten
res14: List[Int] = List(1, 2, 3, 4, 5)
scala> fruit.map(_.toCharArray).flatten
res15: List[Char] = List(a, p, p, l, e, s, o, r, a, n, g, e,
s, p, e, a, r, s)
```

It can only be applied to lists whose elements are all lists. Trying to flatten any other list will give a compilation error:

```
scala> List(1, 2, 3).flatten
<console>:8: error: No implicit view available from Int =>
scala.collection.GenTraversableOnce[B].
           List(1, 2, 3).flatten
                    ^
```

## Zipping lists: zip and unzip

The zip operation takes two lists and forms a list of pairs:

```
scala> abcde.indices zip abcde
res17: scala.collection.immutable.IndexedSeq[(Int, Char)] =
Vector((0,a), (1,b), (2,c), (3,d), (4,e))
```

If the two lists are of different length, any unmatched elements are dropped:

```
scala> val zipped = abcde zip List(1, 2, 3)
zipped: List[(Char, Int)] = List((a,1), (b,2), (c,3))
```

A useful special case is to zip a list with its index. This is done most efficiently with the zipWithIndex method, which pairs every element of a list with the position where it appears in the list.

```
scala> abcde.zipWithIndex
res18: List[(Char, Int)] = List((a,0), (b,1), (c,2), (d,3),
    (e,4))
```

Any list of tuples can also be changed back to a tuple of lists by using the unzip method:

```
scala> zipped.unzip
res19: (List[Char], List[Int])
  = (List(a, b, c),List(1, 2, 3))
```

The zip and unzip methods provide one way to operate on multiple lists together. See Section 16.9 for a more concise way to do this.

**Displaying lists: `toString` and `mkString`**

The `toString` operation returns the canonical string representation of a list:

```
scala> abcde.toString
res20: String = List(a, b, c, d, e)
```

If you want a different representation you can use the `mkString` method. The operation `xs mkString (pre, sep, post)` involves four operands: the list `xs` to be displayed, a prefix string `pre` to be displayed in front of all elements, a separator string `sep` to be displayed between successive elements, and a postfix string `post` to be displayed at the end.

The result of the operation is the string:

$$pre + xs(0) + sep + \ldots + sep + xs(xs.length - 1) + post$$

The `mkString` method has two overloaded variants that let you drop some or all of its arguments. The first variant only takes a separator string:

$$xs \, mkString \, sep \quad equals \quad xs \, mkString \, ("", sep, "")$$

The second variant lets you omit all arguments:

$$xs.mkString \quad equals \quad xs \, mkString \, ""$$

Here are some examples:

```
scala> abcde mkString ("[", ",", "]")
res21: String = [a,b,c,d,e]

scala> abcde mkString ""
res22: String = abcde

scala> abcde.mkString
res23: String = abcde

scala> abcde mkString ("List(", ", ", ")")
res24: String = List(a, b, c, d, e)
```

There are also variants of the `mkString` methods called `addString` which append the constructed string to a `StringBuilder` object,[6] rather than returning them as a result:

---

[6]This is class `scala.StringBuilder`, not `java.lang.StringBuilder`.

```
scala> val buf = new StringBuilder
buf: StringBuilder =

scala> abcde addString (buf, "(", ";", ")")
res25: StringBuilder = (a;b;c;d;e)
```

The `mkString` and `addString` methods are inherited from `List`'s super trait `Traversable`, so they are applicable to all other collections as well.

**Converting lists: `iterator`, `toArray`, `copyToArray`**

To convert data between the flat world of arrays and the recursive world of lists, you can use method `toArray` in class `List` and `toList` in class `Array`:

```
scala> val arr = abcde.toArray
arr: Array[Char] = Array(a, b, c, d, e)

scala> arr.toList
res26: List[Char] = List(a, b, c, d, e)
```

There's also a method `copyToArray`, which copies list elements to successive array positions within some destination array. The operation:

```
xs copyToArray (arr, start)
```

copies all elements of the list `xs` to the array `arr`, beginning with position `start`. You must ensure that the destination array `arr` is large enough to hold the list in full. Here's an example:

```
scala> val arr2 = new Array[Int](10)
arr2: Array[Int] = Array(0, 0, 0, 0, 0, 0, 0, 0, 0, 0)

scala> List(1, 2, 3) copyToArray (arr2, 3)

scala> arr2
res28: Array[Int] = Array(0, 0, 0, 1, 2, 3, 0, 0, 0, 0)
```

Finally, if you need to access list elements via an iterator, you can use the `iterator` method:

```
scala> val it = abcde.iterator
it: Iterator[Char] = non-empty iterator
```

```
scala> it.next
res29: Char = a

scala> it.next
res30: Char = b
```

### Example: Merge sort

The insertion sort presented earlier is concise to write, but it is not very efficient. Its average complexity is proportional to the square of the length of the input list. A more efficient algorithm is *merge sort*.

> **The fast track**
>
> This example provides another illustration of the divide and conquer principle and currying, as well as a useful discussion of algorithmic complexity. If you prefer to move a bit faster on your first pass through this book, however, you can safely skip to Section 16.7.

Merge sort works as follows: First, if the list has zero or one elements, it is already sorted, so the list can be returned unchanged. Longer lists are split into two sub-lists, each containing about half the elements of the original list. Each sub-list is sorted by a recursive call to the sort function, and the resulting two sorted lists are then combined in a merge operation.

For a general implementation of merge sort, you want to leave open the type of list elements to be sorted and the function to be used for the comparison of elements. You obtain a function of maximal generality by passing these two items as parameters. This leads to the implementation shown in Listing 16.1.

The complexity of `msort` is order $(n \, log(n))$, where $n$ is the length of the input list. To see why, note that splitting a list in two and merging two sorted lists each take time proportional to the length of the argument list(s). Each recursive call of `msort` halves the number of elements in its input, so there are about $log(n)$ consecutive recursive calls until the base case of lists of length 1 is reached. However, for longer lists each call spawns off two further calls. Adding everything up, we obtain at each of the $log(n)$ call levels, every element of the original lists takes part in one split operation and one merge operation.

Hence, every call level has a total cost proportional to $n$. Since there are $log(n)$ call levels, we obtain an overall cost proportional to $n \, log(n)$. That

```
def msort[T](less: (T, T) => Boolean)
    (xs: List[T]): List[T] = {

  def merge(xs: List[T], ys: List[T]): List[T] =
    (xs, ys) match {
      case (Nil, _) => ys
      case (_, Nil) => xs
      case (x :: xs1, y :: ys1) =>
        if (less(x, y)) x :: merge(xs1, ys)
        else y :: merge(xs, ys1)
    }

  val n = xs.length / 2
  if (n == 0) xs
  else {
    val (ys, zs) = xs splitAt n
    merge(msort(less)(ys), msort(less)(zs))
  }
}
```

Listing 16.1 · A merge sort function for Lists.

cost does not depend on the initial distribution of elements in the list, so the worst case cost is the same as the average case cost. This property makes merge sort an attractive algorithm for sorting lists.

Here is an example of how msort is used:

```
scala> msort((x: Int, y: Int) => x < y)(List(5, 7, 1, 3))
res31: List[Int] = List(1, 3, 5, 7)
```

The msort function is a classical example of the currying concept discussed in Section 9.3. Currying makes it easy to specialize the function for particular comparison functions. Here's an example:

```
scala> val intSort = msort((x: Int, y: Int) => x < y) _
intSort: List[Int] => List[Int] = <function1>
```

The intSort variable refers to a function that takes a list of integers and sorts them in numerical order. As described in Section 8.6, an underscore stands for a missing argument list. In this case, the missing argument is the list that should be sorted. As another example, here's how you could define a function that sorts a list of integers in reverse numerical order:

```
scala> val reverseIntSort = msort((x: Int, y: Int) => x > y) _
reverseIntSort: (List[Int]) => List[Int] = <function>
```

Because you provided the comparison function already via currying, you now need only provide the list to sort when you invoke the intSort or reverseIntSort functions. Here are some examples:

```
scala> val mixedInts = List(4, 1, 9, 0, 5, 8, 3, 6, 2, 7)
mixedInts: List[Int] = List(4, 1, 9, 0, 5, 8, 3, 6, 2, 7)

scala> intSort(mixedInts)
res0: List[Int] = List(0, 1, 2, 3, 4, 5, 6, 7, 8, 9)

scala> reverseIntSort(mixedInts)
res1: List[Int] = List(9, 8, 7, 6, 5, 4, 3, 2, 1, 0)
```

## 16.7   Higher-order methods on class List

Many operations over lists have a similar structure. Several patterns appear time and time again. Some examples are: transforming every element of a list in some way, verifying whether a property holds for all elements of a list, extracting from a list elements satisfying a certain criterion, or combining the elements of a list using some operator. In Java, such patterns would usually be expressed by idiomatic combinations of for or while loops. In Scala, they can be expressed more concisely and directly using higher-order operators,[7] which are implemented as methods in class List. These higher-order operators are discussed in this section.

### Mapping over lists: map, flatMap and foreach

The operation xs map f takes as operands a list xs of type List[T] and a function f of type T => U. It returns the list that results from applying the function f to each list element in xs. For instance:

---

[7]By *higher-order operators*, we mean higher-order functions used in operator notation. As mentioned in Section 9.1, a function is "higher-order" if it takes one or more other functions as a parameters.

```
scala> List(1, 2, 3) map (_ + 1)
res32: List[Int] = List(2, 3, 4)

scala> val words = List("the", "quick", "brown", "fox")
words: List[String] = List(the, quick, brown, fox)

scala> words map (_.length)
res33: List[Int] = List(3, 5, 5, 3)

scala> words map (_.toList.reverse.mkString)
res34: List[String] = List(eht, kciuq, nworb, xof)
```

The flatMap operator is similar to map, but it takes a function returning a list of elements as its right operand. It applies the function to each list element and returns the concatenation of all function results. The difference between map and flatMap is illustrated in the following example:

```
scala> words map (_.toList)
res35: List[List[Char]] = List(List(t, h, e), List(q, u, i,
    c, k), List(b, r, o, w, n), List(f, o, x))

scala> words flatMap (_.toList)
res36: List[Char] = List(t, h, e, q, u, i, c, k, b, r, o, w,
    n, f, o, x)
```

You see that where map returns a list of lists, flatMap returns a single list in which all element lists are concatenated.

The differences and interplay between map and flatMap are also demonstrated by the following expression, which constructs a list of all pairs $(i, j)$ such that $1 \le j < i < 5$:

```
scala> List.range(1, 5) flatMap (
          i => List.range(1, i) map (j => (i, j))
        )
res37: List[(Int, Int)] = List((2,1), (3,1), (3,2), (4,1),
    (4,2), (4,3))
```

List.range is a utility method that creates a list of all integers in some range. It is used twice in this example: once to generate a list of integers from 1 (including) until 5 (excluding), and a second time to generate a list of integers from 1 until $i$, for each value of $i$ taken from the first list. The map in this expression generates a list of tuples $(i, j)$ where $j < i$. The outer flatMap

325

in this example generates this list for each i between 1 and 5, and then concatenates all the results. Alternatively, the same list can be constructed with a for expression:

```
for (i <- List.range(1, 5); j <- List.range(1, i)) yield (i, j)
```

You'll learn more about the interplay of for expressions and list operations in Chapter 23.

The third map-like operation is foreach. Unlike map and flatMap, however, foreach takes a procedure (a function with result type Unit) as right operand. It simply applies the procedure to each list element. The result of the operation itself is again Unit; no list of results is assembled. As an example, here is a concise way of summing up all numbers in a list:

```
scala> var sum = 0
sum: Int = 0

scala> List(1, 2, 3, 4, 5) foreach (sum += _)

scala> sum
res39: Int = 15
```

### Filtering lists: filter, partition, find, takeWhile, dropWhile, and span

The operation "xs filter p" takes as operands a list xs of type List[T] and a predicate function p of type T => Boolean. It yields the list of all elements x in xs for which p(x) is true. For instance:

```
scala> List(1, 2, 3, 4, 5) filter (_ % 2 == 0)
res40: List[Int] = List(2, 4)

scala> words filter (_.length == 3)
res41: List[String] = List(the, fox)
```

The partition method is like filter but returns a pair of lists. One list contains all elements for which the predicate is true, while the other contains all elements for which the predicate is false. It is defined by the equality:

xs partition p    *equals*    (xs filter p, xs filter (!p(_)))

Here's an example:

```
scala> List(1, 2, 3, 4, 5) partition (_ % 2 == 0)
res42: (List[Int], List[Int]) = (List(2, 4),List(1, 3, 5))
```

The find method is also similar to filter, but it returns the first element
satisfying a given predicate, rather than all such elements. The operation
xs find p takes a list xs and a predicate p as operands. It returns an optional
value. If there is an element x in xs for which p(x) is true, Some(x) is
returned. Otherwise, p is false for all elements, and None is returned. Here
are some examples:

```
scala>  List(1, 2, 3, 4, 5) find (_ % 2 == 0)
res43: Option[Int] = Some(2)

scala>  List(1, 2, 3, 4, 5) find (_ <= 0)
res44: Option[Int] = None
```

The takeWhile and dropWhile operators also take a predicate as their right
operand. The operation xs takeWhile p takes the longest prefix of list xs
such that every element in the prefix satisfies p. Analogously, the operation
xs dropWhile p removes the longest prefix from list xs such that every
element in the prefix satisfies p. Here are some examples:

```
scala> List(1, 2, 3, -4, 5) takeWhile (_ > 0)
res45: List[Int] = List(1, 2, 3)

scala> words dropWhile (_ startsWith "t")
res46: List[String] = List(quick, brown, fox)
```

The span method combines takeWhile and dropWhile in one operation,
just like splitAt combines take and drop. It returns a pair of two lists,
defined by the equality:

$$xs \; span \; p \quad equals \quad (xs \; takeWhile \; p, \; xs \; dropWhile \; p)$$

Like splitAt, span avoids traversing the list xs twice:

```
scala> List(1, 2, 3, -4, 5) span (_ > 0)
res47: (List[Int], List[Int]) = (List(1, 2, 3),List(-4, 5))
```

**Predicates over lists: `forall` and `exists`**

The operation `xs forall p` takes as arguments a list `xs` and a predicate `p`. Its result is `true` if all elements in the list satisfy `p`. Conversely, the operation `xs exists p` returns `true` if there is an element in `xs` that satisfies the predicate `p`. For instance, to find out whether a matrix represented as a list of lists has a row with only zeroes as elements:

```
scala> def hasZeroRow(m: List[List[Int]]) =
         m exists (row => row forall (_ == 0))
hasZeroRow: (m: List[List[Int]])Boolean

scala> hasZeroRow(diag3)
res48: Boolean = false
```

**Folding lists: `/:` and `:\`**

Another common kind of operation combines the elements of a list with some operator. For instance:

$$\text{sum}(\text{List}(a, b, c)) \quad \textit{equals} \quad 0 + a + b + c$$

This is a special instance of a fold operation:

```
scala> def sum(xs: List[Int]): Int = (0 /: xs) (_ + _)
sum: (xs: List[Int])Int
```

Similarly:

$$\text{product}(\text{List}(a, b, c)) \quad \textit{equals} \quad 1 * a * b * c$$

is a special instance of this fold operation:

```
scala> def product(xs: List[Int]): Int = (1 /: xs) (_ * _)
product: (xs: List[Int])Int
```

A *fold left* operation "`(z /: xs) (op)`" involves three objects: a start value z, a list xs, and a binary operation op. The result of the fold is op applied between successive elements of the list prefixed by z. For instance:

$$(z \mathbin{/:} \text{List}(a, b, c)) \, (op) \quad \textit{equals} \quad op(op(op(z, a), b), c)$$

Or, graphically:

Here's another example that illustrates how /: is used. To concatenate all words in a list of strings with spaces between them and in front, you can write this:

```
scala>  ("" /: words) (_ + " " + _)
res49: String = " the quick brown fox"
```

This gives an extra space at the beginning. To remove the space, you can use this slight variation:

```
scala> (words.head /: words.tail)  (_ + " " + _)
res50: String = the quick brown fox
```

The /: operator produces left-leaning operation trees (its syntax with the slash rising forward is intended to be a reflection of that). The operator has :\ as an analog that produces right-leaning trees. For instance:

$$(\text{List}(a, b, c) :\backslash z) (op) \quad \textit{equals} \quad op(a, op(b, op(c, z)))$$

Or, graphically:

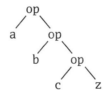

The :\ operator is pronounced *fold right*. It involves the same three operands as fold left, but the first two appear in reversed order: The first operand is the list to fold, the second is the start value.

For associative operations, fold left and fold right are equivalent, but there might be a difference in efficiency. Consider for instance an operation corresponding to the flatten method, which concatenates all elements in a list of lists. This could be implemented with either fold left or fold right:

```
def flattenLeft[T](xss: List[List[T]]) =
    (List[T]() /: xss) (_ ::: _)
def flattenRight[T](xss: List[List[T]]) =
    (xss :\ List[T]()) (_ ::: _)
```

Because list concatenation, xs ::: ys, takes time proportional to its first argument xs, the implementation in terms of fold right in flattenRight is more efficient than the fold left implementation in flattenLeft. The problem is that flattenLeft(xss) copies the first element list xss.head $n - 1$ times, where $n$ is the length of the list xss.

Note that both versions of flatten require a type annotation on the empty list that is the start value of the fold. This is due to a limitation in Scala's type inferencer, which fails to infer the correct type of the list automatically. If you try to leave out the annotation, you get the following:

```
scala> def flattenRight[T](xss: List[List[T]]) =
            (xss :\ List()) (_ ::: _)
<console>:8: error: type mismatch;
  found   : List[T]
  required: List[Nothing]
            (xss :\ List()) (_ ::: _)
                                ^
```

To find out why the type inferencer goes wrong, you'll need to know about the types of the fold methods and how they are implemented. More on this in Section 16.10.

Lastly, although the /: and :\ operators have the advantage that the direction of the slash resembles the graphical depiction of their respective left or right-leaning trees, and the associativity of the colon character places the start value in the same position in the expression as it is in the tree, some may find the resulting code less than intuitive. If you prefer, you can use the methods named foldLeft and foldRight instead, which are also defined on class List.

### Example: List reversal using fold

Earlier in the chapter you saw an implementation of method reverse, named rev, whose running time was quadratic in the length of the list to be reversed. Here is a different implementation of reverse that has linear cost. The idea is to use a fold left operation based on the following scheme:

```
def reverseLeft[T](xs: List[T]) = (startvalue /: xs)(operation)
```

What remains is to fill in the *startvalue* and *operation* parts. In fact, you can try to deduce these parts from some simple examples. To deduce the correct value of *startvalue*, you can start with the smallest possible list, List(), and calculate as follows:

List()
  *equals* (by the properties of reverseLeft)
reverseLeft(List())
  *equals* (by the template for reverseLeft)
(*startvalue* /: List())(*operation*)
  *equals* (by the definition of /:)
*startvalue*

Hence, *startvalue* must be List(). To deduce the second operand, you can pick the next smallest list as an example case. You know already that *startvalue* is List(), so you can calculate as follows:

List(x)
  *equals* (by the properties of reverseLeft)
reverseLeft(List(x))
  *equals* (by the template for reverseLeft, with *startvalue* = List())
(List() /: List(x)) (*operation*)
  *equals* (by the definition of /:)
*operation*(List(), x)

Hence, *operation*(List(), x) equals List(x), which can also be written as x :: List(). This suggests taking as *operation* the :: operator with its operands exchanged. (This operation is sometimes called "snoc," in reference to ::, which is called cons.) We arrive then at the following implementation for reverseLeft:

```
def reverseLeft[T](xs: List[T]) =
  (List[T]() /: xs) {(ys, y) => y :: ys}
```

Again, the type annotation in List[T]() is needed to make the type inferencer work. If you analyze the complexity of reverseLeft, you'll find that it applies a constant-time operation ("snoc") *n* times, where *n* is the length of the argument list. Thus, the complexity of reverseLeft is linear.

**Sorting lists: sortWith**

The operation xs sortWith before, where "xs" is a list and "before" is a function that can be used to compare two elements, sorts the elements of list xs. The expression x before y should return true if x should come before y in the intended ordering for the sort. For instance:

```
scala> List(1, -3, 4, 2, 6) sortWith (_ < _)
res51: List[Int] = List(-3, 1, 2, 4, 6)

scala> words sortWith (_.length > _.length)
res52: List[String] = List(quick, brown, the, fox)
```

Note that sortWith performs a merge sort similar to the msort algorithm shown in the last section. But sortWith is a method of class List, whereas msort is defined outside lists.

## 16.8 Methods of the List object

So far, all operations you have seen in this chapter are implemented as methods of class List, so you invoke them on individual list objects. There are also a number of methods in the globally accessible object scala.List, which is the companion object of class List. Some of these operations are factory methods that create lists. Others are operations that work on lists of some specific shape. Both kinds of methods will be presented in this section.

**Creating lists from their elements: List.apply**

You've already seen on several occasions list literals such as List(1, 2, 3). There's nothing special about their syntax. A literal like List(1, 2, 3) is simply the application of the object List to the elements 1, 2, 3. That is, it is equivalent to List.apply(1, 2, 3):

```
scala> List.apply(1, 2, 3)
res53: List[Int] = List(1, 2, 3)
```

### Creating a range of numbers: List.range

The range method, which you saw briefly earlier in the discussion of map and flatmap, creates a list consisting of a range of numbers. Its simplest form is List.range(from, until), which creates a list of all numbers starting at from and going up to until minus one. So the end value, until, does not form part of the range.

There's also a version of range that takes a step value as third parameter. This operation will yield list elements that are step values apart, starting at from. The step can be positive or negative:

```scala
scala> List.range(1, 5)
res54: List[Int] = List(1, 2, 3, 4)

scala> List.range(1, 9, 2)
res55: List[Int] = List(1, 3, 5, 7)

scala> List.range(9, 1, -3)
res56: List[Int] = List(9, 6, 3)
```

### Creating uniform lists: List.fill

The fill method creates a list consisting of zero or more copies of the same element. It takes two parameters: the length of the list to be created, and the element to be repeated. Each parameter is given in a separate list:

```scala
scala> List.fill(5)('a')
res57: List[Char] = List(a, a, a, a, a)

scala> List.fill(3)("hello")
res58: List[String] = List(hello, hello, hello)
```

If fill is given more than two arguments, then it will make multi-dimensional lists. That is, it will make lists of lists, lists of lists of lists, *etc.* The additional arguments go in the first argument list.

```scala
scala> List.fill(2, 3)('b')
res59: List[List[Char]] = List(List(b, b, b), List(b, b, b))
```

### Tabulating a function: `List.tabulate`

The `tabulate` method creates a list whose elements are computed according to a supplied function. Its arguments are just like those of `List.fill`: the first argument list gives the dimensions of the list to create, and the second describes the elements of the list. The only difference is that instead of the elements being fixed, they are computed from a function:

```
scala> val squares = List.tabulate(5)(n => n * n)
squares: List[Int] = List(0, 1, 4, 9, 16)
scala> val multiplication = List.tabulate(5,5)(_ * _)
multiplication: List[List[Int]] = List(List(0, 0, 0, 0, 0),
    List(0, 1, 2, 3, 4), List(0, 2, 4, 6, 8),
    List(0, 3, 6, 9, 12), List(0, 4, 8, 12, 16))
```

### Concatenating multiple lists: `List.concat`

The `concat` method concatenates a number of element lists. The lists to be concatenated are supplied as direct arguments to `concat`:

```
scala> List.concat(List('a', 'b'), List('c'))
res60: List[Char] = List(a, b, c)
```

```
scala> List.concat(List(), List('b'), List('c'))
res61: List[Char] = List(b, c)
```

```
scala> List.concat()
res62: List[Nothing] = List()
```

## 16.9   Processing multiple lists together

The `zipped` method on tuples generalizes several common operations to work on multiple lists instead of just one. One such operation is `map`. The map method for two zipped lists maps pairs of elements rather than individual elements. One pair is for the first element of each list, another pair is for the second element of each list, and so on—as many pairs as the lists are long. Here is an example of its use:

```
scala> (List(10, 20), List(3, 4, 5)).zipped.map(_ * _)
res63: List[Int] = List(30, 80)
```

Notice that the third element of the second list is discarded. The `zipped` method zips up only as many elements as appear in all the lists together. Any extra elements on the end are discarded.

There are also zipped analogs to `exists` and `forall`. They are just like the single-list versions of those methods except they operate on elements from multiple lists instead of just one:

```
scala> (List("abc", "de"), List(3, 2)).zipped.
          forall(_.length == _)
res64: Boolean = true
scala> (List("abc", "de"), List(3, 2)).zipped.
          exists(_.length != _)
res65: Boolean = false
```

> **The fast track**
>
> In the next (and final) section of this chapter, we provide insight into Scala's type inference algorithm. If you're not interested in such details right now, you can skip the entire section and go straight to the conclusion on page 339.

## 16.10  Understanding Scala's type inference algorithm

One difference between the previous uses of `sortWith` and `msort` concerns the admissible syntactic forms of the comparison function.

Compare:

```
scala> msort((x: Char, y: Char) => x > y)(abcde)
res66: List[Char] = List(e, d, c, b, a)
```

with:

```
scala> abcde sortWith (_ > _)
res67: List[Char] = List(e, d, c, b, a)
```

The two expressions are equivalent, but the first uses a longer form of comparison function with named parameters and explicit types. The second uses the concise form, (_ > _), where named parameters are replaced by underscores. Of course, you could also use the first, longer form of comparison with `sortWith`.

However, the short form cannot be used with msort.

```
scala> msort(_ > _)(abcde)
<console>:12: error: missing parameter type for expanded
function ((x$1, x$2) => x$1.$greater(x$2))
        msort(_ > _)(abcde)
              ^
```

To understand why, you need to know some details of Scala's type inference algorithm. Type inference in Scala is flow based. In a method application m(args), the inferencer first checks whether the method m has a known type. If it does, that type is used to infer the expected type of the arguments. For instance, in abcde.sortWith(_ > _), the type of abcde is List[Char]. Hence, sortWith is known to be a method that takes an argument of type (Char, Char) => Boolean and produces a result of type List[Char]. Since the parameter types of the function arguments are known, they need not be written explicitly. With what it knows about sortWith, the inferencer can deduce that (_ > _) should expand to ((x: Char, y: Char) => x > y) where x and y are some arbitrary fresh names.

Now consider the second case, msort(_ > _)(abcde). The type of msort is a curried, polymorphic method type that takes an argument of type (T, T) => Boolean to a function from List[T] to List[T] *where T is some as-yet unknown type*. The msort method needs to be instantiated with a type parameter before it can be applied to its arguments.

Because the precise instance type of msort in the application is not yet known, it cannot be used to infer the type of its first argument. The type inferencer changes its strategy in this case; it first type checks method arguments to determine the proper instance type of the method. However, when tasked to type check the short-hand function literal, (_ > _), it fails because it has no information about the types of the implicit function parameters that are indicated by underscores.

One way to resolve the problem is to pass an explicit type parameter to msort, as in:

```
scala> msort[Char](_ > _)(abcde)
res68: List[Char] = List(e, d, c, b, a)
```

Because the correct instance type of msort is now known, it can be used to infer the type of the arguments. Another possible solution is to rewrite the msort method so that its parameters are swapped:

```
def msortSwapped[T](xs: List[T])(less:
    (T, T) => Boolean): List[T] = {

  // same implementation as msort,
  // but with arguments swapped
}
```

Now type inference would succeed:

```
scala> msortSwapped(abcde)(_ > _)
res69: List[Char] = List(e, d, c, b, a)
```

What has happened is that the inferencer used the known type of the first parameter abcde to determine the type parameter of msortSwapped. Once the precise type of msortSwapped was known, it could be used in turn to infer the type of the second parameter, (_ > _).

Generally, when tasked to infer the type parameters of a polymorphic method, the type inferencer consults the types of all value arguments in the first parameter list but no arguments beyond that. Since msortSwapped is a curried method with two parameter lists, the second argument (*i.e.*, the function value) did not need to be consulted to determine the type parameter of the method.

This inference scheme suggests the following library design principle: When designing a polymorphic method that takes some non-function arguments and a function argument, place the function argument last in a curried parameter list on its own. That way, the method's correct instance type can be inferred from the non-function arguments, and that type can in turn be used to type check the function argument. The net effect is that users of the method will be able to give less type information and write function literals in more compact ways.

Now to the more complicated case of a *fold* operation. Why is there the need for an explicit type parameter in an expression like the body of the flattenRight method shown on page 330?

```
(xss :\ List[T]()) (_ ::: _)
```

The type of the fold-right operation is polymorphic in two type variables. Given an expression:

```
(xs :\ z) (op)
```

The type of xs must be a list of some arbitrary type A, say xs: List[A]. The start value z can be of some other type B. The operation op must then take two arguments of type A and B, and return a result of type B, *i.e.*, op: (A, B) => B. Because the type of z is not related to the type of the list xs, type inference has no context information for z.

Now consider the expression in the erroneous version of flattenRight, also shown on page 330:

```
(xss :\ List()) (_ ::: _)  // this won't compile
```

The start value z in this fold is an empty list, List(), so without additional type information its type is inferred to be a List[Nothing]. Hence, the inferencer will infer that the B type of the fold is List[Nothing]. Therefore, the operation (_ ::: _) of the fold is expected to be of the following type:

```
(List[T], List[Nothing]) => List[Nothing]
```

This is indeed a possible type for the operation in that fold but it is not a very useful one! It says that the operation always takes an empty list as second argument and always produces an empty list as result.

In other words, the type inference settled too early on a type for List(); it should have waited until it had seen the type of the operation op. So the (otherwise very useful) rule to only consider the first argument section in a curried method application for determining the method's type is at the root of the problem here. On the other hand, even if that rule were relaxed, the inferencer still could not come up with a type for op because its parameter types are not given. Hence, there is a Catch-22 situation that can only be resolved by an explicit type annotation from the programmer.

This example highlights some limitations of the local, flow-based type inference scheme of Scala. It is not present in the more global Hindley-Milner style of type inference used in functional languages, such as ML or Haskell. However, Scala's local type inference deals much more gracefully with object-oriented subtyping than the Hindley-Milner style does. Fortunately, the limitations show up only in some corner cases, and are usually easily fixed by adding an explicit type annotation.

Adding type annotations is also a useful debugging technique when you get confused by type error messages related to polymorphic methods. If you are unsure what caused a particular type error, just add some type arguments or other type annotations, which you think are correct. Then you should be able to quickly see where the real problem is.

## 16.11 Conclusion

Now you have seen many ways to work with lists. You have seen the basic operations like `head` and `tail`, the first-order operations like `reverse`, the higher-order operations like `map`, and the utility methods in the `List` object. Along the way, you learned a bit about how Scala's type inference works.

Lists are a real work horse in Scala, so you will benefit from knowing how to use them. For that reason, this chapter has delved deeply into how to use lists. Lists are just one kind of collection that Scala supports, however. The next chapter is broad, rather than deep, and shows you how to use a variety of Scala's collection types.

# Chapter 17

# Working with Other Collections

Scala has a rich collection library. This chapter gives you a tour of the most commonly used collection types and operations, showing just the parts you will use most frequently. Chapter 24 will provide a more comprehensive tour of what's available, and Chapter 25 will show how Scala's composition constructs are used to provide such a rich API.

## 17.1  Sequences

Sequence types let you work with groups of data lined up in order. Because the elements are ordered, you can ask for the first element, second element, 103rd element, and so on. In this section, we'll give you a quick tour of the most important sequences.

### Lists

Perhaps the most important sequence type to know about is class List, the immutable linked-list described in detail in the previous chapter. Lists support fast addition and removal of items to the beginning of the list, but they do not provide fast access to arbitrary indexes because the implementation must iterate through the list linearly.

This combination of features might sound odd, but they hit a sweet spot that works well for many algorithms. The fast addition and removal of initial elements means that pattern matching works well, as described in Chapter 15. The immutability of lists helps you develop correct, efficient algorithms because you never need to make copies of a list.

Here's a short example showing how to initialize a list, and access its head and tail:

```
scala> val colors = List("red", "blue", "green")
colors: List[String] = List(red, blue, green)

scala> colors.head
res0: String = red

scala> colors.tail
res1: List[String] = List(blue, green)
```

For a refresher on lists, see Step 8 in Chapter 3. You can find details on using lists in Chapter 16. Lists will also be discussed in Chapter 22, which provides insight into how lists are implemented in Scala.

**Arrays**

Arrays allow you to hold a sequence of elements and efficiently access an element at an arbitrary position, either to get or update the element, with a zero-based index. Here's how you create an array whose size you know, but for which you don't yet know the element values:

```
scala> val fiveInts = new Array[Int](5)
fiveInts: Array[Int] = Array(0, 0, 0, 0, 0)
```

Here's how you initialize an array when you do know the element values:

```
scala> val fiveToOne = Array(5, 4, 3, 2, 1)
fiveToOne: Array[Int] = Array(5, 4, 3, 2, 1)
```

As mentioned previously, arrays are accessed in Scala by placing an index in parentheses, not square brackets as in Java. Here's an example of both accessing and updating an array element:

```
scala> fiveInts(0) = fiveToOne(4)

scala> fiveInts
res3: Array[Int] = Array(1, 0, 0, 0, 0)
```

Scala arrays are represented in the same way as Java arrays. So, you can seamlessly use existing Java methods that return arrays.[1]

---

[1] The difference in variance of Scala's and Java's arrays—*i.e.*, whether Array[String] is a subtype of Array[AnyRef]—will be discussed in Section 19.3.

You have seen arrays in action many times in previous chapters. The basics are in Step 7 in Chapter 3. Several examples of iterating through the elements of an array with a for expression are shown in Section 7.3. Arrays also figure prominently in the two-dimensional layout library of Chapter 10.

## List buffers

Class List provides fast access to the head of the list, but not the end. Thus, when you need to build a list by appending to the end, consider building the list backwards by prepending elements to the front. Then when you're done, call reverse to get the elements in the order you need.

Another alternative, which avoids the reverse operation, is to use a ListBuffer. A ListBuffer is a mutable object (contained in package scala.collection.mutable), which can help you build lists more efficiently when you need to append. ListBuffer provides constant time append and prepend operations. You append elements with the += operator, and prepend them with the +=: operator. When you're done building, you can obtain a List by invoking toList on the ListBuffer. Here's an example:

```
scala> import scala.collection.mutable.ListBuffer
import scala.collection.mutable.ListBuffer

scala> val buf = new ListBuffer[Int]
buf: scala.collection.mutable.ListBuffer[Int] = ListBuffer()

scala> buf += 1
res4: buf.type = ListBuffer(1)

scala> buf += 2
res5: buf.type = ListBuffer(1, 2)

scala> buf
res6: scala.collection.mutable.ListBuffer[Int] =
    ListBuffer(1, 2)

scala> 3 +=: buf
res7: buf.type = ListBuffer(3, 1, 2)

scala> buf.toList
res8: List[Int] = List(3, 1, 2)
```

Another reason to use ListBuffer instead of List is to prevent the potential for stack overflow. If you can build a list in the desired order by prepending, but the recursive algorithm that would be required is not tail recursive, you can use a for expression or while loop and a ListBuffer instead. You'll see ListBuffer being used in this way in Section 22.2.

**Array buffers**

An ArrayBuffer is like an array, except that you can additionally add and remove elements from the beginning and end of the sequence. All Array operations are available, though they are a little slower due to a layer of wrapping in the implementation. The new addition and removal operations are constant time on average, but occasionally require linear time due to the implementation needing to allocate a new array to hold the buffer's contents.

To use an ArrayBuffer, you must first import it from the mutable collections package:

```
scala> import scala.collection.mutable.ArrayBuffer
import scala.collection.mutable.ArrayBuffer
```

When you create an ArrayBuffer, you must specify a type parameter, but you don't need to specify a length. The ArrayBuffer will adjust the allocated space automatically as needed:

```
scala> val buf = new ArrayBuffer[Int]()
buf: scala.collection.mutable.ArrayBuffer[Int] =
ArrayBuffer()
```

You can append to an ArrayBuffer using the += method:

```
scala> buf += 12
res9: buf.type = ArrayBuffer(12)

scala> buf += 15
res10: buf.type = ArrayBuffer(12, 15)

scala> buf
res11: scala.collection.mutable.ArrayBuffer[Int] =
    ArrayBuffer(12, 15)
```

All the normal array methods are available. For example, you can ask an ArrayBuffer its length or you can retrieve an element by its index:

```
scala> buf.length
res12: Int = 2

scala> buf(0)
res13: Int = 12
```

**Strings (via StringOps)**

One other sequence to be aware of is StringOps, which implements many sequence methods. Because Predef has an implicit conversion from String to StringOps, you can treat any string like a sequence. Here's an example:

```
scala> def hasUpperCase(s: String) = s.exists(_.isUpper)
hasUpperCase: (s: String)Boolean

scala> hasUpperCase("Robert Frost")
res14: Boolean = true

scala> hasUpperCase("e e cummings")
res15: Boolean = false
```

In this example, the exists method is invoked on the string named s in the hasUpperCase method body. Because no method named "exists" is declared in class String itself, the Scala compiler will implicitly convert s to StringOps, which has the method. The exists method treats the string as a sequence of characters, and will return true if any of the characters are upper case.[2]

## 17.2 Sets and maps

You have already seen the basics of sets and maps in previous chapters, starting with Step 10 in Chapter 3. In this section, we'll offer more insight into their use and show you a few more examples.

As mentioned previously, the Scala collections library offers both mutable and immutable versions of sets and maps. The hierarchy for sets is shown in Figure 3.2 on page 48, and the hierarchy for maps is shown in Figure 3.3 on page 50. As these diagrams show, the simple names Set and Map are used by three traits each, residing in different packages.

---

[2]The code given on page 15 of Chapter 1 presents a similar example.

By default when you write "Set" or "Map" you get an immutable object. If you want the mutable variant, you need to do an explicit import. Scala gives you easier access to the immutable variants, as a gentle encouragement to prefer them over their mutable counterparts. The easy access is provided via the Predef object, which is implicitly imported into every Scala source file. Listing 17.1 shows the relevant definitions:

```
object Predef {
  type Map[A, +B] = collection.immutable.Map[A, B]
  type Set[A] = collection.immutable.Set[A]
  val Map = collection.immutable.Map
  val Set = collection.immutable.Set
  // ...
}
```

Listing 17.1 · Default map and set definitions in Predef.

The "type" keyword is used in Predef to define Set and Map as aliases for the longer fully qualified names of the immutable set and map traits.[3] The vals named Set and Map are initialized to refer to the singleton objects for the immutable Set and Map. So Map is the same as Predef.Map, which is defined to be the same as scala.collection.immutable.Map. This holds both for the Map type and Map object.

If you want to use both mutable and immutable sets or maps in the same source file, one approach is to import the name of the package that contains the mutable variants:

```
scala> import scala.collection.mutable
import scala.collection.mutable
```

You can continue to refer to the immutable set as Set, as before, but can now refer to the mutable set as mutable.Set. Here's an example:

```
scala> val mutaSet = mutable.Set(1, 2, 3)
mutaSet: scala.collection.mutable.Set[Int] = Set(1, 2, 3)
```

---

[3]The type keyword will be explained in more detail in Section 20.6.

**Using sets**

The key characteristic of sets is that they will ensure that at most one of each object, as determined by ==, will be contained in the set at any one time. As an example, we'll use a set to count the number of different words in a string.

The split method on String can separate a string into words, if you specify spaces and punctuation as word separators. The regular expression "[ !,.]+" will suffice: It indicates the string should be split at each place that one or more space and/or punctuation characters exist.

```
scala> val text = "See Spot run. Run, Spot. Run!"
text: String = See Spot run. Run, Spot. Run!

scala> val wordsArray = text.split("[ !,.]+")
wordsArray: Array[String]
  = Array(See, Spot, run, Run, Spot, Run)
```

To count the distinct words, you can convert them to the same case and then add them to a set. Because sets exclude duplicates, each distinct word will appear exactly one time in the set.

First, you can create an empty set using the empty method provided on the Set companion objects:

```
scala>  val words = mutable.Set.empty[String]
words: scala.collection.mutable.Set[String] = Set()
```

Then, just iterate through the words with a for expression, convert each word to lower case, and add it to the mutable set with the += operator:

```
scala> for (word <- wordsArray)
         words += word.toLowerCase

scala> words
res17: scala.collection.mutable.Set[String] =
    Set(see, run, spot)
```

Thus, the text contained exactly three distinct words: spot, run, and see. The most commonly used methods on both mutable and immutable sets are shown in Table 17.1.

Table 17.1 · Common operations for sets

| What it is | What it does |
|---|---|
| val nums = Set(1, 2, 3) | Creates an immutable set (nums.toString returns Set(1, 2, 3)) |
| nums + 5 | Adds an element (returns Set(1, 2, 3, 5)) |
| nums - 3 | Removes an element (returns Set(1, 2)) |
| nums ++ List(5, 6) | Adds multiple elements (returns Set(1, 2, 3, 5, 6)) |
| nums -- List(1, 2) | Removes multiple elements (returns Set(3)) |
| nums & Set(1, 3, 5, 7) | Takes the intersection of two sets (returns Set(1, 3)) |
| nums.size | Returns the size of the set (returns 3) |
| nums.contains(3) | Checks for inclusion (returns true) |
| import scala.collection.mutable | Makes the mutable collections easy to access |
| val words = mutable.Set.empty[String] | Creates an empty, mutable set (words.toString returns Set()) |
| words += "the" | Adds an element (words.toString returns Set(the)) |
| words -= "the" | Removes an element, if it exists (words.toString returns Set()) |
| words ++= List("do", "re", "mi") | Adds multiple elements (words.toString returns Set(do, re, mi)) |
| words --= List("do", "re") | Removes multiple elements (words.toString returns Set(mi)) |
| words.clear | Removes all elements (words.toString returns Set()) |

## Using maps

Maps let you associate a value with each element of a set. Using a map looks similar to using an array, except instead of indexing with integers counting from 0, you can use any kind of key. If you import the mutable package name, you can create an empty mutable map like this:

```
scala> val map = mutable.Map.empty[String, Int]
map: scala.collection.mutable.Map[String,Int] = Map()
```

Note that when you create a map, you must specify two types. The first type is for the *keys* of the map, the second for the *values*. In this case, the keys are strings and the values are integers. Setting entries in a map looks similar to setting entries in an array:

```
scala> map("hello") = 1

scala> map("there") = 2

scala> map
res20: scala.collection.mutable.Map[String,Int] =
    Map(hello -> 1, there -> 2)
```

Likewise, reading a map is similar to reading an array:

```
scala> map("hello")
res21: Int = 1
```

Putting it all together, here is a method that counts the number of times each word occurs in a string:

```
scala> def countWords(text: String) = {
         val counts = mutable.Map.empty[String, Int]
         for (rawWord <- text.split("[ ,!.]+")) {
           val word = rawWord.toLowerCase
           val oldCount =
             if (counts.contains(word)) counts(word)
             else 0
           counts += (word -> (oldCount + 1))
         }
         counts
       }
```

349

```
countWords: (text:
String)scala.collection.mutable.Map[String,Int]
```

```
scala> countWords("See Spot run! Run, Spot. Run!")
res22: scala.collection.mutable.Map[String,Int] =
    Map(spot -> 2, see -> 1, run -> 3)
```

Given these counts, you can see that this text talks a lot about running, but not so much about seeing.

The way this code works is that a mutable map, named `counts`, maps each word to the number of times it occurs in the text. For each word in the text, the word's old count is looked up, that count is incremented by one, and the new count is saved back into `counts`. Note the use of `contains` to check whether a word has been seen yet or not. If `counts.contains(word)` is not true, then the word has not yet been seen and zero is used for the count.

Many of the most commonly used methods on both mutable and immutable maps are shown in Table 17.2.

Table 17.2 · Common operations for maps

| What it is | What it does |
|---|---|
| val nums = Map("i" -> 1, "ii" -> 2) | Creates an immutable map (nums.toString returns Map(i -> 1, ii -> 2)) |
| nums + ("vi" -> 6) | Adds an entry (returns Map(i -> 1, ii -> 2, vi -> 6)) |
| nums - "ii" | Removes an entry (returns Map(i -> 1)) |
| nums ++ List("iii" -> 3, "v" -> 5) | Adds multiple entries (returns Map(i -> 1, ii -> 2, iii -> 3, v -> 5)) |
| nums -- List("i", "ii") | Removes multiple entries (returns Map()) |
| nums.size | Returns the size of the map (returns 2) |
| nums.contains("ii") | Checks for inclusion (returns true) |
| nums("ii") | Retrieves the value at a specified key (returns 2) |
| nums.keys | Returns the keys (returns an Iterable over the strings "i" and "ii") |

## Table 17.2 · continued

| | |
|---|---|
| `nums.keySet` | Returns the keys as a set (returns `Set(i, ii)`) |
| `nums.values` | Returns the values (returns an `Iterable` over the integers 1 and 2) |
| `nums.isEmpty` | Indicates whether the map is empty (returns false) |
| `import scala.collection.mutable` | Makes the mutable collections easy to access |
| `val words =`<br>`mutable.Map.empty[String, Int]` | Creates an empty, mutable map |
| `words += ("one" -> 1)` | Adds a map entry from `"one"` to 1 (`words.toString` returns `Map(one -> 1)`) |
| `words -= "one"` | Removes a map entry, if it exists (`words.toString` returns `Map()`) |
| `words ++= List("one" -> 1,`<br>`"two" -> 2, "three" -> 3)` | Adds multiple map entries (`words.toString` returns `Map(one -> 1, two -> 2, three -> 3)`) |
| `words --= List("one", "two")` | Removes multiple objects (`words.toString` returns `Map(three -> 3)`) |

### Default sets and maps

For most uses, the implementations of mutable and immutable sets and maps provided by the `Set()`, `scala.collection.mutable.Map()`, *etc.*, factories will likely be sufficient. The implementations provided by these factories use a fast lookup algorithm, usually involving a hash table, so they can quickly decide whether or not an object is in the collection.

The `scala.collection.mutable.Set()` factory method, for example, returns a `scala.collection.mutable.HashSet`, which uses a hash table internally. Similarly, the `scala.collection.mutable.Map()` factory returns a `scala.collection.mutable.HashMap`.

The story for immutable sets and maps is a bit more involved. The class

351

Table 17.3 · Default immutable set implementations

| Number of elements | Implementation |
|---|---|
| 0 | scala.collection.immutable.EmptySet |
| 1 | scala.collection.immutable.Set1 |
| 2 | scala.collection.immutable.Set2 |
| 3 | scala.collection.immutable.Set3 |
| 4 | scala.collection.immutable.Set4 |
| 5 or more | scala.collection.immutable.HashSet |

returned by the `scala.collection.immutable.Set()` factory method, for example, depends on how many elements you pass to it, as shown in Table 17.3. For sets with fewer than five elements, a special class devoted exclusively to sets of each particular size is used to maximize performance. Once you request a set that has five or more elements in it, however, the factory method will return an implementation that uses hash tries.

Similarly, the `scala.collection.immutable.Map()` factory method will return a different class depending on how many key-value pairs you pass to it, as shown in Table 17.4. As with sets, for immutable maps with fewer than five elements, a special class devoted exclusively to maps of each particular size is used to maximize performance. Once a map has five or more key-value pairs in it, however, an immutable HashMap is used.

The default immutable implementation classes shown in Tables 17.3 and 17.4 work together to give you maximum performance. For example, if you add an element to an EmptySet, it will return a Set1. If you add an element to that Set1, it will return a Set2. If you then remove an element from the Set2, you'll get another Set1.

Table 17.4 · Default immutable map implementations

| Number of elements | Implementation |
|---|---|
| 0 | scala.collection.immutable.EmptyMap |
| 1 | scala.collection.immutable.Map1 |
| 2 | scala.collection.immutable.Map2 |
| 3 | scala.collection.immutable.Map3 |
| 4 | scala.collection.immutable.Map4 |
| 5 or more | scala.collection.immutable.HashMap |

**Sorted sets and maps**

On occasion you may need a set or map whose iterator returns elements in a particular order. For this purpose, the Scala collections library provides traits SortedSet and SortedMap. These traits are implemented by classes TreeSet and TreeMap, which use a red-black tree to keep elements (in the case of TreeSet) or keys (in the case of TreeMap) in order. The order is determined by the Ordered trait, which the element type of the set, or key type of the map, must either mix in or be implicitly convertible to. These classes only come in immutable variants. Here are some TreeSet examples:

```
scala> import scala.collection.immutable.TreeSet
import scala.collection.immutable.TreeSet

scala> val ts = TreeSet(9, 3, 1, 8, 0, 2, 7, 4, 6, 5)
ts: scala.collection.immutable.TreeSet[Int] =
    TreeSet(0, 1, 2, 3, 4, 5, 6, 7, 8, 9)

scala> val cs = TreeSet('f', 'u', 'n')
cs: scala.collection.immutable.TreeSet[Char] =
    TreeSet(f, n, u)
```

And here are a few TreeMap examples:

```
scala> import scala.collection.immutable.TreeMap
import scala.collection.immutable.TreeMap

scala> var tm = TreeMap(3 -> 'x', 1 -> 'x', 4 -> 'x')
tm: scala.collection.immutable.TreeMap[Int,Char] =
    Map(1 -> x, 3 -> x, 4 -> x)

scala> tm += (2 -> 'x')

scala> tm
res30: scala.collection.immutable.TreeMap[Int,Char] =
    Map(1 -> x, 2 -> x, 3 -> x, 4 -> x)
```

# 17.3 Selecting mutable versus immutable collections

For some problems, mutable collections work better, while for others, immutable collections work better. When in doubt, it is better to start with an immutable collection and change it later, if you need to, because immutable collections can be easier to reason about than mutable ones.

Also, it can be worthwhile to go the opposite way sometimes. If you find some code that uses mutable collections becoming complicated and hard to reason about, consider whether it would help to change some of the collections to immutable alternatives. In particular, if you find yourself worrying about making copies of mutable collections in just the right places, or thinking a lot about who "owns" or "contains" a mutable collection, consider switching some of the collections to their immutable counterparts.

Besides being potentially easier to reason about, immutable collections can usually be stored more compactly than mutable ones if the number of elements stored in the collection is small. For instance an empty mutable map in its default representation of HashMap takes up about 80 bytes, and about 16 more are added for each entry that's added to it. An empty immutable Map is a single object that's shared between all references, so referring to it essentially costs just a single pointer field.

What's more, the Scala collections library currently stores immutable maps and sets with up to four entries in a single object, which typically takes up between 16 and 40 bytes, depending on the number of entries stored in the collection.[4] So for small maps and sets, the immutable versions are much more compact than the mutable ones. Given that many collections are small, switching them to be immutable can bring important space savings and performance advantages.

To make it easier to switch from immutable to mutable collections, and vice versa, Scala provides some syntactic sugar. Even though immutable sets and maps do not support a true += method, Scala gives a useful alternate interpretation to +=. Whenever you write a += b, and a does not support a method named +=, Scala will try interpreting it as a = a + b.

For example, immutable sets do not support a += operator:

```
scala> val people = Set("Nancy", "Jane")
people: scala.collection.immutable.Set[String] =
    Set(Nancy, Jane)

scala> people += "Bob"
<console>:14: error: value += is not a member of
```

---

[4]The "single object" is an instance of Set1 through Set4, or Map1 through Map4, as shown in Tables 17.3 and 17.4.

```
scala.collection.immutable.Set[String]
                people += "Bob"
                      ^
```

However, if you declare people as a var, instead of a val, then the collection can be "updated" with a += operation, even though it is immutable. First, a new collection will be created, and then people will be reassigned to refer to the new collection:

```
scala> var people = Set("Nancy", "Jane")
people: scala.collection.immutable.Set[String] =
    Set(Nancy, Jane)

scala> people += "Bob"

scala> people
res34: scala.collection.immutable.Set[String] =
    Set(Nancy, Jane, Bob)
```

After this series of statements, the people variable refers to a new immutable set, which contains the added string, "Bob". The same idea applies to any method ending in =, not just the += method. Here's the same syntax used with the -= operator, which removes an element from a set, and the ++= operator, which adds a collection of elements to a set:

```
scala> people -= "Jane"

scala> people ++= List("Tom", "Harry")

scala> people
res37: scala.collection.immutable.Set[String] =
    Set(Nancy, Bob, Tom, Harry)
```

To see how this is useful, consider again the following Map example from Section 1.1:

```
var capital = Map("US" -> "Washington", "France" -> "Paris")
capital += ("Japan" -> "Tokyo")
println(capital("France"))
```

This code uses immutable collections. If you want to try using mutable collections instead, all that is necessary is to import the mutable version of Map, thus overriding the default import of the immutable Map:

355

```
import scala.collection.mutable.Map   // only change needed!
var capital = Map("US" -> "Washington", "France" -> "Paris")
capital += ("Japan" -> "Tokyo")
println(capital("France"))
```

Not all examples are quite that easy to convert, but the special treatment of methods ending in an equals sign will often reduce the amount of code that needs changing.

By the way, this syntactic treatment works on any kind of value, not just collections. For example, here it is being used on floating-point numbers:

```
scala> var roughlyPi = 3.0
roughlyPi: Double = 3.0

scala> roughlyPi += 0.1

scala> roughlyPi += 0.04

scala> roughlyPi
res40: Double = 3.14
```

The effect of this expansion is similar to Java's assignment operators (+=, -=, *=, *etc.*), but it is more general because every operator ending in = can be converted.

## 17.4   Initializing collections

As you've seen previously, the most common way to create and initialize a collection is to pass the initial elements to a factory method on the companion object of your chosen collection. You just place the elements in parentheses after the companion object name, and the Scala compiler will transform that to an invocation of an `apply` method on that companion object:

```
scala> List(1, 2, 3)
res41: List[Int] = List(1, 2, 3)

scala> Set('a', 'b', 'c')
res42: scala.collection.immutable.Set[Char] = Set(a, b, c)

scala> import scala.collection.mutable
import scala.collection.mutable
```

```
scala> mutable.Map("hi" -> 2, "there" -> 5)
res43: scala.collection.mutable.Map[String,Int] =
    Map(hi -> 2, there -> 5)·

scala> Array(1.0, 2.0, 3.0)
res44: Array[Double] = Array(1.0, 2.0, 3.0)
```

Although most often you can let the Scala compiler infer the element type of a collection from the elements passed to its factory method, sometimes you may want to create a collection but specify a different type from the one the compiler would choose. This is especially an issue with mutable collections. Here's an example:

```
scala> import scala.collection.mutable
import scala.collection.mutable

scala> val stuff = mutable.Set(42)
stuff: scala.collection.mutable.Set[Int] = Set(42)

scala> stuff += "abracadabra"
<console>:16: error: type mismatch;
  found   : String("abracadabra")
  required: Int
              stuff += "abracadabra"
                     ^
```

The problem here is that stuff was given an element type of Int. If you want it to have an element type of Any, you need to say so explicitly by putting the element type in square brackets, like this:

```
scala> val stuff = mutable.Set[Any](42)
stuff: scala.collection.mutable.Set[Any] = Set(42)
```

Another special situation is if you want to initialize a collection with another collection. For example, imagine you have a list, but you want a TreeSet containing the elements in the list. Here's the list:

```
scala> val colors = List("blue", "yellow", "red", "green")
colors: List[String] = List(blue, yellow, red, green)
```

You cannot pass the colors list to the factory method for TreeSet:

```
scala> import scala.collection.immutable.TreeSet
import scala.collection.immutable.TreeSet

scala> val treeSet = TreeSet(colors)
<console>:16: error: No implicit Ordering defined for
List[String].
       val treeSet = TreeSet(colors)
                            ^
```

Instead, you'll need to create an empty TreeSet[String] and add to it the elements of the list with the TreeSet's ++ operator:

```
scala> val treeSet = TreeSet[String]() ++ colors
treeSet: scala.collection.immutable.TreeSet[String] =
    TreeSet(blue, green, red, yellow)
```

### Converting to array or list

If you need to initialize a list or array with another collection, on the other hand, it is quite straightforward. As you've seen previously, to initialize a new list with another collection, simply invoke toList on that collection:

```
scala> treeSet.toList
res50: List[String] = List(blue, green, red, yellow)
```

Or, if you need an array, invoke toArray:

```
scala> treeSet.toArray
res51: Array[String] = Array(blue, green, red, yellow)
```

Note that although the original colors list was not sorted, the elements in the list produced by invoking toList on the TreeSet are in alphabetical order. When you invoke toList or toArray on a collection, the order of the elements in the resulting list or array will be the same as the order of elements produced by an iterator obtained by invoking elements on that collection. Because a TreeSet[String]'s iterator will produce strings in alphabetical order, those strings will appear in alphabetical order in the list resulting from invoking toList on that TreeSet.

Keep in mind, however, that conversion to lists or arrays usually requires copying all of the elements of the collection, and thus may be slow for large collections. Sometimes you need to do it, though, due to an existing API.

Further, many collections only have a few elements anyway, in which case there is only a small speed penalty.

**Converting between mutable and immutable sets and maps**

Another situation that arises occasionally is the need to convert a mutable set or map to an immutable one, or *vice versa*. To accomplish this, you can use the technique shown on the previous page to initialize a TreeSet with the elements of a list. Create a collection of the new type using the empty method and then add the new elements using either ++ or ++=, whichever is appropriate for the target collection type. Here's how you'd convert the immutable TreeSet from the previous example to a mutable set, and back again to an immutable one:

```
scala> import scala.collection.mutable
import scala.collection.mutable

scala> treeSet
res52: scala.collection.immutable.TreeSet[String] =
    TreeSet(blue, green, red, yellow)

scala> val mutaSet = mutable.Set.empty ++= treeSet
mutaSet: scala.collection.mutable.Set[String] =
    Set(red, blue, green, yellow)

scala> val immutaSet = Set.empty ++ mutaSet
immutaSet: scala.collection.immutable.Set[String] =
    Set(red, blue, green, yellow)
```

You can use the same technique to convert between mutable and immutable maps:

```
scala> val muta = mutable.Map("i" -> 1, "ii" -> 2)
muta: scala.collection.mutable.Map[String,Int] =
    Map(ii -> 2,i -> 1)

scala> val immu = Map.empty ++ muta
immu: scala.collection.immutable.Map[String,Int] =
    Map(ii -> 2, i -> 1)
```

## 17.5   Tuples

As described in Step 9 in Chapter 3, a tuple combines a fixed number of items together so that they can be passed around as a whole. Unlike an array or list, a tuple can hold objects with different types. Here is an example of a tuple holding an integer, a string, and the console:

```
(1, "hello", Console)
```

Tuples save you the tedium of defining simplistic data-heavy classes. Even though defining a class is already easy, it does require a certain minimum effort, which sometimes serves no purpose. Tuples save you the effort of choosing a name for the class, choosing a scope to define the class in, and choosing names for the members of the class. If your class simply holds an integer and a string, there is no clarity added by defining a class named AnIntegerAndAString.

Because tuples can combine objects of different types, tuples do not inherit from Traversable. If you find yourself wanting to group exactly one integer and exactly one string, then you want a tuple, not a List or Array.

A common application of tuples is returning multiple values from a method. For example, here is a method that finds the longest word in a collection and also returns its index:

```
def longestWord(words: Array[String]) = {
  var word = words(0)
  var idx = 0
  for (i <- 1 until words.length)
    if (words(i).length > word.length) {
      word = words(i)
      idx = i
    }
  (word, idx)
}
```

Here is an example use of the method:

```
scala> val longest =
          longestWord("The quick brown fox".split(" "))
longest: (String, Int) = (quick,1)
```

The longestWord function here computes two items: word, the longest word in the array, and idx, the index of that word. To keep things simple, the function assumes there is at least one word in the list, and it breaks ties by choosing the word that comes earlier in the list. Once the function has chosen which word and index to return, it returns both of them together using the tuple syntax (word, idx).

To access elements of a tuple, you can use method _1 to access the first element, _2 to access the second, and so on:

```
scala> longest._1
res53: String = quick

scala> longest._2
res54: Int = 1
```

Additionally, you can assign each element of the tuple to its own variable,[5] like this:

```
scala> val (word, idx) = longest
word: String = quick
idx: Int = 1

scala> word
res55: String = quick
```

By the way, if you leave off the parentheses you get a different result:

```
scala> val word, idx = longest
word: (String, Int) = (quick,1)
idx: (String, Int) = (quick,1)
```

This syntax gives *multiple definitions* of the same expression. Each variable is initialized with its own evaluation of the expression on the right-hand side. That the expression evaluates to a tuple in this case does not matter. Both variables are initialized to the tuple in its entirety. See Chapter 18 for some examples where multiple definitions are convenient.

As a note of warning, tuples are almost too easy to use. Tuples are great when you combine data that has no meaning beyond "an A and a B." However, whenever the combination has some meaning, or you want to add some

---

[5]This syntax is actually a special case of *pattern matching*, as described in detail in Section 15.7.

methods to the combination, it is better to go ahead and create a class. For example, do not use a 3-tuple for the combination of a month, a day, and a year. Make a `Date` class. It makes your intentions explicit, which both clears up the code for human readers and gives the compiler and language opportunities to help you catch mistakes.

## 17.6 Conclusion

This chapter has given an overview of the Scala collections library and the most important classes and traits in it. With this foundation you should be able to work effectively with Scala collections, and know where to look in Scaladoc when you need more information. For more detailed information about Scala collections, look ahead to Chapter 24 and Chapter 25. For now, in the next chapter, we'll turn our attention from the Scala library back to the language and discuss Scala's support for mutable objects.

# Chapter 18

# Mutable Objects

In previous chapters, we put the spotlight on functional (immutable) objects. We did so because the idea of objects without any mutable state deserves to be better known. However, it is also perfectly possible to define objects with mutable state in Scala. Such mutable objects often come up naturally when you want to model objects in the real world that change over time.

This chapter explains what mutable objects are and what Scala provides in terms of syntax to express them. We will also introduce a larger case study on discrete event simulation, which involves mutable objects, as well as building an internal DSL for defining digital circuits to simulate.

## 18.1   What makes an object mutable?

You can observe the principal difference between a purely functional object and a mutable one even without looking at the object's implementation. When you invoke a method or dereference a field on some purely functional object, you will always get the same result.

For instance, given a list of characters:

```
val cs = List('a', 'b', 'c')
```

an application of cs.head will always return 'a'. This is the case even if there is an arbitrary number of operations on the list cs between the point where it is defined and the point where the access cs.head is made.

For a mutable object, on the other hand, the result of a method call or field access may depend on what operations were previously performed on the object. A good example of a mutable object is a bank account. Listing 18.1 shows a simplified implementation of bank accounts:

```scala
class BankAccount {

  private var bal: Int = 0

  def balance: Int = bal

  def deposit(amount: Int) = {
    require(amount > 0)
    bal += amount
  }

  def withdraw(amount: Int): Boolean =
    if (amount > bal) false
    else {
      bal -= amount
      true
    }
}
```

Listing 18.1 · A mutable bank account class.

The BankAccount class defines a private variable, bal, and three public methods: balance returns the current balance; deposit adds a given amount to bal; and withdraw tries to subtract a given amount from bal while assuring that the remaining balance won't be negative. The return value of withdraw is a Boolean indicating whether the requested funds were successfully withdrawn.

Even if you know nothing about the inner workings of the BankAccount class, you can still tell that BankAccounts are mutable objects:

```scala
scala> val account = new BankAccount
account: BankAccount = BankAccount@21cf775d

scala> account deposit 100

scala> account withdraw 80
res1: Boolean = true

scala> account withdraw 80
res2: Boolean = false
```

Note that the two final withdrawals in the previous interaction returned different results. The first withdraw operation returned `true` because the bank account contained sufficient funds to allow the withdrawal. The second operation, although the same as the first one, returned `false` because the balance of the account had been reduced so that it no longer covered the requested funds. So, clearly, bank accounts have mutable state, because the same operation can return different results at different times.

You might think that the mutability of `BankAccount` is immediately apparent because it contains a `var` definition. Mutation and `var`s usually go hand in hand, but things are not always so clear cut. For instance, a class might be mutable without defining or inheriting any `var`s because it forwards method calls to other objects that have mutable state. The reverse is also possible: A class might contain `var`s and still be purely functional. An example would be a class that caches the result of an expensive operation in a field for optimization purposes. To pick an example, assume the following unoptimized class Keyed with an expensive operation `computeKey`:

```
class Keyed {
  def computeKey: Int = ... // this will take some time
  ...
}
```

Provided that `computeKey` neither reads nor writes any `var`s, you can make Keyed more efficient by adding a cache:

```
class MemoKeyed extends Keyed {
  private var keyCache: Option[Int] = None
  override def computeKey: Int = {
    if (!keyCache.isDefined) keyCache = Some(super.computeKey)
    keyCache.get
  }
}
```

Using MemoKeyed instead of Keyed can speed things up because the second time the result of the `computeKey` operation is requested, the value stored in the keyCache field can be returned instead of running `computeKey` once again. But except for this speed gain, the behavior of class Keyed and MemoKeyed is exactly the same. Consequently, if Keyed is purely functional, then so is MemoKeyed, even though it contains a reassignable variable.

## 18.2 Reassignable variables and properties

You can perform two fundamental operations on a reassignable variable: get its value or set it to a new value. In libraries such as JavaBeans, these operations are often encapsulated in separate getter and setter methods, which need to be defined explicitly.

In Scala, every var that is a non-private member of some object implicitly defines a getter and a setter method with it. These getters and setters are named differently from the Java convention, however. The getter of a var x is just named "x", while its setter is named "x_=".

For example, if it appears in a class, the var definition:

```
var hour = 12
```

generates a getter, "hour", and setter, "hour_=", in addition to a reassignable field. The field is always marked private[this], which means it can be accessed only from the object that contains it. The getter and setter, on the other hand, get the same visibility as the original var. If the var definition is public, so are its getter and setter. If it is protected, they are also protected, and so on.

For instance, consider the class Time shown in Listing 18.2, which defines two public vars named hour and minute:

```
class Time {
  var hour = 12
  var minute = 0
}
```

Listing 18.2 · A class with public vars.

This implementation is exactly equivalent to the class definition shown in Listing 18.3. In the definitions shown in Listing 18.3, the names of the local fields h and m are arbitrarily chosen so as not to clash with any names already in use.

An interesting aspect about this expansion of vars into getters and setters is that you can also choose to define a getter and a setter directly, instead of defining a var. By defining these access methods directly you can interpret the operations of variable access and variable assignment as you like. For in-

```
class Time {

  private[this] var h = 12
  private[this] var m = 0

  def hour: Int = h
  def hour_=(x: Int) = { h = x }

  def minute: Int = m
  def minute_=(x: Int) = { m = x }
}
```

Listing 18.3 · How public vars are expanded into getter and setter methods.

stance, the variant of class Time shown in Listing 18.4 contains requirements that catch all assignments to hour and minute with illegal values.

```
class Time {

  private[this] var h = 12
  private[this] var m = 0

  def hour: Int = h
  def hour_= (x: Int) = {
    require(0 <= x && x < 24)
    h = x
  }

  def minute = m
  def minute_= (x: Int) = {
    require(0 <= x && x < 60)
    m = x
  }
}
```

Listing 18.4 · Defining getter and setter methods directly.

Some languages have a special syntactic construct for these variable-like quantities that are not plain variables in that their getter or setter can be redefined. For instance, C# has properties, which fulfill this role. In effect, Scala's convention of always interpreting a variable as a pair of setter and getter methods gives you the same capabilities as C# properties without requiring special syntax.

Properties can serve many different purposes. In the example shown in Listing 18.4, the setters enforced an invariant, thus protecting the variable from being assigned illegal values. You could also use a property to log all accesses to getters or setters of a variable. Or you could integrate variables with events, for instance by notifying some subscriber methods each time a variable is modified (you'll see examples of this in Chapter 35).

It's also possible, and sometimes useful, to define a getter and a setter without an associated field. For example, Listing 18.5 shows a Thermometer class, which encapsulates a temperature variable that can be read and updated. Temperatures can be expressed in Celsius or Fahrenheit degrees. This class allows you to get and set the temperature in either measure.

```scala
class Thermometer {

  var celsius: Float = _

  def fahrenheit = celsius * 9 / 5 + 32
  def fahrenheit_= (f: Float) = {
    celsius = (f - 32) * 5 / 9
  }
  override def toString = fahrenheit + "F/" + celsius + "C"

}
```

Listing 18.5 · Defining a getter and setter without an associated field.

The first line in the body of this class defines a var, celsius, which will contain the temperature in degrees Celsius. The celsius variable is initially set to a default value by specifying '_' as the "initializing value" of the variable. More precisely, an initializer "= _" of a field assigns a zero value to that field. The zero value depends on the field's type. It is 0 for numeric types, false for booleans, and null for reference types. This is the same as if the same variable was defined in Java without an initializer.

Note that you cannot simply leave off the "= _" initializer in Scala. If you had written:

```scala
var celsius: Float
```

this would declare an abstract variable, not an uninitialized one.[1]

---

[1] Abstract variables will be explained in Chapter 20.

The celsius variable definition is followed by a getter, "fahrenheit", and a setter, "fahrenheit_=", which access the same temperature, but in degrees Fahrenheit. There is no separate field that contains the current temperature value in Fahrenheit. Instead the getter and setter methods for Fahrenheit values automatically convert from and to degrees Celsius, respectively. Here's an example of interacting with a Thermometer object:

```
scala> val t = new Thermometer
t: Thermometer = 32.0F/0.0C

scala> t.celsius = 100
t.celsius: Float = 100.0

scala> t
res3: Thermometer = 212.0F/100.0C

scala> t.fahrenheit = -40
t.fahrenheit: Float = -40.0

scala> t
res4: Thermometer = -40.0F/-40.0C
```

## 18.3 Case study: Discrete event simulation

The rest of this chapter shows by way of an extended example how mutable objects can be combined with first-class function values in interesting ways. You'll see the design and implementation of a simulator for digital circuits. This task is broken down into several subproblems, each of which is interesting individually.

First, you'll see a little language for digital circuits. The definition of this language will highlight a general method for embedding domain-specific languages (DSL) in a host language like Scala. Second, we'll present a simple but general framework for discrete event simulation. Its main task will be to keep track of actions that are performed in simulated time. Finally, we'll show how discrete simulation programs can be structured and built. The idea of such simulations is to model physical objects by simulated objects, and use the simulation framework to model physical time.

The example is taken from the classic textbook *Structure and Interpretation of Computer Programs* by Abelson and Sussman [Abe96]. What's different here is that the implementation language is Scala instead of Scheme,

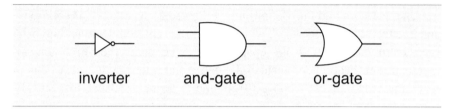

Figure 18.1 · Basic gates.

and that the various aspects of the example are structured into four software layers: one for the simulation framework, another for the basic circuit simulation package, a third for a library of user-defined circuits, and the last layer for each simulated circuit itself. Each layer is expressed as a class, and more specific layers inherit from more general ones.

**The fast track**

Understanding the discrete event simulation example presented in this chapter will take some time. If you feel you want to get on with learning more Scala instead, it's safe to skip ahead to the next chapter.

## 18.4 A language for digital circuits

We'll start with a "little language" to describe digital circuits. A digital circuit is built from *wires* and *function boxes*. Wires carry *signals*, which are transformed by function boxes. Signals are represented by booleans: true for signal-on and false for signal-off.

Figure 18.1 shows three basic function boxes (or *gates*):

- An *inverter*, which negates its signal.

- An *and-gate*, which sets its output to the conjunction of its inputs.

- An *or-gate*, which sets its output to the disjunction of its inputs.

These gates are sufficient to build all other function boxes. Gates have *delays*, so an output of a gate will change only some time after its inputs change.

We'll describe the elements of a digital circuit by the following set of Scala classes and functions. First, there is a class Wire for wires. We can construct wires like this:

```
val a = new Wire
val b = new Wire
val c = new Wire
```

or, equivalent but shorter, like this:

```
val a, b, c = new Wire
```

Second, there are three procedures which "make" the basic gates we need:

```
def inverter(input: Wire, output: Wire)
def andGate(a1: Wire, a2: Wire, output: Wire)
def orGate(o1: Wire, o2: Wire, output: Wire)
```

What's unusual, given the functional emphasis of Scala, is that these procedures construct the gates as a side effect, instead of returning the constructed gates as a result. For instance, an invocation of inverter(a, b) places an inverter between the wires a and b. It turns out that this side-effecting construction makes it easier to construct complicated circuits gradually. Also, although methods most often have verb names, these have noun names that indicate which gate they are making. This reflects the declarative nature of the DSL: it should describe a circuit, not the actions of making one.

More complicated function boxes can be built from the basic gates. For instance, the method shown in Listing 18.6 constructs a half-adder. The halfAdder method takes two inputs, a and b, and produces a sum, s, defined by "s = (a + b) % 2" and a carry, c, defined by "c = (a + b) / 2". A diagram of the half-adder is shown in Figure 18.2.

```
def halfAdder(a: Wire, b: Wire, s: Wire, c: Wire) = {
  val d, e = new Wire
  orGate(a, b, d)
  andGate(a, b, c)
  inverter(c, e)
  andGate(d, e, s)
}
```

Listing 18.6 · The halfAdder method.

Note that halfAdder is a parameterized function box just like the three methods that construct the primitive gates. You can use the halfAdder

371

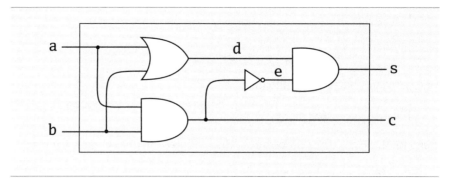

Figure 18.2 · A half-adder circuit.

method to construct more complicated circuits. For instance, Listing 18.7 defines a full, one-bit adder, shown in Figure 18.3, which takes two inputs, a and b, as well as a carry-in, cin, and which produces a sum output defined by "sum = (a + b + cin) % 2" and a carry-out output defined by "cout = (a + b + cin) / 2".

```
def fullAdder(a: Wire, b: Wire, cin: Wire,
    sum: Wire, cout: Wire) = {

  val s, c1, c2 = new Wire
  halfAdder(a, cin, s, c1)
  halfAdder(b, s, sum, c2)
  orGate(c1, c2, cout)
}
```

Listing 18.7 · The fullAdder method.

Class Wire and functions inverter, andGate, and orGate represent a little language with which users can define digital circuits. It's a good example of an *internal* DSL, a domain-specific language defined as a library in a host language instead of being implemented on its own.

The implementation of the circuit DSL still needs to be worked out. Since the purpose of defining a circuit in the DSL is simulating the circuit, it makes sense to base the DSL implementation on a general API for discrete event simulation. The next two sections will present first the simulation API and then the implementation of the circuit DSL on top of it.

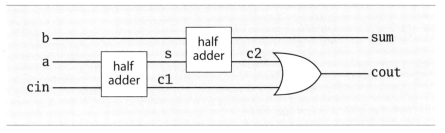

Figure 18.3 · A full-adder circuit.

# 18.5   The Simulation API

The simulation API is shown in Listing 18.8. It consists of class `Simulation` in package `org.stairwaybook.simulation`. Concrete simulation libraries inherit this class and augment it with domain-specific functionality. The elements of the `Simulation` class are presented in this section.

A discrete event simulation performs user-defined *actions* at specified *times*. The actions, which are defined by concrete simulation subclasses, all share a common type:

```
type Action = () => Unit
```

This statement defines `Action` to be an alias of the type of procedure that takes an empty parameter list and returns `Unit`. `Action` is a *type member* of class `Simulation`. You can think of it as a more readable name for type `() => Unit`. Type members will be described in detail in Section 20.6.

The time at which an action is performed is simulated time; it has nothing to do with the actual "wall clock" time. Simulated times are represented simply as integers. The current simulated time is kept in a private variable:

```
private var curtime: Int = 0
```

The variable has a public accessor method, which retrieves the current time:

```
def currentTime: Int = curtime
```

This combination of private variable with public accessor is used to make sure that the current time cannot be modified outside the `Simulation` class. After all, you don't usually want your simulation objects to manipulate the current time, except possibly if your simulation models time travel.

373

```
abstract class Simulation {

  type Action = () => Unit

  case class WorkItem(time: Int, action: Action)

  private var curtime = 0
  def currentTime: Int = curtime

  private var agenda: List[WorkItem] = List()

  private def insert(ag: List[WorkItem],
      item: WorkItem): List[WorkItem] = {

    if (ag.isEmpty || item.time < ag.head.time) item :: ag
    else ag.head :: insert(ag.tail, item)
  }

  def afterDelay(delay: Int)(block: => Unit) = {
    val item = WorkItem(currentTime + delay, () => block)
    agenda = insert(agenda, item)
  }

  private def next() = {
    (agenda: @unchecked) match {
      case item :: rest =>
        agenda = rest
        curtime = item.time
        item.action()
    }
  }

  def run() = {
    afterDelay(0) {
      println("*** simulation started, time = " +
          currentTime + " ***")
    }
    while (!agenda.isEmpty) next()
  }
}
```

Listing 18.8 · The Simulation class.

An action that needs to be executed at a specified time is called a *work item*. Work items are implemented by the following class:

```
case class WorkItem(time: Int, action: Action)
```

We made the WorkItem class a case class because of the syntactic conveniences this entails: You can use the factory method, WorkItem, to create instances of the class, and you get accessors for the constructor parameters time and action for free. Note also that class WorkItem is nested inside class Simulation. Nested classes in Scala are treated similarly to Java. Section 20.7 will give more details.

The Simulation class keeps an *agenda* of all remaining work items that have not yet been executed. The work items are sorted by the simulated time at which they have to be run:

```
private var agenda: List[WorkItem] = List()
```

The agenda list will be kept in the proper sorted order by the insert method, which updates it. You can see insert being called from afterDelay, which is the only way to add a work item to the agenda:

```
def afterDelay(delay: Int)(block: => Unit) = {
  val item = WorkItem(currentTime + delay, () => block)
  agenda = insert(agenda, item)
}
```

As the name implies, this method inserts an action (given by block) into the agenda so that it is scheduled for execution delay time units after the current simulation time. For instance, the following invocation would create a new work item to be executed at the simulated time, currentTime + delay:

```
afterDelay(delay) { count += 1 }
```

The code to be executed is contained in the method's second argument. The formal parameter for this argument has type "=> Unit" (*i.e.*, it is a computation of type Unit which is passed by name). Recall that by-name parameters are not evaluated when passed to a method. So in the call above, count would be incremented only when the simulation framework calls the action stored in the work item. Note that afterDelay is a curried function. It's a good example of the principle set forward in Section 9.5 that currying can be used to make method calls look more like built-in syntax.

375

The created work item still needs to be inserted into the agenda. This is done by the insert method, which maintains the invariant that the agenda is time-sorted:

```
private def insert(ag: List[WorkItem],
    item: WorkItem): List[WorkItem] = {

  if (ag.isEmpty || item.time < ag.head.time) item :: ag
  else ag.head :: insert(ag.tail, item)
}
```

The core of the Simulation class is defined by the run method:

```
def run() = {
  afterDelay(0) {
    println("*** simulation started, time = " +
        currentTime + " ***")
  }
  while (!agenda.isEmpty) next()
}
```

This method repeatedly takes the first item in the agenda, removes it from the agenda and executes it. It does this until there are no more items left in the agenda to execute. Each step is performed by calling the next method, which is defined as follows:

```
private def next() = {
  (agenda: @unchecked) match {
    case item :: rest =>
      agenda = rest
      curtime = item.time
      item.action()
  }
}
```

The next method decomposes the current agenda with a pattern match into a front item, item, and a remaining list of work items, rest. It removes the front item from the current agenda, sets the simulated time curtime to the work item's time, and executes the work item's action.

Note that next can be called only if the agenda is non-empty. There's no case for an empty list, so you would get a MatchError exception if you tried to run next on an empty agenda.

In fact, the Scala compiler would normally warn you that you missed one of the possible patterns for a list:

```
Simulator.scala:19: warning: match is not exhaustive!
missing combination             Nil

    agenda match {
    ^

one warning found
```

In this case, the missing case is not a problem because you know that next is called only on a non-empty agenda. Therefore, you might want to disable the warning. You saw in Section 15.5 that this can be done by adding an @unchecked annotation to the selector expression of the pattern match. That's why the Simulation code uses "(agenda: @unchecked) match", not "agenda match".

That's it. This might look like surprisingly little code for a simulation framework. You might wonder how this framework could possibly support interesting simulations, if all it does is execute a list of work items? In fact the power of the simulation framework comes from the fact that actions stored in work items can themselves install further work items into the agenda when they are executed. That makes it possible to have long-running simulations evolve from simple beginnings.

## 18.6 Circuit Simulation

The next step is to use the simulation framework to implement the domain-specific language for circuits shown in Section 18.4. Recall that the circuit DSL consists of a class for wires and methods that create and-gates, or-gates, and inverters. These are all contained in a BasicCircuitSimulation class, which extends the simulation framework. This class is shown in Listings 18.9 and 18.10.

Class BasicCircuitSimulation declares three abstract methods that represent the delays of the basic gates: InverterDelay, AndGateDelay, and OrGateDelay. The actual delays are not known at the level of this class

```scala
package org.stairwaybook.simulation

abstract class BasicCircuitSimulation extends Simulation {

  def InverterDelay: Int
  def AndGateDelay: Int
  def OrGateDelay: Int

  class Wire {

    private var sigVal = false
    private var actions: List[Action] = List()

    def getSignal = sigVal

    def setSignal(s: Boolean) =
      if (s != sigVal) {
        sigVal = s
        actions foreach (_ ())
      }

    def addAction(a: Action) = {
      actions = a :: actions
      a()
    }
  }

  def inverter(input: Wire, output: Wire) = {
    def invertAction() = {
      val inputSig = input.getSignal
      afterDelay(InverterDelay) {
        output setSignal !inputSig
      }
    }
    input addAction invertAction
  }
  // continued in Listing 18.10...
```

Listing 18.9 · The first half of the BasicCircuitSimulation class.

```
// ...continued from Listing 18.9
def andGate(a1: Wire, a2: Wire, output: Wire) = {
  def andAction() = {
    val a1Sig = a1.getSignal
    val a2Sig = a2.getSignal
    afterDelay(AndGateDelay) {
      output setSignal (a1Sig & a2Sig)
    }
  }
  a1 addAction andAction
  a2 addAction andAction
}

def orGate(o1: Wire, o2: Wire, output: Wire) = {
  def orAction() = {
    val o1Sig = o1.getSignal
    val o2Sig = o2.getSignal
    afterDelay(OrGateDelay) {
      output setSignal (o1Sig | o2Sig)
    }
  }
  o1 addAction orAction
  o2 addAction orAction
}

def probe(name: String, wire: Wire) = {
  def probeAction() = {
    println(name + " " + currentTime +
        " new-value = " + wire.getSignal)
  }
  wire addAction probeAction
}
}
```

Listing 18.10 · The second half of the `BasicCircuitSimulation` class.

because they depend on the technology of circuits that are simulated. That's why the delays are left abstract in class BasicCircuitSimulation, so that their concrete definition is delegated to a subclass.[2] The implementation of class BasicCircuitSimulation's other members is described next.

### The Wire class

A wire needs to support three basic actions:

getSignal: Boolean: returns the current signal on the wire.

setSignal(sig: Boolean): sets the wire's signal to sig.

addAction(p: Action): attaches the specified procedure p to the *actions* of the wire. The idea is that all action procedures attached to some wire will be executed every time the signal of the wire changes. Typically actions are added to a wire by components connected to the wire. An attached action is executed once at the time it is added to a wire, and after that, every time the signal of the wire changes.

Here is the implementation of the Wire class:

```
class Wire {

  private var sigVal = false
  private var actions: List[Action] = List()

  def getSignal = sigVal

  def setSignal(s: Boolean) =
    if (s != sigVal) {
      sigVal = s
      actions foreach (_ ())
    }

  def addAction(a: Action) = {
    actions = a :: actions
    a()
  }
}
```

---

[2]The names of these "delay" methods start with a capital letter because they represent constants. They are methods so they can be overridden in subclasses. You'll find out how to do the same thing with vals in Section 20.3.

Two private variables make up the state of a wire. The variable `sigVal` represents the current signal, and the variable `actions` represents the action procedures currently attached to the wire. The only interesting method implementation is the one for `setSignal`: When the signal of a wire changes, the new value is stored in the variable `sigVal`. Furthermore, all actions attached to a wire are executed. Note the shorthand syntax for doing this: "`actions foreach (_ ())`" applies the function, "`_ ()`", to each element in the `actions` list. As described in Section 8.5, the function "`_ ()`" is a shorthand for "`f => f ()`"—*i.e.*, it takes a function (we'll call it `f`) and applies it to the empty parameter list.

### The `inverter` method

The only effect of creating an inverter is that an action is installed on its input wire. This action is invoked once at the time the action is installed, and thereafter every time the signal on the input changes. The effect of the action is that the value of the inverter's output value is set (via `setSignal`) to the inverse of its input value. Since inverter gates have delays, this change should take effect only `InverterDelay` units of simulated time after the input value has changed and the action was executed. This suggests the following implementation:

```
def inverter(input: Wire, output: Wire) = {
  def invertAction() = {
    val inputSig = input.getSignal
    afterDelay(InverterDelay) {
      output setSignal !inputSig
    }
  }
  input addAction invertAction
}
```

The effect of the `inverter` method is to add `invertAction` to the `input` wire. This action, when invoked, gets the input signal and installs another action that inverts the `output` signal into the simulation agenda. This other action is to be executed after `InverterDelay` units of simulated time. Note how the method uses the `afterDelay` method of the simulation framework to create a new work item that's going to be executed in the future.

## The andGate and orGate methods

The implementation of and-gates is analogous to the implementation of inverters. The purpose of an and-gate is to output the conjunction of its input signals. This should happen at AndGateDelay simulated time units after any one of its two inputs changes. Hence, the following implementation:

```
def andGate(a1: Wire, a2: Wire, output: Wire) = {
  def andAction() = {
    val a1Sig = a1.getSignal
    val a2Sig = a2.getSignal
    afterDelay(AndGateDelay) {
      output setSignal (a1Sig & a2Sig)
    }
  }
  a1 addAction andAction
  a2 addAction andAction
}
```

The effect of the andGate method is to add andAction to both of its input wires a1 and a2. This action, when invoked, gets both input signals and installs another action that sets the output signal to the conjunction of both input signals. This other action is to be executed after AndGateDelay units of simulated time. Note that the output has to be recomputed if either of the input wires changes. That's why the same andAction is installed on each of the two input wires a1 and a2. The orGate method is implemented similarly, except it performs a logical-or instead of a logical-and.

## Simulation output

To run the simulator, you need a way to inspect changes of signals on wires. To accomplish this, you can simulate the action of putting a probe on a wire:

```
def probe(name: String, wire: Wire) = {
  def probeAction() = {
    println(name + " " + currentTime +
        " new-value = " + wire.getSignal)
  }
  wire addAction probeAction
}
```

The effect of the probe procedure is to install a `probeAction` on a given wire. As usual, the installed action is executed every time the wire's signal changes. In this case it simply prints the name of the wire (which is passed as first parameter to `probe`), as well as the current simulated time and the wire's new value.

## Running the simulator

After all these preparations, it's time to see the simulator in action. To define a concrete simulation, you need to inherit from a simulation framework class. To see something interesting, we'll create an abstract simulation class that extends `BasicCircuitSimulation` and contains method definitions for half-adders and full-adders as they were presented earlier in this chapter in Listings 18.6 and 18.7, respectively. This class, which we'll call `CircuitSimulation`, is shown in Listing 18.11.

```
package org.stairwaybook.simulation

abstract class CircuitSimulation
  extends BasicCircuitSimulation {

  def halfAdder(a: Wire, b: Wire, s: Wire, c: Wire) = {
    val d, e = new Wire
    orGate(a, b, d)
    andGate(a, b, c)
    inverter(c, e)
    andGate(d, e, s)
  }

  def fullAdder(a: Wire, b: Wire, cin: Wire,
      sum: Wire, cout: Wire) = {

    val s, c1, c2 = new Wire
    halfAdder(a, cin, s, c1)
    halfAdder(b, s, sum, c2)
    orGate(c1, c2, cout)
  }
}
```

Listing 18.11 · The `CircuitSimulation` class.

A concrete circuit simulation will be an object that inherits from class CircuitSimulation. The object still needs to fix the gate delays according to the circuit implementation technology that's simulated. Finally, you will also need to define the concrete circuit that's going to be simulated.

You can do these steps interactively in the Scala interpreter:

```
scala> import org.stairwaybook.simulation._
import org.stairwaybook.simulation._
```

First, the gate delays. Define an object (call it MySimulation) that provides some numbers:

```
scala> object MySimulation extends CircuitSimulation {
         def InverterDelay = 1
         def AndGateDelay = 3
         def OrGateDelay = 5
       }
defined module MySimulation
```

Because you are going to access the members of the MySimulation object repeatedly, an import of the object keeps the subsequent code shorter:

```
scala> import MySimulation._
import MySimulation._
```

Next, the circuit. Define four wires, and place probes on two of them:

```
scala> val input1, input2, sum, carry = new Wire
input1: MySimulation.Wire =
    BasicCircuitSimulation$Wire@111089b
input2: MySimulation.Wire =
    BasicCircuitSimulation$Wire@14c352e
sum: MySimulation.Wire =
    BasicCircuitSimulation$Wire@37a04c
carry: MySimulation.Wire =
    BasicCircuitSimulation$Wire@1fd10fa

scala> probe("sum", sum)
sum 0 new-value = false

scala> probe("carry", carry)
carry 0 new-value = false
```

Note that the probes immediately print an output. This is because every action installed on a wire is executed a first time when the action is installed.

Now define a half-adder connecting the wires:

```scala
scala> halfAdder(input1, input2, sum, carry)
```

Finally, set the signals, one after another, on the two input wires to true and run the simulation:

```scala
scala> input1 setSignal true

scala> run()
*** simulation started, time = 0 ***
sum 8 new-value = true

scala> input2 setSignal true

scala> run()
*** simulation started, time = 8 ***
carry 11 new-value = true
sum 15 new-value = false
```

## 18.7 Conclusion

This chapter brought together two techniques that seem disparate at first: mutable state and higher-order functions. Mutable state was used to simulate physical entities whose state changes over time. Higher-order functions were used in the simulation framework to execute actions at specified points in simulated time. They were also used in the circuit simulations as *triggers* that associate actions with state changes. Along the way, you saw a simple way to define a domain-specific language as a library. That's probably enough for one chapter!

If you feel like staying a bit longer, you might want to try more simulation examples. You can combine half-adders and full-adders to create larger circuits, or design new circuits from the basic gates defined so far and simulate them. In the next chapter, you'll learn about type parameterization in Scala, and see another example in which a combination of functional and imperative approaches yields a good solution.

Chapter 19

# Type Parameterization

In this chapter, we'll explain the details of type parameterization in Scala. Along the way we'll demonstrate some of the techniques for information hiding introduced in Chapter 13 by using a concrete example: the design of a class for purely functional queues. We're presenting type parameterization and information hiding together, because information hiding can be used to obtain more general type parameterization variance annotations.

Type parameterization allows you to write generic classes and traits. For example, sets are generic and take a type parameter: they are defined as Set[T]. As a result, any particular set instance might be a Set[String], a Set[Int], *etc.*, but it must be a set of *something*. Unlike Java, which allows raw types, Scala requires that you specify type parameters. Variance defines inheritance relationships of parameterized types, such as whether a Set[String], for example, is a subtype of Set[AnyRef].

The chapter contains three parts. The first part develops a data structure for purely functional queues. The second part develops techniques to hide internal representation details of this structure. The final part explains variance of type parameters and how it interacts with information hiding.

## 19.1 Functional queues

A functional queue is a data structure with three operations:

| | |
|---|---|
| head | returns the first element of the queue |
| tail | returns a queue without its first element |
| enqueue | returns a new queue with a given element appended at the end |

Unlike a mutable queue, a functional queue does not change its contents
when an element is appended. Instead, a new queue is returned that contains
the element. The goal of this chapter will be to create a class, which we'll
name Queue, that works like this:

```
scala> val q = Queue(1, 2, 3)
q: Queue[Int] = Queue(1, 2, 3)

scala> val q1 = q enqueue 4
q1: Queue[Int] = Queue(1, 2, 3, 4)

scala> q
res0: Queue[Int] = Queue(1, 2, 3)
```

If Queue were a mutable implementation, the enqueue operation in the second input line above would affect the contents of q; in fact both the result,
q1, and the original queue, q, would contain the sequence 1, 2, 3, 4 after
the operation. But for a functional queue, the appended value shows up only
in the result, q1, not in the queue, q, being operated on.

Purely functional queues also have some similarity with lists. Both are
so called *fully persistent* data structures, where old versions remain available
even after extensions or modifications. Both support head and tail operations. But where a list is usually extended at the front, using a :: operation,
a queue is extended at the end, using enqueue.

How can this be implemented efficiently? Ideally, a functional (immutable) queue should not have a fundamentally higher overhead than an
imperative (mutable) one. That is, all three operations, head, tail, and
enqueue, should operate in constant time.

One simple approach to implement a functional queue would be to use a
list as representation type. Then head and tail would just translate into the
same operations on the list, whereas enqueue would be concatenation.

This would give the following implementation:

```
class SlowAppendQueue[T](elems: List[T]) { // Not efficient
  def head = elems.head
  def tail = new SlowAppendQueue(elems.tail)
  def enqueue(x: T) = new SlowAppendQueue(elems ::: List(x))
}
```

The problem with this implementation is in the enqueue operation. It takes
time proportional to the number of elements stored in the queue. If you want

388

constant time append, you could also try to reverse the order of the elements in the representation list, so that the last element that's appended comes first in the list. This would lead to the following implementation:

```scala
class SlowHeadQueue[T](smele: List[T]) { // Not efficient
  // smele is elems reversed
  def head = smele.last
  def tail = new SlowHeadQueue(smele.init)
  def enqueue(x: T) = new SlowHeadQueue(x :: smele)
}
```

Now enqueue is constant time, but head and tail are not. They now take time proportional to the number of elements stored in the queue.

Looking at these two examples, it does not seem easy to come up with an implementation that's constant time for all three operations. In fact, it looks doubtful that this is even possible! However, by combining the two operations you can get very close. The idea is to represent a queue by two lists, called leading and trailing. The leading list contains elements towards the front, whereas the trailing list contains elements towards the back of the queue in reversed order. The contents of the whole queue are at each instant equal to "leading ::: trailing.reverse".

Now, to append an element, you just cons it to the trailing list using the :: operator, so enqueue is constant time. This means that, when an initially empty queue is constructed from successive enqueue operations, the trailing list will grow whereas the leading list will stay empty. Then, before the first head or tail operation is performed on an empty leading list, the whole trailing list is copied to leading, reversing the order of the elements. This is done in an operation called mirror. Listing 19.1 shows an implementation of queues that uses this approach.

What is the complexity of this implementation of queues? The mirror operation might take time proportional to the number of queue elements, but only if list leading is empty. It returns directly if leading is non-empty. Because head and tail call mirror, their complexity might be linear in the size of the queue, too. However, the longer the queue gets, the less often mirror is called.

Indeed, assume a queue of length $n$ with an empty leading list. Then mirror has to reverse-copy a list of length $n$. However, the next time mirror will have to do any work is once the leading list is empty again, which will

```
class Queue[T](
  private val leading: List[T],
  private val trailing: List[T]
) {
  private def mirror =
    if (leading.isEmpty)
      new Queue(trailing.reverse, Nil)
    else
      this

  def head = mirror.leading.head

  def tail = {
    val q = mirror
    new Queue(q.leading.tail, q.trailing)
  }

  def enqueue(x: T) =
    new Queue(leading, x :: trailing)
}
```

Listing 19.1 · A basic functional queue.

be the case after *n* `tail` operations. This means you can "charge" each of these *n* `tail` operations with one *n*'th of the complexity of `mirror`, which means a constant amount of work. Assuming that head, `tail`, and enqueue operations appear with about the same frequency, the *amortized* complexity is hence constant for each operation. So functional queues are asymptotically just as efficient as mutable ones.

Now, there are some caveats that need to be attached to this argument. First, the discussion was only about asymptotic behavior; the constant factors might well be somewhat different. Second, the argument rested on the fact that head, `tail` and enqueue are called with about the same frequency. If head is called much more often than the other two operations, the argument is not valid, as each call to head might involve a costly re-organization of the list with `mirror`. The second caveat can be avoided; it is possible to design functional queues so that in a sequence of successive head operations only the first one might require a re-organization. You will find out at the end of this chapter how this is done.

## 19.2    Information hiding

The implementation of Queue shown in Listing 19.1 is now quite good with regards to efficiency. You might object, though, that this efficiency is paid for by exposing a needlessly detailed implementation. The Queue constructor, which is globally accessible, takes two lists as parameters, where one is reversed—hardly an intuitive representation of a queue. What's needed is a way to hide this constructor from client code. In this section, we'll show you some ways to accomplish this in Scala.

### Private constructors and factory methods

In Java, you can hide a constructor by making it private. In Scala, the primary constructor does not have an explicit definition; it is defined implicitly by the class parameters and body. Nevertheless, it is still possible to hide the primary constructor by adding a private modifier in front of the class parameter list, as shown in Listing 19.2:

```scala
class Queue[T] private (
  private val leading: List[T],
  private val trailing: List[T]
)
```

Listing 19.2 · Hiding a primary constructor by making it private.

The private modifier between the class name and its parameters indicates that the constructor of Queue is private: it can be accessed only from within the class itself and its companion object. The class name Queue is still public, so you can use it as a type, but you cannot call its constructor:

```scala
scala> new Queue(List(1, 2), List(3))
<console>:9: error: constructor Queue in class Queue cannot
be accessed in object $iw
              new Queue(List(1, 2), List(3))
              ^
```

Now that the primary constructor of class Queue can no longer be called from client code, there needs to be some other way to create new queues. One possibility is to add an auxiliary constructor, like this:

```scala
def this() = this(Nil, Nil)
```

The auxiliary constructor shown in the previous example builds an empty queue. As a refinement, the auxiliary constructor could take a list of initial queue elements:

```
def this(elems: T*) = this(elems.toList, Nil)
```

Recall that T* is the notation for repeated parameters, as described in Section 8.8.

Another possibility is to add a factory method that builds a queue from such a sequence of initial elements. A neat way to do this is to define an object Queue that has the same name as the class being defined and contains an apply method, as shown in Listing 19.3:

```
object Queue {
  // constructs a queue with initial elements 'xs'
  def apply[T](xs: T*) = new Queue[T](xs.toList, Nil)
}
```

Listing 19.3 · An apply factory method in a companion object.

By placing this object in the same source file as class Queue, you make the object a companion object of the class. You saw in Section 13.5 that a companion object has the same access rights as its class. Because of this, the apply method in object Queue can create a new Queue object, even though the constructor of class Queue is private.

Note that, because the factory method is called apply, clients can create queues with an expression such as Queue(1, 2, 3). This expression expands to Queue.apply(1, 2, 3) since Queue is an object instead of a function. As a result, Queue looks to clients as if it was a globally defined factory method. In reality, Scala has no globally visible methods; every method must be contained in an object or a class. However, using methods named apply inside global objects, you can support usage patterns that look like invocations of global methods.

**An alternative: private classes**

Private constructors and private members are one way to hide the initialization and representation of a class. Another more radical way is to hide the class itself and only export a trait that reveals the public interface of the

392

```scala
trait Queue[T] {
  def head: T
  def tail: Queue[T]
  def enqueue(x: T): Queue[T]
}
object Queue {

  def apply[T](xs: T*): Queue[T] =
    new QueueImpl[T](xs.toList, Nil)

  private class QueueImpl[T](
    private val leading: List[T],
    private val trailing: List[T]
  ) extends Queue[T] {

    def mirror =
      if (leading.isEmpty)
        new QueueImpl(trailing.reverse, Nil)
      else
        this

    def head: T = mirror.leading.head

    def tail: QueueImpl[T] = {
      val q = mirror
      new QueueImpl(q.leading.tail, q.trailing)
    }

    def enqueue(x: T) =
      new QueueImpl(leading, x :: trailing)
  }
}
```

Listing 19.4 · Type abstraction for functional queues.

class. The code in Listing 19.4 implements this design. There's a trait Queue, which declares the methods head, tail, and enqueue. All three methods are implemented in a subclass QueueImpl, which is itself a private inner class of object Queue. This exposes to clients the same information as before, but using a different technique. Instead of hiding individual constructors and methods, this version hides the whole implementation class.

## 19.3   Variance annotations

Queue, as defined in Listing 19.4, is a trait, but not a type. Queue is not a type because it takes a type parameter.

As a result, you cannot create variables of type Queue:

```
scala> def doesNotCompile(q: Queue) = {}
<console>:8: error: class Queue takes type parameters
       def doesNotCompile(q: Queue) = {}
                             ^
```

Instead, trait Queue enables you to specify *parameterized* types, such as Queue[String], Queue[Int], or Queue[AnyRef]:

```
scala> def doesCompile(q: Queue[AnyRef]) = {}
doesCompile: (q: Queue[AnyRef])Unit
```

Thus, Queue is a trait and Queue[String] is a type. Queue is also called a *type constructor* because you can construct a type with it by specifying a type parameter. (This is analogous to constructing an object instance with a plain-old constructor by specifying a value parameter.) The type constructor Queue "generates" a family of types, which includes Queue[Int], Queue[String], and Queue[AnyRef].

You can also say that Queue is a *generic* trait. (Classes and traits that take type parameters are "generic," but the types they generate are "parameterized," not generic.) The term "generic" means that you are defining many specific types with one generically written class or trait. For example, trait Queue in Listing 19.4 defines a generic queue. Queue[Int] and Queue[String], *etc.*, would be the specific queues.

The combination of type parameters and subtyping poses some interesting questions. For example, are there any special subtyping relationships between members of the family of types generated by Queue[T]? More specifically, should a Queue[String] be considered a subtype of Queue[AnyRef]?

Or more generally, if S is a subtype of type T, then should Queue[S] be considered a subtype of Queue[T]? If so, you could say that trait Queue is *covariant* (or "flexible") in its type parameter T. Or, since it just has one type parameter, you could say simply that Queues are covariant. Covariant Queues would mean, for example, that you could pass a Queue[String] to the doesCompile method shown previously, which takes a value parameter of type Queue[AnyRef].

Intuitively, all this seems OK, since a queue of Strings looks like a special case of a queue of AnyRefs. In Scala, however, generic types have by default *nonvariant* (or "rigid") subtyping. That is, with Queue defined as in Listing 19.4, queues with different element types would never be in a subtype relationship. A Queue[String] would not be usable as a Queue[AnyRef]. However, you can demand covariant (flexible) subtyping of queues by changing the first line of the definition of class Queue like this:

```
trait Queue[+T] { ... }
```

Prefixing a formal type parameter with a + indicates that subtyping is covariant (flexible) in that parameter. By adding this single character, you are telling Scala that you want Queue[String], for example, to be considered a subtype of Queue[AnyRef]. The compiler will check that Queue is defined in a way that this subtyping is sound.

Besides +, there is also a prefix –, which indicates *contravariant* subtyping. If Queue were defined like this:

```
trait Queue[-T] { ... }
```

then if T is a subtype of type S, this would imply that Queue[S] is a subtype of Queue[T] (which in the case of queues would be rather surprising!). Whether a type parameter is covariant, contravariant, or nonvariant is called the parameter's *variance* . The + and – symbols you can place next to type parameters are called *variance annotations*.

In a purely functional world, many types are naturally covariant (flexible). However, the situation changes once you introduce mutable data. To find out why, consider the simple type of one-element cells that can be read or written, shown in Listing 19.5.

The Cell type of Listing 19.5 is declared nonvariant (rigid). For the sake of argument, assume for a moment that Cell was declared covariant instead—*i.e.*, it was declared class Cell[+T]—and that this passed the

```
class Cell[T](init: T) {
  private[this] var current = init
  def get = current
  def set(x: T) = { current = x }
}
```

Listing 19.5 · A nonvariant (rigid) Cell class.

Scala compiler. (It doesn't, and we'll explain why shortly.) Then you could construct the following problematic statement sequence:

```
val c1 = new Cell[String]("abc")
val c2: Cell[Any] = c1
c2.set(1)
val s: String = c1.get
```

Seen by itself, each of these four lines looks OK. The first line creates a cell of strings and stores it in a val named c1. The second line defines a new val, c2, of type Cell[Any], which initialized with c1. This is OK since Cells are assumed to be covariant. The third line sets the value of cell c2 to 1. This is also OK because the assigned value 1 is an instance of c2's element type Any. Finally, the last line assigns the element value of c1 into a string. Nothing strange here, as both the sides are of the same type. But taken together, these four lines end up assigning the integer 1 to the string s. This is clearly a violation of type soundness.

Which operation is to blame for the runtime fault? It must be the second one, which uses covariant subtyping. The other statements are too simple and fundamental. Thus, a Cell of String is *not* also a Cell of Any, because there are things you can do with a Cell of Any that you cannot do with a Cell of String. You cannot use set with an Int argument on a Cell of String, for example.

In fact, were you to pass the covariant version of Cell to the Scala compiler, you would get a compile-time error:

```
Cell.scala:7: error: covariant type T occurs in
contravariant position in type T of value x
    def set(x: T) = current = x
                ^
```

**Variance and arrays**

It's interesting to compare this behavior with arrays in Java. In principle, arrays are just like cells except that they can have more than one element. Nevertheless, arrays are treated as covariant in Java.

You can try an example analogous to the cell interaction described here with Java arrays:

```
// this is Java
String[] a1 = { "abc" };
Object[] a2 = a1;
a2[0] = new Integer(17);
String s = a1[0];
```

If you try out this example, you will find that it compiles. But executing the program will cause an `ArrayStore` exception to be thrown when a2[0] is assigned to an `Integer`:

```
Exception in thread "main" java.lang.ArrayStoreException:
java.lang.Integer
        at JavaArrays.main(JavaArrays.java:8)
```

What happens here is that Java stores the element type of the array at runtime. Then, every time an array element is updated, the new element value is checked against the stored type. If it is not an instance of that type, an `ArrayStore` exception is thrown.

You might ask why Java adopted this design, which seems both unsafe and expensive. When asked this question, James Gosling, the principal inventor of the Java language, answered that they wanted to have a simple means to treat arrays generically. For instance, they wanted to be able to write a method to sort all elements of an array, using a signature like the following that takes an array of `Object`:

```
void sort(Object[] a, Comparator cmp) { ... }
```

Covariance of arrays was needed so that arrays of arbitrary reference types could be passed to this `sort` method. Of course, with the arrival of Java generics, such a `sort` method can now be written with a type parameter, so the covariance of arrays is no longer necessary. For compatibility reasons, though, it has persisted in Java to this day.

Scala tries to be purer than Java in not treating arrays as covariant. Here's what you get if you translate the first two lines of the array example to Scala:

```
scala> val a1 = Array("abc")
a1: Array[String] = Array(abc)

scala> val a2: Array[Any] = a1
<console>:8: error: type mismatch;
 found    : Array[String]
 required: Array[Any]
       val a2: Array[Any] = a1
                            ^
```

What happened here is that Scala treats arrays as nonvariant (rigid), so an Array[String] is not considered to conform to an Array[Any]. However, sometimes it is necessary to interact with legacy methods in Java that use an Object array as a means to emulate a generic array. For instance, you might want to call a sort method like the one described previously with an array of Strings as argument. To make this possible, Scala lets you cast an array of Ts to an array of any supertype of T:

```
scala> val a2: Array[Object] =
          a1.asInstanceOf[Array[Object]]
a2: Array[Object] = Array(abc)
```

The cast is always legal at compile-time, and it will always succeed at run-time because the JVM's underlying run-time model treats arrays as covariant, just as Java the language does. But you might get ArrayStore exceptions afterwards, again just as you would in Java.

## 19.4   Checking variance annotations

Now that you have seen some examples where variance is unsound, you may be wondering which kind of class definitions need to be rejected and which can be accepted. So far, all violations of type soundness involved some re-assignable field or array element. The purely functional implementation of queues, on the other hand, looks like a good candidate for covariance. However, the following example shows that you can "engineer" an unsound situation even if there is no reassignable field.

To set up the example, assume that queues as defined in Listing 19.4 are covariant. Then, create a subclass of queues that specializes the element type to Int and overrides the enqueue method:

```
class StrangeIntQueue extends Queue[Int] {
  override def enqueue(x: Int) = {
    println(math.sqrt(x))
    super.enqueue(x)
  }
}
```

The enqueue method in StrangeIntQueue prints out the square root of its (integer) argument before doing the append proper.

Now, you can write a counterexample in two lines:

```
val x: Queue[Any] = new StrangeIntQueue
x.enqueue("abc")
```

The first of these two lines is valid because StrangeIntQueue is a subclass of Queue[Int] and, assuming covariance of queues, Queue[Int] is a subtype of Queue[Any]. The second line is valid because you can append a String to a Queue[Any]. However, taken together, these two lines have the effect of applying a square root method to a string, which makes no sense.

Clearly it's not just mutable fields that make covariant types unsound. The problem is more general. It turns out that as soon as a generic parameter type appears as the type of a method parameter, the containing class or trait may not be covariant in that type parameter.

For queues, the enqueue method violates this condition:

```
class Queue[+T] {
  def enqueue(x: T) =
    ...
}
```

Running a modified queue class like the one above through a Scala compiler would yield:

```
Queues.scala:11: error: covariant type T occurs in
contravariant position in type T of value x
  def enqueue(x: T) =
              ^
```

Reassignable fields are a special case of the rule that disallows type parameters annotated with + from being used as method parameter types. As mentioned in Section 18.2, a reassignable field, "var x: T", is treated in Scala as a getter method, "def x: T", and a setter method, "def x_=(y: T)". As you can see, the setter method has a parameter of the field's type T. So that type may not be covariant.

> **The fast track**
>
> In the rest of this section, we'll describe the mechanism by which the Scala compiler checks variance annotations. If you're not interested in such detail right now, you can safely skip to Section 19.5. The most important thing to understand is that the Scala compiler will check any variance annotations you place on type parameters. For example, if you try to declare a type parameter to be covariant (by adding a +), but that could lead to potential runtime errors, your program won't compile.

To verify correctness of variance annotations, the Scala compiler classifies all positions in a class or trait body as *positive*, *negative* or *neutral*. A "position" is any location in the class or trait (but from now on we'll just write "class") body where a type parameter may be used. For example, every method value parameter is a position because a method value parameter has a type. Therefore a type parameter could appear in that position.

The compiler checks each use of each of the class's type parameters. Type parameters annotated with + may only be used in positive positions, while type parameters annotated with – may only be used in negative positions. A type parameter with no variance annotation may be used in any position, and is, therefore, the only kind of type parameter that can be used in neutral positions of the class body.

To classify the positions, the compiler starts from the declaration of a type parameter and then moves inward through deeper nesting levels. Positions at the top level of the declaring class are classified as positive. By default, positions at deeper nesting levels are classified the same as that at enclosing levels, but there are a handful of exceptions where the classification changes. Method value parameter positions are classified to the *flipped* classification relative to positions outside the method, where the flip of a positive classification is negative, the flip of a negative classification is positive, and the flip of a neutral classification is still neutral.

Besides method value parameter positions, the current classification is also flipped at the type parameters of methods. A classification is sometimes flipped at the type argument position of a type, such as the Arg in C[Arg], depending on the variance of the corresponding type parameter. If C's type parameter is annotated with a + then the classification stays the same. If C's type parameter is annotated with a -, then the current classification is flipped. If C's type parameter has no variance annotation then the current classification is changed to neutral.

As a somewhat contrived example, consider the following class definition, where several positions are annotated with their classifications, $^+$ (for positive) or $^-$ (for negative):

```
abstract class Cat[-T, +U] {
  def meow[W⁻](volume: T⁻, listener: Cat[U⁺, T⁻]⁻)
    : Cat[Cat[U⁺, T⁻]⁻, U⁺]⁺
}
```

The positions of the type parameter, W, and the two value parameters, volume and listener, are all negative. Looking at the result type of meow, the position of the first Cat[U, T] argument is negative because Cat's first type parameter, T, is annotated with a -. The type U inside this argument is again in positive position (two flips), whereas the type T inside that argument is still in negative position.

You see from this discussion that it's quite hard to keep track of variance positions. That's why it's a welcome relief that the Scala compiler does this job for you.

Once the classifications are computed, the compiler checks that each type parameter is only used in positions that are classified appropriately. In this case, T is only used in negative positions, and U is only used in positive positions. So class Cat is type correct.

## 19.5 Lower bounds

Back to the Queue class. You saw that the previous definition of Queue[T] shown in Listing 19.4 cannot be made covariant in T because T appears as a type of a parameter of the enqueue method, and that's a negative position.

Fortunately, there's a way to get unstuck: you can generalize enqueue by making it polymorphic (*i.e.*, giving the enqueue method itself a type parameter) and using a *lower bound* for its type parameter. Listing 19.6 shows a new formulation of Queue that implements this idea.

```
class Queue[+T] (private val leading: List[T],
    private val trailing: List[T] ) {
  def enqueue[U >: T](x: U) =
    new Queue[U](leading, x :: trailing) // ...
}
```

Listing 19.6 · A type parameter with a lower bound.

The new definition gives enqueue a type parameter U, and with the syntax, "U >: T", defines T as the lower bound for U. As a result, U is required to be a supertype of T.[1] The parameter to enqueue is now of type U instead of type T, and the return value of the method is now Queue[U] instead of Queue[T].

For example, suppose there is a class Fruit with two subclasses, Apple and Orange. With the new definition of class Queue, it is possible to append an Orange to a Queue[Apple]. The result will be a Queue[Fruit].

This revised definition of enqueue is type correct. Intuitively, if T is a more specific type than expected (for example, Apple instead of Fruit), a call to enqueue will still work because U (Fruit) will still be a supertype of T (Apple).[2]

The new definition of enqueue is arguably better than the old, because it is more general. Unlike the old version, the new definition allows you to append an arbitrary supertype U of the queue element type T. The result is then a Queue[U]. Together with queue covariance, this gives the right kind of flexibility for modeling queues of different element types in a natural way.

This shows that variance annotations and lower bounds play well together. They are a good example of *type-driven design*, where the types of an interface guide its detailed design and implementation. In the case of queues, it's likely you would not have thought of the refined implementation of enqueue with a lower bound. But you might have decided to make queues covariant, in which case, the compiler would have pointed out the variance error for enqueue. Correcting the variance error by adding a lower bound makes enqueue more general and queues as a whole more usable.

---

[1] Supertype and subtype relationships are reflexive, which means a type is both a supertype and a subtype of itself. Even though T is a lower bound for U, you could still pass in a T to enqueue.

[2] Technically, what happens is a flip occurs for lower bounds. The type parameter U is in a negative position (1 flip), while the lower bound (>: T) is in a positive position (2 flips).

This observation is also the main reason that Scala prefers declaration-site variance over use-site variance as it is found in Java's wildcards. With use-site variance, you are on your own designing a class. It will be the clients of the class that need to put in the wildcards, and if they get it wrong, some important instance methods will no longer be applicable. Variance being a tricky business, users usually get it wrong, and they come away thinking that wildcards and generics are overly complicated. With definition-side variance, you express your intent to the compiler, and the compiler will double check that the methods you want available will indeed be available.

## 19.6 Contravariance

So far in this chapter, all examples you've seen were either covariant or non-variant. But there are also cases where contravariance is natural. For instance, consider the trait of output channels shown in Listing 19.7:

```
trait OutputChannel[-T] {
  def write(x: T)
}
```

Listing 19.7 · A contravariant output channel.

Here, OutputChannel is defined to be contravariant in T. So an output channel of AnyRefs, say, is a subtype of an output channel of Strings. Although it may seem non-intuitive, it actually makes sense. To see why, consider what you can do with an OutputChannel[String]. The only supported operation is writing a String to it. The same operation can also be done on an OutputChannel[AnyRef]. So it is safe to substitute an OutputChannel[AnyRef] for an OutputChannel[String]. By contrast, it would not be safe to substitute an OutputChannel[String] where an OutputChannel[AnyRef] is required. After all, you can send any object to an OutputChannel[AnyRef], whereas an OutputChannel[String] requires that the written values are all strings.

This reasoning points to a general principle in type system design: It is safe to assume that a type T is a subtype of a type U if you can substitute a value of type T wherever a value of type U is required. This is called the *Liskov Substitution Principle*. The principle holds if T supports the same operations as U, and all of T's operations require less and provide more

```
trait Function1[-S, +T] {
  def apply(x: S): T
}
```

Listing 19.8 · Covariance and contravariance of Function1s.

than the corresponding operations in U. In the case of output channels, an OutputChannel[AnyRef] can be a subtype of an OutputChannel[String] because the two support the same write operation, and this operation requires less in OutputChannel[AnyRef] than in OutputChannel[String]. "Less" means the argument is only required to be an AnyRef in the first case, whereas it is required to be a String in the second case.

Sometimes covariance and contravariance are mixed in the same type. A prominent example is Scala's function traits. For instance, whenever you write the function type A => B, Scala expands this to Function1[A, B]. The definition of Function1 in the standard library uses both covariance and contravariance: the Function1 trait is contravariant in the function argument type S and covariant in the result type T, as shown in Listing 19.8. This satisfies the Liskov Substitution Principle because arguments are something that's required, whereas results are something that's provided.

As an example, consider the application shown in Listing 19.9. Here, class Publication contains one parametric field, title, of type String. Class Book extends Publication and forwards its string title parameter to the constructor of its superclass. The Library singleton object defines a set of books and a method printBookList, which takes a function, named info, of type Book => AnyRef. In other words, the type of the lone parameter to printBookList is a function that takes one Book argument and returns an AnyRef. The Customer application defines a method, getTitle, which takes a Publication as its lone parameter and returns a String, the title of the passed Publication.

Now take a look at the last line in Customer. This line invokes Library's printBookList method and passes getTitle, wrapped in a function value:

```
Library.printBookList(getTitle)
```

This line of code type checks even though String, the function's result type, is a subtype of AnyRef, the result type of printBookList's info parameter. This code passes the compiler because function result types are de-

```
class Publication(val title: String)
class Book(title: String) extends Publication(title)

object Library {
 val books: Set[Book] =
   Set(
     new Book("Programming in Scala"),
     new Book("Walden")
   )
 def printBookList(info: Book => AnyRef) = {
   for (book <- books) println(info(book))
 }
}

object Customer extends App {
 def getTitle(p: Publication): String = p.title
 Library.printBookList(getTitle)
}
```

Listing 19.9 · Demonstration of function type parameter variance.

clared to be covariant (the +T in Listing 19.8). If you look inside the body of printBookList, you can get a glimpse of why this makes sense.

The printBookList method iterates through its book list and invokes the passed function on each book. It passes the AnyRef result returned by info to println, which invokes toString on it and prints the result. This activity will work with String as well as any other subclass of AnyRef, which is what covariance of function result types means.

Now consider the parameter type of the function being passed to the printBookList method. Although printBookList's parameter type is declared as Book, the getTitle we're passing in takes a Publication, a *supertype* of Book. The reason this works is that since printBookList's parameter type is Book, the body of the printBookList method will only be allowed to pass a Book into the function. And because getTitle's parameter type is Publication, the body of that function will only be able to access on its parameter, p, members that are declared in class Publication. Because any method declared in Publication is also available on its subclass Book, everything should work, which is what contravariance of function parameter

405

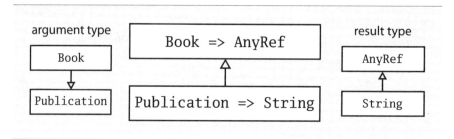

Figure 19.1 · Covariance and contravariance in function type parameters.

types means. You can see all this graphically in Figure 19.1.

The code in Listing 19.9 compiles because Publication => String is a subtype of Book => AnyRef, as shown in the center of the Figure 19.1. Because the result type of a Function1 is defined as covariant, the inheritance relationship of the two result types, shown at the right of the diagram, is in the same direction as that of the two functions shown in the center. By contrast, because the parameter type of a Function1 is defined as contravariant, the inheritance relationship of the two parameter types, shown at the left of the diagram, is in the opposite direction as that of the two functions.

## 19.7   Object private data

The Queue class seen so far has a problem in that the mirror operation will repeatedly copy the trailing into the leading list if head is called several times in a row on a list where leading is empty. The wasteful copying could be avoided by adding some judicious side effects. Listing 19.10 presents a new implementation of Queue, which performs at most one trailing to leading adjustment for any sequence of head operations.

What's different with respect to the previous version is that now leading and trailing are reassignable variables, and mirror performs the reverse copy from trailing to leading as a side effect on the current queue instead of returning a new queue. This side effect is purely internal to the implementation of the Queue operation; since leading and trailing are private variables, the effect is not visible to clients of Queue. So by the terminology established in Chapter 18, the new version of Queue still defines purely functional objects, in spite of the fact that they now contain reassignable fields.

```scala
class Queue[+T] private (
  private[this] var leading: List[T],
  private[this] var trailing: List[T]
) {

  private def mirror() =
    if (leading.isEmpty) {
      while (!trailing.isEmpty) {
        leading = trailing.head :: leading
        trailing = trailing.tail
      }
    }

  def head: T = {
    mirror()
    leading.head
  }

  def tail: Queue[T] = {
    mirror()
    new Queue(leading.tail, trailing)
  }

  def enqueue[U >: T](x: U) =
    new Queue[U](leading, x :: trailing)
}
```

Listing 19.10 · An optimized functional queue.

You might wonder whether this code passes the Scala type checker. After all, queues now contain two reassignable fields of the covariant parameter type T. Is this not a violation of the variance rules? It would be indeed, except for the detail that leading and trailing have a private[this] modifier, and are thus declared to be object private.

As mentioned in Section 13.5, object private members can be accessed only from within the object in which they are defined. It turns out that accesses to variables from the same object in which they are defined do not cause problems with variance. The intuitive explanation is that, in order to construct a case where variance would lead to type errors, you need to have a reference to a containing object that has a statically weaker type than the type

407

the object was defined with. For accesses to object private values, however, this is impossible.

Scala's variance checking rules contain a special case for object private definitions. Such definitions are omitted when it is checked that a type parameter with either a + or − annotation occurs only in positions that have the same variance classification. Therefore, the code in Listing 19.10 compiles without error. On the other hand, if you had left out the [this] qualifiers from the two private modifiers, you would see two type errors:

```
Queues.scala:1: error: covariant type T occurs in
contravariant position in type List[T] of parameter of
setter leading_=
class Queue[+T] private (private var leading: List[T],
                                           ^

Queues.scala:1: error: covariant type T occurs in
contravariant position in type List[T] of parameter of
setter trailing_=
                        private var trailing: List[T]) {
                                        ^
```

## 19.8   Upper bounds

In Listing 16.1 on page 323, we showed a merge sort function for lists that took a comparison function as a first argument and a list to sort as a second, curried argument. Another way you might want to organize such a sort function is by requiring the type of the list to mix in the Ordered trait. As mentioned in Section 12.4, by mixing Ordered into a class and implementing Ordered's one abstract method, compare, you enable clients to compare instances of that class with <, >, <=, and >=. For example, Listing 19.11 shows Ordered being mixed into a Person class.

As a result, you can compare two people like this:

```
scala> val robert = new Person("Robert", "Jones")
robert: Person = Robert Jones

scala> val sally = new Person("Sally", "Smith")
sally: Person = Sally Smith

scala> robert < sally
res0: Boolean = true
```

408

```
class Person(val firstName: String, val lastName: String)
    extends Ordered[Person] {

  def compare(that: Person) = {
    val lastNameComparison =
      lastName.compareToIgnoreCase(that.lastName)
    if (lastNameComparison != 0)
      lastNameComparison
    else
      firstName.compareToIgnoreCase(that.firstName)
  }

  override def toString = firstName + " " + lastName
}
```

Listing 19.11 · A Person class that mixes in the Ordered trait.

```
def orderedMergeSort[T <: Ordered[T]](xs: List[T]): List[T] = {
  def merge(xs: List[T], ys: List[T]): List[T] =
    (xs, ys) match {
      case (Nil, _) => ys
      case (_, Nil) => xs
      case (x :: xs1, y :: ys1) =>
        if (x < y) x :: merge(xs1, ys)
        else y :: merge(xs, ys1)
    }
  val n = xs.length / 2
  if (n == 0) xs
  else {
    val (ys, zs) = xs splitAt n
    merge(orderedMergeSort(ys), orderedMergeSort(zs))
  }
}
```

Listing 19.12 · A merge sort function with an upper bound.

To require that the type of the list passed to your new sort function mixes in Ordered, you need to use an *upper bound*. An upper bound is specified similar to a lower bound, except instead of the >: symbol used for lower bounds, you use a <: symbol, as shown in Listing 19.12.

With the "T <: Ordered[T]" syntax, you indicate that the type parameter, T, has an upper bound, Ordered[T]. This means that the element type of the list passed to orderedMergeSort must be a subtype of Ordered. Thus, you could pass a List[Person] to orderedMergeSort because Person mixes in Ordered.

For example, consider this list:

```
scala> val people = List(
         new Person("Larry", "Wall"),
         new Person("Anders", "Hejlsberg"),
         new Person("Guido", "van Rossum"),
         new Person("Alan", "Kay"),
         new Person("Yukihiro", "Matsumoto")
       )
people: List[Person] = List(Larry Wall, Anders Hejlsberg,
    Guido van Rossum, Alan Kay, Yukihiro Matsumoto)
```

Because the element type of this list, Person, mixes in (and is therefore a subtype of) Ordered[People], you can pass the list to orderedMergeSort:

```
scala> val sortedPeople = orderedMergeSort(people)
sortedPeople: List[Person] = List(Anders Hejlsberg, Alan Kay,
    Yukihiro Matsumoto, Guido van Rossum, Larry Wall)
```

Now, although the sort function shown in Listing 19.12 serves as a useful illustration of upper bounds, it isn't actually the most general way in Scala to design a sort function that takes advantage of the Ordered trait.

For example, you couldn't use the orderedMergeSort function to sort a list of integers, because class Int is not a subtype of Ordered[Int]:

```
scala> val wontCompile = orderedMergeSort(List(3, 2, 1))
<console>:5: error: inferred type arguments [Int] do
   not conform to method orderedMergeSort's type
      parameter bounds [T <: Ordered[T]]
         val wontCompile = orderedMergeSort(List(3, 2, 1))
                           ^
```

In Section 21.6, we'll show you how to use *implicit parameters* and *context bounds* to achieve a more general solution.

## 19.9 Conclusion

In this chapter you saw several techniques for information hiding: private constructors, factory methods, type abstraction, and object private members. You also learned how to specify data type variance and what it implies for class implementation. Finally, you saw two techniques which help in obtaining flexible variance annotations: lower bounds for method type parameters and `private[this]` annotations for local fields and methods.

Chapter 20

# Abstract Members

A member of a class or trait is *abstract* if the member does not have a complete definition in the class. Abstract members are intended to be implemented in subclasses of the class in which they are declared. This idea is found in many object-oriented languages. For instance, Java lets you declare abstract methods. Scala also lets you declare such methods, as you saw in Section 10.2. But Scala goes beyond that and implements the idea in its full generality: Besides methods, you can declare abstract fields and even abstract types as members of classes and traits.

In this chapter we'll describe all four kinds of abstract member: vals, vars, methods, and types. Along the way we'll discuss pre-initialized fields, lazy vals, path-dependent types, and enumerations.

## 20.1 A quick tour of abstract members

The following trait declares one of each kind of abstract member: an abstract type (T), method (transform), val (initial), and var (current):

```
trait Abstract {
  type T
  def transform(x: T): T
  val initial: T
  var current: T
}
```

A concrete implementation of Abstract needs to fill in definitions for each of its abstract members. Here is an example implementation that provides these definitions:

```
class Concrete extends Abstract {
  type T = String
  def transform(x: String) = x + x
  val initial = "hi"
  var current = initial
}
```

The implementation gives a concrete meaning to the type name T by defining it as an alias of type String. The transform operation concatenates a given string with itself, and the initial and current values are both set to "hi".

This example gives you a rough first idea of what kinds of abstract members exist in Scala. The remainder of the chapter will present the details and explain what the new forms of abstract members, as well as type members in general, are good for.

## 20.2  Type members

As you can see from the example in the previous section, the term *abstract type* in Scala means a type declared (with the "type" keyword) to be a member of a class or trait, without specifying a definition. Classes themselves may be abstract, and traits are by definition abstract, but neither of these are what are referred to as *abstract types* in Scala. An abstract type in Scala is always a member of some class or trait, such as type T in trait Abstract.

You can think of a non-abstract (or "concrete") type member, such as type T in class Concrete, as a way to define a new name, or *alias*, for a type. In class Concrete, for example, the type String is given the alias T. As a result, anywhere T appears in the definition of class Concrete, it means String. This includes the parameter and result types of transform, initial, and current, which mention T when they are declared in super-trait Abstract. Thus, when class Concrete implements these methods, those Ts are interpreted to mean String.

One reason to use a type member is to define a short, descriptive alias for a type whose real name is more verbose, or less obvious in meaning, than the alias. Such type members can help clarify the code of a class or trait. The other main use of type members is to declare abstract types that must be defined in subclasses. This use, which was demonstrated in the previous section, will be described in detail later in this chapter.

## 20.3  Abstract vals

An abstract `val` declaration has a form like:

```
val initial: String
```

It gives a name and type for a `val`, but not its value. This value has to be provided by a concrete `val` definition in a subclass. For instance, class `Concrete` implemented the `val` using:

```
val initial = "hi"
```

You use an abstract `val` declaration in a class when you do not know the correct value in the class, but you do know that the variable will have an unchangeable value in each instance of the class.

An abstract `val` declaration resembles an abstract parameterless method declaration such as:

```
def initial: String
```

Client code would refer to both the `val` and the method in exactly the same way (*i.e.*, `obj.initial`). However, if `initial` is an abstract `val`, the client is guaranteed that `obj.initial` will yield the same value every time it is referenced. If `initial` were an abstract method, that guarantee would not hold because, in that case, `initial` could be implemented by a concrete method that returns a different value every time it's called.

In other words, an abstract `val` constrains its legal implementation: Any implementation must be a `val` definition; it may not be a `var` or a `def`. Abstract method declarations, on the other hand, may be implemented by both concrete method definitions and concrete `val` definitions. Given the abstract class `Fruit` shown in Listing 20.1, class `Apple` would be a legal subclass implementation, but class `BadApple` would not.

```
abstract class Fruit {
  val v: String // 'v' for value
  def m: String // 'm' for method
}

abstract class Apple extends Fruit {
  val v: String
  val m: String // OK to override a 'def' with a 'val'
}

abstract class BadApple extends Fruit {
  def v: String // ERROR: cannot override a 'val' with a 'def'
  def m: String
}
```

Listing 20.1 · Overriding abstract vals and parameterless methods.

## 20.4 Abstract vars

Like an abstract val, an abstract var declares just a name and a type, but not an initial value. For instance, Listing 20.2 shows a trait AbstractTime, which declares two abstract variables named hour and minute:

```
trait AbstractTime {
  var hour: Int
  var minute: Int
}
```

Listing 20.2 · Declaring abstract vars.

What is the meaning of abstract vars like hour and minute? You saw in Section 18.2 that vars declared as members of classes come equipped with getter and setter methods. This holds for abstract vars as well. If you declare an abstract var named hour, for example, you implicitly declare an abstract getter method, hour, and an abstract setter method, hour_=. There's no reassignable field to be defined—that will come in subclasses that define the concrete implementation of the abstract var. For instance, the definition of AbstractTime shown in Listing 20.2 is exactly equivalent to the definition shown in Listing 20.3.

```
trait AbstractTime {
  def hour: Int           // getter for 'hour'
  def hour_=(x: Int)      // setter for 'hour'
  def minute: Int         // getter for 'minute'
  def minute_=(x: Int)    // setter for 'minute'
}
```

Listing 20.3 · How abstract vars are expanded into getters and setters.

## 20.5  Initializing abstract vals

Abstract vals sometimes play a role analogous to superclass parameters: they let you provide details in a subclass that are missing in a superclass. This is particularly important for traits, because traits don't have a constructor to which you could pass parameters. So the usual notion of parameterizing a trait works via abstract vals that are implemented in subclasses.

As an example, consider a reformulation of class Rational from Chapter 6, as shown in Listing 6.5 on page 113, as a trait:

```
trait RationalTrait {
  val numerArg: Int
  val denomArg: Int
}
```

The Rational class from Chapter 6 had two parameters: n for the numerator of the rational number, and d for the denominator. The RationalTrait trait given here defines instead two abstract vals: numerArg and denomArg. To instantiate a concrete instance of that trait, you need to implement the abstract val definitions. Here's an example:

```
new RationalTrait {
  val numerArg = 1
  val denomArg = 2
}
```

Here the keyword new appears in front of a trait name, RationalTrait, which is followed by a class body in curly braces. This expression yields an instance of an *anonymous class* that mixes in the trait and is defined by the

417

body. This particular anonymous class instantiation has an effect analogous to the instance creation new Rational(1, 2).

The analogy is not perfect, however. There's a subtle difference concerning the order in which expressions are initialized. When you write:

```
new Rational(expr1, expr2)
```

the two expressions, expr1 and expr2, are evaluated before class Rational is initialized, so the values of expr1 and expr2 are available for the initialization of class Rational.

For traits, the situation is the opposite. When you write:

```
new RationalTrait {
  val numerArg = expr1
  val denomArg = expr2
}
```

the expressions, expr1 and expr2, are evaluated as part of the initialization of the anonymous class, but the anonymous class is initialized *after* the RationalTrait. So the values of numerArg and denomArg are not available during the initialization of RationalTrait (more precisely, a selection of either value would yield the default value for type Int, 0). For the definition of RationalTrait given previously, this is not a problem, because the trait's initialization does not make use of values numerArg or denomArg. However, it becomes a problem in the variant of RationalTrait shown in Listing 20.4, which defines normalized numerators and denominators.

If you try to instantiate this trait with some numerator and denominator expressions that are not simple literals, you'll get an exception:

```
scala> val x = 2
x: Int = 2

scala> new RationalTrait {
          val numerArg = 1 * x
          val denomArg = 2 * x
       }
java.lang.IllegalArgumentException: requirement failed
  at scala.Predef$.require(Predef.scala:207)
  at RationalTrait$class.$init$(<console>:10)
  ... 28 elided
```

```
trait RationalTrait {
  val numerArg: Int
  val denomArg: Int
  require(denomArg != 0)
  private val g = gcd(numerArg, denomArg)
  val numer = numerArg / g
  val denom = denomArg / g
  private def gcd(a: Int, b: Int): Int =
    if (b == 0) a else gcd(b, a % b)
  override def toString = numer + "/" + denom
}
```

Listing 20.4 · A trait that uses its abstract vals.

The exception in this example was thrown because denomArg still had its default value of 0 when class RationalTrait was initialized, which caused the require invocation to fail.

This example demonstrates that initialization order is not the same for class parameters and abstract fields. A class parameter argument is evaluated *before* it is passed to the class constructor (unless the parameter is by-name). An implementing val definition in a subclass, by contrast, is evaluated only *after* the superclass has been initialized.

Now that you understand why abstract vals behave differently from parameters, it would be good to know what can be done about this. Is it possible to define a RationalTrait that can be initialized robustly, without fearing errors due to uninitialized fields? In fact, Scala offers two alternative solutions to this problem, *pre-initialized fields* and *lazy vals*. They are presented in the remainder of this section.

## Pre-initialized fields

The first solution, pre-initialized fields, lets you initialize a field of a subclass before the superclass is called. To do this, simply place the field definition in braces before the superclass constructor call. As an example, Listing 20.5 shows another attempt to create an instance of RationalTrait. As you see from this example, the initialization section comes before the mention of the supertrait RationalTrait. Both are separated by a with.

419

```
scala> new {
         val numerArg = 1 * x
         val denomArg = 2 * x
       } with RationalTrait
res1: RationalTrait = 1/2
```

Listing 20.5 · Pre-initialized fields in an anonymous class expression.

Pre-initialized fields are not restricted to anonymous classes; they can also be used in objects or named subclasses. Two examples are shown in Listings 20.6 and 20.7. As you can see from these examples, the pre-initialization section comes in each case after the extends keyword of the defined object or class. Class RationalClass, shown in Listing 20.7, exemplifies a general schema of how class parameters can be made available for the initialization of a supertrait.

```
object twoThirds extends {
  val numerArg = 2
  val denomArg = 3
} with RationalTrait
```

Listing 20.6 · Pre-initialized fields in an object definition.

```
class RationalClass(n: Int, d: Int) extends {
  val numerArg = n
  val denomArg = d
} with RationalTrait {
  def + (that: RationalClass) = new RationalClass(
    numer * that.denom + that.numer * denom,
    denom * that.denom
  )
}
```

Listing 20.7 · Pre-initialized fields in a class definition.

Because pre-initialized fields are initialized before the superclass constructor is called, their initializers cannot refer to the object that's being constructed. Consequently, if such an initializer refers to this, the reference goes to the object containing the class or object that's being constructed, not the constructed object itself.

Here's an example:

```
scala> new {
          val numerArg = 1
          val denomArg = this.numerArg * 2
      } with RationalTrait
<console>:11: error: value numerArg is not a member of object
  $iw
              val denomArg = this.numerArg * 2
                                  ^
```

The example did not compile because the reference this.numerArg was looking for a numerArg field in the object containing the new (which in this case was the synthetic object named $iw, into which the interpreter puts user input lines). Once more, pre-initialized fields behave in this respect like class constructor arguments.

## Lazy vals

You can use pre-initialized fields to simulate precisely the initialization behavior of class constructor arguments. Sometimes, however, you might prefer to let the system itself sort out how things should be initialized. This can be achieved by making your val definitions *lazy*. If you prefix a val definition with a lazy modifier, the initializing expression on the right-hand side will only be evaluated the first time the val is used.

For an example, define an object Demo with a val as follows:

```
scala> object Demo {
          val x = { println("initializing x"); "done" }
       }
defined object Demo
```

Now, first refer to Demo, then to Demo.x:

```
scala> Demo
initializing x
res3: Demo.type = Demo$@2129a843
```

```
scala> Demo.x
res4: String = done
```

As you can see, the moment you use Demo, its x field becomes initialized. The initialization of x forms part of the initialization of Demo. The situation changes, however, if you define the x field to be lazy:

```
scala> object Demo {
         lazy val x = { println("initializing x"); "done" }
       }
defined object Demo

scala> Demo
res5: Demo.type = Demo$@5b1769c

scala> Demo.x
initializing x
res6: String = done
```

Now, initializing Demo does not involve initializing x. The initialization of x will be deferred until the first time x is used. This is similar to the situation where x is defined as a parameterless method, using a def. However, unlike a def, a lazy val is never evaluated more than once. In fact, after the first evaluation of a lazy val the result of the evaluation is stored, to be reused when the same val is used subsequently.

Looking at this example, it seems that objects like Demo themselves behave like lazy vals, in that they are also initialized on demand, the first time they are used. This is correct. In fact an object definition can be seen as a shorthand for the definition of a lazy val with an anonymous class that describes the object's contents.

Using lazy vals, you could reformulate RationalTrait as shown in Listing 20.8. In the new trait definition, all concrete fields are defined lazy. Another change with respect to the previous definition of RationalTrait, shown in Listing 20.4, is that the require clause was moved from the body of the trait to the initializer of the private field, g, which computes the greatest common divisor of numerArg and denomArg. With these changes, there's nothing that remains to be done when LazyRationalTrait is initialized; all initialization code is now part of the right-hand side of a lazy val. Thus, it is safe to initialize the abstract fields of LazyRationalTrait after the class is defined.

```
trait LazyRationalTrait {
  val numerArg: Int
  val denomArg: Int
  lazy val numer = numerArg / g
  lazy val denom = denomArg / g
  override def toString = numer + "/" + denom
  private lazy val g = {
    require(denomArg != 0)
    gcd(numerArg, denomArg)
  }
  private def gcd(a: Int, b: Int): Int =
    if (b == 0) a else gcd(b, a % b)
}
```

Listing 20.8 · Initializing a trait with lazy vals.

Here's an example:

```
scala> val x = 2
x: Int = 2
scala> new LazyRationalTrait {
         val numerArg = 1 * x
         val denomArg = 2 * x
       }
res7: LazyRationalTrait = 1/2
```

No pre-initialization is needed. It's instructive to trace the sequence of initializations that lead to the string 1/2 to be printed in the code above:

1. A fresh instance of LazyRationalTrait gets created and the initialization code of LazyRationalTrait is run. This initialization code is empty; none of the fields of LazyRationalTrait is initialized yet.

2. Next, the primary constructor of the anonymous subclass defined by the new expression is executed. This involves the initialization of numerArg with 2 and denomArg with 4.

3. Next, the toString method is invoked on the constructed object by the interpreter, so that the resulting value can be printed.

4. Next, the numer field is accessed for the first time by the toString method in trait LazyRationalTrait, so its initializer is evaluated.

5. The initializer of numer accesses the private field, g, so g is evaluated next. This evaluation accesses numerArg and denomArg, which were defined in Step 2.

6. Next, the toString method accesses the value of denom, which causes denom's evaluation. The evaluation of denom accesses the values of denomArg and g. The initializer of the g field is not re-evaluated, because it was already evaluated in Step 5.

7. Finally, the result string "1/2" is constructed and printed.

Note that the definition of g comes textually after the definitions of numer and denom in class LazyRationalTrait. Nevertheless, because all three values are lazy, g gets initialized before the initialization of numer and denom is completed.

This shows an important property of lazy vals: The textual order of their definitions does not matter because values get initialized on demand. Thus, lazy vals can free you as a programmer from having to think hard how to arrange val definitions to ensure that everything is defined when it is needed.

However, this advantage holds only as long as the initialization of lazy vals neither produces side effects nor depends on them. In the presence of side effects, initialization order starts to matter. And then it can be quite difficult to trace in what order initialization code is run, as the previous example has demonstrated. So lazy vals are an ideal complement to functional objects, where the order of initializations does not matter, as long as everything gets initialized eventually. They are less well suited for code that's predominantly imperative.

---

### Lazy functional languages

Scala is by no means the first language to have exploited the perfect match of lazy definitions and functional code. In fact, there is a category of "lazy functional programming languages" in which *every* value and parameter is initialized lazily. The best known member of this class of languages is Haskell [SPJ02].

---

## 20.6  Abstract types

In the beginning of this chapter, you saw, "type T", an abstract type declaration. The rest of this chapter discusses what such an abstract type declaration means and what it's good for. Like all other abstract declarations, an abstract type declaration is a placeholder for something that will be defined concretely in subclasses. In this case, it is a type that will be defined further down the class hierarchy. So T above refers to a type that is as yet unknown at the point where it is declared. Different subclasses can provide different realizations of T.

Here is a well-known example where abstract types show up naturally. Suppose you are given the task of modeling the eating habits of animals. You might start with a class Food and a class Animal with an eat method:

```
class Food
abstract class Animal {
  def eat(food: Food)
}
```

You might then attempt to specialize these two classes to a class of Cows that eat Grass:

```
class Grass extends Food
class Cow extends Animal {
  override def eat(food: Grass) = {} // This won't compile
}
```

However, if you tried to compile the new classes, you'd get the following compilation errors:

```
BuggyAnimals.scala:7: error: class Cow needs to be
abstract, since method eat in class Animal of type
    (Food)Unit is not defined
class Cow extends Animal {
      ^
BuggyAnimals.scala:8: error: method eat overrides nothing
  override def eat(food: Grass) = {}
               ^
```

425

What happened is that the eat method in class Cow did not override the eat method in class Animal because its parameter type is different: it's Grass in class Cow vs. Food in class Animal.

Some people have argued that the type system is unnecessarily strict in refusing these classes. They have said that it should be OK to specialize a parameter of a method in a subclass. However, if the classes were allowed as written, you could get yourself in unsafe situations very quickly.

For instance, the following script would pass the type checker:

```
class Food
abstract class Animal {
  def eat(food: Food)
}
class Grass extends Food
class Cow extends Animal {
  override def eat(food: Grass) = {} // This won't compile,
}                                    // but if it did,...
class Fish extends Food
val bessy: Animal = new Cow
bessy eat (new Fish)      // ...you could feed fish to cows.
```

The program would compile if the restriction were eased, because Cows are Animals and Animals do have an eat method that accepts any kind of Food, including Fish. But surely it would do a cow no good to eat a fish!

What you need to do instead is apply some more precise modeling. Animals do eat Food, but what kind of Food each Animal eats depends on the Animal. This can be neatly expressed with an abstract type, as shown in Listing 20.9:

```
class Food
abstract class Animal {
  type SuitableFood <: Food
  def eat(food: SuitableFood)
}
```

Listing 20.9 · Modeling suitable food with an abstract type.

With the new class definition, an Animal can eat only food that's suitable. What food is suitable cannot be determined at the level of the Animal class.

426

That's why SuitableFood is modeled as an abstract type. The type has an upper bound, Food, which is expressed by the "<: Food" clause. This means that any concrete instantiation of SuitableFood (in a subclass of Animal) must be a subclass of Food. For example, you would not be able to instantiate SuitableFood with class IOException.

With Animal defined, you can now progress to cows, as shown in Listing 20.10. Class Cow fixes its SuitableFood to be Grass and also defines a concrete eat method for this kind of food.

```
class Grass extends Food
class Cow extends Animal {
  type SuitableFood = Grass
  override def eat(food: Grass) = {}
}
```

Listing 20.10 · Implementing an abstract type in a subclass.

These new class definitions compile without errors. If you tried to run the "cows-that-eat-fish" counterexample with the new class definitions, you would get the following compiler error:

```
scala> class Fish extends Food
defined class Fish

scala> val bessy: Animal = new Cow
bessy: Animal = Cow@1515d8a6

scala> bessy eat (new Fish)
<console>:14: error: type mismatch;
 found   : Fish
 required: bessy.SuitableFood
              bessy eat (new Fish)
                        ^
```

## 20.7  Path-dependent types

Have a look at the last error message again. What's interesting about it is the type required by the eat method: bessy.SuitableFood. This type consists of an object reference, bessy, followed by a type field, SuitableFood, of

427

the object. So this shows that objects in Scala can have types as members. The meaning of bessy.SuitableFood is "the type SuitableFood that is a member of the object referenced from bessy" or, alternatively, the type of food that's suitable for bessy.

A type like bessy.SuitableFood is called a *path-dependent type*. The word "path" here means a reference an object. It could be a single name, such as bessy, or a longer access path, such as farm.barn.bessy, where each of farm, barn, and bessy are variables (or singleton object names) that refer to objects.

As the term "path-dependent type" implies, the type depends on the path; in general, different paths give rise to different types. For instance, say you defined classes DogFood and Dog, like this:

```scala
class DogFood extends Food
class Dog extends Animal {
  type SuitableFood = DogFood
  override def eat(food: DogFood) = {}
}
```

If you attempted to feed a dog with food fit for a cow, your code would not compile:

```scala
scala> val bessy = new Cow
bessy: Cow = Cow@713e7e09

scala> val lassie = new Dog
lassie: Dog = Dog@6eaf2c57

scala> lassie eat (new bessy.SuitableFood)
<console>:16: error: type mismatch;
 found    : Grass
 required: DogFood
               lassie eat (new bessy.SuitableFood)
                           ^
```

The problem here is that the type of the SuitableFood object passed to the eat method, bessy.SuitableFood, is incompatible with the parameter type of eat, lassie.SuitableFood.

The case would be different for two Dogs. Because Dog's SuitableFood type is defined to be an alias for class DogFood, the SuitableFood types of two Dogs are in fact the same. As a result, the Dog instance named lassie could actually eat the suitable food of a different Dog instance (which we'll name bootsie):

```
scala> val bootsie = new Dog
bootsie: Dog = Dog@13a7c48c

scala> lassie eat (new bootsie.SuitableFood)
```

A path-dependent type resembles the syntax for an inner class type in Java, but there is a crucial difference: a path-dependent type names an outer *object*, whereas an inner class type names an outer *class*. Java-style inner class types can also be expressed in Scala, but they are written differently. Consider these two classes, Outer and Inner:

```
class Outer {
  class Inner
}
```

In Scala, the inner class is addressed using the expression Outer#Inner instead of Java's Outer.Inner. The '.' syntax is reserved for objects. For example, imagine you instantiate two objects of type Outer, like this:

```
val o1 = new Outer
val o2 = new Outer
```

Here o1.Inner and o2.Inner are two path-dependent types (and they are different types). Both of these types conform to (are subtypes of) the more general type Outer#Inner, which represents the Inner class with an *arbitrary* outer object of type Outer. By contrast, type o1.Inner refers to the Inner class with a *specific* outer object (the one referenced from o1). Likewise, type o2.Inner refers to the Inner class with a different, specific outer object (the one referenced from o2).

In Scala, as in Java, inner class instances hold a reference to an enclosing outer class instance. This allows an inner class, for example, to access members of its outer class. Thus you can't instantiate an inner class without in some way specifying an outer class instance. One way to do this is to instantiate the inner class inside the body of the outer class. In this case, the current outer class instance (referenced from this) will be used.

Another way is to use a path-dependent type. For example, because the type, o1.Inner, names a specific outer object, you can instantiate it:

```
scala> new o1.Inner
res11: o1.Inner = Outer$Inner@1ae1e03f
```

The resulting inner object will contain a reference to its outer object, the object referenced from o1. By contrast, because the type Outer#Inner does not name any specific instance of Outer, you can't create an instance of it:

```
scala> new Outer#Inner
<console>:9: error: Outer is not a legal prefix for a
constructor
            new Outer#Inner
                  ^
```

## 20.8   Refinement types

When a class inherits from another, the first class is said to be a *nominal* subtype of the other one. It's a *nominal* subtype because each type has a *name*, and the names are explicitly declared to have a subtyping relationship. Scala additionally supports *structural* subtyping, where you get a subtyping relationship simply because two types have compatible members. To get structural subtyping in Scala, use Scala's *refinement types*.

Nominal subtyping is usually more convenient, so you should try nominal types first with any new design. A name is a single short identifier and thus is more concise than an explicit listing of member types. Further, structural subtyping is often more flexible than you want. A widget can draw(), and a Western cowboy can draw(), but they aren't really substitutable. You'd typically prefer to get a compilation error if you tried to substitute a cowboy for a widget.

Nonetheless, structural subtyping has its own advantages. One is that sometimes there really is no more to a type than its members. For example, suppose you want to define a Pasture class that can contain animals that eat grass. One option would be to define a trait AnimalThatEatsGrass and mix it into every class where it applies. It would be verbose, however. Class Cow has already declared that it's an animal and that it eats grass, and now it would have to declare that it is also an animal-that-eats-grass.

Instead of defining AnimalThatEatsGrass, you can use a refinement type. Simply write the base type, Animal, followed by a sequence of members listed in curly braces. The members in the curly braces further specify— or refine, if you will—the types of members from the base class.

Here is how you write the type, "animal that eats grass":

```
Animal { type SuitableFood = Grass }
```

Given this type, you can now write the pasture class like this:

```
class Pasture {
  var animals: List[Animal { type SuitableFood = Grass }] = Nil
  // ...
}
```

## 20.9   Enumerations

An interesting application of path-dependent types is found in Scala's support for enumerations. Some other languages, including Java and C#, have enumerations as a built-in language construct to define new types. Scala does not need special syntax for enumerations. Instead, there's a class in its standard library, scala.Enumeration.

To create a new enumeration, you define an object that extends this class, as in the following example, which defines a new enumeration of Colors:

```
object Color extends Enumeration {
  val Red = Value
  val Green = Value
  val Blue = Value
}
```

Scala lets you also shorten several successive val or var definitions with the same right-hand side. Equivalently to the above you could write:

```
object Color extends Enumeration {
  val Red, Green, Blue = Value
}
```

This object definition provides three values: Color.Red, Color.Green, and Color.Blue. You could also import everything in Color with:

```
import Color._
```

and then just use Red, Green, and Blue. But what is the type of these values?

Enumeration defines an inner class named Value, and the same-named parameterless Value method returns a fresh instance of that class. In other words, a value such as Color.Red is of type Color.Value; Color.Value is the type of all enumeration values defined in object Color. It's a path-dependent type, with Color being the path and Value being the dependent type. What's significant about this is that it is a completely new type, different from all other types.

In particular, if you define another enumeration, such as:

```scala
object Direction extends Enumeration {
  val North, East, South, West = Value
}
```

then Direction.Value would be different from Color.Value because the path parts of the two types differ.

Scala's Enumeration class also offers many other features found in the enumeration designs of other languages. You can associate names with enumeration values by using a different overloaded variant of the Value method:

```scala
object Direction extends Enumeration {
  val North = Value("North")
  val East = Value("East")
  val South = Value("South")
  val West = Value("West")
}
```

You can iterate over the values of an enumeration via the set returned by the enumeration's values method:

```scala
scala> for (d <- Direction.values) print(d + " ")
North East South West
```

Values of an enumeration are numbered from 0, and you can find out the number of an enumeration value by its id method:

```scala
scala> Direction.East.id
res14: Int = 1
```

It's also possible to go the other way, from a non-negative integer number to the value that has this number as id in an enumeration:

432

```
scala> Direction(1)
res15: Direction.Value = East
```

This should be enough to get you started with enumerations. You can find
more information in the Scaladoc comments of class `scala.Enumeration`.

## 20.10   Case study: Currencies

The rest of this chapter presents a case study that explains how abstract types
can be used in Scala. The task is to design a class `Currency`. A typical
instance of `Currency` would represent an amount of money in dollars, euros,
yen, or some other currency. It should be possible to do some arithmetic on
currencies. For instance, you should be able to add two amounts of the same
currency. Or you should be able to multiply a currency amount by a factor
representing an interest rate.

These thoughts lead to the following first design for a currency class:

```
// A first (faulty) design of the Currency class
abstract class Currency {
  val amount: Long
  def designation: String
  override def toString = amount + " " + designation
  def + (that: Currency): Currency = ...
  def * (x: Double): Currency = ...
}
```

The `amount` of a currency is the number of currency units it represents. This
is a field of type Long so that very large amounts of money, such as the mar-
ket capitalization of Google or Apple, can be represented. It's left abstract
here, waiting to be defined when a subclass talks about concrete amounts of
money. The `designation` of a currency is a string that identifies it. The
`toString` method of class `Currency` indicates an amount and a designation.
It would yield results such as:

```
79 USD
11000 Yen
99 Euro
```

433

Finally, there are methods + for adding currencies and * for multiplying a currency with a floating-point number. You can create a concrete currency value by supplying concrete `amount` and `designation` values, like this:

```
new Currency {
  val amount = 79L
  def designation = "USD"
}
```

This design would be OK if all we wanted to model was a single currency, like only dollars or only euros. But it fails if we need to deal with several currencies. Assume that you model dollars and euros as two subclasses of class currency:

```
abstract class Dollar extends Currency {
  def designation = "USD"
}
abstract class Euro extends Currency {
  def designation = "Euro"
}
```

At first glance this looks reasonable. But it would let you add dollars to euros. The result of such an addition would be of type `Currency`. But it would be a funny currency that was made up of a mix of euros and dollars. What you want instead is a more specialized version of the + method. When implemented in class `Dollar`, it should take `Dollar` arguments and yield a `Dollar` result; when implemented in class `Euro`, it should take `Euro` arguments and yield a `Euro` result. So the type of the addition method would change depending on which class you are in. Nonetheless, you would like to write the addition method just once, not each time a new currency is defined.

In Scala, there's a simple technique to deal with situations like this. If something is not known at the point where a class is defined, make it abstract in the class. This applies to both values and types. In the case of currencies, the exact argument and result type of the addition method are not known, so it is a good candidate for an abstract type.

This would lead to the following sketch of class `AbstractCurrency`:

```
// A second (still imperfect) design of the Currency class
abstract class AbstractCurrency {
  type Currency <: AbstractCurrency
  val amount: Long
  def designation: String
  override def toString = amount + " " + designation
  def + (that: Currency): Currency = ...
  def * (x: Double): Currency = ...
}
```

The only differences from the previous situation are that the class is now
called AbstractCurrency, and that it contains an abstract type Currency,
which represents the real currency in question. Each concrete subclass of
AbstractCurrency would need to fix the Currency type to refer to the
concrete subclass itself, thereby "tying the knot."

For instance, here is a new version of class Dollar, which now extends
class AbstractCurrency:

```
abstract class Dollar extends AbstractCurrency {
  type Currency = Dollar
  def designation = "USD"
}
```

This design is workable, but it is still not perfect. One problem is hidden by
the ellipses that indicate the missing method definitions of + and * in class
AbstractCurrency. In particular, how should addition be implemented
in this class? It's easy enough to calculate the correct amount of the new
currency as this.amount + that.amount, but how would you convert the
amount into a currency of the right type?

You might try something like:

```
def + (that: Currency): Currency = new Currency {
  val amount = this.amount + that.amount
}
```

However, this would not compile:

```
error: class type required
   def + (that: Currency): Currency = new Currency {
                                          ^
```

One of the restrictions of Scala's treatment of abstract types is that you can neither create an instance of an abstract type nor have an abstract type as a supertype of another class.[1] So the compiler would refuse the example code here that attempted to instantiate Currency.

However, you can work around this restriction using a *factory method*. Instead of creating an instance of an abstract type directly, declare an abstract method that does it. Then, wherever the abstract type is fixed to be some concrete type, you also need to give a concrete implementation of the factory method. For class AbstractCurrency, this would look as follows:

```
abstract class AbstractCurrency {
  type Currency <: AbstractCurrency // abstract type
  def make(amount: Long): Currency  // factory method
  ...                               // rest of class
}
```

A design like this could be made to work, but it looks rather suspicious. Why place the factory method *inside* class AbstractCurrency? This looks dubious for at least two reasons. First, if you have some amount of currency (say, one dollar), you also hold in your hand the ability to make more of the same currency, using code such as:

```
myDollar.make(100)   // here are a hundred more!
```

In the age of color copying this might be a tempting scenario, but hopefully not one which you would be able to do for very long without being caught. The second problem with this code is that you can make more Currency objects if you already have a reference to a Currency object. But how do you get the first object of a given Currency? You'd need another creation method, which does essentially the same job as make. So you have a case of code duplication, which is a sure sign of a code smell.

The solution, of course, is to move the abstract type and the factory method outside class AbstractCurrency. You need to create another class that contains the AbstractCurrency class, the Currency type, and the make factory method.

We'll call this a CurrencyZone:

---

[1] There's some promising recent research on *virtual classes*, which would allow this, but virtual classes are not currently supported in Scala.

```
abstract class CurrencyZone {
  type Currency <: AbstractCurrency
  def make(x: Long): Currency
  abstract class AbstractCurrency {
    val amount: Long
    def designation: String
    override def toString = amount + " " + designation
    def + (that: Currency): Currency =
      make(this.amount + that.amount)
    def * (x: Double): Currency =
      make((this.amount * x).toLong)
  }
}
```

An example concrete `CurrencyZone` is the US, which could be defined as:

```
object US extends CurrencyZone {
  abstract class Dollar extends AbstractCurrency {
    def designation = "USD"
  }
  type Currency = Dollar
  def make(x: Long) = new Dollar { val amount = x }
}
```

Here, US is an object that extends `CurrencyZone`. It defines a class `Dollar`, which is a subclass of `AbstractCurrency`. So the type of money in this zone is `US.Dollar`. The US object also fixes the type `Currency` to be an alias for `Dollar`, and it gives an implementation of the make factory method to return a dollar amount.

This is a workable design. There are only a few refinements to be added. The first refinement concerns subunits. So far, every currency was measured in a single unit: dollars, euros, or yen. However, most currencies have subunits: For instance, in the US, it's dollars and cents. The most straightforward way to model cents is to have the amount field in US.Currency represent cents instead of dollars. To convert back to dollars, it's useful to introduce a field `CurrencyUnit` into class `CurrencyZone`, which contains the amount of one standard unit in that currency:

```
class CurrencyZone {
  ...
  val CurrencyUnit: Currency
}
```

As shown in Listing 20.11, The US object could define the quantities Cent, Dollar, and CurrencyUnit. This definition is just like the previous definition of the US object, except that it adds three new fields. The field Cent represents an amount of 1 US.Currency. It's an object analogous to a one-cent coin. The field Dollar represents an amount of 100 US.Currency. So the US object now defines the name Dollar in two ways. The *type* Dollar (defined by the abstract inner class named Dollar) represents the generic name of the Currency valid in the US currency zone. By contrast, the *value* Dollar (referenced from the val field named Dollar) represents a single US dollar, analogous to a one-dollar bill. The third field definition of CurrencyUnit specifies that the standard currency unit in the US zone is the Dollar (*i.e.*, the value Dollar, referenced from the field, not the type Dollar).

The toString method in class Currency also needs to be adapted to take subunits into account. For instance, the sum of ten dollars and twenty three cents should print as a decimal number: 10.23 USD. To achieve this, you could implement Currency's toString method as follows:

```
override def toString =
  ((amount.toDouble / CurrencyUnit.amount.toDouble)
    formatted ("%." + decimals(CurrencyUnit.amount) + "f")
    + " " + designation)
```

Here, formatted is a method that Scala makes available on several classes, including Double.[2] The formatted method returns the string that results from formatting the original string on which formatted was invoked according to a format string passed as the formatted method's right-hand operand. The syntax of format strings passed to formatted is the same as that of Java's String.format method.

For instance, the format string %.2f formats a number with two decimal digits. The format string used in the toString shown previously is assembled by calling the decimals method on CurrencyUnit.amount. This

---

[2]Scala uses rich wrappers, described in Section 5.10, to make formatted available.

```
object US extends CurrencyZone {
  abstract class Dollar extends AbstractCurrency {
    def designation = "USD"
  }
  type Currency = Dollar
  def make(cents: Long) = new Dollar {
    val amount = cents
  }
  val Cent = make(1)
  val Dollar = make(100)
  val CurrencyUnit = Dollar
}
```

Listing 20.11 · The US currency zone.

method returns the number of decimal digits of a decimal power minus one. For instance, decimals(10) is 1, decimals(100) is 2, and so on. The decimals method is implemented by a simple recursion:

```
private def decimals(n: Long): Int =
  if (n == 1) 0 else 1 + decimals(n / 10)
```

Listing 20.12 shows some other currency zones. As another refinement, you can add a currency conversion feature to the model. First, you could write a Converter object that contains applicable exchange rates between currencies, as shown in Listing 20.13. Then, you could add a conversion method, from, to class Currency, which converts from a given source currency into the current Currency object:

```
def from(other: CurrencyZone#AbstractCurrency): Currency =
  make(math.round(
    other.amount.toDouble * Converter.exchangeRate
      (other.designation)(this.designation)))
```

The from method takes an arbitrary currency as argument. This is expressed by its formal parameter type, CurrencyZone#AbstractCurrency, which indicates that the argument passed as other must be an AbstractCurrency type in some arbitrary and unknown CurrencyZone. It produces its result

```
object Europe extends CurrencyZone {
  abstract class Euro extends AbstractCurrency {
    def designation = "EUR"
  }
  type Currency = Euro
  def make(cents: Long) = new Euro {
    val amount = cents
  }
  val Cent = make(1)
  val Euro = make(100)
  val CurrencyUnit = Euro
}
object Japan extends CurrencyZone {
  abstract class Yen extends AbstractCurrency {
    def designation = "JPY"
  }
  type Currency = Yen
  def make(yen: Long) = new Yen {
    val amount = yen
  }
  val Yen = make(1)
  val CurrencyUnit = Yen
}
```

Listing 20.12 · Currency zones for Europe and Japan.

by multiplying the amount of the other currency with the exchange rate between the other and the current currency.[3]

The final version of the CurrencyZone class is shown in Listing 20.14. You can test the class in the Scala command shell. We'll assume that the CurrencyZone class and all concrete CurrencyZone objects are defined in a package org.stairwaybook.currencies. The first step is to import "org.stairwaybook.currencies._" into the command shell. Then you can do some currency conversions:

---

[3]By the way, in case you think you're getting a bad deal on Japanese yen, the exchange rates convert currencies based on their CurrencyZone amounts. Thus, 1.211 is the exchange rate between US cents and Japanese yen.

440

```scala
object Converter {
  var exchangeRate = Map(
    "USD" -> Map("USD" -> 1.0   , "EUR" -> 0.7596,
                 "JPY" -> 1.211 , "CHF" -> 1.223),
    "EUR" -> Map("USD" -> 1.316 , "EUR" -> 1.0   ,
                 "JPY" -> 1.594 , "CHF" -> 1.623),
    "JPY" -> Map("USD" -> 0.8257, "EUR" -> 0.6272,
                 "JPY" -> 1.0   , "CHF" -> 1.018),
    "CHF" -> Map("USD" -> 0.8108, "EUR" -> 0.6160,
                 "JPY" -> 0.982 , "CHF" -> 1.0   )
  )
}
```

Listing 20.13 · A converter object with an exchange rates map.

```scala
scala> Japan.Yen from US.Dollar * 100
res16: Japan.Currency = 12110 JPY

scala> Europe.Euro from res16
res17: Europe.Currency = 75.95 EUR

scala> US.Dollar from res17
res18: US.Currency = 99.95 USD
```

The fact that we obtain almost the same amount after three conversions implies that these are some pretty good exchange rates! You can also add up values of the same currency:

```scala
scala> US.Dollar * 100 + res18
res19: US.Currency = 199.95 USD
```

On the other hand, you cannot add amounts of different currencies:

```scala
scala> US.Dollar + Europe.Euro
<console>:12: error: type mismatch;
 found    : Europe.Euro
 required: US.Currency
    (which expands to)  US.Dollar
            US.Dollar + Europe.Euro
                      ^
```

441

```
abstract class CurrencyZone {

  type Currency <: AbstractCurrency
  def make(x: Long): Currency

  abstract class AbstractCurrency {

    val amount: Long
    def designation: String

    def + (that: Currency): Currency =
      make(this.amount + that.amount)
    def * (x: Double): Currency =
      make((this.amount * x).toLong)
    def - (that: Currency): Currency =
      make(this.amount - that.amount)
    def / (that: Double) =
      make((this.amount / that).toLong)
    def / (that: Currency) =
      this.amount.toDouble / that.amount

    def from(other: CurrencyZone#AbstractCurrency): Currency =
      make(math.round(
        other.amount.toDouble * Converter.exchangeRate
          (other.designation)(this.designation)))

    private def decimals(n: Long): Int =
      if (n == 1) 0 else 1 + decimals(n / 10)

    override def toString =
      ((amount.toDouble / CurrencyUnit.amount.toDouble)
        formatted ("%." + decimals(CurrencyUnit.amount) + "f")
      + " " + designation)
  }

  val CurrencyUnit: Currency
}
```

Listing 20.14 · The full code of class CurrencyZone.

442

By preventing the addition of two values with different units (in this case, currencies), the type abstraction has done its job. It prevents us from performing calculations that are unsound. Failures to convert correctly between different units may seem like trivial bugs, but they have caused many serious systems faults. An example is the crash of the Mars Climate Orbiter spacecraft on September 23, 1999, which was caused because one engineering team used metric units while another used English units. If units had been coded in the same way as currencies are coded in this chapter, this error would have been detected by a simple compilation run. Instead, it caused the crash of the orbiter after a near ten-month voyage.

## 20.11 Conclusion

Scala offers systematic and very general support for object-oriented abstraction. It enables you to not only abstract over methods, but also over values, variables, and types. This chapter has shown how to take advantage of abstract members. They support a simple yet effective principle for systems structuring: when designing a class, make everything that is not yet known into an abstract member. The type system will then drive the development of your model, just as you saw with the currency case study. It does not matter whether the unknown is a type, method, variable or value. In Scala, all of these can be declared abstract.

# Chapter 21

# Implicit Conversions and Parameters

There's a fundamental difference between your own code and other people's libraries: You can change or extend your own code as you wish, but if you want to use someone else's libraries, you usually have to take them as they are. A number of constructs have sprung up in programming languages to alleviate this problem. Ruby has modules, and Smalltalk lets packages add to each other's classes. These are very powerful but also dangerous, in that you can modify the behavior of a class for an entire application, some parts of which you might not know. C# 3.0 has static extension methods, which are more local but also more restrictive, in that you can only add methods, not fields, to a class, and you can't make a class implement new interfaces.

Scala's answer is implicit conversions and parameters. These can make existing libraries much more pleasant to deal with by letting you leave out tedious, obvious details that obscure the interesting parts of your code. Used tastefully, this results in code that is focused on the interesting, non-trivial parts of your program. This chapter shows you how implicits work, and it presents some of the most common ways they are used.

## 21.1 Implicit conversions

Before delving into the details of implicit conversions, take a look at a typical example of their use. Implicit conversions are often helpful for working with two bodies of software that were developed without each other in mind. Each library has its own way to encode a concept that is essentially the same thing. Implicit conversions help by reducing the number of explicit conversions that are needed from one type to another.

445

Java includes a library named Swing for implementing cross-platform user interfaces. One of the things Swing does is process events from the operating system, convert them to platform-independent event objects, and pass those events to parts of an application called event listeners.

If Swing had been written with Scala in mind, event listeners would probably have been represented by a function type. Callers could then use the function literal syntax as a lightweight way to specify what should happen for a certain class of events. Since Java doesn't have function literals, Swing uses the next best thing, an inner class that implements a one-method interface. In the case of action listeners, the interface is `ActionListener`.

Without the use of implicit conversions, a Scala program that uses Swing must use inner classes just like in Java. Here's an example that creates a button and hooks up an action listener to it. The action listener is invoked whenever the button is pressed, at which point it prints the string `"pressed!"`:

```
val button = new JButton
button.addActionListener(
  new ActionListener {
    def actionPerformed(event: ActionEvent) = {
      println("pressed!")
    }
  }
)
```

This code has a lot of information-free boilerplate. The fact that this listener is an `ActionListener`, the fact that the callback method is named `actionPerformed`, and the fact that the argument is an `ActionEvent` are all implied for any argument to `addActionListener`. The only new information here is the code to be performed, namely the call to `println`. This new information is drowned out by the boilerplate. Someone reading this code will need to have an eagle's eye to pick through the noise and find the informative part.

A more Scala-friendly version would take a function as an argument, greatly reducing the amount of boilerplate:

```
button.addActionListener( // Type mismatch!
  (_: ActionEvent) => println("pressed!")
)
```

As written so far, this code doesn't work.[1] The addActionListener method wants an action listener but is getting a function. With implicit conversions, however, this code can be made to work.

The first step is to write an implicit conversion between the two types. Here is an implicit conversion from functions to action listeners:

```
implicit def function2ActionListener(f: ActionEvent => Unit) =
  new ActionListener {
    def actionPerformed(event: ActionEvent) = f(event)
  }
```

This is a one-argument method that takes a function and returns an action listener. Like any other one-argument method, it can be called directly and have its result passed on to another expression:

```
button.addActionListener(
  function2ActionListener(
    (_: ActionEvent) => println("pressed!")
  )
)
```

This is already an improvement on the version with the inner class. Note how arbitrary amounts of boilerplate end up replaced by a function literal and a call to a method. It gets better, though, with implicit conversions. Because function2ActionListener is marked as implicit, it can be left out and the compiler will insert it automatically. Here is the result:

```
// Now this works
button.addActionListener(
  (_: ActionEvent) => println("pressed!")
)
```

The way this code works is that the compiler first tries to compile it as is, but it sees a type error. Before giving up, it looks for an implicit conversion that can repair the problem. In this case, it finds function2ActionListener. It tries that conversion method, sees that it works, and moves on. The compiler works hard here so that the developer can ignore one more fiddly detail.

---

[1] As will be explained in Section 31.5, it does work in Scala 2.12.

Action listener? Action event function? Either one will work—use the one that's more convenient.

In this section, we illustrated some some of the power of implicit conversions and how they let you dress up existing libraries. In the next sections, you'll learn the rules that determine when implicit conversions are tried and how they are found.

## 21.2  Rules for implicits

Implicit definitions are those that the compiler is allowed to insert into a program in order to fix any of its type errors. For example, if x + y does not type check, then the compiler might change it to convert(x) + y, where convert is some available implicit conversion. If convert changes x into something that has a + method, then this change might fix a program so that it type checks and runs correctly. If convert really is just a simple conversion function, then leaving it out of the source code can be a clarification.

Implicit conversions are governed by the following general rules:

**Marking rule: Only definitions marked `implicit` are available.**  The `implicit` keyword is used to mark which declarations the compiler may use as implicits. You can use it to mark any variable, function, or object definition. Here's an example of an implicit function definition:[2]

```
implicit def intToString(x: Int) = x.toString
```

The compiler will only change x + y to convert(x) + y if convert is marked as `implicit`. This way, you avoid the confusion that would result if the compiler picked random functions that happen to be in scope and inserted them as "conversions." The compiler will only select among the definitions you have explicitly marked as implicit.

**Scope rule: An inserted implicit conversion must be in scope as a single identifier, or be associated with the source or target type of the conversion.**  The Scala compiler will only consider implicit conversions that are in scope. To make an implicit conversion available, therefore, you must in

---

[2]Variables and singleton objects marked implicit can be used as *implicit parameters*. This use case will be described later in this chapter.

some way bring it into scope. Moreover, with one exception, the implicit conversion must be in scope *as a single identifier*. The compiler will not insert a conversion of the form someVariable.convert. For example, it will not expand x + y to someVariable.convert(x) + y. If you want to make someVariable.convert available as an implicit, you would need to import it, which would make it available as a single identifier. Once imported, the compiler would be free to apply it as convert(x) + y. In fact, it is common for libraries to include a Preamble object including a number of useful implicit conversions. Code that uses the library can then do a single "import Preamble._" to access the library's implicit conversions.

There's one exception to the "single identifier" rule. The compiler will also look for implicit definitions in the companion object of the source or expected target types of the conversion. For example, if you're attempting to pass a Dollar object to a method that takes a Euro, the source type is Dollar and the target type is Euro. You could, therefore, package an implicit conversion from Dollar to Euro in the companion object of either class, Dollar or Euro.

Here's an example in which the implicit definition is placed in Dollar's companion object:

```
object Dollar {
  implicit def dollarToEuro(x: Dollar): Euro = ...
}
class Dollar { ... }
```

In this case, the conversion dollarToEuro is said to be *associated* to the type Dollar. The compiler will find such an associated conversion every time it needs to convert from an instance of type Dollar. There's no need to import the conversion separately into your program.

The Scope Rule helps with modular reasoning. When you read code in a file, the only things you need to consider from other files are those that are either imported or are explicitly referenced through a fully qualified name. This benefit is at least as important for implicits as for explicitly written code. If implicits took effect system-wide, then to understand a file you would have to know about every implicit introduced anywhere in the program!

**One-at-a-time rule: Only one implicit is inserted.** The compiler will never rewrite x + y to convert1(convert2(x)) + y. Doing so would cause compile times to increase dramatically on erroneous code, and it would increase the difference between what the programmer writes and what the program actually does. For sanity's sake, the compiler does not insert further implicit conversions when it is already in the middle of trying another implicit. However, it's possible to circumvent this restriction by having implicits take implicit parameters, which will be described later in this chapter.

**Explicits-first rule: Whenever code type checks as it is written, no implicits are attempted.** The compiler will not change code that already works. A corollary of this rule is that you can always replace implicit identifiers by explicit ones, thus making the code longer but with less apparent ambiguity. You can trade between these choices on a case-by-case basis. Whenever you see code that seems repetitive and verbose, implicit conversions can help you decrease the tedium. Whenever code seems terse to the point of obscurity, you can insert conversions explicitly. The amount of implicits you leave the compiler to insert is ultimately a matter of style.

### Naming an implicit conversion

Implicit conversions can have arbitrary names. The name of an implicit conversion matters only in two situations: If you want to write it explicitly in a method application and for determining which implicit conversions are available at any place in the program. To illustrate the second point, say you have an object with two implicit conversions:

```
object MyConversions {
  implicit def stringWrapper(s: String):
      IndexedSeq[Char] = ...
  implicit def intToString(x: Int): String = ...
}
```

In your application, you want to make use of the `stringWrapper` conversion, but you don't want integers to be converted automatically to strings by means of the `intToString` conversion. You can achieve this by importing only one conversion, but not the other:

```
import MyConversions.stringWrapper
... // code making use of stringWrapper
```

In this example, it was important that the implicit conversions had names, because only that way could you selectively import one and not the other.

**Where implicits are tried**

There are three places implicits are used in the language: conversions to an expected type, conversions of the receiver of a selection, and implicit parameters. Implicit conversions to an expected type let you use one type in a context where a different type is expected. For example, you might have a String and want to pass it to a method that requires an IndexedSeq[Char]. Conversions of the receiver let you adapt the receiver of a method call (*i.e.*, the object on which a method is invoked), if the method is not applicable on the original type. An example is "abc".exists, which is converted to stringWrapper("abc").exists because the exists method is not available on Strings but is available on IndexedSeqs. Implicit parameters, on the other hand, are usually used to provide more information to the called function about what the caller wants. Implicit parameters are especially useful with generic functions, where the called function might otherwise know nothing at all about the type of one or more arguments. We will examine each of these three kinds of implicits in the next sections.

## 21.3   Implicit conversion to an expected type

Implicit conversion to an expected type is the first place the compiler will use implicits. The rule is simple. Whenever the compiler sees an X, but needs a Y, it will look for an implicit function that converts X to Y. For example, normally a double cannot be used as an integer because it loses precision:

```
scala> val i: Int = 3.5
<console>:7: error: type mismatch;
 found   : Double(3.5)
 required: Int
       val i: Int = 3.5
                    ^
```

However, you can define an implicit conversion to smooth this over:

```
scala> implicit def doubleToInt(x: Double) = x.toInt
doubleToInt: (x: Double)Int
```

```
scala> val i: Int = 3.5
i: Int = 3
```

What happens here is that the compiler sees a `Double`, specifically `3.5`, in a context where it requires an `Int`. So far, the compiler is looking at an ordinary type error. Before giving up, though, it searches for an implicit conversion from `Double` to `Int`. In this case, it finds one: `doubleToInt`, because `doubleToInt` is in scope as a single identifier. (Outside the interpreter, you might bring `doubleToInt` into scope via an `import` or possibly through inheritance.) The compiler then inserts a call to `doubleToInt` automatically. Behind the scenes, the code becomes:

```
val i: Int = doubleToInt(3.5)
```

This is literally an *implicit* conversion. You did not explicitly ask for conversion. Instead, you marked `doubleToInt` as an available implicit conversion by bringing it into scope as a single identifier, and then the compiler automatically used it when it needed to convert from a `Double` to an `Int`.

Converting `Double`s to `Int`s might raise some eyebrows because, it's a dubious idea to have something that causes a loss in precision happen invisibly. So this is not really a conversion we recommend. It makes much more sense to go the other way, from some more constrained type to a more general one. For instance, an `Int` can be converted without loss of precision to a `Double`, so an implicit conversion from `Int` to `Double` makes sense. In fact, that's exactly what happens. The `scala.Predef` object, which is implicitly imported into every Scala program, defines implicit conversions that convert "smaller" numeric types to "larger" ones. For instance, you will find in `Predef` the following conversion:

```
implicit def int2double(x: Int): Double = x.toDouble
```

That's why in Scala `Int` values can be stored in variables of type `Double`. There's no special rule in the type system for this; it's just an implicit conversion that gets applied.[3]

---

[3]The Scala compiler backend will treat the conversion specially, however, translating it to a special "i2d" bytecode. So the compiled image is the same as in Java.

## 21.4  Converting the receiver

Implicit conversions also apply to the receiver of a method call, the object on which the method is invoked. This kind of implicit conversion has two main uses. First, receiver conversions allow smoother integration of a new class into an existing class hierarchy. And second, they support writing domain-specific languages (DSLs) within the language.

To see how it works, suppose you write down `obj.doIt`, and `obj` does not have a member named `doIt`. The compiler will try to insert conversions before giving up. In this case, the conversion needs to apply to the receiver, `obj`. The compiler will act as if the expected "type" of `obj` was "has a member named `doIt`." This "has a `doIt`" type is not a normal Scala type, but it is there conceptually and is why the compiler will insert an implicit conversion in this case.

### Interoperating with new types

As mentioned previously, one major use of receiver conversions is allowing smoother integration of new types with existing types. In particular, they allow you to enable client programmers to use instances of existing types as if they were instances of your new type. Take, for example, class `Rational` shown in Listing 6.5 on page 113. Here's a snippet of that class again:

```
class Rational(n: Int, d: Int) {
  ...
  def + (that: Rational): Rational = ...
  def + (that: Int): Rational = ...
}
```

Class `Rational` has two overloaded variants of the `+` method, which take `Rational`s and `Int`s, respectively, as arguments. So you can either add two rational numbers or a rational number and an integer:

```
scala> val oneHalf = new Rational(1, 2)
oneHalf: Rational = 1/2

scala> oneHalf + oneHalf
res0: Rational = 1/1

scala> oneHalf + 1
res1: Rational = 3/2
```

What about an expression like 1 + oneHalf? This expression is tricky because the receiver, 1, does not have a suitable + method. So the following gives an error:

```
scala> 1 + oneHalf
<console>:6: error: overloaded method value + with
alternatives (Double)Double <and> ... cannot be applied
to (Rational)
       1 + oneHalf
         ^
```

To allow this kind of mixed arithmetic, you need to define an implicit conversion from Int to Rational:

```
scala> implicit def intToRational(x: Int) =
         new Rational(x, 1)
intToRational: (x: Int)Rational
```

With the conversion in place, converting the receiver does the trick:

```
scala> 1 + oneHalf
res2: Rational = 3/2
```

What happens behind the scenes here is that the Scala compiler first tries to type check the expression 1 + oneHalf as it is. This fails because Int has several + methods, but none that takes a Rational argument. Next, the compiler searches for an implicit conversion from Int to another type that has a + method which can be applied to a Rational. It finds your conversion and applies it, which yields:

```
intToRational(1) + oneHalf
```

In this case, the compiler found the implicit conversion function because you entered its definition into the interpreter, which brought it into scope for the remainder of the interpreter session.

### Simulating new syntax

The other major use of implicit conversions is to simulate adding new syntax. Recall that you can make a Map using syntax like this:

```
Map(1 -> "one", 2 -> "two", 3 -> "three")
```

454

Have you wondered how the -> is supported? It's not syntax! Instead, -> is a method of the class ArrowAssoc, a class defined inside the standard Scala preamble (scala.Predef). The preamble also defines an implicit conversion from Any to ArrowAssoc. When you write 1 -> "one", the compiler inserts a conversion from 1 to ArrowAssoc so that the -> method can be found. Here are the relevant definitions:

```
package scala
object Predef {
  class ArrowAssoc[A](x: A) {
    def -> [B](y: B): Tuple2[A, B] = Tuple2(x, y)
  }
  implicit def any2ArrowAssoc[A](x: A): ArrowAssoc[A] =
    new ArrowAssoc(x)
  ...
}
```

This "rich wrappers" pattern is common in libraries that provide syntax-like extensions to the language, so you should be ready to recognize the pattern when you see it. Whenever you see someone calling methods that appear not to exist in the receiver class, they are probably using implicits. Similarly, if you see a class named RichSomething (*e.g.*, RichInt or RichBoolean), that class is likely adding syntax-like methods to type Something.

You have already seen this rich wrappers pattern for the basic types described in Chapter 5. As you can now see, these rich wrappers apply more widely, often letting you get by with an internal DSL defined as a library where programmers in other languages might feel the need to develop an external DSL.

**Implicit classes**

Implicit classes were added in Scala 2.10 to make it easier to write rich wrapper classes. An implicit class is a class that is preceded by the implicit keyword. For any such class, the compiler generates an implicit conversion from the class's constructor parameter to the class itself. Such a conversion is just what you need if you plan to use the class for the rich wrappers pattern.

For example, suppose you have a class named Rectangle for representing the width and height of a rectangle on the screen:

```
case class Rectangle(width: Int, height: Int)
```

If you use this class very frequently, you might want to use the rich wrappers pattern so you can more easily construct it. Here's one way to do so.

```
implicit class RectangleMaker(width: Int) {
  def x(height: Int) = Rectangle(width, height)
}
```

The above definition defines a RectangleMaker class in the usual manner. In addition, it causes the following conversion to be automatically generated:

```
// Automatically generated
implicit def RectangleMaker(width: Int) =
  new RectangleMaker(width)
```

As a result, you can create points by putting an x in between two integers:

```
scala> val myRectangle = 3 x 4
  myRectangle: Rectangle = Rectangle(3,4)
```

This is how it works: Since type Int has no method named x, the compiler will look for an implicit conversion from Int to something that does. It will find the generated RectangleMaker conversion, and RectangleMaker does have a method named x. The compiler inserts a call to this conversion, after which the call to x type checks and does what is desired.

As a warning to the adventurous, it might be tempting to think that any class can have implicit put in front of it. It's not so. An implicit class cannot be a case class, and its constructor must have exactly one parameter. Also, an implicit class must be located within some other object, class, or trait. In practice, so long as you use implicit classes as rich wrappers to add a few methods onto an existing class, these restrictions should not matter.

## 21.5   Implicit parameters

The remaining place the compiler inserts implicits is within argument lists. The compiler will sometimes replace someCall(a) with someCall(a)(b), or new SomeClass(a) with new SomeClass(a)(b), thereby adding a missing parameter list to complete a function call. It is the entire last curried parameter list that's supplied, not just the last parameter. For example, if someCall's missing last parameter list takes three parameters, the compiler might replace someCall(a) with someCall(a)(b, c, d). For this usage, not only must the inserted identifiers, such as b, c, and d in (b, c, d), be marked implicit where they are defined, but also the last parameter list in someCall's or someClass's definition must be marked implicit.

Here's a simple example. Suppose you have a class PreferredPrompt, which encapsulates a shell prompt string (such as, say "$ " or "> ") that is preferred by a user:

```
class PreferredPrompt(val preference: String)
```

Also, suppose you have a Greeter object with a greet method, which takes two parameter lists. The first parameter list takes a string user name, and the second parameter list takes a PreferredPrompt:

```
object Greeter {
  def greet(name: String)(implicit prompt: PreferredPrompt) = {
    println("Welcome, " + name + ". The system is ready.")
    println(prompt.preference)
  }
}
```

The last parameter list is marked implicit, which means it can be supplied implicitly. But you can still provide the prompt explicitly, like this:

```
scala> val bobsPrompt = new PreferredPrompt("relax> ")
bobsPrompt: PreferredPrompt = PreferredPrompt@714d36d6

scala> Greeter.greet("Bob")(bobsPrompt)
Welcome, Bob. The system is ready.
relax>
```

To let the compiler supply the parameter implicitly, you must first define a variable of the expected type, which in this case is PreferredPrompt. You could do this, for example, in a preferences object:

```
object JoesPrefs {
  implicit val prompt = new PreferredPrompt("Yes, master> ")
}
```

457

Note that the val itself is marked implicit. If it wasn't, the compiler would not use it to supply the missing parameter list. It will also not use it if it isn't in scope as a single identifier, as shown in this example:

```
scala> Greeter.greet("Joe")
<console>:13: error: could not find implicit value for
parameter prompt: PreferredPrompt
              Greeter.greet("Joe")
                      ^
```

Once you bring it into scope via an import, however, it will be used to supply the missing parameter list:

```
scala> import JoesPrefs._
import JoesPrefs._

scala> Greeter.greet("Joe")
Welcome, Joe. The system is ready.
Yes, master>
```

Note that the implicit keyword applies to an entire parameter list, not to individual parameters. Listing 21.1 shows an example in which the last parameter list of Greeter's greet method, which is again marked implicit, has two parameters: prompt (of type PreferredPrompt) and drink (of type PreferredDrink).

Singleton object JoesPrefs declares two implicit vals, prompt of type PreferredPrompt and drink of type PreferredDrink. As before, however, so long as these are not in scope as single identifiers, they won't be used to fill in a missing parameter list to greet:

```
scala> Greeter.greet("Joe")
<console>:19: error: could not find implicit value for
parameter prompt: PreferredPrompt
              Greeter.greet("Joe")
                      ^
```

You can bring both implicit vals into scope with an import:

```
scala> import JoesPrefs._
import JoesPrefs._
```

Because both prompt and drink are now in scope as single identifiers, you can use them to supply the last parameter list explicitly, like this:

```
class PreferredPrompt(val preference: String)
class PreferredDrink(val preference: String)

object Greeter {
  def greet(name: String)(implicit prompt: PreferredPrompt,
      drink: PreferredDrink) = {

    println("Welcome, " + name + ". The system is ready.")
    print("But while you work, ")
    println("why not enjoy a cup of " + drink.preference + "?")
    println(prompt.preference)
  }
}

object JoesPrefs {
  implicit val prompt = new PreferredPrompt("Yes, master> ")
  implicit val drink = new PreferredDrink("tea")
}
```

Listing 21.1 · An implicit parameter list with multiple parameters.

```
scala> Greeter.greet("Joe")(prompt, drink)
Welcome, Joe. The system is ready.
But while you work, why not enjoy a cup of tea?
Yes, master>
```

And because all the rules for implicit parameters are now met, you can alternatively let the Scala compiler supply prompt and drink for you by leaving off the last parameter list:

```
scala> Greeter.greet("Joe")
Welcome, Joe. The system is ready.
But while you work, why not enjoy a cup of tea?
Yes, master>
```

One thing to note about the previous examples is that we didn't use String as the type of prompt or drink, even though ultimately it was a String that each of them provided through their preference fields. Because the compiler selects implicit parameters by matching types of parameters against types of values in scope, implicit parameters usually have "rare" or "special" enough types that accidental matches are unlikely. For example, the types PreferredPrompt and PreferredDrink in Listing 21.1 were defined solely to serve as implicit parameter types. As a result, it is unlikely that implicit variables of these types will be in scope if they aren't intended to be used as implicit parameters to Greeter.greet.

Another thing to know about implicit parameters is that they are perhaps most often used to provide information about a type mentioned *explicitly* in an earlier parameter list, similar to the type classes of Haskell.

As an example, consider the maxListOrdering function shown in Listing 21.2, which returns the maximum element of the passed list.

```
def maxListOrdering[T](elements: List[T])
      (ordering: Ordering[T]): T =
   elements match {
     case List() =>
       throw new IllegalArgumentException("empty list!")
     case List(x) => x
     case x :: rest =>
       val maxRest = maxListOrdering(rest)(ordering)
       if (ordering.gt(x, maxRest)) x
       else maxRest
   }
```

Listing 21.2 · A function with an upper bound.

The signature of maxListOrdering is similar to that of orderedMergeSort, shown in Listing 19.12 on page 409: It takes a List[T] as its argument, and now it takes an additional argument of type Ordering[T]. This additional argument specifies which ordering to use when comparing elements of type T. As such, this version can be used for types that don't have a built-in ordering. Additionally, this version can be used for types that *do* have a built-in ordering, but for which you occasionally want to use some other ordering.

This version is more general, but it's also more cumbersome to use. Now a caller must specify an explicit ordering even if T is something like String or Int that has an obvious default ordering. To make the new method more convenient, it helps to make the second argument implicit. This approach is shown in Listing 21.3.

The ordering parameter in this example is used to describe the ordering

```
def maxListImpParm[T](elements: List[T])
      (implicit ordering: Ordering[T]): T =

  elements match {
    case List() =>
      throw new IllegalArgumentException("empty list!")
    case List(x) => x
    case x :: rest =>
      val maxRest = maxListImpParm(rest)(ordering)
      if (ordering.gt(x, maxRest)) x
      else maxRest
  }
```

Listing 21.3 · A function with an implicit parameter.

of Ts. In the body of maxListImpParm, this ordering is used in two places: a recursive call to maxListImpParm, and an if expression that checks whether the head of the list is larger than the maximum element of the rest of the list.

The maxListImpParm function is an example of an implicit parameter used to provide more information about a type mentioned explicitly in an earlier parameter list. To be specific, the implicit parameter ordering, of type Ordering[T], provides more information about type T—in this case, how to order Ts. Type T is mentioned in List[T], the type of parameter elements, which appears in the earlier parameter list. Because elements must always be provided explicitly in any invocation of maxListImpParm, the compiler will know T at compile time and can therefore determine whether an implicit definition of type Ordering[T] is available. If so, it can pass in the second parameter list, ordering, implicitly.

This pattern is so common that the standard Scala library provides implicit "ordering" methods for many common types. You could therefore use this maxListImpParm method with a variety of types:

```
scala> maxListImpParm(List(1,5,10,3))
res9: Int = 10

scala> maxListImpParm(List(1.5, 5.2, 10.7, 3.14159))
res10: Double = 10.7

scala> maxListImpParm(List("one", "two", "three"))
res11: String = two
```

In the first case, the compiler inserted an `ordering` for `Ints`; in the second case, for `Doubles`; in the third case, for `Strings`.

**A style rule for implicit parameters**

As a style rule, it is best to use a custom named type in the types of implicit parameters. For example, the types of `prompt` and `drink` in the previous example was not `String`, but `PreferredPrompt` and `PreferredDrink`, respectively. As a counterexample, consider that the `maxListImpParm` function could just as well have been written with the following type signature:

```
def maxListPoorStyle[T](elements: List[T])
    (implicit orderer: (T, T) => Boolean): T
```

To use this version of the function, though, the caller would have to supply an `orderer` parameter of type `(T, T) => Boolean`. This is a fairly generic type that includes any function from two `Ts` to a `Boolean`. It does not indicate anything at all about what the type is for; it could be an equality test, a less-than test, a greater-than test, or something else entirely.

The actual code for `maxListImpParm`, given in Listing 21.3, shows better style. It uses an `ordering` parameter of type `Ordering[T]`. The word `Ordering` in this type indicates exactly what the implicit parameter is used for: it is for ordering elements of T. Because this `ordering` type is more explicit, it's no trouble to add implicit providers for this type in the standard library. To contrast, imagine the chaos that would ensue if you added an implicit of type `(T, T) => Boolean` in the standard library, and the compiler started sprinkling it around in people's code. You would end up with code that compiles and runs, but that does fairly arbitrary tests against pairs of items! Thus the style rule: Use at least one role-determining name within the type of an implicit parameter.

## 21.6 Context bounds

The previous example showed an opportunity to use an implicit but did not. Note that when you use `implicit` on a parameter, not only will the compiler try to *supply* that parameter with an implicit value, but the compiler will also *use* that parameter as an available implicit in the body of the method! Thus, the first use of `ordering` within the body of the method can be left out.

```
def maxList[T](elements: List[T])
    (implicit ordering: Ordering[T]): T =

  elements match {
    case List() =>
      throw new IllegalArgumentException("empty list!")
    case List(x) => x
    case x :: rest =>
      val maxRest = maxList(rest)       // (ordering) is implicit
      if (ordering.gt(x, maxRest)) x  // this ordering is
      else maxRest                    // still explicit
  }
```

Listing 21.4 · A function that uses an implicit parameter internally.

When the compiler examines the code in Listing 21.4, it will see that the types do not match up. The expression maxList(rest) only supplies one parameter list, but maxList requires two. Since the second parameter list is implicit, the compiler does not give up type checking immediately. Instead, it looks for an implicit parameter of the appropriate type, in this case Ordering[T]. In this case, it finds one and rewrites the call to maxList(rest)(ordering), after which the code type checks.

There is also a way to eliminate the second use of ordering. It involves the following method defined in the standard library:

```
def implicitly[T](implicit t: T) = t
```

The effect of calling implicitly[Foo] is that the compiler will look for an implicit definition of type Foo. It will then call the implicitly method with that object, which in turn returns the object right back. Thus you can write implicitly[Foo] whenever you want to find an implicit object of type Foo in the current scope. For example, Listing 21.5 shows a use of implicitly[Ordering[T]] to retrieve the ordering parameter by its type.

Look closely at this last version of maxList. There is not a single mention of the ordering parameter in the text of the method. The second parameter could just as well be named "comparator":

```
def maxList[T](elements: List[T])
      (implicit ordering: Ordering[T]): T =

  elements match {
    case List() =>
      throw new IllegalArgumentException("empty list!")
    case List(x) => x
    case x :: rest =>
      val maxRest = maxList(rest)
      if (implicitly[Ordering[T]].gt(x, maxRest)) x
      else maxRest
  }
```

Listing 21.5 · A function that uses implicitly.

```
def maxList[T](elements: List[T])
      (implicit comparator: Ordering[T]): T = // same body...
```

For that matter, this version works as well:

```
def maxList[T](elements: List[T])
      (implicit iceCream: Ordering[T]): T = // same body...
```

Because this pattern is common, Scala lets you leave out the name of this parameter and shorten the method header by using a *context bound*. Using a context bound, you would write the signature of maxList as shown in Listing 21.6. The syntax [T : Ordering] is a context bound, and it does two things. First, it introduces a type parameter T as normal. Second, it adds an implicit parameter of type Ordering[T]. In previous versions of maxList, that parameter was called ordering, but when using a context bound you don't know what the parameter will be called. As shown earlier, you often don't need to know what the parameter is called.

Intuitively, you can think of a context bound as saying something *about* a type parameter. When you write [T <: Ordered[T]] you are saying that a T *is* an Ordered[T]. To contrast, when you write [T : Ordering] you are not so much saying what T is; rather, you are saying that there is some form of ordering associated with T. Thus, a context bound is quite flexible. It allows you to use code that requires orderings—or any other property of a type—without having to change the definition of that type.

464

```scala
def maxList[T : Ordering](elements: List[T]): T =
  elements match {
    case List() =>
      throw new IllegalArgumentException("empty list!")
    case List(x) => x
    case x :: rest =>
      val maxRest = maxList(rest)
      if (implicitly[Ordering[T]].gt(x, maxRest)) x
      else maxRest
  }
```

Listing 21.6 · A function with a context bound.

## 21.7  When multiple conversions apply

It can happen that multiple implicit conversions are in scope and each would
work. For the most part, Scala refuses to insert a conversion in such a case.
Implicits work well when the conversion left out is completely obvious and
pure boilerplate. If multiple conversions apply, then the choice isn't so obvious after all.

Here's a simple example. There is a method that takes a sequence, a
conversion that turns an integer into a range, and a conversion that turns an
integer into a list of digits:

```scala
scala> def printLength(seq: Seq[Int]) = println(seq.length)
printLength: (seq: Seq[Int])Unit

scala> implicit def intToRange(i: Int) = 1 to i
intToRange: (i:
Int)scala.collection.immutable.Range.Inclusive

scala> implicit def intToDigits(i: Int) =
         i.toString.toList.map(_.toInt)
intToDigits: (i: Int)List[Int]

scala> printLength(12)
<console>:26: error: type mismatch;
 found    : Int(12)
 required: Seq[Int]
Note that implicit conversions are not applicable because
```

465

```
they are ambiguous:
 both method intToRange of type (i:
Int)scala.collection.immutable.Range.Inclusive
 and method intToDigits of type (i: Int)List[Int]
 are possible conversion functions from Int(12) to Seq[Int]
            printLength(12)
                  ^
```

The ambiguity here is real. Converting an integer to a sequence of digits is completely different from converting it to a range. In this case, the programmer should specify which one is intended and be explicit. Up through Scala 2.7, that was the end of the story. Whenever multiple implicit conversions applied, the compiler refused to choose between them. The situation was just as with method overloading. If you try to call foo(null) and there are two different foo overloads that accept null, the compiler will refuse. It will say that the method call's target is ambiguous.

Scala 2.8 loosened this rule. If one of the available conversions is strictly *more specific* than the others, then the compiler will choose the more specific one. The idea is that whenever there is a reason to believe a programmer would always choose one of the conversions over the others, don't require the programmer to write it explicitly. After all, method overloading has the same relaxation. Continuing the previous example, if one of the available foo methods takes a String while the other takes an Any, then choose the String version. It's clearly more specific.

To be more precise, one implicit conversion is *more specific* than another if one of the following applies:

- The argument type of the former is a subtype of the latter's.

- Both conversions are methods, and the enclosing class of the former extends the enclosing class of the latter.

The motivation to revisit this issue and revise the rule was to improve interoperation between Java collections, Scala collections, and strings. Here's a simple example:

```
val cba = "abc".reverse
```

What is the type inferred for cba? Intuitively, the type should be String. Reversing a string should yield another string, right? However, in Scala 2.7, what happened was that "abc" was converted to a Scala collection. Reversing a Scala collection yields a Scala collection, so the type of cba would be a collection. There's also an implicit conversion back to a string, but that didn't patch up every problem. For example, in versions prior to Scala 2.8, "abc" == "abc".reverse.reverse was false!

With Scala 2.8, the type of cba is String. The old implicit conversion to a Scala collection (now named WrappedString) is retained. However, there is a more specific conversion supplied from String to a new type called StringOps. StringOps has many methods such as reverse, but instead of returning a collection, they return a String. The conversion to StringOps is defined directly in Predef, whereas the conversion to a Scala collection is defined in a new class, LowPriorityImplicits, which is extended by Predef. Whenever a choice exists between these two conversions, the compiler chooses the conversion to StringOps, because it's defined in a subclass of the class where the other conversion is defined.

## 21.8   Debugging implicits

Implicits are a powerful feature in Scala, but one that's sometimes difficult to get right. This section contains a few tips for debugging implicits.

Sometimes you might wonder why the compiler did not find an implicit conversion that you think should apply. In that case it helps to write the conversion out explicitly. If that also gives an error message, you then know why the compiler could not apply your implicit.

For instance, assume that you mistakenly took wrapString to be a conversion from Strings to Lists, instead of IndexedSeqs. You would wonder why the following code does not work:

```
scala> val chars: List[Char] = "xyz"
<console>:24: error: type mismatch;
 found    : String("xyz")
 required: List[Char]
        val chars: List[Char] = "xyz"
                                 ^
```

Again, it helps to write the wrapString conversion explicitly to find out what went wrong:

```
scala> val chars: List[Char] = wrapString("xyz")
<console>:24: error: type mismatch;
 found    : scala.collection.immutable.WrappedString
 required: List[Char]
       val chars: List[Char] = wrapString("xyz")
                               ^
```

With this, you have found the cause of the error: `wrapString` has the wrong
return type. On the other hand, it's also possible that inserting the conversion
explicitly will make the error go away. In that case you know that one of the
other rules (such as the Scope Rule) was preventing the implicit conversion
from being applied.

When you are debugging a program, it can sometimes help to see what
implicit conversions the compiler is inserting. The `-Xprint:typer` option to
the compiler is useful for this. If you run `scalac` with this option, the compiler will show you what your code looks like after all implicit conversions
have been added by the type checker. An example is shown in Listing 21.7
and Listing 21.8. If you look at the last statement in each of these listings,
you'll see that the second parameter list to `enjoy`, which was left off in the
code in Listing 21.7, "`enjoy("reader")`," was inserted by the compiler, as
shown in Listing 21.8:

```
Mocha.this.enjoy("reader")(Mocha.this.pref)
```

If you are brave, try `scala -Xprint:typer` to get an interactive shell
that prints out the post-typing source code it uses internally. If you do so, be
prepared to see an enormous amount of boilerplate surrounding the meat of
your code.

## 21.9   Conclusion

Implicits are a powerful, code-condensing feature of Scala. This chapter has
shown you Scala's rules about implicits and several common programming
situations where you can profit from using implicits.

As a word of warning, implicits can make code confusing if they are
used too frequently. Thus, before adding a new implicit conversion, first
ask whether you can achieve a similar effect through other means, such as
inheritance, mixin composition, or method overloading. If all of these fail,
however, and you feel like a lot of your code is still tedious and redundant,
then implicits might just be able to help you out.

```scala
object Mocha extends App {

  class PreferredDrink(val preference: String)

  implicit val pref = new PreferredDrink("mocha")

  def enjoy(name: String)(implicit drink: PreferredDrink) = {
    print("Welcome, " + name)
    print(". Enjoy a ")
    print(drink.preference)
    println("!")
  }

  enjoy("reader")
}
```

Listing 21.7 · Sample code that uses an implicit parameter.

```
$ scalac -Xprint:typer mocha.scala
[[syntax trees at end of typer]]// Scala source: mocha.scala
package <empty> {
  final object Mocha extends java.lang.Object with Application
      with ScalaObject {
    // ...
    private[this] val pref: Mocha.PreferredDrink =
      new Mocha.this.PreferredDrink("mocha");
    implicit <stable> <accessor>
      def pref: Mocha.PreferredDrink = Mocha.this.pref;
    def enjoy(name: String)
        (implicit drink: Mocha.PreferredDrink): Unit = {
      scala.this.Predef.print("Welcome, ".+(name));
      scala.this.Predef.print(". Enjoy a ");
      scala.this.Predef.print(drink.preference);
      scala.this.Predef.println("!")
    };
    Mocha.this.enjoy("reader")(Mocha.this.pref)
  }
}
```

Listing 21.8 · Sample code after type checking and insertion of implicits.

# Chapter 22

# Implementing Lists

Lists have been ubiquitous in this book. Class List is probably the most commonly used structured data type in Scala. Chapter 16 showed you how to use lists. This chapter "opens up the covers" and explains a bit about how lists are implemented in Scala.

Knowing the internal workings of the List class is useful for several reasons. You gain a better idea of the relative efficiency of list operations, which will help you in writing fast and compact code using lists. You also gain a toolbox of techniques that you can apply in the design of your own libraries. Finally, the List class is a sophisticated application of Scala's type system in general and its genericity concepts in particular. So studying class List will deepen your knowledge in these areas.

## 22.1  The List class in principle

Lists are not "built-in" as a language construct in Scala; they are defined by an abstract class List in the scala package, which comes with two subclasses for :: and Nil. In this chapter we will present a quick tour through class List. This section presents a somewhat simplified account of the class, compared to its real implementation in the Scala standard library, which is covered in Section 22.3.

```
package scala
abstract class List[+T] {
```

List is an abstract class, so you cannot define elements by calling the empty List constructor. For instance the expression "new List" would be ille-

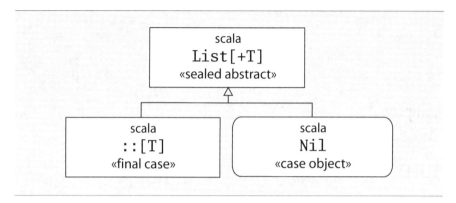

Figure 22.1 · Class hierarchy for Scala lists.

gal. The class has a type parameter T. The + in front of this type parameter specifies that lists are covariant, as discussed in Chapter 19.

Because of this property, you can assign a value of type List[Int] to a variable of type List[Any]:

```
scala> val xs = List(1, 2, 3)
xs: List[Int] = List(1, 2, 3)

scala> var ys: List[Any] = xs
ys: List[Any] = List(1, 2, 3)
```

All list operations can be defined in terms of three basic methods:

```
def isEmpty: Boolean
def head: T
def tail: List[T]
```

These three methods are all abstract in class List. They are defined in the subobject Nil and the subclass ::. The hierarchy for List is shown in Figure 22.1.

### The Nil object

The Nil object defines an empty list. Its definition is shown in Listing 22.1. The Nil object inherits from type List[Nothing]. Because of covariance, this means that Nil is compatible with every instance of the List type.

```
case object Nil extends List[Nothing] {
  override def isEmpty = true
  def head: Nothing =
    throw new NoSuchElementException("head of empty list")
  def tail: List[Nothing] =
    throw new NoSuchElementException("tail of empty list")
}
```

Listing 22.1 · The definition of the Nil singleton object.

The three abstract methods of class List are implemented in the Nil object in a straightforward way: The isEmpty method returns true, and the head and tail methods both throw an exception. Note that throwing an exception is not only reasonable, but practically the only possible thing to do for head: Because Nil is a List of Nothing, the result type of head must be Nothing. Since there is no value of this type, this means that head cannot return a normal value. It has to return abnormally by throwing an exception.[1]

**The :: class**

Class ::, pronounced "cons" for "construct," represents non-empty lists. It's named that way in order to support pattern matching with the infix ::. You have seen in Section 16.5 that every infix operation in a pattern is treated as a constructor application of the infix operator to its arguments. So the pattern x :: xs is treated as ::(x, xs) where :: is a case class.

Here is the definition of the :: class:

```
final case class ::[T](hd: T, tl: List[T]) extends List[T] {
  def head = hd
  def tail = tl
  override def isEmpty: Boolean = false
}
```

The implementation of the :: class is straightforward. It takes two parameters hd and tl, representing the head and the tail of the list to be constructed.

---

[1]To be precise, the types would also permit for head to always go into an infinite loop instead of throwing an exception, but this is clearly not what's wanted.

473

The definitions of the head and `tail` method simply return the corresponding parameter. In fact, this pattern can be abbreviated by letting the parameters directly implement the head and `tail` methods of the superclass `List`,
as in the following equivalent but shorter definition of the `::` class:

```
final case class ::[T](head: T, tail: List[T])
    extends List[T] {

  override def isEmpty: Boolean = false
}
```

This works because every case class parameter is implicitly also a field of the
class (it's like the parameter declaration was prefixed with `val`). Recall from
Section 20.3 that Scala allows you to implement an abstract parameterless
method such as head or `tail` with a field. So the code above directly uses
the parameters head and `tail` as implementations of the abstract methods
head and `tail` that were inherited from class `List`.

### Some more methods

All other `List` methods can be written using the basic three. For instance:

```
def length: Int =
  if (isEmpty) 0 else 1 + tail.length
```

or:

```
def drop(n: Int): List[T] =
  if (isEmpty) Nil
  else if (n <= 0) this
  else tail.drop(n - 1)
```

or:

```
def map[U](f: T => U): List[U] =
  if (isEmpty) Nil
  else f(head) :: tail.map(f)
```

**List construction**

The list construction methods :: and ::: are special. Because they end in a colon, they are bound to their right operand. That is, an operation such as x :: xs is treated as the method call xs.::(x), not x.::(xs). In fact, x.::(xs) would not make sense, as x is of the list element type, which can be arbitrary, so we cannot assume that this type would have a :: method.

For this reason, the :: method should take an element value and yield a new list. What is the required type of the element value? You might be tempted to say it should be the same as the list's element type, but in fact this is more restrictive than necessary.

To see why, consider this class hierarchy:

```scala
abstract class Fruit
class Apple extends Fruit
class Orange extends Fruit
```

Listing 22.2 shows what happens when you construct lists of fruit:

```scala
scala> val apples = new Apple :: Nil
apples: List[Apple] = List(Apple@e885c6a)

scala> val fruits = new Orange :: apples
fruits: List[Fruit] = List(Orange@3f51b349, Apple@e885c6a)
```

Listing 22.2 · Prepending a supertype element to a subtype list.

The apples value is treated as a List of Apples, as expected. However, the definition of fruits shows that it's still possible to add an element of a different type to that list. The element type of the resulting list is Fruit, which is the most precise common supertype of the original list element type (*i.e.*, Apple) and the type of the element to be added (*i.e.*, Orange). This flexibility is obtained by defining the :: method (cons) as shown in Listing 22.3:

```scala
def ::[U >: T](x: U): List[U] = new scala.::(x, this)
```

Listing 22.3 · The definition of method :: (cons) in class List.

Note that the method is itself polymorphic—it takes a type parameter named U. Furthermore, U is constrained in [U >: T] to be a supertype of the

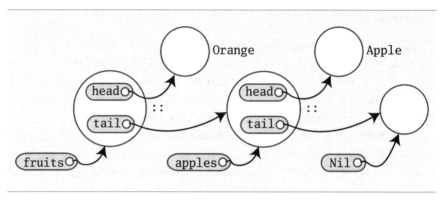

Figure 22.2 · The structure of the Scala lists shown in Listing 22.2.

list element type T. The element to be added is required to be of type U and the result is a List[U].

With the formulation of :: shown in Listing 22.3, you can check how the definition of fruits shown in Listing 22.2 works out type-wise: In that definition, the type parameter U of :: is instantiated to Fruit. The lower-bound constraint of U is satisfied because the list apples has type List[Apple] and Fruit is a supertype of Apple. The argument to the :: is new Orange, which conforms to type Fruit. Therefore, the method application is type-correct with result type List[Fruit]. Figure 22.2 illustrates the structure of the lists that result from executing the code shown in Listing 22.2.

In fact, the polymorphic definition of :: with the lower bound T is not only convenient, it is also necessary to render the definition of class List type-correct. This is because Lists are defined to be covariant.

Assume for a moment that we had defined :: like this:

```
// A thought experiment (which wouldn't work)
def ::(x: T): List[T] = new scala.::(x, this)
```

You saw in Chapter 19 that method parameters count as contravariant positions, so the list element type T is in contravariant position in the definition above. But then List cannot be declared covariant in T. The lower bound [U >: T] thus kills two birds with one stone: It removes a typing problem and leads to a :: method that's more flexible to use. The list concatenation method ::: is defined in a similar way to ::, as shown in Listing 22.4.

```
def :::[U >: T](prefix: List[U]): List[U] =
  if (prefix.isEmpty) this
  else prefix.head :: prefix.tail ::: this
```

Listing 22.4 · The definition of method ::: in class List.

Like cons, concatenation is polymorphic. The result type is "widened" as necessary to include the types of all list elements. Note again that the order of the arguments is swapped between an infix operation and an explicit method call. Because both ::: and :: end in a colon, they both bind to the right and are both right associative. For instance, the else part of the definition of ::: shown in Listing 22.4 contains infix operations of both :: and :::.

These infix operations can be expanded to equivalent method calls as follows:

prefix.head :: prefix.tail ::: **this**
    *equals* (because :: and ::: are right-associative)

prefix.head :: (prefix.tail ::: **this**)
    *equals* (because :: binds to the right)

(prefix.tail ::: **this**).::(prefix.head)
    *equals* (because ::: binds to the right)

**this**.:::(prefix.tail).::(prefix.head)

## 22.2  The ListBuffer class

The typical access pattern for a list is recursive. For instance, to increment every element of a list without using map you could write:

```
def incAll(xs: List[Int]): List[Int] = xs match {
  case List() => List()
  case x :: xs1 => x + 1 :: incAll(xs1)
}
```

One shortcoming of this program pattern is that it is not tail recursive. Note that the recursive call to incAll above occurs inside a :: operation. Therefore each recursive call requires a new stack frame.

On today's virtual machines this means that you cannot apply incAll to lists of much more than about 30,000 to 50,000 elements. This is a pity. How do you write a version of incAll that can work on lists of arbitrary size (as much as heap-capacity allows)?

One approach is to use a loop:

```
for (x <- xs) // ??
```

But what should go in the loop body? Note that where incAll constructs the list by prepending elements to the result of the recursive call, the loop needs to append new elements at the end of the result list. A very inefficient possibility is to use : : :, the list append operator:

```
var result = List[Int]()      // a very inefficient approach
for (x <- xs) result = result ::: List(x + 1)
result
```

This is terribly inefficient. Because : : : takes time proportional to the length of its first operand, the whole operation takes time proportional to the square of the length of the list. This is clearly unacceptable.

A better alternative is to use a list buffer. List buffers let you accumulate the elements of a list. To do this, you use an operation such as "buf += elem", which appends the element elem at the end of the list buffer buf. Once you are done appending elements, you can turn the buffer into a list using the toList operation.

ListBuffer is a class in package scala.collection.mutable. To use the simple name only, you can import ListBuffer from its package:

```
import scala.collection.mutable.ListBuffer
```

Using a list buffer, the body of incAll can now be written as follows:

```
val buf = new ListBuffer[Int]
for (x <- xs) buf += x + 1
buf.toList
```

This is a very efficient way to build lists. In fact, the list buffer implementation is organized so that both the append operation (+=) and the toList operation take (very short) constant time.

## 22.3    The List class in practice

The implementations of list methods given in Section 22.1 are concise and clear, but suffer from the same stack overflow problem as the non-tail recursive implementation of incAll. Therefore, most methods in the real implementation of class List avoid recursion and use loops with list buffers instead. For example, Listing 22.5 shows the real implementation of map in class List:

```scala
final override def map[U](f: T => U): List[U] = {
  val b = new ListBuffer[U]
  var these = this
  while (!these.isEmpty) {
    b += f(these.head)
    these = these.tail
  }
  b.toList
}
```

Listing 22.5 · The definition of method map in class List.

This revised implementation traverses the list with a simple loop, which is highly efficient. A tail recursive implementation would be similarly efficient, but a general recursive implementation would be slower and less scalable. But what about the operation b.toList at the end? What is its complexity? In fact, the call to the toList method takes only a small number of cycles, which is independent of the length of the list.

To understand why, take a second look at class ::, which constructs non-empty lists. In practice, this class does not quite correspond to its idealized definition given previously in Section 22.1. The real definition is shown in Listing 22.6. As you can see, there's one peculiarity: the tl argument is a var! This means that it is possible to modify the tail of a list after the list is constructed. However, because the variable tl has the modifier private[scala], it can be accessed only from within package scala. Client code outside this package can neither read nor write tl.

Since the ListBuffer class is contained in a subpackage of package scala, scala.collection.mutable, ListBuffer can access the tl field of a cons cell. In fact the elements of a list buffer are represented as a list and

479

```
final case class ::[U](hd: U,
    private[scala] var tl: List[U]) extends List[U] {
  def head = hd
  def tail = tl
  override def isEmpty: Boolean = false
}
```

Listing 22.6 · The definition of the :: subclass of List.

appending new elements involves a modification of the tl field of the last ::
cell in that list. Here's the start of class ListBuffer:

```
package scala.collection.immutable
final class ListBuffer[T] extends Buffer[T] {
  private var start: List[T] = Nil
  private var last0: ::[T] = _
  private var exported: Boolean = false
  ...
```

You see three private fields that characterize a ListBuffer:

|  |  |
|---|---|
| start | points to the list of all elements stored in the buffer |
| last0 | points to the last :: cell in that list |
| exported | indicates whether the buffer has been turned into a list using a toList operation |

The toList operation is very simple:

```
override def toList: List[T] = {
  exported = !start.isEmpty
  start
}
```

It returns the list of elements referred to by start and also sets exported
to true if that list is nonempty. So toList is very efficient because it does
not copy the list which is stored in a ListBuffer. But what happens if the
list is further extended after the toList operation? Of course, once a list
is returned from toList, it must be immutable. However, appending to the
last0 element will modify the list which is referred to by start.

480

To maintain the correctness of the list buffer operations, you need to work on a fresh list instead. This is achieved by the first line in the implementation of the += operation:

```
override def += (x: T) = {
  if (exported) copy()
  if (start.isEmpty) {
    last0 = new scala.::(x, Nil)
    start = last0
  } else {
    val last1 = last0
    last0 = new scala.::(x, Nil)
    last1.tl = last0
  }
}
```

You see that += copies the list pointed to by start if exported is true. So, in the end, there is no free lunch. If you want to go from lists which can be extended at the end to immutable lists, there needs to be some copying. However, the implementation of ListBuffer is such that copying is necessary only for list buffers that are further extended after they have been turned into lists. This case is quite rare in practice. Most use cases of list buffers add elements incrementally and then do one toList operation at the end. In such cases, no copying is necessary.

## 22.4   Functional on the outside

In the previous section, we showed key elements of the implementation of Scala's List and ListBuffer classes. You saw that lists are purely functional on the "outside" but have an imperative implementation using list buffers on the "inside." This is a typical strategy in Scala programming—trying to combine purity with efficiency by carefully delimiting the effects of impure operations.

But you might ask, Why insist on purity? Why not just open up the definition of lists, making the tail field, and maybe also the head field, mutable? The disadvantage of such an approach is that it would make programs much more fragile. Note that constructing lists with :: re-uses the tail of the constructed list.

481

So when you write:

```
val ys = 1 :: xs
val zs = 2 :: xs
```

the tails of lists ys and zs are shared; they point to the same data structure. This is essential for efficiency; if the list xs was copied every time you added a new element onto it, this would be much slower. Because sharing is pervasive, changing list elements, if it were possible, would be quite dangerous. For instance, taking the code above, if you wanted to truncate list ys to its first two elements by writing:

```
ys.drop(2).tail = Nil  // can't do this in Scala!
```

you would also truncate lists zs and xs as a side effect.

Clearly, it would be quite difficult to keep track of what gets changed. That's why Scala opts for pervasive sharing and no mutation for lists. The ListBuffer class still allows you to build up lists imperatively and incrementally, if you wish. But since list buffers are not lists, the types keep mutable buffers and immutable lists separate.

The design of Scala's List and ListBuffer is quite similar to what's done in Java's pair of classes String and StringBuffer. This is no coincidence. In both situations the designers wanted to maintain a pure immutable data structure but also provide an efficient way to construct this structure incrementally. For Java and Scala strings, StringBuffers (or, in Java 5, StringBuilders) provide a way to construct a string incrementally. For Scala's lists, you have a choice: You can either construct lists incrementally by adding elements to the beginning of a list using ::, or you use a list buffer for adding elements to the end. Which one is preferable depends on the situation. Usually, :: lends itself well to recursive algorithms in the divide-and-conquer style. List buffers are often used in a more traditional loop-based style.

## 22.5  Conclusion

In this chapter, you saw how lists are implemented in Scala. List is one of the most heavily used data structures in Scala, and it has a refined implementation. List's two subclasses, Nil and ::, are both case classes. Instead of recursing through this structure, however, many core list methods are implemented using a ListBuffer. ListBuffer, in turn, is carefully implemented so that it can efficiently build lists without allocating extraneous memory. It is functional on the outside, but uses mutation internally to speed up the common case where a buffer is discarded after toList has been called. After studying all of this, you now know the list classes inside and out, and you might have learned an implementation trick or two.

# Chapter 23

# For Expressions Revisited

Chapter 16 demonstrated that higher-order functions, such as `map`, `flatMap`, and `filter`, provide powerful constructions for dealing with lists. But sometimes the level of abstraction required by these functions makes a program a bit hard to understand.

Here's an example. Say you are given a list of persons, each defined as an instance of a class `Person`. Class `Person` has fields indicating the person's name, whether he or she is male, and his or her children.

Here's the class definition:

```
scala> case class Person(name: String,
                         isMale: Boolean,
                         children: Person*)
```

Here's a list of some sample `persons`:

```
val lara = Person("Lara", false)
val bob = Person("Bob", true)
val julie = Person("Julie", false, lara, bob)
val persons = List(lara, bob, julie)
```

Now, say you want to find out the names of all pairs of mothers and their children in that list. Using `map`, `flatMap` and `filter`, you can formulate the following query:

```
scala> persons filter (p => !p.isMale) flatMap (p =>
            (p.children map (c => (p.name, c.name))))
res0: List[(String, String)] = List((Julie,Lara),
    (Julie,Bob))
```

You could optimize this example a bit by using a `withFilter` call instead of `filter`. This would avoid the creation of an intermediate data structure for female persons:

```
scala> persons withFilter (p => !p.isMale) flatMap (p =>
           (p.children map (c => (p.name, c.name)))))
res1: List[(String, String)] = List((Julie,Lara),
    (Julie,Bob))
```

These queries do their job, but they are not exactly trivial to write or understand. Is there a simpler way? In fact, there is. Remember the for expressions in Section 7.3? Using a for expression, the same example can be written as follows:

```
scala> for (p <- persons; if !p.isMale; c <- p.children)
         yield (p.name, c.name)
res2: List[(String, String)] = List((Julie,Lara),
    (Julie,Bob))
```

The result of this expression is exactly the same as the result of the previous expression. What's more, most readers of the code would likely find the for expression much clearer than the previous query, which used the higher-order functions, map, flatMap, and withFilter.

However, the last two queries are not as dissimilar as it might seem. In fact, it turns out that the Scala compiler will translate the second query into the first one. More generally, all for expressions that yield a result are translated by the compiler into combinations of invocations of the higher-order methods map, flatMap, and withFilter. All for loops without yield are translated into a smaller set of higher-order functions: just withFilter and foreach.

In this chapter, you'll find out first about the precise rules of writing for expressions. After that, you'll see how they can make combinatorial problems easier to solve. Finally, you'll learn how for expressions are translated, and how as a result, for expressions can help you "grow" the Scala language into new application domains.

## 23.1   For expressions

Generally, a for expression is of the form:

```
for ( seq ) yield expr
```

Here, *seq* is a sequence of *generators*, *definitions*, and *filters*, with semi-colons between successive elements. An example is the for expression:

```
for (p <- persons; n = p.name; if (n startsWith "To"))
yield' n
```

This for expression contains one generator, one definition, and one filter. As mentioned in Section 7.3 on page 125, you can also enclose the sequence in braces instead of parentheses. Then the semicolons become optional:

```
for {
  p <- persons          // a generator
  n = p.name            // a definition
  if (n startsWith "To") // a filter
} yield n
```

A *generator* is of the form:

```
pat <- expr
```

The expression *expr* typically returns a list, even though you will see later that this can be generalized. The pattern *pat* gets matched one-by-one against all elements of that list. If the match succeeds, the variables in the pattern get bound to the corresponding parts of the element, just the way it is described in Chapter 15. But if the match fails, no MatchError is thrown. Instead, the element is simply discarded from the iteration.

In the most common case, the pattern *pat* is just a variable $x$, as in $x$ <- *expr*. In that case, the variable $x$ simply iterates over all elements returned by *expr*.

A *definition* is of the form:

```
pat = expr
```

This definition binds the pattern *pat* to the value of *expr*, so it has the same effect as a val definition:

```
val x = expr
```

The most common case is again where the pattern is a simple variable x (*e.g.*, $x$ = *expr*). This defines $x$ as a name for the value *expr*.

A *filter* is of the form:

```
if expr
```

Here, *expr* is an expression of type Boolean. The filter drops from the iteration all elements for which *expr* returns false.

Every for expression starts with a generator. If there are several generators in a for expression, later generators vary more rapidly than earlier ones.
You can verify this easily with the following simple test:

```
scala> for (x <- List(1, 2); y <- List("one", "two"))
       yield (x, y)
res3: List[(Int, String)] =
  List((1,one), (1,two), (2,one), (2,two))
```

## 23.2   The n-queens problem

A particularly suitable application area of for expressions are combinatorial
puzzles. An example of such a puzzle is the 8-queens problem: Given a
standard chess-board, place eight queens such that no queen is in check from
any other (a queen can check another piece if they are on the same column,
row, or diagonal). To find a solution to this problem, it's actually simpler to
generalize it to chess-boards of arbitrary size. Hence, the problem is to place
$N$ queens on a chess-board of $N \times N$ squares, where the size $N$ is arbitrary.
We'll start numbering cells at one, so the upper-left cell of an $N \times N$ board
has coordinate $(1,1)$ and the lower-right cell has coordinate $(N,N)$.

To solve the N-queens problem, note that you need to place a queen in
each row. So you could place queens in successive rows, each time checking
that a newly placed queen is not in check from any other queens that have
already been placed. In the course of this search, it might happen that a
queen that needs to be placed in row $k$ would be in check in all fields of that
row from queens in row 1 to $k-1$. In that case, you need to abort that part
of the search in order to continue with a different configuration of queens in
columns 1 to $k-1$.

An imperative solution to this problem would place queens one by one,
moving them around on the board. But it looks difficult to come up with a
scheme that really tries all possibilities. A more functional approach represents a solution directly, as a value. A solution consists of a list of coordinates, one for each queen placed on the board. Note, however, that a full

solution can not be found in a single step. It needs to be built up gradually, by occupying successive rows with queens.

This suggests a recursive algorithm. Assume you have already generated all solutions of placing $k$ queens on a board of size $N \times N$, where $k$ is less than $N$. Each such solution can be presented by a list of length $k$ of coordinates (row, column), where both row and column numbers range from 1 to $N$. It's convenient to treat these partial solution lists as stacks, where the coordinates of the queen in row $k$ come first in the list, followed by the coordinates of the queen in row $k-1$, and so on. The bottom of the stack is the coordinate of the queen placed in the first row of the board. All solutions together are represented as a list of lists, with one element for each solution.

Now, to place the next queen in row $k+1$, generate all possible extensions of each previous solution by one more queen. This yields another list of solution lists, this time of length $k+1$. Continue the process until you have obtained all solutions of the size of the chess-board $N$.

This algorithmic idea is embodied in function placeQueens below:

```
def queens(n: Int): List[List[(Int, Int)]] = {
  def placeQueens(k: Int): List[List[(Int, Int)]] =
    if (k == 0)
      List(List())
    else
      for {
        queens <- placeQueens(k - 1)
        column <- 1 to n
        queen = (k, column)
        if isSafe(queen, queens)
      } yield queen :: queens

  placeQueens(n)
}
```

The outer function queens in the program above simply calls placeQueens with the size of the board n as its argument. The task of the function application placeQueens(k) is to generate all partial solutions of length k in a list. Every element of the list is one solution, represented by a list of length k. So placeQueens returns a list of lists.

If the parameter k to placeQueens is 0, this means that it needs to generate all solutions of placing zero queens on zero rows. There is only one such solution: place no queen at all. This solution is represented by the empty list. So if k is zero, placeQueens returns List(List()), a list consisting of a single element that is the empty list. Note that this is quite different from the empty list List(). If placeQueens returns List(), this means *no solutions*, instead of a single solution consisting of no placed queens.

In the other case, where k is not zero, all the work of placeQueens is done in a for expression. The first generator of that for expression iterates through all solutions of placing k - 1 queens on the board. The second generator iterates through all possible columns on which the k'th queen might be placed. The third part of the for expression defines the newly considered queen position to be the pair consisting of row k and each produced column. The fourth part of the for expression is a filter which checks with isSafe whether the new queen is safe from check by all previous queens (the definition of isSafe will be discussed a bit later).

If the new queen is not in check from any other queens, it can form part of a partial solution, so placeQueens generates with queen :: queens a new solution. If the new queen is not safe from check, the filter returns false, so no solution is generated.

The only remaining bit is the isSafe method, which is used to check whether a given queen is in check from any other element in a list of queens. Here is its definition:

```
def isSafe(queen: (Int, Int), queens: List[(Int, Int)]) =
  queens forall (q => !inCheck(queen, q))

def inCheck(q1: (Int, Int), q2: (Int, Int)) =
  q1._1 == q2._1 ||  // same row
  q1._2 == q2._2 ||  // same column
  (q1._1 - q2._1).abs == (q1._2 - q2._2).abs // on diagonal
```

The isSafe method expresses that a queen is safe with respect to some other queens if it is not in check from any other queen. The inCheck method expresses that queens q1 and q2 are mutually in check.

It returns true in one of three cases:

1. If the two queens have the same row coordinate,

2. If the two queens have the same column coordinate,

3. If the two queens are on the same diagonal (*i.e.*, the difference between their rows and the difference between their columns are the same).

The first case—that the two queens have the same row coordinate—cannot happen in the application because `placeQueens` already takes care to place each queen in a different row. So you could remove the test without changing the functionality of the program.

## 23.3 Querying with for expressions

The for notation is essentially equivalent to common operations of database query languages. For instance, say you are given a database named books, represented as a list of books, where Book is defined as follows:

```
case class Book(title: String, authors: String*)
```

Here is a small example database represented as an in-memory list:

```
val books: List[Book] =
  List(
    Book(
      "Structure and Interpretation of Computer Programs",
      "Abelson, Harold", "Sussman, Gerald J."
    ),
    Book(
      "Principles of Compiler Design",
      "Aho, Alfred", "Ullman, Jeffrey"
    ),
    Book(
      "Programming in Modula-2",
      "Wirth, Niklaus"
    ),
    Book(
      "Elements of ML Programming",
      "Ullman, Jeffrey"
    ),
    Book(
      "The Java Language Specification", "Gosling, James",
      "Joy, Bill", "Steele, Guy", "Bracha, Gilad"
    )
  )
```

To find the titles of all books whose author's last name is "Gosling":

```
scala> for (b <- books; a <- b.authors
            if a startsWith "Gosling")
        yield b.title
res4: List[String] = List(The Java Language Specification)
```

Or to find the titles of all books that have the string "Program" in their title:

```
scala> for (b <- books if (b.title indexOf "Program") >= 0)
        yield b.title
res5: List[String] = List(Structure and Interpretation of
  Computer Programs, Programming in Modula-2, Elements of ML
    Programming)
```

Or to find the names of all authors who have written at least two books in the database:

```
scala> for (b1 <- books; b2 <- books if b1 != b2;
            a1 <- b1.authors; a2 <- b2.authors if a1 == a2)
        yield a1
res6: List[String] = List(Ullman, Jeffrey, Ullman, Jeffrey)
```

The last solution is still not perfect because authors will appear several times in the list of results. You still need to remove duplicate authors from result lists. This can be achieved with the following function:

```
scala> def removeDuplicates[A](xs: List[A]): List[A] = {
          if (xs.isEmpty) xs
          else
            xs.head :: removeDuplicates(
              xs.tail filter (x => x != xs.head)
            )
        }
removeDuplicates: [A](xs: List[A])List[A]

scala> removeDuplicates(res6)
res7: List[String] = List(Ullman, Jeffrey)
```

It's worth noting that the last expression in method removeDuplicates can be equivalently expressed using a for expression:

```
xs.head :: removeDuplicates(
  for (x <- xs.tail if x != xs.head) yield x
)
```

## 23.4 Translation of for expressions

Every for expression can be expressed in terms of the three higher-order functions map, flatMap, and withFilter. This section describes the translation scheme, which is also used by the Scala compiler.

**Translating for expressions with one generator**

First, assume you have a simple for expression:

> for (*x* <- *expr₁*) yield *expr₂*

where *x* is a variable. Such an expression is translated to:

> *expr₁*.map(*x* => *expr₂*)

**Translating for expressions starting with a generator and a filter**

Now, consider for expressions that combine a leading generator with some other elements. A for expression of the form:

> for (*x* <- *expr₁* if *expr₂*) yield *expr₃*

is translated to:

> for (*x* <- *expr₁* withFilter (*x* => *expr₂*)) yield *expr₃*

This translation gives another for expression that is shorter by one element than the original, because an if element is transformed into an application of withFilter on the first generator expression. The translation then continues with this second expression, so in the end you obtain:

> *expr₁* withFilter (*x* => *expr₂*) map (*x* => *expr₃*)

The same translation scheme also applies if there are further elements following the filter. If *seq* is an arbitrary sequence of generators, definitions, and filters, then:

493

**for** ($x$ <- $expr_1$ **if** $expr_2$; $seq$) **yield** $expr_3$

is translated to:

**for** ($x$ <- $expr_1$ withFilter $expr_2$; $seq$) **yield** $expr_3$

Then translation continues with the second expression, which is again shorter by one element than the original one.

### Translating for expressions starting with two generators

The next case handles for expressions that start with two generators, as in:

**for** ($x$ <- $expr_1$; $y$ <- $expr_2$; $seq$) **yield** $expr_3$

Again, assume that *seq* is an arbitrary sequence of generators, definitions, and filters. In fact, *seq* might also be empty, and in that case there would not be a semicolon after $expr_2$. The translation scheme stays the same in each case. The for expression above is translated to an application of flatMap:

$expr_1$.flatMap($x$ => **for** ($y$ <- $expr_2$; $seq$) **yield** $expr_3$)

This time, there is another for expression in the function value passed to flatMap. That for expression (which is again simpler by one element than the original) is in turn translated with the same rules.

The three translation schemes given so far are sufficient to translate all for expressions that contain just generators and filters, and where generators bind only simple variables. Take, for instance, the query, "find all authors who have published at least two books," from Section 23.3:

```
for (b1 <- books; b2 <- books if b1 != b2;
     a1 <- b1.authors; a2 <- b2.authors if a1 == a2)
yield a1
```

This query translates to the following map/flatMap/filter combination:

```
books flatMap (b1 =>
  books withFilter (b2 => b1 != b2) flatMap (b2 =>
    b1.authors flatMap (a1 =>
      b2.authors withFilter (a2 => a1 == a2) map (a2 =>
        a1))))
```

The translation scheme presented so far does not yet handle generators that bind whole patterns instead of simple variables. It also does not yet cover definitions. These two aspects will be explained in the next two sub-sections.

### Translating patterns in generators

The translation scheme becomes more complicated if the left hand side of generator is a pattern, *pat*, other than a simple variable. The case where the for expression binds a tuple of variables is still relatively easy to handle. In that case, almost the same scheme as for single variables applies.

A for expression of the form:

```
for ((x₁, ..., xₙ) <- expr₁) yield expr₂
```

translates to:

```
expr₁.map { case (x₁, ..., xₙ) => expr₂ }
```

Things become a bit more involved if the left hand side of the generator is an arbitrary pattern *pat* instead of a single variable or a tuple.

In this case:

```
for (pat <- expr₁) yield expr₂
```

translates to:

```
expr₁ withFilter {
  case pat => true
  case _ => false
} map {
  case pat => expr₂
}
```

That is, the generated items are first filtered and only those that match *pat* are mapped. Therefore, it's guaranteed that a pattern-matching generator will never throw a `MatchError`.

The scheme here only treated the case where the for expression contains a single pattern-matching generator. Analogous rules apply if the for expression contains other generators, filters or definitions. Because these additional rules don't add much new insight, they are omitted from discussion here. If you are interested, you can look them up in the *Scala Language Specification* [Ode11].

495

## Translating definitions

The last missing situation is where a for expression contains embedded definitions. Here's a typical case:

```
for (x <- expr₁; y = expr₂; seq) yield expr₃
```

Assume again that *seq* is a (possibly empty) sequence of generators, definitions, and filters. This expression is translated to this one:

```
for ((x, y) <- for (x <- expr₁) yield (x, expr₂); seq)
  yield expr₃
```

So you see that *expr₂* is evaluated each time there is a new *x* value being generated. This re-evaluation is necessary because *expr₂* might refer to *x* and so needs to be re-evaluated for changing values of *x*. For you as a programmer, the conclusion is that it's probably not a good idea to have definitions embedded in for expressions that do not refer to variables bound by some preceding generator, because re-evaluating such expressions would be wasteful. For instance, instead of:

```
for (x <- 1 to 1000; y = expensiveComputationNotInvolvingX)
  yield x * y
```

it's usually better to write:

```
val y = expensiveComputationNotInvolvingX
for (x <- 1 to 1000) yield x * y
```

## Translating for loops

The previous subsections showed how for expressions that contain a yield are translated. What about for loops that simply perform a side effect without returning anything? Their translation is similar, but simpler than for expressions. In principle, wherever the previous translation scheme used a map or a flatMap in the translation, the translation scheme for for loops uses just a foreach.

For instance, the expression:

```
for (x <- expr₁) body
```

translates to:

```
expr₁ foreach (x => body)
```

A larger example is the expression:

**for** (x <- expr₁; **if** expr₂; y <- expr₃) body

This expression translates to:

```
expr₁ withFilter (x => expr₂) foreach (x =>
  expr₃ foreach (y => body))
```

For example, the following expression sums up all elements of a matrix represented as a list of lists:

```
var sum = 0
for (xs <- xss; x <- xs) sum += x
```

This loop is translated into two nested `foreach` applications:

```
var sum = 0
xss foreach (xs =>
  xs foreach (x =>
    sum += x))
```

## 23.5 Going the other way

The previous section showed that `for` expressions can be translated into applications of the higher-order functions `map`, `flatMap`, and `withFilter`. In fact, you could equally go the other way: Every application of a `map`, `flatMap`, or `filter` can be represented as a `for` expression.

Here are implementations of the three methods in terms of `for` expressions. The methods are contained in an object Demo to distinguish them from the standard operations on Lists. To be concrete, the three functions all take a List as parameter, but the translation scheme would work just as well with other collection types:

```scala
object Demo {
  def map[A, B](xs: List[A], f: A => B): List[B] =
    for (x <- xs) yield f(x)

  def flatMap[A, B](xs: List[A], f: A => List[B]): List[B] =
    for (x <- xs; y <- f(x)) yield y

  def filter[A](xs: List[A], p: A => Boolean): List[A] =
    for (x <- xs if p(x)) yield x
}
```

Not surprisingly, the translation of the for expression used in the body of Demo.map will produce a call to map in class List. Similarly, Demo.flatMap and Demo.filter translate to flatMap and withFilter in class List. So this little demonstration shows that for expressions really are equivalent in their expressiveness to applications of the three functions map, flatMap, and withFilter.

## 23.6   Generalizing for

Because the translation of for expressions only relies on the presence of methods map, flatMap, and withFilter, it is possible to apply the for notation to a large class of data types.

You have already seen for expressions over lists and arrays. These are supported because lists, as well as arrays, define operations map, flatMap, and withFilter. Because they define a foreach method as well, for loops over these data types are also possible.

Besides lists and arrays, there are many other types in the Scala standard library that support the same four methods and therefore allow for expressions. Examples are ranges, iterators, streams, and all implementations of sets. It's also perfectly possible for your own data types to support for expressions by defining the necessary methods. To support the full range of for expressions and for loops, you need to define map, flatMap, withFilter, and foreach as methods of your data type. But it's also possible to define a subset of these methods, and thereby support a subset of all possible for expressions or loops.

Here are the precise rules:

- If your type defines just map, it allows for expressions consisting of a single generator.

- If it defines flatMap as well as map, it allows for expressions consisting of several generators.

- If it defines foreach, it allows for loops (both with single and multiple generators).

- If it defines withFilter, it allows for filter expressions starting with an if in the for expression.

The translation of for expressions happens before type checking. This allows for maximum flexibility because the only requirement is that the result of expanding a for expression type checks. Scala defines no typing rules for the for expressions themselves, and does not require that methods map, flatMap, withFilter, or foreach have any particular type signatures.

Nevertheless, there is a typical setup that captures the most common intention of the higher order methods to which for expressions translate. Say you have a parameterized class, C, which typically would stand for some sort of collection. Then it's quite natural to pick the following type signatures for map, flatMap, withFilter, and foreach:

```
abstract class C[A] {
  def map[B](f: A => B): C[B]
  def flatMap[B](f: A => C[B]): C[B]
  def withFilter(p: A => Boolean): C[A]
  def foreach(b: A => Unit): Unit
}
```

That is, the map function takes a function from the collection's element type A to some other type B. It produces a new collection of the same kind C, but with B as the element type. The flatMap method takes a function f from A to some C-collection of Bs and produces a C-collection of Bs. The withFilter method takes a predicate function from the collection's element type A to Boolean. It produces a collection of the same type as the one on which it is invoked. Finally, the foreach method takes a function from A to Unit and produces a Unit result:

In class C above, the withFilter method produces a new collection of the same class. That means that every invocation of withFilter creates a new C object, just the same as filter would work. Now, in the translation of for expressions, any calls to withFilter are always followed by calls to one of the other three methods. Therefore, the object created by `withFilter` will be taken apart by one of the other methods immediately afterwards. If objects of class C are large (think long sequences), you might want to avoid the creation of such an intermediate object. A standard technique is to let `withFilter` return not a C object but just a wrapper object that "remembers" that elements need to be filtered before being processed further.

Concentrating on just the first three functions of class C, the following facts are noteworthy. In functional programming, there's a general concept called a *monad*, which can explain a large number of types with computations, ranging from collections, to computations with state and I/O, backtracking computations, and transactions, to name a few. You can formulate functions `map`, `flatMap`, and `withFilter` on a monad, and, if you do, they end up having exactly the types given here.

Furthermore, you can characterize every monad by `map`, `flatMap`, and `withFilter`, plus a "unit" constructor that produces a monad from an element value. In an object-oriented language, this "unit" constructor is simply an instance constructor or a factory method. Therefore, `map`, `flatMap`, and `withFilter` can be seen as an object-oriented version of the functional concept of monad. Because for expressions are equivalent to applications of these three methods, they can be seen as syntax for monads.

All this suggests that the concept of for expression is more general than just iteration over a collection, and indeed it is. For instance, for expressions also play an important role in asynchronous I/O, or as an alternative notation for optional values. Watch out in the Scala libraries for occurrences of `map`, `flatMap`, and `withFilter`—when they are present, for expressions suggest themselves as a concise way of manipulating elements of the type.

## 23.7   Conclusion

In this chapter, you were given a peek under the hood of for expressions and for loops. You learned that they translate into applications of a standard set of higher-order methods. As a result, you saw that for expressions are really much more general than mere iterations over collections, and that you can design your own classes to support them.

# Chapter 24

# Collections in Depth

Scala includes an elegant and powerful collection library. Even though the collections API is subtle at first glance, the changes it can provoke in your programming style can be profound. Quite often it's as if you work on a higher level with the basic building blocks of a program being whole collections instead of their elements. This new style of programming requires some adaptation. Fortunately, the adaptation is helped by several nice properties of Scala collections. They are easy to use, concise, safe, fast, and universal.

**Easy to use:** A small vocabulary of twenty to fifty methods is enough to solve most collection problems in a couple of operations. No need to wrap your head around complicated looping structures or recursions. Persistent collections and side-effect-free operations mean that you need not worry about accidentally corrupting existing collections with new data. Interference between iterators and collection updates is eliminated.

**Concise:** You can achieve with a single word what used to take one or several loops. You can express functional operations with lightweight syntax and combine operations effortlessly, so that the result feels like a custom algebra.

**Safe:** This one has to be experienced to sink in. The statically typed and functional nature of Scala's collections means that the overwhelming majority of errors you might make are caught at compile-time. The reason is that (1) the collection operations themselves are heavily used and therefore well tested. (2) the usages of the collection operation make inputs and output explicit as function parameters and results. (3) These explicit inputs and outputs are subject to static type checking. The bottom line is that the large majority of misuses will manifest themselves as type errors. It's not at all uncommon to have programs of several hundred lines run at first try.

**Fast:** Collection operations are tuned and optimized in the libraries. As a result, using collections is typically quite efficient. You might be able to do a little bit better with carefully hand-tuned data structures and operations, but you might also do a lot worse by making some suboptimal implementation decisions along the way. What's more, collections are have been adapted to parallel execution on multi-cores. Parallel collections support the same operations as sequential ones, so no new operations need to be learned and no code needs to be rewritten. You can turn a sequential collection into a parallel one simply by invoking the par method.

**Universal:** Collections provide the same operations on any type where it makes sense to do so. So you can achieve a lot with a fairly small vocabulary of operations. For instance, a string is conceptually a sequence of characters. Consequently, in Scala collections, strings support all sequence operations. The same holds for arrays.

This chapter describes in depth the APIs of the Scala collection classes from a user perspective. You've already seen a quick tour of the collections library, in Chapter 17. This chapter takes you on a more detailed tour, showing all the collection classes and all the methods they define, so it includes everything you need to know to use Scala collections. Looking ahead, Chapter 25 will concentrate on the architecture and extensibility aspects of the library, for people implementing new collection types.

## 24.1 Mutable and immutable collections

As is now familiar to you, Scala collections systematically distinguish between mutable and immutable collections. A mutable collection can be updated or extended in place. This means you can change, add, or remove elements of a collection as a side effect. Immutable collections, by contrast, never change. You still have operations that simulate additions, removals, or updates, but those operations will in each case return a new collection and leave the old collection unchanged.

All collection classes are found in the package `scala.collection` or one of its subpackages: `mutable`, `immutable`, and `generic`. Most collection classes needed by client code exist in three variants, each of which has different characteristics with respect to mutability. The three variants are located in packages `scala.collection`, `scala.collection.immutable`, and `scala.collection.mutable`.

A collection in package `scala.collection.immutable` is guaranteed to be immutable for everyone. Such a collection will never change after it is created. Therefore, you can rely on the fact that accessing the same collection value repeatedly at different points in time will always yield a collection with the same elements.

A collection in package `scala.collection.mutable` is known to have some operations that change the collection in place. These operations let you write code to mutate the collection yourself. However, you must be careful to understand and defend against any updates performed by other parts of the code base.

A collection in package `scala.collection` can be either mutable or immutable. For instance, `scala.collection.IndexedSeq[T]` is a supertrait of both `scala.collection.immutable.IndexedSeq[T]` and its mutable sibling `scala.collection.mutable.IndexedSeq[T]`. Generally, the root collections in package `scala.collection` define the same interface as the immutable collections. And typically, the mutable collections in package `scala.collection.mutable` add some side-effecting modification operations to this immutable interface.

The difference between root collections and immutable collections is that clients of an immutable collection have a guarantee that nobody can mutate the collection, whereas clients of a root collection only know that they can't change the collection themselves. Even though the static type of such a collection provides no operations for modifying the collection, it might still be possible that the run-time type is a mutable collection that can be changed by other clients.

By default, Scala always picks immutable collections. For instance, if you just write `Set` without any prefix or without having imported anything, you get an immutable set, and if you write `Iterable` you get an immutable iterable, because these are the default bindings imported from the `scala` package. To get the mutable default versions, you need to write explicitly collection.mutable.Set, or collection.mutable.Iterable.

The last package in the collection hierarchy is collection.generic. This package contains building blocks for implementing collections. Typically, collection classes defer the implementations of some of their operations to classes in generic. Everyday users of the collection framework on the other hand should need to refer to classes in generic only in exceptional circumstances.

## 24.2 Collections consistency

The most important collection classes are shown in Figure 24.1. There is quite a bit of commonality shared by all these classes. For instance, every kind of collection can be created by the same uniform syntax, writing the collection class name followed by its elements:

```
Traversable(1, 2, 3)
Iterable("x", "y", "z")
Map("x" -> 24, "y" -> 25, "z" -> 26)
Set(Color.Red, Color.Green, Color.Blue)
SortedSet("hello", "world")
Buffer(x, y, z)
IndexedSeq(1.0, 2.0)
LinearSeq(a, b, c)
```

The same principle also applies for specific collection implementations:

```
List(1, 2, 3)
HashMap("x" -> 24, "y" -> 25, "z" -> 26)
```

The toString methods for all collections produce output written as above, with a type name followed by the elements of the collection in parentheses. All collections support the API provided by Traversable, but their methods all return their own class rather than the root class Traversable. For instance, the map method on List has a return type of List, whereas the map method on Set has a return type of Set. Thus the static return type of these methods is fairly precise:

```
Traversable
    Iterable
        Seq
            IndexedSeq
                Vector
                ResizableArray
                GenericArray
            LinearSeq
                MutableList
                List
                Stream
            Buffer
                ListBuffer
                ArrayBuffer
        Set
            SortedSet
                TreeSet
            HashSet (mutable)
            LinkedHashSet
            HashSet (immutable)
            BitSet
            EmptySet, Set1, Set2, Set3, Set4
        Map
            SortedMap
                TreeMap
            HashMap (mutable)
            LinkedHashMap (mutable)
            HashMap (immutable)
            EmptyMap, Map1, Map2, Map3, Map4
```

Figure 24.1 · Collection hierarchy.

```
scala> List(1, 2, 3) map (_ + 1)
res0: List[Int] = List(2, 3, 4)

scala> Set(1, 2, 3) map (_ * 2)
res1: scala.collection.immutable.Set[Int] = Set(2, 4, 6)
```

Equality is also organized uniformly for all collection classes; more on this in Section 24.13.

Most of the classes in Figure 24.1 exist in three variants: root, mutable, and immutable. The only exception is the `Buffer` trait, which only exists as a mutable collection.

In the remainder of this chapter, we will review these classes one by one.

## 24.3  Trait Traversable

At the top of the collection hierarchy is trait `Traversable`. Its only abstract operation is `foreach`:

```
def foreach[U](f: Elem => U)
```

Collection classes implementing `Traversable` just need to define this method; all other methods can be inherited from `Traversable`.

The `foreach` method is meant to traverse all elements of the collection, and apply the given operation, `f`, to each element. The type of the operation is `Elem => U`, where `Elem` is the type of the collection's elements and `U` is an arbitrary result type. The invocation of `f` is done for its side effect only; in fact any function result of `f` is discarded by `foreach`.

`Traversable` also defines many concrete methods, which are all listed in Table 24.1 on page 508. These methods fall into the following categories:

*Addition* ++, which appends two traversables together, or appends all elements of an iterator to a traversable.

*Map operations* map, flatMap, and collect, which produce a new collection by applying some function to collection elements.

*Conversions* toIndexedSeq, toIterable, toStream, toArray, toList, toSeq, toSet, and toMap, which turn a `Traversable` collection into a more specific collection. All these conversions return the receiver object if it already matches the demanded collection type. For instance, applying toList to a list will yield the list itself.

506

***Copying operations*** copyToBuffer and copyToArray. As their names imply, these copy collection elements to a buffer or array, respectively.

***Size operations*** isEmpty, nonEmpty, size, and hasDefiniteSize. Collections that are traversable can be finite or infinite. An example of an infinite traversable collection is the stream of natural numbers Stream.from(0). The method hasDefiniteSize indicates whether a collection is possibly infinite. If hasDefiniteSize returns true, the collection is certainly finite. If it returns false, the collection might be infinite, in which case size will emit an error or not return.

***Element retrieval operations*** head, last, headOption, lastOption, and find. These select the first or last element of a collection, or else the first element matching a condition. Note, however, that not all collections have a well-defined meaning of what "first" and "last" means. For instance, a hash set might store elements according to their hash keys, which might change from run to run. In that case, the "first" element of a hash set could also be different for different runs of a program. A collection is *ordered* if it always yields its elements in the same order. Most collections are ordered, but some (such as hash sets) are not—dropping the ordering provides a little bit of extra efficiency. Ordering is often essential to give reproducible tests and help in debugging. That's why Scala collections provide ordered alternatives for all collection types. For instance, the ordered alternative for HashSet is LinkedHashSet.

***Subcollection retrieval operations*** takeWhile, tail, init, slice, take, drop, filter, dropWhile, filterNot, and withFilter. These all return some subcollection identified by an index range or a predicate.

***Subdivision operations*** splitAt, span, partition, and groupBy, which split the elements of this collection into several subcollections.

***Element tests*** exists, forall, and count, which test collection elements with a given predicate.

***Folds*** foldLeft, foldRight, /:, :\, reduceLeft, reduceRight, which apply a binary operation to successive elements.

***Specific folds*** sum, product, min, and max, which work on collections of specific types (numeric or comparable).

507

***String operations*** mkString, addString, and stringPrefix, which provide alternative ways of converting a collection to a string.

***View operations*** consisting of two overloaded variants of the view method. A view is a collection that's evaluated lazily. You'll learn more about views in Section 24.14.

Table 24.1 · Operations in trait Traversable

| What it is | What it does |
| --- | --- |
| **Abstract method:** | |
| xs foreach f | Executes function f for every element of xs. |
| **Addition:** | |
| xs ++ ys | A collection consisting of the elements of both xs and ys. ys is a TraversableOnce collection, *i.e.*, either a Traversable or an Iterator. |
| **Maps:** | |
| xs map f | The collection obtained from applying the function f to every element in xs. |
| xs flatMap f | The collection obtained from applying the collection-valued function f to every element in xs and concatenating the results. |
| xs collect f | The collection obtained from applying the partial function f to every element in xs for which it is defined and collecting the results. |
| **Conversions:** | |
| xs.toArray | Converts the collection to an array. |
| xs.toList | Converts the collection to a list. |
| xs.toIterable | Converts the collection to an iterable. |
| xs.toSeq | Converts the collection to a sequence. |
| xs.toIndexedSeq | Converts the collection to an indexed sequence. |
| xs.toStream | Converts the collection to a stream (a lazily computed sequence). |
| xs.toSet | Converts the collection to a set. |
| xs.toMap | Converts a collection of key/value pairs to a map. |

## Table 24.1 · continued

**Copying:**

| | |
|---|---|
| xs copyToBuffer buf | Copies all elements of the collection to buffer buf. |
| xs copyToArray(arr, s, len) | Copies at most len elements of arr, starting at index s. The last two arguments are optional. |

**Size info:**

| | |
|---|---|
| xs.isEmpty | Tests whether the collection is empty. |
| xs.nonEmpty | Tests whether the collection contains elements. |
| xs.size | The number of elements in the collection. |
| xs.hasDefiniteSize | True if xs is known to have finite size. |

**Element retrieval:**

| | |
|---|---|
| xs.head | The first element of the collection (or, some element, if no order is defined). |
| xs.headOption | The first element of xs in an option value, or None if xs is empty. |
| xs.last | The last element of the collection (or, some element, if no order is defined). |
| xs.lastOption | The last element of xs in an option value, or None if xs is empty. |
| xs find p | An option containing the first element in xs that satisfies p, or None if no element qualifies. |

**Subcollections:**

| | |
|---|---|
| xs.tail | The rest of the collection except xs.head. |
| xs.init | The rest of the collection except xs.last. |
| xs slice (from, to) | A collection consisting of elements in some index range of xs (from from, up to and excluding to). |
| xs take n | A collection consisting of the first n elements of xs (or, some arbitrary n elements, if no order is defined). |
| xs drop n | The rest of the collection except xs take n. |
| xs takeWhile p | The longest prefix of elements in the collection that all satisfy p. |
| xs dropWhile p | The collection without the longest prefix of elements that all satisfy p. |

## Table 24.1 · continued

| | |
|---|---|
| xs filter p | The collection consisting of those elements of xs that satisfy the predicate p. |
| xs withFilter p | A non-strict filter of this collection. All operations on the resulting filter will only apply to those elements of xs for which the condition p is true. |
| xs filterNot p | The collection consisting of those elements of xs that do not satisfy the predicate p. |

**Subdivisions:**

| | |
|---|---|
| xs splitAt n | Splits xs at a position, giving the pair of collections (xs take n, xs drop n). |
| xs span p | Splits xs according to a predicate, giving the pair of collections (xs takeWhile p, xs.dropWhile p). |
| xs partition p | Splits xs into a pair of collections; one with elements that satisfy the predicate p, the other with elements that do not, giving the pair of collections (xs filter p, xs.filterNot p). |
| xs groupBy f | Partitions xs into a map of collections according to a discriminator function f. |

**Element conditions:**

| | |
|---|---|
| xs forall p | A boolean indicating whether the predicate p holds for all elements of xs. |
| xs exists p | A boolean indicating whether the predicate p holds for some element in xs. |
| xs count p | The number of elements in xs that satisfy the predicate p. |

**Folds:**

| | |
|---|---|
| (z /: xs)(op) | Applies binary operation op between successive elements of xs, going left to right, starting with z. |
| (xs :\ z)(op) | Applies binary operation op between successive elements of xs, going right to left, starting with z. |
| xs.foldLeft(z)(op) | Same as (z /: xs)(op). |
| xs.foldRight(z)(op) | Same as (xs :\ z)(op). |

Table 24.1 · continued

| | |
|---|---|
| xs reduceLeft op | Applies binary operation op between successive elements of non-empty collection xs, going left to right. |
| xs reduceRight op | Applies binary operation op between successive elements of non-empty collection xs, going right to left. |

**Specific folds:**

| | |
|---|---|
| xs.sum | The sum of the numeric element values of collection xs. |
| xs.product | The product of the numeric element values of collection xs. |
| xs.min | The minimum of the ordered element values of collection xs. |
| xs.max | The maximum of the ordered element values of collection xs. |

**Strings:**

| | |
|---|---|
| xs addString (b, start, sep, end) | Adds a string to StringBuilder b that shows all elements of xs between separators sep enclosed in strings start and end. start, sep, and end are all optional. |
| xs mkString (start, sep, end) | Converts the collection to a string that shows all elements of xs between separators sep enclosed in strings start and end. start, sep, and end are all optional. |
| xs.stringPrefix | The collection name at the beginning of the string returned from xs.toString. |

**Views:**

| | |
|---|---|
| xs.view | Produces a view over xs. |
| xs view (from, to) | Produces a view that represents the elements in some index range of xs. |

## 24.4  Trait Iterable

The next trait from the top in Figure 24.1 is Iterable. All methods in this trait are defined in terms of an abstract method, iterator, which yields the collection's elements one by one. The abstract foreach method inherited

511

from trait `Traversable` is implemented in `Iterable` in terms of `iterator`. Here is the actual implementation:

```
def foreach[U](f: Elem => U): Unit = {
  val it = iterator
  while (it.hasNext) f(it.next())
}
```

Quite a few subclasses of `Iterable` override this standard implementation of `foreach` in `Iterable`, because they can provide a more efficient implementation. Remember that `foreach` is the basis of the implementation of all operations in `Traversable`, so its performance matters.

Two more methods exist in `Iterable` that return iterators: grouped and sliding. These iterators, however, do not return single elements but whole subsequences of elements of the original collection. The maximal size of these subsequences is given as an argument to these methods. The grouped method chunks its elements into increments, whereas `sliding` yields a sliding window over the elements. The difference between the two should become clear by looking at the following interpreter interaction:

```
scala> val xs = List(1, 2, 3, 4, 5)
xs: List[Int] = List(1, 2, 3, 4, 5)

scala> val git = xs grouped 3
git: Iterator[List[Int]] = non-empty iterator

scala> git.next()
res2: List[Int] = List(1, 2, 3)

scala> git.next()
res3: List[Int] = List(4, 5)

scala> val sit = xs sliding 3
sit: Iterator[List[Int]] = non-empty iterator

scala> sit.next()
res4: List[Int] = List(1, 2, 3)

scala> sit.next()
res5: List[Int] = List(2, 3, 4)

scala> sit.next()
res6: List[Int] = List(3, 4, 5)
```

Trait `Iterable` also adds some other methods to `Traversable` that can be implemented efficiently only if an iterator is available. They are summarized in Table 24.2:

Table 24.2 · Operations in trait `Iterable`

| What it is | What it does |
| --- | --- |
| **Abstract method:** | |
| xs.iterator | An iterator that yields every element in xs, in the same order as `foreach` traverses elements |
| **Other iterators:** | |
| xs grouped size | An iterator that yields fixed-sized "chunks" of this collection |
| xs sliding size | An iterator that yields a sliding fixed-sized window of elements in this collection |
| **Subcollections:** | |
| xs takeRight n | A collection consisting of the last n elements of xs (or, some arbitrary n elements, if no order is defined) |
| xs dropRight n | The rest of the collection except xs `takeRight n` |
| **Zippers:** | |
| xs zip ys | An iterable of pairs of corresponding elements from xs and ys |
| xs zipAll (ys, x, y) | An iterable of pairs of corresponding elements from xs and ys, where the shorter sequence is extended to match the longer one by appending elements x or y |
| xs.zipWithIndex | An iterable of pairs of elements from xs with their indicies |
| **Comparison:** | |
| xs sameElements ys | Tests whether xs and ys contain the same elements in the same order |

**Why have both Traversable and Iterable?**

You might wonder why the extra trait `Traversable` is above `Iterable`. Can we not do everything with an `iterator`? So what's the point of having a more abstract trait that defines its methods in terms of `foreach` instead

513

of `iterator`? One reason for having `Traversable` is that sometimes it is easier or more efficient to provide an implementation of `foreach` than to provide an implementation of `iterator`. Here's a simple example. Let's say you want a class hierarchy for binary trees that have integer elements at the leaves. You might design this hierarchy like this:

```
sealed abstract class Tree
case class Branch(left: Tree, right: Tree) extends Tree
case class Node(elem: Int) extends Tree
```

Now assume you want to make trees traversable. To do this, have `Tree` inherit from `Traversable[Int]` and define a `foreach` method like this:

```
sealed abstract class Tree extends Traversable[Int] {
  def foreach[U](f: Int => U) = this match {
   case Node(elem) => f(elem)
   case Branch(l, r) => l foreach f; r foreach f
  }
}
```

That's not too hard, and it is also very efficient—traversing a balanced tree takes time proportional to the number of elements in the tree. To see this, consider that for a balanced tree with N leaves you will have N − 1 interior nodes of class `Branch`. So the total number of steps to traverse the tree is N + N − 1.

Now, compare this with making trees iterable. To do this, have `Tree` inherit from `Iterable[Int]` and define an `iterator` method like this:

```
sealed abstract class Tree extends Iterable[Int] {
  def iterator: Iterator[Int] = this match {
    case Node(elem) => Iterator.single(elem)
    case Branch(l, r) => l.iterator ++ r.iterator
  }
}
```

At first glance, this looks no harder than the `foreach` solution. However, there's an efficiency problem that has to do with the implementation of the iterator concatenation method, `++`. Every time an element is produced by a concatenated iterator such as `l.iterator ++ r.iterator`, the computation needs to follow one indirection to get at the right iterator (either `l.iterator`, or r.iterator). Overall, that makes log($N$) indirections to get at a leaf of a balanced tree with N leaves. So the cost of visiting all elements of a tree went up from about 2N for the foreach traversal method to N log(N) for the traversal with iterator. If the tree has a million elements that means about two million steps for foreach and about twenty million steps for iterator. So the foreach solution has a clear advantage.

**Subcategories of Iterable**

In the inheritance hierarchy below Iterable you find three traits: Seq, Set, and Map. A common aspect of these three traits is that they all implement the PartialFunction trait[1] with its apply and isDefinedAt methods. However, the way each trait implements PartialFunction differs.

For sequences, apply is positional indexing, where elements are always numbered from 0. That is, Seq(1, 2, 3)(1) == 2. For sets, apply is a membership test. For instance, Set('a', 'b', 'c')('b') == true whereas Set()('a') == false. Finally for maps, apply is a selection. For instance, Map('a' -> 1, 'b' -> 10, 'c' -> 100)('b') == 10.

In the following three sections, we will explain each of the three kinds of collections in more detail.

## 24.5 The sequence traits Seq, IndexedSeq, and LinearSeq

The Seq trait represents sequences. A sequence is a kind of iterable that has a length and whose elements have fixed index positions, starting from 0.

The operations on sequences, summarized in Figure 24.3, fall into the following categories:

*Indexing and length operations* apply, isDefinedAt, length, indices, and lengthCompare. For a Seq, the apply operation means indexing; hence a sequence of type Seq[T] is a partial function that takes an Int argument (an index) and yields a sequence element of type T. In other words Seq[T] extends PartialFunction[Int, T]. The elements of a sequence are indexed from zero up to the length of the sequence minus one. The length method on sequences is an alias of

---

[1] Partial functions were described in Section 15.7.

the `size` method of general collections. The `lengthCompare` method allows you to compare the lengths of two sequences even if one of the sequences has infinite length.

***Index search operations*** `indexOf`, `lastIndexOf`, `indexOfSlice`, `lastIndexOfSlice`, `indexWhere`, `lastIndexWhere`, `segmentLength`, and `prefixLength`, which return the index of an element equal to a given value or matching some predicate.

***Addition operations*** `+:`, `:+`, and `padTo`, which return new sequences obtained by adding elements at the front or the end of a sequence.

***Update operations*** `updated` and `patch`, which return a new sequence obtained by replacing some elements of the original sequence.

***Sorting operations*** `sorted`, `sortWith`, and `sortBy`, which sort sequence elements according to various criteria.

***Reversal operations*** `reverse`, `reverseIterator`, and `reverseMap`, which yield or process sequence elements in reverse order, from last to first.

***Comparison operations*** `startsWith`, `endsWith`, `contains`, `corresponds`, and `containsSlice`, which relate two sequences or search an element in a sequence.

***Multiset operations*** `intersect`, `diff`, `union`, and `distinct`, which perform set-like operations on the elements of two sequences or remove duplicates.

If a sequence is mutable, it offers in addition a side-effecting `update` method, which lets sequence elements be updated. Recall from Chapter 3 that syntax like `seq(idx) = elem` is just a shorthand for `seq.update(idx, elem)`. Note the difference between `update` and `updated`. The `update` method changes a sequence element in place, and is only available for mutable sequences. The `updated` method is available for all sequences and always returns a new sequence instead of modifying the original.

Table 24.3 · Operations in trait Seq

| What it is | What it does |
|---|---|
| **Indexing and length:** | |
| xs(i) | (or, written out, xs apply i) The element of xs at index i. |
| xs isDefinedAt i | Tests whether i is contained in xs.indices. |
| xs.length | The length of the sequence (same as size). |
| xs.lengthCompare ys | Returns −1 if xs is shorter than ys, +1 if it is longer, and 0 is they have the same length. Works even if one of the sequences is infinite. |
| xs.indices | The index range of xs, extending from 0 to xs.length − 1. |
| **Index search:** | |
| xs indexOf x | The index of the first element in xs equal to x (several variants exist). |
| xs lastIndexOf x | The index of the last element in xs equal to x (several variants exist). |
| xs indexOfSlice ys | The first index of xs such that successive elements starting from that index form the sequence ys. |
| xs lastIndexOfSlice ys | The last index of xs such that successive elements starting from that index form the sequence ys. |
| xs indexWhere p | The index of the first element in xs that satisfies p (several variants exist). |
| xs segmentLength (p, i) | The length of the longest uninterrupted segment of elements in xs, starting with xs(i), that all satisfy the predicate p. |
| xs prefixLength p | The length of the longest prefix of elements in xs that all satisfy the predicate p. |
| **Additions:** | |
| x +: xs | A new sequence consisting of x prepended to xs. |
| xs :+ x | A new sequence that consists of x append to xs. |
| xs padTo (len, x) | The sequence resulting from appending the value x to xs until length len is reached. |
| **Updates:** | |

## Table 24.3 · continued

| | |
|---|---|
| xs patch (i, ys, r) | The sequence resulting from replacing r elements of xs starting with i by the patch ys. |
| xs updated (i, x) | A copy of xs with the element at index i replaced by x. |
| xs(i) = x | (or, written out, xs.update(i, x), only available for mutable.Seqs) Changes the element of xs at index i to y. |

**Sorting:**

| | |
|---|---|
| xs.sorted | A new sequence obtained by sorting the elements of xs using the standard ordering of the element type of xs. |
| xs sortWith lessThan | A new sequence obtained by sorting the elements of xs, using lessThan as comparison operation. |
| xs sortBy f | A new sequence obtained by sorting the elements of xs. Comparison between two elements proceeds by mapping the function f over both and comparing the results. |

**Reversals:**

| | |
|---|---|
| xs.reverse | A sequence with the elements of xs in reverse order. |
| xs.reverseIterator | An iterator yielding all the elements of xs in reverse order. |
| xs reverseMap f | A sequence obtained by mapping f over the elements of xs in reverse order. |

**Comparisons:**

| | |
|---|---|
| xs startsWith ys | Tests whether xs starts with sequence ys (several variants exist). |
| xs endsWith ys | Tests whether xs ends with sequence ys (several variants exist). |
| xs contains x | Tests whether xs has an element equal to x. |
| xs containsSlice ys | Tests whether xs has a contiguous subsequence equal to ys. |
| (xs corresponds ys)(p) | Tests whether corresponding elements of xs and ys satisfy the binary predicate p. |

**Multiset operations:**

## Table 24.3 · continued

| | |
|---|---|
| xs intersect ys | The multi-set intersection of sequences xs and ys that preserves the order of elements in xs. |
| xs diff ys | The multi-set difference of sequences xs and ys that preserves the order of elements in xs. |
| xs union ys | Multiset union; same as xs ++ ys. |
| xs.distinct | A subsequence of xs that contains no duplicated element. |

Each Seq trait has two subtraits, LinearSeq and IndexedSeq. These do not add any new operations, but each offers different performance characteristics. A linear sequence has efficient head and tail operations, whereas an indexed sequence has efficient apply, length, and (if mutable) update operations. List is a frequently used linear sequence, as is Stream. Two frequently used indexed sequences are Array and ArrayBuffer. The Vector class provides an interesting compromise between indexed and linear access. It has both effectively constant time indexing overhead and constant time linear access overhead. Because of this, vectors are a good foundation for mixed access patterns where both indexed and linear accesses are used. More on vectors in Section 24.8.

### Buffers

An important sub-category of mutable sequences is buffers. Buffers allow not only updates of existing elements but also element insertions, element removals, and efficient additions of new elements at the end of the buffer. The principal new methods supported by a buffer are += and ++=, for element addition at the end, +=: and ++=: for addition at the front, insert and insertAll for element insertions, as well as remove and -= for element removal. These operations are summarized in Table 24.4.

Two Buffer implementations that are commonly used are ListBuffer and ArrayBuffer. As the name implies, a ListBuffer is backed by a List and supports efficient conversion of its elements to a List, whereas an ArrayBuffer is backed by an array, and can be quickly converted into one. You saw a glimpse of the implementation of ListBuffer in Section 22.2.

Table 24.4 · Operations in trait `Buffer`

| What it is | What it does |
|---|---|
| **Additions:** | |
| buf += x | Appends element x to buffer buf, and returns buf itself as result |
| buf += (x, y, z) | Appends given elements to buffer |
| buf ++= xs | Appends all elements in xs to buffer |
| x +=: buf | Prepends element x to buffer |
| xs ++=: buf | Prepends all elements in xs to buffer |
| buf insert (i, x) | Inserts element x at index i in buffer |
| buf insertAll (i, xs) | Inserts all elements in xs at index i in buffer |
| **Removals:** | |
| buf -= x | Removes element x from buffer |
| buf remove i | Removes element at index i from buffer |
| buf remove (i, n) | Removes n elements starting at index i from buffer |
| buf trimStart n | Removes first n elements from buffer |
| buf trimEnd n | Removes last n elements from buffer |
| buf.clear() | Removes all elements from buffer |
| **Cloning:** | |
| buf.clone | A new buffer with the same elements as buf |

## 24.6   Sets

Sets are `Iterables` that contain no duplicate elements. The operations on sets are summarized in Table 24.5 for general sets and Table 24.6 for mutable sets. They fall into the following categories:

*Tests* contains, apply, and subsetOf. The contains method indicates whether a set contains a given element. The apply method for a set is the same as contains, so set(elem) is the same as set contains elem. That means sets can also be used as test functions that return true for the elements they contain. For example:

```
scala> val fruit = Set("apple", "orange", "peach", "banana")
fruit: scala.collection.immutable.Set[String] =
  Set(apple, orange, peach, banana)

scala> fruit("peach")
res7: Boolean = true

scala> fruit("potato")
res8: Boolean = false
```

*Additions*  + and ++, which add one or more elements to a set, yielding a new
set as a result.

*Removals*  – and --, which remove one or more elements from a set, yielding
a new set.

*Set operations*  for union, intersection, and set difference. These set operations exist in two forms: alphabetic and symbolic. The alphabetic
versions are intersect, union, and diff, whereas the symbolic ver-
sions are &, |, and &~. The ++ that Set inherits from Traversable
can be seen as yet another alias of union or |, except that ++ takes a
Traversable argument whereas union and | take sets.

Table 24.5 · Operations in trait Set

| **What it is** | **What it does** |
| --- | --- |
| **Tests:** | |
| xs contains x | Tests whether x is an element of xs |
| xs(x) | Same as xs contains x |
| xs subsetOf ys | Tests whether xs is a subset of ys |
| **Additions:** | |
| xs + x | The set containing all elements of xs as well as x |
| xs + (x, y, z) | The set containing all elements of xs as well as the given additional elements |
| xs ++ ys | The set containing all elements of xs as well as all elements of ys |
| **Removals:** | |

521

## Table 24.5 · continued

| | |
|---|---|
| xs - x | The set containing all elements of xs except x |
| xs - (x, y, z) | The set containing all elements of xs except the given elements |
| xs -- ys | The set containing all elements of xs except the elements of ys |
| xs.empty | An empty set of the same class as xs |
| **Binary operations:** | |
| xs & ys | The set intersection of xs and ys |
| xs intersect ys | Same as xs & ys |
| xs \| ys | The set union of xs and ys |
| xs union ys | Same as xs \| ys |
| xs &~ ys | The set difference of xs and ys |
| xs diff ys | Same as xs &~ ys |

Mutable sets have methods that add, remove, or update elements, which are summarized in Table 24.6:

## Table 24.6 · Operations in trait `mutable.Set`

| What it is | What it does |
|---|---|
| **Additions:** | |
| xs += x | Adds element x to set xs as a side effect and returns xs itself |
| xs += (x, y, z) | Adds the given elements to set xs as a side effect and returns xs itself |
| xs ++= ys | Adds all elements in ys to set xs as a side effect and returns xs itself |
| xs add x | Adds element x to xs and returns `true` if x was not previously contained in the set, `false` if it was previously contained |
| **Removals:** | |
| xs -= x | Removes element x from set xs as a side effect and returns xs itself |
| xs -= (x, y, z) | Removes the given elements from set xs as a side effect and returns xs itself |

Table 24.6 · continued

| | |
|---|---|
| xs ---= ys | Removes all elements in ys from set xs as a side effect and returns xs itself |
| xs remove x | Removes element x from xs and returns true if x was previously contained in the set, false if it was not previously contained |
| xs retain p | Keeps only those elements in xs that satisfy predicate p |
| xs.clear() | Removes all elements from xs |
| **Update:** | |
| xs(x) = b | (or, written out, xs.update(x, b)) If boolean argument b is true, adds x to xs, otherwise removes x from xs |
| **Cloning:** | |
| xs.clone | A new mutable set with the same elements as xs |

Just like an immutable set, a mutable set offers the + and ++ operations for element additions and the - and -- operations for element removals. But these are less often used for mutable sets since they involve copying the set. As a more efficient alternative, mutable sets offer the update methods += and -=. The operation s += elem adds elem to the set s as a side effect, and returns the mutated set as a result. Likewise, s -= elem removes elem from the set, and returns the mutated set as a result. Besides += and -= there are also the bulk operations ++= and --=, which add or remove all elements of a traversable or an iterator.

The choice of the method names += and -= means that very similar code can work with either mutable or immutable sets. Consider first the following interpreter dialogue that uses an immutable set s:

```scala
scala> var s = Set(1, 2, 3)
s: scala.collection.immutable.Set[Int] = Set(1, 2, 3)

scala> s += 4; s -= 2

scala> s
res10: scala.collection.immutable.Set[Int] = Set(1, 3, 4)
```

In this example, we used += and -= on a var of type immutable.Set. As was explained in Step 10 in Chapter 3, a statement such as s += 4 is an abbreviation for s = s + 4. So this invokes the addition method + on the set s and then assigns the result back to the s variable. Consider now an analogous interaction with a mutable set:

```
scala> val s = collection.mutable.Set(1, 2, 3)
s: scala.collection.mutable.Set[Int] = Set(1, 2, 3)

scala> s += 4
res11: s.type = Set(1, 2, 3, 4)

scala> s -= 2
res12: s.type = Set(1, 3, 4)
```

The end effect is very similar to the previous interaction; we start with a Set(1, 2, 3) and end up with a Set(1, 3, 4). However, even though the statements look the same as before, they do something different. The s += 4 statement now invokes the += method on the mutable set value s, changing the set in place. Likewise, the s -= 2 statement now invokes the -= method on the same set.

Comparing the two interactions shows an important principle. You often can replace a mutable collection stored in a val by an immutable collection stored in a var, and *vice versa*. This works at least as long as there are no alias references to the collection through which you can observe whether it was updated in place or a new collection was created.

Mutable sets also provide add and remove as variants of += and -=. The difference is that add and remove return a boolean result indicating whether the operation had an effect on the set.

The current default implementation of a mutable set uses a hash table to store the set's elements. The default implementation of an immutable set uses a representation that adapts to the number of elements of the set. An empty set is represented by just a singleton object. Sets of sizes up to four are represented by a single object that stores all elements as fields. Beyond that size, immutable sets are implemented as hash tries.[2]

A consequence of these representation choices is that for sets of small sizes, up to about four, immutable sets are more compact and more efficient than mutable sets. So if you expect the size of a set to be small, try to make it immutable.

---

[2]Hash tries are described in Section 24.8.

## 24.7 Maps

Maps are `Iterables` of pairs of keys and values (also named *mappings* or *associations*). As explained in Section 21.4, Scala's `Predef` class offers an implicit conversion that lets you write key -> value as an alternate syntax for the pair (key, value). Therefore, Map("x" -> 24, "y" -> 25, "z" -> 26) means exactly the same as Map(("x", 24), ("y", 25), ("z", 26)), but reads better.

The fundamental operations on maps, summarized in Table 24.7, are similar to those on sets. Mutable maps additionally support the operations shown in Table 24.8. Map operations fall into the following categories:

***Lookups*** apply, get, getOrElse, contains, and isDefinedAt. These operations turn maps into partial functions from keys to values. The fundamental lookup method for a map is:

```
def get(key): Option[Value]
```

The operation "m get key" tests whether the map contains an association for the given key. If so, it returns the associated value in a Some. If no key is defined in the map, get returns None. Maps also define an apply method that returns the value associated with a given key directly, without wrapping it in an Option. If the key is not defined in the map, an exception is raised.

***Additions and updates*** +, ++, and updated, which let you add new bindings to a map or change existing bindings.

***Removals*** - and --, which remove bindings from a map.

***Subcollection producers*** keys, keySet, keysIterator, valuesIterator, and values, which return a map's keys and values separately in various forms.

***Transformations*** filterKeys and mapValues, which produce a new map by filtering and transforming bindings of an existing map.

Table 24.7 · Operations in trait Map

| What it is | What it does |
| --- | --- |
| **Lookups:** | |
| ms get k | The value associated with key k in map ms as an option, or None if not found |
| ms(k) | (or, written out, ms apply k) The value associated with key k in map ms, or a thrown exception if not found |
| ms getOrElse (k, d) | The value associated with key k in map ms, or the default value d if not found |
| ms contains k | Tests whether ms contains a mapping for key k |
| ms isDefinedAt k | Same as contains |
| **Additions and updates:** | |
| ms + (k -> v) | The map containing all mappings of ms as well as the mapping k -> v from key k to value v |
| ms + (k -> v, l -> w) | The map containing all mappings of ms as well as the given key/value pairs |
| ms ++ kvs | The map containing all mappings of ms as well as all key/value pairs of kvs |
| ms updated (k, v) | Same as ms + (k -> v) |
| **Removals:** | |
| ms – k | The map containing all mappings of ms except for any mapping of key k |
| ms – (k, l, m) | The map containing all mappings of ms except for any mapping with the given keys |
| ms -- ks | The map containing all mappings of ms except for any mapping with a key in ks |
| **Subcollections:** | |
| ms.keys | An iterable containing each key in ms |
| ms.keySet | A set containing each key in ms |
| ms.keysIterator | An iterator yielding each key in ms |
| ms.values | An iterable containing each value associated with a key in ms |
| ms.valuesIterator | An iterator yielding each value associated with a key in ms |

Table 24.7 · continued

**Transformation:**

| | |
|---|---|
| ms filterKeys p | A map view containing only those mappings in ms where the key satisfies predicate p |
| ms mapValues f | A map view resulting from applying function f to each value associated with a key in ms |

Table 24.8 · Operations in trait `mutable.Map`

| **What it is** | **What it does** |
|---|---|
| **Additions and updates:** | |
| ms(k) = v | (or, written out, ms.update(k, v)) Adds mapping from key k to value v to map ms as a side effect, overwriting any previous mapping of k |
| ms += (k -> v) | Adds mapping from key k to value v to map ms as a side effect and returns ms itself |
| ms += (k -> v, l -> w) | Adds the given mappings to ms as a side effect and returns ms itself |
| ms ++= kvs | Adds all mappings in kvs to ms as a side effect and returns ms itself |
| ms put (k, v) | Adds mapping from key k to value v to ms and returns any value previously associated with k as an option |
| ms getOrElseUpdate (k, d) | If key k is defined in map ms, returns its associated value. Otherwise, updates ms with the mapping k -> d and returns d |
| **Removals:** | |
| ms -= k | Removes mapping with key k from ms as a side effect and returns ms itself |
| ms -= (k, l, m) | Removes mappings with the given keys from ms as a side effect and returns ms itself |
| ms --= ks | Removes all keys in ks from ms as a side effect and returns ms itself |
| ms remove k | Removes any mapping with key k from ms and returns any value previously associated with k as an option |

527

## Table 24.8 · continued

| | |
|---|---|
| ms retain p | Keeps only those mappings in ms that have a key satisfying predicate p. |
| ms.clear() | Removes all mappings from ms |
| **Transformation and cloning:** | |
| ms transform f | Transforms all associated values in map ms with function f |
| ms.clone | Returns a new mutable map with the same mappings as ms |

The addition and removal operations for maps mirror those for sets. As for sets, mutable maps also support the non-destructive addition operations +, −, and updated, but they are used less frequently because they involve a copying of the mutable map. Instead, a mutable map m is usually updated "in place," using the two variants m(key) = value or m += (key -> value). There is also the variant m put (key, value), which returns an Option value that contains the value previously associated with key, or None if the key did not exist in the map before.

The getOrElseUpdate is useful for accessing maps that act as caches. Say you have an expensive computation triggered by invoking a function f:

```scala
scala> def f(x: String) = {
         println("taking my time."); Thread.sleep(100)
         x.reverse }
f: (x: String)String
```

Assume further that f has no side-effects, so invoking it again with the same argument will always yield the same result. In that case you could save time by storing previously computed bindings of argument and results of f in a map, and only computing the result of f if a result of an argument was not found there. You could say the map is a *cache* for the computations of the function f.

```scala
scala> val cache = collection.mutable.Map[String, String]()
cache: scala.collection.mutable.Map[String,String] = Map()
```

You can now create a more efficient caching version of the f function:

```scala
scala> def cachedF(s: String) = cache.getOrElseUpdate(s, f(s))
cachedF: (s: String)String

scala> cachedF("abc")
taking my time.
res16: String = cba

scala> cachedF("abc")
res17: String = cba
```

Note that the second argument to getOrElseUpdate is "by-name," so the computation of f("abc") above is only performed if getOrElseUpdate requires the value of its second argument, which is precisely if its first argument is not found in the cache map. You could also have implemented cachedF directly, using just basic map operations, but it would have have taken more code to do so:

```scala
def cachedF(arg: String) = cache get arg match {
  case Some(result) => result
  case None =>
    val result = f(arg)
    cache(arg) = result
    result
}
```

## 24.8 Concrete immutable collection classes

Scala provides many concrete immutable collection classes for you to choose from. They differ in the traits they implement (maps, sets, sequences), whether they can be infinite, and the speed of various operations. We'll start by reviewing the most common immutable collection types.

### Lists

Lists are finite immutable sequences. They provide constant-time access to their first element as well as the rest of the list, and they have a constant-time cons operation for adding a new element to the front of the list. Many other operations take linear time. See Chapters 16 and 22 for extensive discussions about lists.

**Streams**

A stream is like a list except that its elements are computed lazily. Because of this, a stream can be infinitely long. Only those elements requested will be computed. Otherwise, streams have the same performance characteristics as lists.

Whereas lists are constructed with the `::` operator, streams are constructed with the similar-looking `#::`. Here is a simple example of a stream containing the integers 1, 2, and 3:

```
scala> val str = 1 #:: 2 #:: 3 #:: Stream.empty
str: scala.collection.immutable.Stream[Int] = Stream(1, ?)
```

The head of this stream is 1, and the tail of it has 2 and 3. The tail is not printed here, though, because it hasn't been computed yet! Streams are required to compute lazily, and the `toString` method of a stream is careful not to force any extra evaluation.

Below is a more complex example. It computes a stream that contains a Fibonacci sequence starting with the given two numbers. A Fibonacci sequence is one where each element is the sum of the previous two elements in the series:

```
scala> def fibFrom(a: Int, b: Int): Stream[Int] =
         a #:: fibFrom(b, a + b)
fibFrom: (a: Int, b: Int)Stream[Int]
```

This function is deceptively simple. The first element of the sequence is clearly a, and the rest of the sequence is the Fibonacci sequence starting with b followed by a + b. The tricky part is computing this sequence without causing an infinite recursion. If the function used `::` instead of `#::`, then every call to the function would result in another call, thus causing an infinite recursion. Since it uses `#::`, though, the right-hand side is not evaluated until it is requested.

Here are the first few elements of the Fibonacci sequence starting with two ones:

```
scala> val fibs = fibFrom(1, 1).take(7)
fibs: scala.collection.immutable.Stream[Int] = Stream(1, ?)

scala> fibs.toList
res23: List[Int] = List(1, 1, 2, 3, 5, 8, 13)
```

**Vectors**

Lists are very efficient when the algorithm processing them is careful to only process their heads. Accessing, adding, and removing the head of a list takes only constant time, whereas accessing or modifying elements later in the list takes time linear in the depth into the list.

Vectors are a collection type that give efficient access to elements beyond the head. Access to any elements of a vector take only "effectively constant time," as defined below. It's a larger constant than for access to the head of a list or for reading an element of an array, but it's a constant nonetheless. As a result, algorithms using vectors do not have to be careful about accessing just the head of the sequence. They can access and modify elements at arbitrary locations, and thus they can be much more convenient to write.

Vectors are built and modified just like any other sequence:

```
scala> val vec = scala.collection.immutable.Vector.empty
vec: scala.collection.immutable.Vector[Nothing] = Vector()

scala> val vec2 = vec :+ 1 :+ 2
vec2: scala.collection.immutable.Vector[Int] = Vector(1, 2)

scala> val vec3 = 100 +: vec2
vec3: scala.collection.immutable.Vector[Int]
  = Vector(100, 1, 2)

scala> vec3(0)
res24: Int = 100
```

Vectors are represented as broad, shallow trees. Every tree node contains up to 32 elements of the vector or contains up to 32 other tree nodes. Vectors with up to 32 elements can be represented in a single node. Vectors with up to 32 * 32 = 1024 elements can be represented with a single indirection. Two hops from the root of the tree to the final element node are sufficient for vectors with up to $2^{15}$ elements, three hops for vectors with $2^{20}$, four hops for vectors with $2^{25}$ elements and five hops for vectors with up to $2^{30}$ elements. So for all vectors of reasonable size, an element selection involves up to five primitive array selections. This is what we meant when we wrote that element access is "effectively constant time."

Vectors are immutable, so you cannot change an element of a vector in place. However, with the `updated` method you can create a new vector that differs from a given vector only in a single element:

```
scala> val vec = Vector(1, 2, 3)
vec: scala.collection.immutable.Vector[Int] = Vector(1, 2, 3)

scala> vec updated (2, 4)
res25: scala.collection.immutable.Vector[Int] = Vector(1, 2, 4)

scala> vec
res26: scala.collection.immutable.Vector[Int] = Vector(1, 2, 3)
```

As the last line above shows, a call to updated has no effect on the original vector vec. Like selection, functional vector updates are also "effectively constant time." Updating an element in the middle of a vector can be done by copying the node that contains the element, and every node that points to it, starting from the root of the tree. This means that a functional update creates between one and five nodes that each contain up to 32 elements or subtrees. This is certainly more expensive than an in-place update in a mutable array, but still a lot cheaper than copying the whole vector.

Because vectors strike a good balance between fast random selections and fast random functional updates, they are currently the default implementation of immutable indexed sequences:

```
scala> collection.immutable.IndexedSeq(1, 2, 3)
res27: scala.collection.immutable.IndexedSeq[Int]
    = Vector(1, 2, 3)
```

**Immutable stacks**

If you need a last-in-first-out sequence, you can use a Stack. You push an element onto a stack with push, pop an element with pop, and peek at the top of the stack without removing it with top. All of these operations are constant time.

Here are some simple operations performed on a stack:

```
scala> val stack = scala.collection.immutable.Stack.empty
stack: scala.collection.immutable.Stack[Nothing] = Stack()

scala> val hasOne = stack.push(1)
hasOne: scala.collection.immutable.Stack[Int] = Stack(1)

scala> stack
res28: scala.collection.immutable.Stack[Nothing] = Stack()
```

```
scala> hasOne.top
res29: Int = 1

scala> hasOne.pop
res30: scala.collection.immutable.Stack[Int] = Stack()
```

Immutable stacks are used rarely in Scala programs because their functionality is subsumed by lists: A push on an immutable stack is the same as a :: on a list, and a pop on a stack is the same a tail on a list.

## Immutable queues

A queue is just like a stack except that it is first-in-first-out rather than last-in-first-out. A simplified implementation of immutable queues was discussed in Chapter 19. Here's how you can create an empty immutable queue:

```
scala> val empty = scala.collection.immutable.Queue[Int]()
empty: scala.collection.immutable.Queue[Int] = Queue()
```

You can append an element to an immutable queue with enqueue:

```
scala> val has1 = empty.enqueue(1)
has1: scala.collection.immutable.Queue[Int] = Queue(1)
```

To append multiple elements to a queue, call enqueue with a collection as its argument:

```
scala> val has123 = has1.enqueue(List(2, 3))
has123: scala.collection.immutable.Queue[Int] = Queue(1, 2,
3)
```

To remove an element from the head of the queue, use dequeue:

```
scala> val (element, has23) = has123.dequeue
element: Int = 1
has23: scala.collection.immutable.Queue[Int] = Queue(2, 3)
```

Note that dequeue returns a pair consisting of the element removed and the rest of the queue.

## Ranges

A range is an ordered sequence of integers that are equally spaced apart. For example, "1, 2, 3" is a range, as is "5, 8, 11, 14." To create a range in Scala, use the predefined methods to and by. Here are some examples:

```
scala> 1 to 3
res31: scala.collection.immutable.Range.Inclusive
  = Range(1, 2, 3)
scala> 5 to 14 by 3
res32: scala.collection.immutable.Range = Range(5, 8, 11, 14)
```

If you want to create a range that is exclusive of its upper limit, use the convenience method until instead of to:

```
scala> 1 until 3
res33: scala.collection.immutable.Range = Range(1, 2)
```

Ranges are represented in constant space, because they can be defined by just three numbers: their start, their end, and the stepping value. Because of this representation, most operations on ranges are extremely fast.

## Hash tries

Hash tries[3] are a standard way to implement immutable sets and maps efficiently. Their representation is similar to vectors in that they are also trees where every node has 32 elements or 32 subtrees, but selection is done based on a hash code. For instance, to find a given key in a map, you use the lowest five bits of the hash code of the key to select the first subtree, the next five bits the next subtree, and so on. Selection stops once all elements stored in a node have hash codes that differ from each other in the bits that are selected so far. Thus, not all the bits of the hash code are necessarily used.

Hash tries strike a nice balance between reasonably fast lookups and reasonably efficient functional insertions (+) and deletions (-). That's why they underlie Scala's default implementations of immutable maps and sets. In fact, Scala has a further optimization for immutable sets and maps that contain less than five elements. Sets and maps with one to four elements are stored as single objects that just contain the elements (or key/value pairs in

---

[3]"Trie" comes from the word "re*trie*val" and is pronounced *tree* or *try*.

the case of a map) as fields. The empty immutable set and empty immutable map is in each case a singleton object—there's no need to duplicate storage for those because an empty immutable set or map will always stay empty.

## Red-black trees

Red-black trees are a form of balanced binary trees where some nodes are designated "red" and others "black." Like any balanced binary tree, operations on them reliably complete in time logarithmic to the size of the tree.

Scala provides implementations of sets and maps that use a red-black tree internally. You access them under the names TreeSet and TreeMap:

```
scala> val set = collection.immutable.TreeSet.empty[Int]
set: scala.collection.immutable.TreeSet[Int] = TreeSet()

scala> set + 1 + 3 + 3
res34: scala.collection.immutable.TreeSet[Int] = TreeSet(1, 3)
```

Red-black trees are also the standard implementation of SortedSet in Scala, because they provide an efficient iterator that returns all elements of the set in sorted order.

## Immutable bit sets

A bit set represents a collection of small integers as the bits of a larger integer. For example, the bit set containing 3, 2, and 0 would be represented as the integer 1101 in binary, which is 13 in decimal.

Internally, bit sets use an array of 64-bit Longs. The first Long in the array is for integers 0 through 63, the second is for 64 through 127, and so on. Thus, bit sets are very compact so long as the largest integer in the set is less than a few hundred or so.

Operations on bit sets are very fast. Testing for inclusion takes constant time. Adding an item to the set takes time proportional to the number of Longs in the bit set's array, which is typically a small number. Here are some simple examples of the use of a bit set:

```
scala> val bits = scala.collection.immutable.BitSet.empty
bits: scala.collection.immutable.BitSet = BitSet()

scala> val moreBits = bits + 3 + 4 + 4
moreBits: scala.collection.immutable.BitSet = BitSet(3, 4)
```

```
scala> moreBits(3)
res35: Boolean = true

scala> moreBits(0)
res36: Boolean = false
```

**List maps**

A list map represents a map as a linked list of key-value pairs. In general, operations on a list map might have to iterate through the entire list. Thus, operations on a list map take time linear in the size of the map. In fact there is little usage for list maps in Scala because standard immutable maps are almost always faster. The only possible difference is if the map is for some reason constructed in such a way that the first elements in the list are selected much more often than the other elements.

```
scala> val map = collection.immutable.ListMap(
         1 -> "one", 2 -> "two")
map: scala.collection.immutable.ListMap[Int,String] = Map(1
-> one, 2 -> two)

scala> map(2)
res37: String = "two"
```

## 24.9   Concrete mutable collection classes

Now that you've seen the most commonly used immutable collection classes that Scala provides in its standard library, take a look at the mutable collection classes.

**Array buffers**

You've already seen array buffers in Section 17.1. An array buffer holds an array and a size. Most operations on an array buffer have the same speed as an array, because the operations simply access and modify the underlying array. Additionally, array buffers can have data efficiently added to the end. Appending an item to an array buffer takes amortized constant time. Thus, array buffers are useful for efficiently building up a large collection whenever the new items are always added to the end. Here are some examples:

```
scala> val buf = collection.mutable.ArrayBuffer.empty[Int]
buf: scala.collection.mutable.ArrayBuffer[Int]
  = ArrayBuffer()

scala> buf += 1
res38: buf.type = ArrayBuffer(1)

scala> buf += 10
res39: buf.type = ArrayBuffer(1, 10)

scala> buf.toArray
res40: Array[Int] = Array(1, 10)
```

**List buffers**

You've also already seen list buffers in Section 17.1. A list buffer is like an array buffer except that it uses a linked list internally instead of an array. If you plan to convert the buffer to a list once it is built up, use a list buffer instead of an array buffer. Here's an example:[4]

```
scala> val buf = collection.mutable.ListBuffer.empty[Int]
buf: scala.collection.mutable.ListBuffer[Int]
  = ListBuffer()

scala> buf += 1
res41: buf.type = ListBuffer(1)

scala> buf += 10
res42: buf.type = ListBuffer(1, 10)

scala> buf.toList
res43: List[Int] = List(1, 10)
```

**String builders**

Just like an array buffer is useful for building arrays, and a list buffer is useful for building lists, a string builder is useful for building strings. String builders are so commonly used that they are already imported into the default namespace. Create them with a simple new `StringBuilder`, like this:

---

[4]The "buf.type" that appears in the interpreter responses in this and several other examples in this section is a *singleton type*. As will be explained in Section 29.6, buf.type means the variable holds exactly the object referred to by buf.

```
scala> val buf = new StringBuilder
buf: StringBuilder =

scala> buf += 'a'
res44: buf.type = a

scala> buf ++= "bcdef"
res45: buf.type = abcdef

scala> buf.toString
res46: String = abcdef
```

**Linked lists**

Linked lists are mutable sequences that consist of nodes that are linked with
next pointers. In most languages null would be picked as the empty linked
list. That does not work for Scala collections, because even empty sequences
must support all sequence methods. LinkedList.empty.isEmpty, in particular, should return true and not throw a NullPointerException. Empty
linked lists are encoded instead in a special way: Their next field points back
to the node itself.

Like their immutable cousins, linked lists are best operated on sequentially. In addition, linked lists make it easy to insert an element or linked list
into another linked list.

**Double linked lists**

DoubleLinkedLists are like the single linked lists described in the previous
subsection, except besides next, they have another mutable field, prev, that
points to the element preceding the current node. The main benefit of that
additional link is that it makes element removal very fast.

**Mutable lists**

A MutableList consists of a single linked list together with a pointer that
refers to the terminal empty node of that list. This makes list append a constant time operation because it avoids having to traverse the list in search for
its terminal node. MutableList is currently the standard implementation of
mutable.LinearSeq in Scala.

## Queues

Scala provides mutable queues in addition to immutable ones. You use a mutable queue similarly to the way you use an immutable one, but instead of enqueue, you use the += and ++= operators to append. Also, on a mutable queue, the dequeue method will just remove the head element from the queue and return it. Here's an example:

```
scala> val queue = new scala.collection.mutable.Queue[String]
queue: scala.collection.mutable.Queue[String] = Queue()

scala> queue += "a"
res47: queue.type = Queue(a)

scala> queue ++= List("b", "c")
res48: queue.type = Queue(a, b, c)

scala> queue
res49: scala.collection.mutable.Queue[String] = Queue(a, b, c)

scala> queue.dequeue
res50: String = a

scala> queue
res51: scala.collection.mutable.Queue[String] = Queue(b, c)
```

## Array sequences

Array sequences are mutable sequences of fixed size that store their elements internally in an `Array[AnyRef]`. They are implemented in Scala by class `ArraySeq`.

You would typically use an `ArraySeq` if you want an array for its performance characteristics, but you also want to create generic instances of the sequence where you do not know the type of the elements and do not have a `ClassTag` to provide it at run-time. You will find out about these issues with arrays shortly, in Section 24.10.

## Stacks

You saw immutable stacks earlier. There is also a mutable version. It works exactly the same as the immutable version except that modifications happen in place. Here's an example:

```
scala> val stack = new scala.collection.mutable.Stack[Int]
stack: scala.collection.mutable.Stack[Int] = Stack()

scala> stack.push(1)
res52: stack.type = Stack(1)

scala> stack
res53: scala.collection.mutable.Stack[Int] = Stack(1)

scala> stack.push(2)
res54: stack.type = Stack(2, 1)

scala> stack
res55: scala.collection.mutable.Stack[Int] = Stack(2, 1)

scala> stack.top
res56: Int = 2

scala> stack
res57: scala.collection.mutable.Stack[Int] = Stack(2, 1)

scala> stack.pop
res58: Int = 2

scala> stack
res59: scala.collection.mutable.Stack[Int] = Stack(1)
```

### Array stacks

ArrayStack is an alternative implementation of a mutable stack, which is backed by an Array that gets resized as needed. It provides fast indexing and is generally slightly more efficient for most operations than a normal mutable stack.

### Hash tables

A hash table stores its elements in an underlying array, placing each item at a position in the array determined by the hash code of that item. Adding an element to a hash table takes only constant time, so long as there isn't already another element in the array that has the same hash code. Hash tables are thus very fast so long as the objects placed in them have a good distribution of hash codes. As a result, the default mutable map and set types in Scala are based on hash tables.

Hash sets and maps are used just like any other set or map. Here are some simple examples:

```
scala> val map = collection.mutable.HashMap.empty[Int,String]
map: scala.collection.mutable.HashMap[Int,String] = Map()

scala> map += (1 -> "make a web site")
res60: map.type = Map(1 -> make a web site)

scala> map += (3 -> "profit!")
res61: map.type = Map(1 -> make a web site, 3 -> profit!)

scala> map(1)
res62: String = make a web site

scala> map contains 2
res63: Boolean = false
```

Iteration over a hash table is not guaranteed to occur in any particular order. Iteration simply proceeds through the underlying array in whichever order it happens to be. To get a guaranteed iteration order, use a *linked* hash map or set instead of a regular one. A linked hash map or set is just like a regular hash map or set except that it also includes a linked list of the elements in the order they were added. Iteration over such a collection is always in the same order that the elements were initially added.

**Weak hash maps**

A weak hash map is a special kind of hash map in which the garbage collector does not follow links from the map to the keys stored in it. This means that a key and its associated value will disappear from the map if there is no other reference to that key. Weak hash maps are useful for tasks such as caching, where you want to re-use an expensive function's result if the function is called again on the same key. If keys and function results are stored in a regular hash map, the map could grow without bounds, and no key would ever become garbage. Using a weak hash map avoids this problem. As soon as a key object becomes unreachable, it's entry is removed from the weak hash map. Weak hash maps in Scala are implemented as a wrapper of an underlying Java implementation, `java.util.WeakHashMap`.

## Concurrent Maps

A concurrent map can be accessed by several threads at once. In addition to the usual Map operations, it provides the following atomic operations:

Table 24.9 · Operations in trait ConcurrentMap

| What it is | What it does |
|---|---|
| m putIfAbsent(k, v) | Adds key/value binding k -> m unless k is already defined in m |
| m remove (k, v) | Removes entry for k if it is currently mapped to v |
| m replace (k, old, new) | Replaces value associated with key k to new, if it was previously bound to old |
| m replace (k, v) | Replaces value associated with key k to v, if it was previously bound to some value |

ConcurrentMap is a trait in the Scala collections library. Currently, its only implementation is Java's java.util.concurrent.ConcurrentMap, which can be converted automatically into a Scala map using the standard Java/Scala collection conversions, which will be described in Section 24.17.

## Mutable bit sets

A mutable bit set is just like an immutable one, except that it can be modified in place. Mutable bit sets are slightly more efficient at updating than immutable ones, because they don't have to copy around Longs that haven't changed. Here is an example:

```
scala> val bits = scala.collection.mutable.BitSet.empty
bits: scala.collection.mutable.BitSet = BitSet()

scala> bits += 1
res64: bits.type = BitSet(1)

scala> bits += 3
res65: bits.type = BitSet(1, 3)

scala> bits
res66: scala.collection.mutable.BitSet = BitSet(1, 3)
```

## 24.10 Arrays

Arrays are a special kind of collection in Scala. One the one hand, Scala arrays correspond one-to-one to Java arrays. That is, a Scala array `Array[Int]` is represented as a Java `int[]`, an `Array[Double]` is represented as a Java `double[]` and an `Array[String]` is represented as a Java `String[]`. But at the same time, Scala arrays offer much more their Java analogues. First, Scala arrays can be *generic*. That is, you can have an `Array[T]`, where T is a type parameter or abstract type. Second, Scala arrays are compatible with Scala sequences—you can pass an `Array[T]` where a `Seq[T]` is required. Finally, Scala arrays also support all sequence operations. Here's an example of this in action:

```
scala> val a1 = Array(1, 2, 3)
a1: Array[Int] = Array(1, 2, 3)

scala> val a2 = a1 map (_ * 3)
a2: Array[Int] = Array(3, 6, 9)

scala> val a3 = a2 filter (_ % 2 != 0)
a3: Array[Int] = Array(3, 9)

scala> a3.reverse
res1: Array[Int] = Array(9, 3)
```

Given that Scala arrays are represented just like Java arrays, how can these additional features be supported in Scala?

The answer lies in systematic use of implicit conversions. An array cannot pretend to *be* a sequence, because the data type representation of a native array is not a subtype of `Seq`. Instead, whenever an array would be used as a `Seq`, implicitly wrap it in a subclass of `Seq`. The name of that subclass is `scala.collection.mutable.WrappedArray`. Here you see it in action:

```
scala> val seq: Seq[Int] = a1
seq: Seq[Int] = WrappedArray(1, 2, 3)

scala> val a4: Array[Int] = seq.toArray
a4: Array[Int] = Array(1, 2, 3)

scala> a1 eq a4
res2: Boolean = true
```

This interaction demonstrates that arrays are compatible with sequences, because there's an implicit conversion from `Array` to `WrappedArray`. To go the other way, from a `WrappedArray` to an `Array`, you can use the `toArray` method defined in `Traversable`. The last interpreter line above shows that wrapping then unwrapping with `toArray` gives you back the same array you started with.

There is yet another implicit conversion that gets applied to arrays. This conversion simply "adds" all sequence methods to arrays but does not turn the array itself into a sequence. "Adding" means that the array is wrapped in another object of type `ArrayOps`, which supports all sequence methods. Typically, this `ArrayOps` object is short-lived; it will usually be inaccessible after the call to the sequence method and its storage can be recycled. Modern VMs often avoid creating this object entirely.

The difference between the two implicit conversions on arrays is demonstrated here:

```
scala> val seq: Seq[Int] = a1
seq: Seq[Int] = WrappedArray(1, 2, 3)

scala> seq.reverse
res2: Seq[Int] = WrappedArray(3, 2, 1)

scala> val ops: collection.mutable.ArrayOps[Int] = a1
ops: scala.collection.mutable.ArrayOps[Int] = [I(1, 2, 3)

scala> ops.reverse
res3: Array[Int] = Array(3, 2, 1)
```

You see that calling `reverse` on `seq`, which is a `WrappedArray`, will give again a `WrappedArray`. That's logical, because wrapped arrays are `Seqs`, and calling `reverse` on any `Seq` will give again a `Seq`. On the other hand, calling `reverse` on the `ops` value of class `ArrayOps` will result in an `Array`, not a `Seq`.

The `ArrayOps` example above was quite artificial, intended only to show the difference to `WrappedArray`. Normally, you'd never define a value of class `ArrayOps`. You'd just call a Seq method on an array:

```
scala> a1.reverse
res4: Array[Int] = Array(3, 2, 1)
```

The ArrayOps object gets inserted automatically by the implicit conversion. So the line above is equivalent to the following line, where intArrayOps was the conversion that was implicitly inserted previously:

```
scala> intArrayOps(a1).reverse
res5: Array[Int] = Array(3, 2, 1)
```

This raises the question how the compiler picked intArrayOps over the other implicit conversion to WrappedArray in the line above. After all, both conversions map an array to a type that supports a reverse method, which is what the input specified. The answer to that question is that the two implicit conversions are prioritized. The ArrayOps conversion has a higher priority than the WrappedArray conversion. The first is defined in the Predef object whereas the second is defined in a class scala.LowPriorityImplicits, which is a superclass of Predef. Implicits in subclasses and subobjects take precedence over implicits in base classes. So if both conversions are applicable, the one in Predef is chosen. A very similar scheme, which was described in Section 21.7, works for strings.

So now you know how arrays can be compatible with sequences and how they can support all sequence operations. What about genericity? In Java you cannot write a T[] where T is a type parameter. How then is Scala's Array[T] represented? In fact a generic array like Array[T] could be at run time any of Java's eight primitive array types byte[], short[], char[], int[], long[], float[], double[], boolean[], or it could be an array of objects. The only common run-time type encompassing all of these types is AnyRef (or, equivalently java.lang.Object), so that's the type to which the Scala compiler maps Array[T]. At run-time, when an element of an array of type Array[T] is accessed or updated there is a sequence of type tests that determine the actual array type, followed by the correct array operation on the Java array. These type tests slow down array operations somewhat. You can expect accesses to generic arrays to be three to four times slower than accesses to primitive or object arrays. This means that if you need maximal performance, you should prefer concrete over generic arrays.

Representing the generic array type is not enough, however, there must also be a way to *create* generic arrays. This is an even harder problem, which requires a little bit of help from you. To illustrate the problem, consider the following attempt to write a generic method that creates an array:

```
// This is wrong!
def evenElems[T](xs: Vector[T]): Array[T] = {
  val arr = new Array[T]((xs.length + 1) / 2)
  for (i <- 0 until xs.length by 2)
    arr(i / 2) = xs(i)
  arr
}
```

The evenElems method returns a new array that consists of all elements of
the argument vector xs that are at even positions in the vector. The first
line of the body of evenElems creates the result array, which has the same
element type as the argument. So depending on the actual type parameter for
T, this could be an Array[Int], or an Array[Boolean], or an array of some
of the other primitive types in Java, or an array of some reference type. But
these types all have different runtime representations, so how is the Scala
runtime going to pick the correct one? In fact, it can't do that based on the
information it is given, because the actual type that corresponds to the type
parameter T is erased at runtime. That's why you will get the following error
message if you attempt to compile the code above:

```
error: cannot find class tag for element type T
  val arr = new Array[T]((arr.length + 1) / 2)
                        ^
```

What's required here is that you help the compiler by providing a runtime
hint of what the actual type parameter of evenElems is. This runtime hint
takes the form of a *class tag* of type scala.reflect.ClassTag. A class tag
describes the *erased type* of a given type, which is all the information needed
to construct an array of that type.

In many cases the compiler can generate a class tag on its own. Such is
the case for a concrete type like Int or String. It's also the case for certain
generic types, like List[T], where enough information is known to predict
the erased type; in this example the erased type would be List.

For fully generic cases, the usual idiom is to pass the class tag using a
context bound, as discussed in Section 21.6. Here is how the above definition
could be fixed by using a context bound:

```
// This works
import scala.reflect.ClassTag
def evenElems[T: ClassTag](xs: Vector[T]): Array[T] = {
```

```
  val arr = new Array[T]((xs.length + 1) / 2)
  for (i <- 0 until xs.length by 2)
    arr(i / 2) = xs(i)
  arr
}
```

In this new definition, when the Array[T] is created, the compiler looks for a class tag for the type parameter T, that is, it will look for an implicit value of type ClassTag[T]. If such a value is found, the class tag is used to construct the right kind of array. Otherwise, you'll see an error message like the one shown previously.

Here is an interpreter interaction that uses the evenElems method:

```
scala> evenElems(Vector(1, 2, 3, 4, 5))
res6: Array[Int] = Array(1, 3, 5)

scala> evenElems(Vector("this", "is", "a", "test", "run"))
res7: Array[java.lang.String] = Array(this, a, run)
```

In both cases, the Scala compiler automatically constructed a class tag for the element type (first Int, then String) and passed it to the implicit parameter of the evenElems method. The compiler can do that for all concrete types, but not if the argument is itself another type parameter without its class tag. For instance, the following fails:

```
scala> def wrap[U](xs: Vector[U]) = evenElems(xs)
<console>:9: error: No ClassTag available for U
     def wrap[U](xs: Vector[U]) = evenElems(xs)
                                            ^
```

What happened here is that the evenElems demands a class tag for the type parameter U, but none was found. The solution in this case is, of course, to demand another implicit class tag for U. So the following works:

```
scala> def wrap[U: ClassTag](xs: Vector[U]) = evenElems(xs)
wrap: [U](xs: Vector[U])(implicit evidence$1:
      scala.reflect.ClassTag[U])Array[U]
```

This example also shows that the context bound in the definition of U is just a shorthand for an implicit parameter named here evidence$1 of type ClassTag[U].

## 24.11 Strings

Like arrays, strings are not directly sequences, but they can be converted to them, and they also support all sequence operations. Here are some examples of operations you can invoke on strings:

```
scala> val str = "hello"
str: java.lang.String = hello

scala> str.reverse
res6: String = olleh

scala> str.map(_.toUpper)
res7: String = HELLO

scala> str drop 3
res8: String = lo

scala> str slice (1, 4)
res9: String = ell

scala> val s: Seq[Char] = str
s: Seq[Char] = WrappedString(h, e, l, l, o)
```

These operations are supported by two implicit conversions, which were explained in Section 21.7. The first, low-priority conversion maps a String to a WrappedString, which is a subclass of immutable.IndexedSeq. This conversion was applied in the last line of the previous example in which a string was converted into a Seq. The other, high-priority conversion maps a string to a StringOps object, which adds all methods on immutable sequences to strings. This conversion was implicitly inserted in the method calls of reverse, map, drop, and slice in the previous example.

## 24.12 Performance characteristics

As the previous explanations have shown, different collection types have different performance characteristics. That's often the primary reason for picking one collection type over another. You can see the performance characteristics of some common operations on collections summarized in two tables, Table 24.10 and Table 24.11.

|              | head | tail | apply | update | prepend | append | insert |
|--------------|------|------|-------|--------|---------|--------|--------|
| **immutable** |      |      |       |        |         |        |        |
| List         | C    | C    | L     | L      | C       | L      | -      |
| Stream       | C    | C    | L     | L      | C       | L      | -      |
| Vector       | eC   | eC   | eC    | eC     | eC      | eC     | -      |
| Stack        | C    | C    | L     | L      | C       | L      | -      |
| Queue        | aC   | aC   | L     | L      | L       | C      | -      |
| Range        | C    | C    | C     | -      | -       | -      | -      |
| String       | C    | L    | C     | L      | L       | L      | -      |
| **mutable**  |      |      |       |        |         |        |        |
| ArrayBuffer  | C    | L    | C     | C      | L       | aC     | L      |
| ListBuffer   | C    | L    | L     | L      | C       | C      | L      |
| StringBuilder| C    | L    | C     | C      | L       | aC     | L      |
| MutableList  | C    | L    | L     | L      | C       | C      | L      |
| Queue        | C    | L    | L     | L      | C       | C      | L      |
| ArraySeq     | C    | L    | C     | C      | -       | -      | -      |
| Stack        | C    | L    | L     | L      | C       | L      | L      |
| ArrayStack   | C    | L    | C     | C      | aC      | L      | L      |
| Array        | C    | L    | C     | C      | -       | -      | -      |

Table 24.10 · Performance characteristics of sequence types

|                   | lookup | add | remove | min     |
|-------------------|--------|-----|--------|---------|
| **immutable**     |        |     |        |         |
| HashSet/HashMap   | eC     | eC  | eC     | L       |
| TreeSet/TreeMap   | Log    | Log | Log    | Log     |
| BitSet            | C      | L   | L      | $eC^a$  |
| ListMap           | L      | L   | L      | L       |
| **mutable**       |        |     |        |         |
| HashSet/HashMap   | eC     | eC  | eC     | L       |
| WeakHashMap       | eC     | eC  | eC     | L       |
| BitSet            | C      | aC  | C      | $eC^a$  |

Table 24.11 · Performance characteristics of set and map types

---

[a] Assuming bits are densely packed.

The entries in these two tables are explained as follows:

C     The operation takes (fast) constant time.

eC    The operation takes effectively constant time, but this might depend on some assumptions such as the maximum length of a vector or the distribution of hash keys.

aC    The operation takes amortized constant time. Some invocations of the operation might take longer, but if many operations are performed on average only constant time per operation is taken.

Log   The operation takes time proportional to the logarithm of the collection size.

L     The operation is linear, that is it takes time proportional to the collection size.

-     The operation is not supported.

Table 24.10 treats sequence types—both immutable and mutable—with the following operations:

head     Selecting the first element of the sequence.

tail      Producing a new sequence that consists of all elements except the first one.

apply    Indexing.

update   Functional update (with `updated`) for immutable sequences, side-effecting update (with `update`) for mutable sequences.

prepend  Adding an element to the front of the sequence. For immutable sequences, this produces a new sequence. For mutable sequences it modifies the existing sequence.

append   Adding an element at the end of the sequence. For immutable sequences, this produces a new sequence. For mutable sequences it modifies the existing sequence.

insert    Inserting an element at an arbitrary position in the sequence. This is only supported directly for mutable sequences.

Table 24.11 treats mutable and immutable sets and maps with the following operations:

lookup    Testing whether an element is contained in set, or selecting a value associated with a key.

add    Adding a new element to a set or a new key/value pair to a map.

remove    Removing an element from a set or a key from a map.

min    The smallest element of the set, or the smallest key of a map.

## 24.13 Equality

The collection libraries have a uniform approach to equality and hashing. The idea is, first, to divide collections into sets, maps, and sequences. Collections in different categories are always unequal. For instance, Set(1, 2, 3) is unequal to List(1, 2, 3) even though they contain the same elements. On the other hand, within the same category, collections are equal if and only if they have the same elements (for sequences: the same elements in the same order). For example, List(1, 2, 3) == Vector(1, 2, 3), and HashSet(1, 2) == TreeSet(2, 1).

It does not matter for the equality check whether a collection is mutable or immutable. For a mutable collection, equality simply depends on the current elements at the time the equality test is performed. This means that a mutable collection might be equal to different collections at different times, depending what elements are added or removed. This is a potential trap when using a mutable collection as a key in a hash map. For example:

```
scala> import collection.mutable.{HashMap, ArrayBuffer}
import collection.mutable.{HashMap, ArrayBuffer}

scala> val buf = ArrayBuffer(1, 2, 3)
buf: scala.collection.mutable.ArrayBuffer[Int] =
ArrayBuffer(1, 2, 3)

scala> val map = HashMap(buf -> 3)
map: scala.collection.mutable.HashMap[scala.collection.
mutable.ArrayBuffer[Int],Int] = Map((ArrayBuffer(1, 2, 3),3))
```

```
scala> map(buf)
res13: Int = 3

scala> buf(0) += 1

scala> map(buf)
java.util.NoSuchElementException: key not found:
  ArrayBuffer(2, 2, 3)
```

In this example, the selection in the last line will most likely fail because the hash code of the array xs has changed in the second-to-last line. Therefore, the hash-code-based lookup will look at a different place than the one in which xs was stored.

## 24.14   Views

Collections have quite a few methods that construct new collections. Some examples are map, filter, and ++. We call such methods *transformers* because they take at least one collection as their receiver object and produce another collection in their result.

Transformers can be implemented in two principal ways: strict and non-strict (or lazy). A strict transformer constructs a new collection with all of its elements. A non-strict, or lazy, transformer constructs only a proxy for the result collection, and its elements are constructed on demand.

As an example of a non-strict transformer, consider the following implementation of a lazy map operation:

```
def lazyMap[T, U](coll: Iterable[T], f: T => U) =
  new Iterable[U] {
    def iterator = coll.iterator map f
  }
```

Note that lazyMap constructs a new Iterable without stepping through all elements of the given collection coll. The given function f is instead applied to the elements of the new collection's iterator as they are demanded.

Scala collections are by default strict in all their transformers, except for Stream, which implements all its transformer methods lazily. However, there is a systematic way to turn every collection into a lazy one and *vice versa*, which is based on collection views. A *view* is a special kind of collection that represents some base collection, but implements all of its transformers lazily.

To go from a collection to its view, you can use the `view` method on the collection. If `xs` is some collection, then `xs.view` is the same collection, but with all transformers implemented lazily. To get back from a view to a strict collection, you can use the `force` method.

As an example, say you have a vector of `Int`s over which you want to map two functions in succession:

```
scala> val v = Vector(1 to 10: _*)
v: scala.collection.immutable.Vector[Int] =
   Vector(1, 2, 3, 4, 5, 6, 7, 8, 9, 10)

scala> v map (_ + 1) map (_ * 2)
res5: scala.collection.immutable.Vector[Int] =
   Vector(4, 6, 8, 10, 12, 14, 16, 18, 20, 22)
```

In the last statement, the expression v map (_ + 1) constructs a new vector that is then transformed into a third vector by the second call to map (_ * 2). In many situations, constructing the intermediate result from the first call to map is a bit wasteful. In the pseudo example, it would be faster to do a single map with the composition of the two functions (_ + 1) and (_ * 2). If you have the two functions available in the same place you can do this by hand. But quite often, successive transformations of a data structure are done in different program modules. Fusing those transformations would then undermine modularity. A more general way to avoid the intermediate results is by turning the vector first into a view, applying all transformations to the view, and finally forcing the view to a vector:

```
scala> (v.view map (_ + 1) map (_ * 2)).force
res12: Seq[Int] = Vector(4, 6, 8, 10, 12, 14, 16, 18, 20, 22)
```

We'll do this sequence of operations again, one by one:

```
scala> val vv = v.view
vv: scala.collection.SeqView[Int,Vector[Int]] =
   SeqView(1, 2, 3, 4, 5, 6, 7, 8, 9, 10)
```

The application `v.view` gives you a `SeqView`, *i.e.*, a lazily evaluated `Seq`. The type `SeqView` has two type parameters. The first, `Int`, shows the type of the view's elements. The second, Vector[Int], shows you the type constructor you get back when forcing the view.

Applying the first map to the view gives you:

```
scala> vv map (_ + 1)
res13: scala.collection.SeqView[Int,Seq[_]] = SeqViewM(...)
```

The result of the map is a value that prints SeqViewM(...). This is in essence a wrapper that records the fact that a map with function (_ + 1) needs to be applied on the vector v. It does not apply that map until the view is forced, however. The "M" after SeqView is an indication that the view encapsulates a map operation. Other letters indicate other delayed operations. For instance "S" indicates a delayed slice operation, and "R" indicates a reverse. We'll now apply the second map to the last result.

```
scala> res13 map (_ * 2)
res14: scala.collection.SeqView[Int,Seq[_]] = SeqViewMM(...)
```

You now get a SeqView that contains two map operations, so it prints with a double "M": SeqViewMM(...). Finally, forcing the last result gives:

```
scala> res14.force
res15: Seq[Int] = Vector(4, 6, 8, 10, 12, 14, 16, 18, 20, 22)
```

Both stored functions get applied as part of the execution of the force operation and a new vector is constructed. That way, no intermediate data structure is needed.

One detail to note is that the static type of the final result is a Seq, not a Vector. Tracing the types back we see that as soon as the first delayed map was applied, the result had static type SeqViewM[Int, Seq[_]]. That is, the "knowledge" that the view was applied to the specific sequence type Vector got lost. The implementation of a view for any particular class requires quite a bit of code, so the Scala collection libraries provide views mostly only for general collection types, not for specific implementations.[5]

There are two reasons why you might want to consider using views. The first is performance. You have seen that by switching a collection to a view the construction of intermediate results can be avoided. These savings can

---

[5]An exception to this is arrays: applying delayed operations on arrays will again give results with static type Array.

be quite important. As another example, consider the problem of finding the first palindrome in a list of words. A palindrome is a word that reads backwards the same as forwards. Here are the necessary definitions:

```
def isPalindrome(x: String) = x == x.reverse
def findPalindrome(s: Seq[String]) = s find isPalindrome
```

Now, assume you have a very long sequence words and you want to find a palindrome in the first million words of that sequence. Can you re-use the definition of findPalindrome? Of course, you could write:

```
findPalindrome(words take 1000000)
```

This nicely separates the two aspects of taking the first million words of a sequence and finding a palindrome in it. But the downside is that it always constructs an intermediary sequence consisting of one million words, even if the first word of that sequence is already a palindrome. So potentially, 999,999 words are copied into the intermediary result without being inspected at all afterwards. Many programmers would give up here and write their own specialized version of finding palindromes in some given prefix of an argument sequence. But with views, you don't have to. Simply write:

```
findPalindrome(words.view take 1000000)
```

This has the same nice separation of concerns, but instead of a sequence of a million elements it will only construct a single lightweight view object. This way, you do not need to choose between performance and modularity.

The second use case applies to views over mutable sequences. Many transformer functions on such views provide a window into the original sequence that can then be used to update selectively some elements of that sequence. To see this in an example, suppose you have an array arr:

```
scala> val arr = (0 to 9).toArray
arr: Array[Int] = Array(0, 1, 2, 3, 4, 5, 6, 7, 8, 9)
```

You can create a subwindow into that array by creating a slice of a view of the array:

```
scala> val subarr = arr.view.slice(3, 6)
subarr: scala.collection.mutable.IndexedSeqView[
  Int,Array[Int]] = IndexedSeqViewS(...)
```

This gives a view, `subarr`, which refers to the elements at positions 3 through 5 of the array `arr`. The view does not copy these elements, it just provides a reference to them. Now, assume you have a method that modifies some elements of a sequence. For instance, the following `negate` method would negate all elements of the sequence of integers it's given:

```
scala> def negate(xs: collection.mutable.Seq[Int]) =
         for (i <- 0 until xs.length) xs(i) = -xs(i)
negate: (xs: scala.collection.mutable.Seq[Int])Unit
```

Assume now you want to negate elements at positions three through five of the array `arr`. Can you use `negate` for this? Using a view, this is simple:

```
scala> negate(subarr)

scala> arr
res4: Array[Int] = Array(0, 1, 2, -3, -4, -5, 6, 7, 8, 9)
```

What happened here is that `negate` changed all elements of `subarr`, which were a slice of the elements of `arr`. Again, you see that views help in keeping things modular. The code above nicely separated the question of what index range to apply a method to from the question what method to apply.

After having seen all these nifty uses of views you might wonder why have strict collections at all? One reason is that performance comparisons do not always favor lazy over strict collections. For smaller collection sizes the added overhead of forming and applying closures in views is often greater than the gain from avoiding the intermediary data structures. A possibly more important reason is that evaluation in views can be very confusing if the delayed operations have side effects.

Here's an example that bit a few users of versions of Scala before 2.8. In these versions the Range type was lazy, so it behaved in effect like a view. People were trying to create a number of actors[6] like this:

```
val actors = for (i <- 1 to 10) yield actor { ... }
```

They were surprised that none of the actors were executing afterwards, even though the `actor` method should create and start an actor from the code that's enclosed in the braces following it. To explain why nothing happened, remember that the `for` expression above is equivalent to an application of the `map` method:

---

[6]The Scala actors library has been deprecated, but this historical example is still relevant.

556

```
val actors = (1 to 10) map (i => actor { ... })
```

Since previously the range produced by (1 to 10) behaved like a view, the result of the map was again a view. That is, no element was computed, and, consequently, no actor was created! Actors would have been created by forcing the range of the whole expression, but it's far from obvious that this is what was required to make the actors do their work.

To avoid surprises like this, the Scala collections gained more regular rules in version 2.8. All collections except streams and views are strict. The only way to go from a strict to a lazy collection is via the view method. The only way to go back is via force. So the actors definition above would behave as expected in Scala 2.8 in that it would create and start ten actors. To get back the surprising previous behavior, you'd have to add an explicit view method call:

```
val actors = for (i <- (1 to 10).view) yield actor { ... }
```

In summary, views are a powerful tool to reconcile concerns of efficiency with concerns of modularity. But in order not to be entangled in aspects of delayed evaluation, you should restrict views to two scenarios. Either you apply views in purely functional code where collection transformations do not have side effects. Or you apply them over mutable collections where all modifications are done explicitly. What's best avoided is a mixture of views and operations that create new collections while also having side effects.

## 24.15 Iterators

An iterator is not a collection, but rather a way to access the elements of a collection one by one. The two basic operations on an iterator it are next and hasNext. A call to it.next() will return the next element of the iterator and advance the state of the iterator. Calling next again on the same iterator will then yield the element one beyond the one returned previously. If there are no more elements to return, a call to next will throw a NoSuchElementException. You can find out whether there are more elements to return using Iterator's hasNext method.

The most straightforward way to "step through" all the elements returned by an iterator is to use a while loop:

```
while (it.hasNext)
  println(it.next())
```

Iterators in Scala also provide analogues of most of the methods that you find in the Traversable, Iterable, and Seq traits. For instance, they provide a foreach method that executes a given procedure on each element returned by an iterator. Using foreach, the loop above could be abbreviated to:

```
it foreach println
```

As always, for expressions can be used as an alternate syntax for expressions involving foreach, map, filter, and flatMap, so yet another way to print all elements returned by an iterator would be:

```
for (elem <- it) println(elem)
```

There's an important difference between the foreach method on iterators and the same method on traversable collections: When called on an iterator, foreach will leave the iterator at its end when it is done. So calling next again on the same iterator will fail with a NoSuchElementException. By contrast, when called on a collection, foreach leaves the number of elements in the collection unchanged (unless the passed function adds or removes elements, but this is discouraged, because it can easily lead to surprising results).

The other operations that Iterator has in common with Traversable have the same property of leaving the iterator at its end when done. For instance, iterators provide a map method, which returns a new iterator:

```
scala> val it = Iterator("a", "number", "of", "words")
it: Iterator[java.lang.String] = non-empty iterator

scala> it.map(_.length)
res1: Iterator[Int] = non-empty iterator

scala> res1 foreach println
1
6
2
5

scala> it.next()
java.util.NoSuchElementException: next on empty iterator
```

As you can see, after the call to map, the it iterator has advanced to its end.

Another example is the dropWhile method, which can be used to find the first element of an iterator that has a certain property. For instance, to find the first word in the iterator shown previously that has at least two characters, you could write:

```
scala> val it = Iterator("a", "number", "of", "words")
it: Iterator[java.lang.String] = non-empty iterator

scala> it dropWhile (_.length < 2)
res4: Iterator[java.lang.String] = non-empty iterator

scala> it.next()
res5: java.lang.String = number
```

Note again that it has changed by the call to dropWhile: it now points to the second word "number" in the list. In fact, it and the result res4 returned by dropWhile will return exactly the same sequence of elements.

There is only one standard operation, duplicate, which allows you to re-use the same iterator:

```
val (it1, it2) = it.duplicate
```

The call to duplicate gives you *two* iterators, which each return exactly the same elements as the iterator it. The two iterators work independently; advancing one does not affect the other. By contrast the original iterator, it, is advanced to its end by duplicate and is thus rendered unusable.

In summary, iterators behave like collections *if you never access an iterator again after invoking a method on it.* The Scala collection libraries make this explicit with an abstraction called TraversableOnce, which is a common supertrait of Traversable and Iterator. As the name implies, TraversableOnce objects can be traversed using foreach, but the state of that object after the traversal is not specified. If the TraversableOnce object is in fact an Iterator, it will be at its end after the traversal, but if it is a Traversable, it will still exist as before. A common use case of TraversableOnce is as an argument type for methods that can take either an iterator or traversable as argument. An example is the appending method ++ in trait Traversable. It takes a TraversableOnce parameter, so you can append elements coming from either an iterator or a traversable collection.

All operations on iterators are summarized in Table 24.12.

Table 24.12 · Operations in trait `Iterator`

| What it is | What it does |
|---|---|
| **Abstract methods:** | |
| `it.next()` | Returns the next element in the iterator and advances past `it`. |
| `it.hasNext` | Returns `true` if `it` can return another element. |
| **Variations:** | |
| `it.buffered` | A buffered iterator returning all elements of `it`. |
| `it grouped size` | An iterator that yields the elements returned by `it` in fixed-sized sequence "chunks." |
| `xs sliding size` | An iterator that yields the elements returned by `it` in sequences representing a sliding fixed-sized window. |
| **Copying:** | |
| `it copyToBuffer buf` | Copies all elements returned by `it` to buffer `buf`. |
| `it copyToArray(arr, s, l)` | Copies at most `l` elements returned by `it` to array `arr` starting at index `s`. The last two arguments are optional. |
| **Duplication:** | |
| `it.duplicate` | A pair of iterators that each independently return all elements of `it`. |
| **Additions:** | |
| `it ++ jt` | An iterator returning all elements returned by iterator `it`, followed by all elements returned by iterator `jt`. |
| `it padTo (len, x)` | The iterator that returns all elements of `it` followed by copies of `x` until length `len` elements are returned overall. |
| **Maps:** | |
| `it map f` | The iterator obtained from applying the function `f` to every element returned from `it`. |
| `it flatMap f` | The iterator obtained from applying the iterator-valued function `f` to every element in `it` and appending the results. |

## Table 24.12 · continued

| | |
|---|---|
| it collect f | The iterator obtained from applying the partial function f to every element in it for which it is defined and collecting the results. |

**Conversions:**

| | |
|---|---|
| it.toArray | Collects the elements returned by it in an array. |
| it.toList | Collects the elements returned by it in a list. |
| it.toIterable | Collects the elements returned by it in an iterable. |
| it.toSeq | Collects the elements returned by it in a sequence. |
| it.toIndexedSeq | Collects the elements returned by it in an indexed sequence. |
| it.toStream | Collects the elements returned by it in a stream. |
| it.toSet | Collects the elements returned by it in a set. |
| it.toMap | Collects the key/value pairs returned by it in a map. |

**Size info:**

| | |
|---|---|
| it.isEmpty | Tests whether the iterator is empty (opposite of hasNext). |
| it.nonEmpty | Tests whether the collection contains elements (alias of hasNext). |
| it.size | The number of elements returned by it. Note: it will be at its end after this operation! |
| it.length | Same as it.size. |
| it.hasDefiniteSize | Returns true if it is known to return finitely many elements (by default the same as isEmpty). |

**Element retrieval index search:**

| | |
|---|---|
| it find p | An option containing the first element returned by it that satisfies p, or None if no element qualifies. Note: The iterator advances to just after the element, or, if none is found, to the end. |
| it indexOf x | The index of the first element returned by it that equals x. Note: The iterator advances past the position of this element. |

## Table 24.12 · continued

| | |
|---|---|
| it indexWhere p | The index of the first element returned by it that satisfies p. Note: The iterator advances past the position of this element. |

**Subiterators:**

| | |
|---|---|
| it take n | An iterator returning of the first n elements of it. Note: it will advance to the position after the n'th element, or to its end, if it contains less than n elements. |
| it drop n | The iterator that starts with the (n + 1)'th element of it. Note: it will advance to the same position. |
| it slice (m, n) | The iterator that returns a slice of the elements returned from it, starting with the m'th element and ending before the n'th element. |
| it takeWhile p | An iterator returning elements from it as long as condition p is true. |
| it dropWhile p | An iterator skipping elements from it as long as condition p is true, and returning the remainder. |
| it filter p | An iterator returning all elements from it that satisfy the condition p. |
| it withFilter p | Same as it filter p. Needed so that iterators can be used in for expressions. |
| it filterNot p | An iterator returning all elements from it that do not satisfy the condition p. |

**Subdivisions:**

| | |
|---|---|
| it partition p | Splits it into a pair of two iterators; one returning all elements from it that satisfy the predicate p, the other returning all elements from it that do not. |

**Element conditions:**

| | |
|---|---|
| it forall p | A boolean indicating whether the predicate p holds for all elements returned by it. |
| it exists p | A boolean indicating whether the predicate p holds for some element in it. |
| it count p | The number of elements in it that satisfy the predicate p. |

**Folds:**

## Table 24.12 · continued

| | |
|---|---|
| (z /: it)(op) | Applies binary operation op between successive elements returned by it, going left to right, starting with z. |
| (it :\ z)(op) | Applies binary operation op between successive elements returned by it, going right to left, starting with z. |
| it.foldLeft(z)(op) | Same as (z /: it)(op). |
| it.foldRight(z)(op) | Same as (it :\ z)(op). |
| it reduceLeft op | Applies binary operation op between successive elements returned by non-empty iterator it, going left to right. |
| it reduceRight op | Applies binary operation op between successive elements returned by non-empty iterator it, going right to left. |

**Specific folds:**

| | |
|---|---|
| it.sum | The sum of the numeric element values returned by iterator it. |
| it.product | The product of the numeric element values returned by iterator it. |
| it.min | The minimum of the ordered element values returned by iterator it. |
| it.max | The maximum of the ordered element values returned by iterator it. |

**Zippers:**

| | |
|---|---|
| it zip jt | An iterator of pairs of corresponding elements returned from iterators it and jt. |
| it zipAll (jt, x, y) | An iterator of pairs of corresponding elements returned from iterators it and jt, where the shorter iterator is extended to match the longer one by appending elements x or y. |
| it.zipWithIndex | An iterator of pairs of elements returned from it with their indicies. |

**Update:**

| | |
|---|---|
| it patch (i, jt, r) | The iterator resulting from it by replacing r elements starting with i by the patch iterator jt. |

**Comparison:**

563

Table 24.12 · continued

| | |
|---|---|
| it sameElements jt | A test whether iterators it and jt return the same elements in the same order. Note: At least one of it and jt will be at its end after this operation. |
| **Strings:** | |
| it addString (b, start, sep, end) | Adds a string to StringBuilder b that shows all elements returned by it between separators sep enclosed in strings start and end. start, sep, and end are all optional. |
| it mkString (start, sep, end) | Converts the iterator to a string that shows all elements returned by it between separators sep enclosed in strings start and end. start, sep, and end are all optional. |

## Buffered iterators

Sometimes you want an iterator that can "look ahead" so that you can inspect the next element to be returned without advancing past that element. Consider, for instance, the task to skip leading empty strings from an iterator that returns a sequence of strings. You might be tempted to write something like the following method:

```
// This won't work
def skipEmptyWordsNOT(it: Iterator[String]) = {
  while (it.next().isEmpty) {}
}
```

But looking at this code more closely, it's clear that this is wrong: the code will indeed skip leading empty strings, but it will also advance it past the first non-empty string!

The solution to this problem is to use a buffered iterator, an instance of trait BufferedIterator. BufferedIterator is a subtrait of Iterator, which provides one extra method, head. Calling head on a buffered iterator will return its first element, but will not advance the iterator. Using a buffered iterator, skipping empty words can be written like this:

```
def skipEmptyWords(it: BufferedIterator[String]) =
  while (it.head.isEmpty) { it.next() }
```

Every iterator can be converted to a buffered iterator by calling its `buffered` method. Here's an example:

```
scala> val it = Iterator(1, 2, 3, 4)
it: Iterator[Int] = non-empty iterator

scala> val bit = it.buffered
bit: java.lang.Object with scala.collection.
  BufferedIterator[Int] = non-empty iterator

scala> bit.head
res10: Int = 1

scala> bit.next()
res11: Int = 1

scala> bit.next()
res11: Int = 2
```

Note that calling `head` on the buffered iterator, `bit`, did not advance it. Therefore, the subsequent call, `bit.next()`, returned again the same value as `bit.head`.

## 24.16 Creating collections from scratch

You have already seen syntax like `List(1, 2, 3)`, which creates a list of three integers, and `Map('A' -> 1, 'C' -> 2)`, which creates a map with two bindings. This is actually a universal feature of Scala collections. You can take any collection name and follow it by a list of elements in parentheses. The result will be a new collection with the given elements. Here are some more examples:

```
Traversable()            // An empty traversable object
List()                   // The empty list
List(1.0, 2.0)           // A list with elements 1.0, 2.0
Vector(1.0, 2.0)         // A vector with elements 1.0, 2.0
Iterator(1, 2, 3)        // An iterator returning three integers.
Set(dog, cat, bird)      // A set of three animals
HashSet(dog, cat, bird)  // A hash set of the same animals
Map('a' -> 7, 'b' -> 0)  // A map from characters to integers
```

"Under the covers" each of the above lines is a call to the `apply` method of some object. For instance, the third line above expands to:

```
List.apply(1.0, 2.0)
```

So this is a call to the `apply` method of the companion object of the `List` class. That method takes an arbitrary number of arguments and constructs a list from them. Every collection class in the Scala library has a companion object with such an `apply` method. It does not matter whether the collection class represents a concrete implementation, like `List`, `Stream`, or `Vector`, or whether it is an trait such as `Seq`, `Set`, or `Traversable`. In the latter case, calling `apply` will produce some default implementation of the trait. Here are some examples:

```
scala> List(1, 2, 3)
res17: List[Int] = List(1, 2, 3)

scala> Traversable(1, 2, 3)
res18: Traversable[Int] = List(1, 2, 3)

scala> mutable.Traversable(1, 2, 3)
res19: scala.collection.mutable.Traversable[Int] =
  ArrayBuffer(1, 2, 3)
```

Besides `apply`, every collection companion object also defines a member `empty`, which returns an empty collection. So instead of `List()` you could write `List.empty`, instead of `Map()`, `Map.empty`, and so on.

Descendants of `Seq` traits also provide other factory operations in their companion objects. These are summarized in Table 24.13. In short, there's:

> `concat`, which concatenates an arbitrary number of traversables together,

> `fill` and `tabulate`, which generate single or multi-dimensional sequences of given dimensions initialized by some expression or tabulating function,

> `range`, which generates integer sequences with some constant step length, and

> `iterate`, which generates the sequence resulting from repeated application of a function to a start element.

Table 24.13 · Factory methods for sequences

| What it is | What it does |
| --- | --- |
| S.empty | The empty sequence |
| S(x, y, z) | A sequence consisting of elements x, y, and z |
| S.concat(xs, ys, zs) | The sequence obtained by concatenating the elements of xs, ys, and zs |
| S.fill(n)(e) | A sequence of length n where each element is computed by expression e |
| S.fill(m, n)(e) | A sequence of sequences of dimension m × n where each element is computed by expression e (exists also in higher dimensions) |
| S.tabulate(n)(f) | A sequence of length n where the element at each index $i$ is computed by f($i$) |
| S.tabulate(m, n)(f) | A sequence of sequences of dimension m×n where the element at each index $(i, j)$ is computed by f($i, j$) (exists also in higher dimensions) |
| S.range(start, end) | The sequence of integers start ... end − 1 |
| S.range(start, end, step) | The sequence of integers starting with start and progressing by step increments up to, and excluding, the end value |
| S.iterate(x, n)(f) | The sequence of length n with elements x, f(x), f(f(x)), ... |

## 24.17   Conversions between Java and Scala collections

Like Scala, Java has a rich collections library. There are many similarities between the two. For instance, both libraries know iterators, iterables, sets, maps, and sequences. But there are also important differences. In particular, the Scala libraries put much more emphasis on immutable collections, and provide many more operations that transform a collection into a new one.

Sometimes you might need to convert from one collection framework to the other. For instance, you might want to access to an existing Java collection, as if it were a Scala collection. Or you might want to pass one of Scala's collections to a Java method that expects the Java counterpart. It is quite easy to do this, because Scala offers implicit conversions between all the major collection types in the JavaConversions object. In particular, you will find bidirectional conversions between the following types:

| | | |
|---|---|---|
| Iterator | ⇔ | java.util.Iterator |
| Iterator | ⇔ | java.util.Enumeration |
| Iterable | ⇔ | java.lang.Iterable |
| Iterable | ⇔ | java.util.Collection |
| mutable.Buffer | ⇔ | java.util.List |
| mutable.Set | ⇔ | java.util.Set |
| mutable.Map | ⇔ | java.util.Map |

To enable these conversions, simply import like this:

```
scala> import collection.JavaConversions._
import collection.JavaConversions._
```

You have now automatic conversions between Scala collections and their corresponding Java collections.

```
scala> import collection.mutable._
import collection.mutable._

scala> val jul: java.util.List[Int] = ArrayBuffer(1, 2, 3)
jul: java.util.List[Int] = [1, 2, 3]

scala> val buf: Seq[Int] = jul
buf: scala.collection.mutable.Seq[Int] = ArrayBuffer(1, 2, 3)

scala> val m: java.util.Map[String, Int] =
          HashMap("abc" -> 1, "hello" -> 2)
m: java.util.Map[String,Int] = {hello=2, abc=1}
```

Internally, these conversion work by setting up a "wrapper" object that forwards all operations to the underlying collection object. So collections are never copied when converting between Java and Scala. An interesting property is that if you do a round-trip conversion from, say, a Java type to its corresponding Scala type, and back to the same Java type, you end up with the identical collection object you started with.

Some other common Scala collections exist that can also be converted to Java types, but for which no corresponding conversion exists in the other direction. These are:

```
Seq              ⇒   java.util.List
mutable.Seq      ⇒   java.util.List
Set              ⇒   java.util.Set
Map              ⇒   java.util.Map
```

Because Java does not distinguish between mutable and immutable collections in their type, a conversion from, say, `collection.immutable.List` will yield a `java.util.List`, on which all attempted mutation operations will throw an `UnsupportedOperationException`. Here's an example:

```
scala> val jul: java.util.List[Int] = List(1, 2, 3)
jul: java.util.List[Int] = [1, 2, 3]

scala> jul.add(7)
java.lang.UnsupportedOperationException
        at java.util.AbstractList.add(AbstractList.java:131)
```

## 24.18  Conclusion

You've now seen how to use Scala's collection in great detail. Scala's collections take the approach of giving you powerful building blocks rather than a number of ad hoc utility methods. Putting together two or three such building blocks allows you to express an enormous number of useful computations. This style of library is especially effective due to Scala having a light syntax for function literals, and due to it providing many collection types that are persistent and immutable.

This chapter has shown collections from the point of view of a programmer using the collection library. The next chapter will show you how the collections are built and how you can add your own collection types.

Chapter 25

# The Architecture of Scala Collections

This chapter describes the architecture of the Scala collections framework in detail. Continuing the theme of Chapter 24, you will find out more about the internal workings of the framework. You will also learn how this architecture helps you define your own collections in a few lines of code, while reusing the overwhelming part of collection functionality from the framework.

Chapter 24 enumerated a large number of collection operations, which exist uniformly on many different collection implementations. Implementing every collection operation anew for every collection type would lead to an enormous amount of code, most of which would be copied from somewhere else. Such code duplication could lead to inconsistencies over time, when an operation is added or modified in one part of the collection library but not in others. The principal design objective of the new collections framework was to avoid any duplication, defining every operation in as few places as possible.[1] The design approach was to implement most operations in collection "templates" that can be flexibly inherited from individual base classes and implementations. In this chapter, we will examine these templates, and other classes and traits that constitute the "building blocks" of the framework, as well as the construction principles they support.

## 25.1 Builders

Almost all collection operations are implemented in terms of *traversals* and *builders*. Traversals are handled by Traversable's foreach method, and

---

[1]Ideally, everything should be defined in one place only, but there are a few exceptions where things needed to be redefined.

building new collections is handled by instances of class Builder. Listing 25.1 presents a slightly abbreviated outline of this class.

```
package scala.collection.generic

class Builder[-Elem, +To] {
  def +=(elem: Elem): this.type
  def result(): To
  def clear()
  def mapResult[NewTo](f: To => NewTo): Builder[Elem, NewTo]
    = ...
}
```

Listing 25.1 · An outline of the Builder class.

You can add an element x to a builder b with b += x. There's also syntax
to add more than one element at once: For instance, b += (x, y) and b ++= xs
work as for buffers. (In fact, buffers are an enriched version of builders.) The
result() method returns a collection from a builder. The state of the builder
is undefined after taking its result, but it can be reset into a new empty state
using clear(). Builders are generic in both the element type, Elem, and in
the type, To, of collections they return.

Often, a builder can refer to some other builder for assembling the elements of a collection, but then would like to transform the result of the other
builder—for example, to give it a different type. This task is simplified by
method mapResult in class Builder. Suppose for instance you have an
array buffer buf. Array buffers are builders for themselves, so taking the
result() of an array buffer will return the same buffer. If you want to use
this buffer to produce a builder that builds arrays, you can use mapResult:

```
scala> val buf = new ArrayBuffer[Int]
buf: scala.collection.mutable.ArrayBuffer[Int] = ArrayBuffer()

scala> val bldr = buf mapResult (_.toArray)
bldr: scala.collection.mutable.Builder[Int,Array[Int]]
  = ArrayBuffer()
```

The result value, bldr, is a builder that uses the array buffer, buf, to collect elements. When a result is demanded from bldr, the result of buf is
computed, which yields the array buffer buf itself. This array buffer is then

mapped with `_.toArray` to an array. So the end result is that `bldr` is a builder for arrays.

## 25.2 Factoring out common operations

The main design objectives of the collection library redesign were to have, at the same time, natural types and maximal sharing of implementation code. In particular, Scala's collections follow the "same-result-type" principle: Wherever possible, a transformation method on a collection will yield a collection of the same type. For instance, the `filter` operation should yield, on every collection type, an instance of the same collection type. Applying `filter` on a `List` should give a `List`; applying it on a `Map` should give a `Map`; and so on. In the rest of this section, you will find out how this is achieved.

> **The fast track**
>
> The material in this section is a bit more dense than usual and might require some time to absorb. If you want to move ahead quickly, you could skip the remainder of this section and move on to Section 25.3 on page 578 where you will learn from concrete examples how to integrate your own collection classes in the framework.

The Scala collection library avoids code duplication and achieves the "same-result-type" principle by using generic builders and traversals over collections in so-called *implementation traits*. These traits are named with a `Like` suffix; for instance, `IndexedSeqLike` is the implementation trait for `IndexedSeq`, and similarly, `TraversableLike` is the implementation trait for `Traversable`. Collection classes such as `Traversable` or `IndexedSeq` inherit all their concrete method implementations from these traits. Implementation traits have two type parameters instead of one for normal collections. They parameterize not only over the collection's element type, but also over the collection's *representation type* (*i.e.*, the type of the underlying collection), such as `Seq[I]` or `List[T]`.

For instance, here is the header of trait `TraversableLike`:

```
trait TraversableLike[+Elem, +Repr] { ... }
```

The type parameter, `Elem`, stands for the element type of the traversable whereas the type parameter `Repr` stands for its representation. There are no constraints on `Repr`. In particular `Repr` might be instantiated to a type

```
package scala.collection

trait TraversableLike[+Elem, +Repr] {
  def newBuilder: Builder[Elem, Repr] // deferred
  def foreach[U](f: Elem => U)         // deferred
         ...
  def filter(p: Elem => Boolean): Repr = {
    val b = newBuilder
    foreach { elem => if (p(elem)) b += elem }
    b.result
  }
}
```

Listing 25.2 · Implementation of filter in TraversableLike.

that is itself not a subtype of Traversable. That way, classes outside the collections hierarchy, such as String and Array, can still make use of all operations defined in a collection implementation trait.

Taking filter as an example, this operation is defined once for all collection classes in the trait TraversableLike. An outline of the relevant code is shown in Listing 25.2. The trait declares two abstract methods, newBuilder and foreach, which are implemented in concrete collection classes. The filter operation is implemented in the same way for all collections using these methods. It first constructs a new builder for the representation type Repr, using newBuilder. It then traverses all elements of the current collection, using foreach. If an element x satisfies the given predicate p—i.e., p(x) is true—it is added with the builder. Finally, the elements collected in the builder are returned as an instance of the Repr collection type by calling the builder's result method.

The map operation on collections is a bit more complicated. For instance, if f is a function from String to Int, and xs is a List[String], then xs map f should give a List[Int]. Likewise, if ys is an Array[String], then ys map f should give a Array[Int]. But how do you achieve that without duplicating the definition of the map method in lists and arrays?

The newBuilder/foreach framework shown in Listing 25.2 is not sufficient for this because it only allows creation of new instances of the same collection *type*, whereas map needs an instance of the same collection *type constructor* but possibly with a different element type. What's more, even

574

the result type constructor of a function like map might depend, in non-trivial ways, on the other argument types. Here is an example:

```
scala> import collection.immutable.BitSet
import collection.immutable.BitSet

scala> val bits = BitSet(1, 2, 3)
bits: scala.collection.immutable.BitSet = BitSet(1, 2, 3)

scala> bits map (_ * 2)
res13: scala.collection.immutable.BitSet = BitSet(2, 4, 6)

scala> bits map (_.toFloat)
res14: scala.collection.immutable.Set[Float] =
   Set(1.0, 2.0, 3.0)
```

If you map the doubling function _ * 2 over a bit set you obtain another bit set. However, if you map the function (_.toFloat) over the same bit set, the result is a general Set[Float]. Of course, it can't be a bit set because bit sets contain Ints, not Floats.

Note that map's result type depends on the type of function that's passed to it. If the result type of that function argument is again an Int, the result of map is a BitSet. But if the result type of the function argument is something else, the result of map is just a Set. You'll find out soon how this type-flexibility is achieved in Scala.

The problem with BitSet is not an isolated case. Here are two more interactions with the interpreter that both map a function over a map:

```
scala> Map("a" -> 1, "b" -> 2) map { case (x, y) => (y, x) }
res3: scala.collection.immutable.Map[Int,java.lang.String] =
   Map(1 -> a, 2 -> b)

scala> Map("a" -> 1, "b" -> 2) map { case (x, y) => y }
res4: scala.collection.immutable.Iterable[Int] =
   List(1, 2)
```

The first function swaps two arguments of a key/value pair. The result of mapping this function is again a map, but now going in the other direction. In fact, the first expression yields the inverse of the original map, provided it is invertible. The second function, however, maps the key/value pair to an integer, namely its value component. In that case, we cannot form a Map from the results, but we still can form an Iterable, a supertrait of Map.

575

You might ask, Why not restrict map so that it can always return the same kind of collection? For instance, on bit sets map could accept only Int-to-Int functions and on maps it could only accept pair-to-pair functions. Not only are such restrictions undesirable from an object-oriented modeling point of view, they are illegal because they would violate the Liskov Substitution Principle: A Map *is* an Iterable. So every operation that's legal on an Iterable must also be legal on a Map.

Scala solves this problem instead with overloading: Not the simple form of overloading inherited by Java (that would not be flexible enough), but the more systematic form of overloading that's provided by implicit parameters.

```
def map[B, That](f: Elem => B)
    (implicit bf: CanBuildFrom[Repr, B, That]): That = {
  val b = bf(this)
  for (x <- this) b += f(x)
  b.result
}
```

Listing 25.3 · Implementation of map in TraversableLike.

Listing 25.3 shows trait TraversableLike's implementation of map. It's quite similar to the implementation of filter shown in Listing 25.2. The principal difference is that where filter used the newBuilder method, which is abstract in class TraversableLike, map uses a *builder factory* that's passed as an additional implicit parameter of type CanBuildFrom.

```
package scala.collection.generic

trait CanBuildFrom[-From, -Elem, +To] {
  // Creates a new builder
  def apply(from: From): Builder[Elem, To]
}
```

Listing 25.4 · The CanBuildFrom trait.

Listing 25.4 shows the definition of the trait CanBuildFrom, which represents builder factories. It has three type parameters: Elem indicates the element type of the collection to be built, To indicates the type of collection to build, and From indicates the type for which this builder factory applies.

By defining the right implicit definitions of builder factories, you can tailor the right typing behavior as needed.

Take class BitSet as an example. Its companion object would contain a builder factory of type CanBuildFrom[BitSet, Int, BitSet]. This means that when operating on a BitSet you can construct another BitSet, provided the type of the collection to build is Int. If this is not the case, you can always fall back to a different implicit builder factory, this time implemented in mutable.Set's companion object. The type of this more general builder factory, where A is a generic type parameter, is:

```
CanBuildFrom[Set[_], A, Set[A]]
```

This means that when operating on an arbitrary Set, expressed by the wildcard type Set[_], you can build a Set again no matter what the element type A is. Given these two implicit instances of CanBuildFrom, you can then rely on Scala's rules for implicit resolution to pick the one that's appropriate and maximally specific.

So implicit resolution provides the correct static types for tricky collection operations, such as map. But what about the dynamic types? Specifically, say you have a list value that has Iterable as its static type, and you map some function over that value:

```
scala> val xs: Iterable[Int] = List(1, 2, 3)
xs: Iterable[Int] = List(1, 2, 3)

scala> val ys = xs map (x => x * x)
ys: Iterable[Int] = List(1, 4, 9)
```

The static type of ys above is Iterable, as expected. But its dynamic type is (and should be) still List! This behavior is achieved by one more indirection. The apply method in CanBuildFrom is passed the source collection as argument. Most builder factories for generic traversables (in fact all except builder factories for leaf classes) forward the call to a method genericBuilder of a collection. The genericBuilder method in turn calls the builder that belongs to the collection in which it is defined. So Scala uses static implicit resolution to resolve constraints on the types of map, and virtual dispatch to pick the best dynamic type that corresponds to these constraints.

## 25.3  Integrating new collections

What needs to be done if you want to integrate a new collection class, so that it can profit from all predefined operations at the right types? In this section we'll show you two examples that do this.

### Integrating sequences

Say you want to create a new sequence type for RNA strands, which are sequences of bases A (adenine), T (thymine), G (guanine), and U (uracil). The definitions for bases are easily set up as shown in Listing 25.5.

Every base is defined as a case object that inherits from a common abstract class `Base`. The `Base` class has a companion object that defines two functions that map between bases and the integers 0 to 3. You can see in the examples two different ways to use collections to implement these functions. The `toInt` function is implemented as a `Map` from `Base` values to integers. The reverse function, `fromInt`, is implemented as an array. This makes use of the fact that both maps and arrays *are* functions because they inherit from the `Function1` trait.

The next task is to define a class for strands of RNA. Conceptually, a strand of RNA is simply a `Seq[Base]`. However, RNA strands can get quite long, so it makes sense to invest some work in a compact representation. Because there are only four bases, a base can be identified with two bits, and you can therefore store sixteen bases as two-bit values in an integer. The idea, then, is to construct a specialized subclass of `Seq[Base]`, which uses this packed representation.

Listing 25.6 presents the first version of this class; it will be refined later. The class `RNA1` has a constructor that takes an array of `Int`s as its first argument. This array contains the packed RNA data, with sixteen bases in each element, except for the last array element, which might be partially filled. The second argument, `length`, specifies the total number of bases on the array (and in the sequence). Class `RNA1` extends `IndexedSeq[Base]`. Trait `IndexedSeq`, which comes from package `scala.collection.immutable`, defines two abstract methods, `length` and `apply`.

These need to be implemented in concrete subclasses. Class `RNA1` implements `length` automatically by defining a parametric field (described in Section 10.6) of the same name. It implements the indexing method `apply` with the code given in Listing 25.6. Essentially, `apply` first extracts an in-

```
abstract class Base
case object A extends Base
case object T extends Base
case object G extends Base
case object U extends Base

object Base {
  val fromInt: Int => Base = Array(A, T, G, U)
  val toInt: Base => Int = Map(A -> 0, T -> 1, G -> 2, U -> 3)
}
```

Listing 25.5 · RNA Bases.

teger value from the groups array, then extracts the correct two-bit number from that integer using right shift (>>) and mask (&). The private constants S, N, and M come from the RNA1 companion object. S specifies the size of each packet (*i.e.*, two); N specifies the number of two-bit packets per integer; and M is a bit mask that isolates the lowest S bits in a word.

Note that the constructor of class RNA1 is `private`. This means that clients cannot create RNA1 sequences by calling new, which makes sense, because it hides the representation of RNA1 sequences in terms of packed arrays from the user. If clients cannot see what the representation details of RNA sequences are, it becomes possible to change these representation details at any point in the future without affecting client code.

In other words, this design achieves a good decoupling of the interface of RNA sequences and its implementation. However, if constructing an RNA sequence with new is impossible, there must be some other way to create new RNA sequences, or else the whole class would be rather useless. There are two alternatives for RNA sequence creation, both provided by the RNA1 companion object. The first way is method `fromSeq`, which converts a given sequence of bases (*i.e.*, a value of type `Seq[Base]`) into an instance of class RNA1. The `fromSeq` method does this by packing all the bases contained in its argument sequence into an array, then calling RNA1's private constructor with that array and the length of the original sequence as arguments. This makes use of the fact that a private constructor of a class is visible in the class's companion object.

The second way to create an RNA1 value is provided by the `apply` method in the RNA1 object. It takes a variable number of Base arguments and simply

579

```
import collection.IndexedSeqLike
import collection.mutable.{Builder, ArrayBuffer}
import collection.generic.CanBuildFrom

final class RNA1 private (val groups: Array[Int],
    val length: Int) extends IndexedSeq[Base] {

  import RNA1._

  def apply(idx: Int): Base = {
    if (idx < 0 || length <= idx)
      throw new IndexOutOfBoundsException
    Base.fromInt(groups(idx / N) >> (idx % N * S) & M)
  }
}

object RNA1 {

  // Number of bits necessary to represent group
  private val S = 2

  // Number of groups that fit in an Int
  private val N = 32 / S

  // Bitmask to isolate a group
  private val M = (1 << S) - 1

  def fromSeq(buf: Seq[Base]): RNA1 = {
    val groups = new Array[Int]((buf.length + N - 1) / N)
    for (i <- 0 until buf.length)
      groups(i / N) |= Base.toInt(buf(i)) << (i % N * S)
    new RNA1(groups, buf.length)
  }

  def apply(bases: Base*) = fromSeq(bases)
}
```

Listing 25.6 · RNA strands class, first version.

forwards them as a sequence to `fromSeq`.

Here are the two creation schemes in action:

```
scala> val xs = List(A, G, T, A)
xs: List[Product with Base] = List(A, G, T, A)

scala> RNA1.fromSeq(xs)
res1: RNA1 = RNA1(A, G, T, A)

scala> val rna1 = RNA1(A, U, G, G, T)
rna1: RNA1 = RNA1(A, U, G, G, T)
```

*Adapting the result type of RNA methods*

Here are some more interactions with the RNA1 abstraction:

```
scala> rna1.length
res2: Int = 5

scala> rna1.last
res3: Base = T

scala> rna1.take(3)
res4: IndexedSeq[Base] = Vector(A, U, G)
```

The first two results are as expected, but the last result of taking the first three elements of `rna1` might not be. In fact, you see an `IndexedSeq[Base]` as static result type and a `Vector` as the dynamic type of the result value. You might have expected to see an RNA1 value instead. But this is not possible because all that was done in Listing 25.6 was make RNA1 extend `IndexedSeq`. Class `IndexedSeq`, on the other hand, has a `take` method that returns an `IndexedSeq`, and that's implemented in terms of `IndexedSeq`'s default implementation, `Vector`.

Now that you understand why things are the way they are, the next question should be what needs to be done to change them? One way to do this would be to override the `take` method in class RNA1, maybe like this:

```
def take(count: Int): RNA1 = RNA1.fromSeq(super.take(count))
```

This would do the job for `take`. But what about `drop`, or `filter`, or `init`? In fact there are over fifty methods on sequences that return again a sequence.

For consistency, all of these would have to be overridden. This looks less and less like an attractive option.

Fortunately, there is a much easier way to achieve the same effect. The RNA class needs to inherit not only from IndexedSeq, but also from its implementation trait IndexedSeqLike. This is shown in Listing 25.7. The new implementation differs from the previous one in only two aspects. First, class RNA2 now also extends from IndexedSeqLike[Base, RNA2]. The IndexedSeqLike trait implements all concrete methods of IndexedSeq in an extensible way.

For instance, the return type of methods like take, drop, filter or init is the second type parameter passed to class IndexedSeqLike (*i.e.*, RNA2 in Listing 25.7). To do this, IndexedSeqLike bases itself on the newBuilder abstraction, which creates a builder of the right kind. Sub-classes of trait IndexedSeqLike have to override newBuilder to return collections of their own kind. In class RNA2, the newBuilder method returns a builder of type Builder[Base, RNA2]. To construct this builder, it first creates an ArrayBuffer, which itself is a Builder[Base, ArrayBuffer]. It then transforms the ArrayBuffer builder by calling its mapResult method to an RNA2 builder. The mapResult method expects a transformation function from ArrayBuffer to RNA2 as its parameter. The function given is simply RNA2.fromSeq, which converts an arbitrary base sequence to an RNA2 value (recall that an array buffer is a kind of sequence, so RNA2.fromSeq can be applied to it).

```
final class RNA2 private (
  val groups: Array[Int],
  val length: Int
) extends IndexedSeq[Base] with IndexedSeqLike[Base, RNA2] {

  import RNA2._

  override def newBuilder: Builder[Base, RNA2] =
    new ArrayBuffer[Base] mapResult fromSeq

  def apply(idx: Int): Base = // as before
}
```

Listing 25.7 · RNA strands class, second version.

If you had left out the newBuilder definition, you would have gotten an error message like the following:

```
RNA2.scala:5:  error:  overriding method newBuilder in trait
TraversableLike of type => scala.collection.mutable.Builder[Base,RNA2];
 method newBuilder in trait GenericTraversableTemplate of type
 => scala.collection.mutable.Builder[Base,IndexedSeq[Base]] has
 incompatible type
class RNA2 private (val groups:  Array[Int], val length:  Int)
          ^
one error found
```

The error message is quite long and complicated, which reflects the intricate way the collection libraries are put together. It's best to ignore the information about where the methods come from, because in this case it detracts more than it helps. What remains is that a method `newBuilder` with result type `Builder[Base, RNA2]` needed to be defined, but a method `newBuilder` with result type `Builder[Base,IndexedSeq[Base]]` was found. The latter does not override the former.

The first method, whose result type is `Builder[Base, RNA2]`, is an abstract method that got instantiated at this type in Listing 25.7 by passing the RNA2 type parameter to `IndexedSeqLike`. The second method, of result type `Builder[Base,IndexedSeq[Base]]`, is what's provided by the inherited `IndexedSeq` class. In other words, the RNA2 class is invalid without a definition of `newBuilder` with the first result type.

With the refined implementation of the RNA class in Listing 25.7, methods like `take`, `drop`, or `filter` work now as expected:

```
scala> val rna2 = RNA2(A, U, G, G, T)
rna2: RNA2 = RNA2(A, U, G, G, T)

scala> rna2 take 3
res5: RNA2 = RNA2(A, U, G)

scala> rna2 filter (U !=)
res6: RNA2 = RNA2(A, G, G, T)
```

### Dealing with map and friends

There is another class of methods in collections that we haven't dealt with yet. These methods do not always return the collection type exactly. They

583

might return the same kind of collection, but with a different element type. The classical example of this is the map method. If s is a Seq[Int], and f is a function from Int to String, then s.map(f) would return a Seq[String]. So the element type changes between the receiver and the result, but the kind of collection stays the same.

There are a number of other methods that behave like map. For some of them you would expect this (*e.g.*, flatMap, collect), but for others you might not. For instance, the append method, ++, also might return a result whose type differs from that of its arguments—appending a list of String to a list of Int would give a list of Any. How should these methods be adapted to RNA strands? Ideally we'd expect that mapping bases to bases over an RNA strand would yield again an RNA strand:

```
scala> val rna = RNA(A, U, G, G, T)
rna: RNA = RNA(A, U, G, G, T)

scala> rna map { case A => T case b => b }
res7: RNA = RNA(T, U, G, G, T)
```

Likewise, appending two RNA strands with ++ should yield again another RNA strand:

```
scala> rna ++ rna
res8: RNA = RNA(A, U, G, G, T, A, U, G, G, T)
```

On the other hand, mapping bases to some other type over an RNA strand cannot yield another RNA strand because the new elements have the wrong type. It has to yield a sequence instead. In the same vein appending elements that are not of type Base to an RNA strand can yield a general sequence, but it cannot yield another RNA strand.

```
scala> rna map Base.toInt
res2: IndexedSeq[Int] = Vector(0, 3, 2, 2, 1)

scala> rna ++ List("missing", "data")
res3: IndexedSeq[java.lang.Object] =
  Vector(A, U, G, G, T, missing, data)
```

This is what you'd expect in the ideal case. But this is not what the RNA2 class as given in Listing 25.7 provides. In fact, if you ran the first two examples above with instances of this class you would obtain:

584

```
scala> val rna2 = RNA2(A, U, G, G, T)
rna2: RNA2 = RNA2(A, U, G, G, T)

scala> rna2 map { case A => T case b => b }
res0: IndexedSeq[Base] = Vector(T, U, G, G, T)

scala> rna2 ++ rna2
res1: IndexedSeq[Base] = Vector(A, U, G, G, T, A, U, G, G, T)
```

So the result of map and ++ is never an RNA strand, even if the element type of the generated collection is a Base. To see how to do better, it pays to have a close look at the signature of the map method (or of ++, which has a similar signature). The map method is originally defined in class scala.collection.TraversableLike with the following signature:

```
def map[B, That](f: Elem => B)
    (implicit cbf: CanBuildFrom[Repr, B, That]): That
```

Here Elem is the type of elements of the collection, and Repr is the type of the collection itself; that is, the second type parameter that gets passed to implementation classes such as TraversableLike and IndexedSeqLike. The map method takes two more type parameters, B and That. The B parameter stands for the result type of the mapping function, which is also the element type of the new collection. The That appears as the result type of map, so it represents the type of the new collection that gets created.

How is the That type determined? It is linked to the other types by an implicit parameter cbf, of type CanBuildFrom[Repr, B, That]. These CanBuildFrom implicits are defined by the individual collection classes. In essence, an implicit value of type CanBuildFrom[From, Elem, To] says: "Here is a way, given a collection of type From, to build with elements of type Elem a collection of type To."

Now the behavior of map and ++ on RNA2 sequences becomes clearer. There is no CanBuildFrom instance that creates RNA2 sequences, so the next best available CanBuildFrom was found in the companion object of the inherited trait IndexedSeq. That implicit creates IndexedSeqs, and that's what you saw when applying map to rna2.

To address this shortcoming, you need to define an implicit instance of CanBuildFrom in the companion object of the RNA class. That instance should have type CanBuildFrom[RNA, Base, RNA]. Hence, this instance states that, given an RNA strand and a new element type Base, you can

585

```
final class RNA private (val groups: Array[Int], val length: Int)
  extends IndexedSeq[Base] with IndexedSeqLike[Base, RNA] {

  import RNA._

  // Mandatory re-implementation of 'newBuilder' in 'IndexedSeq'
  override protected[this] def newBuilder: Builder[Base, RNA] =
    RNA.newBuilder

  // Mandatory implementation of 'apply' in 'IndexedSeq'
  def apply(idx: Int): Base = {
    if (idx < 0 || length <= idx)
      throw new IndexOutOfBoundsException
    Base.fromInt(groups(idx / N) >> (idx % N * S) & M)
  }

  // Optional re-implementation of foreach,
  // to make it more efficient.
  override def foreach[U](f: Base => U): Unit = {
    var i = 0
    var b = 0
    while (i < length) {
      b = if (i % N == 0) groups(i / N) else b >>> S
      f(Base.fromInt(b & M))
      i += 1
    }
  }
}
```

Listing 25.8 · RNA strands class, final version.

build another collection which is again an RNA strand. Listing 25.8 and
Listing 25.9 show the details.

```
object RNA {
  private val S = 2              // number of bits in group
  private val M = (1 << S) - 1   // bitmask to isolate a group
  private val N = 32 / S         // number of groups in an Int

  def fromSeq(buf: Seq[Base]): RNA = {
    val groups = new Array[Int]((buf.length + N - 1) / N)
    for (i <- 0 until buf.length)
      groups(i / N) |= Base.toInt(buf(i)) << (i % N * S)
    new RNA(groups, buf.length)
  }

  def apply(bases: Base*) = fromSeq(bases)

  def newBuilder: Builder[Base, RNA] =
    new ArrayBuffer mapResult fromSeq

  implicit def canBuildFrom: CanBuildFrom[RNA, Base, RNA] =
    new CanBuildFrom[RNA, Base, RNA] {
      def apply(): Builder[Base, RNA] = newBuilder
      def apply(from: RNA): Builder[Base, RNA] = newBuilder
    }
}
```

Listing 25.9 · RNA companion object—final version.

Compared to class RNA2 there are two important differences. First, the
newBuilder implementation has moved from the RNA class to its companion object. The newBuilder method in class RNA simply forwards to this
definition. Second, there is now an implicit CanBuildFrom value in object
RNA. To create such an object you need to define two apply methods in the
CanBuildFrom trait. Both create a new builder for an RNA collection, but
they differ in their argument list. The apply() method simply creates a new
builder of the right type. By contrast, the apply(from) method takes the
original collection as argument. This can be useful to adapt the dynamic type
of builder's return type to be the same as the dynamic type of the receiver. In
the case of RNA this does not come into play because RNA is a final class, so

587

any receiver of static type RNA also has RNA as its dynamic type. That's why `apply(from)` also simply calls `newBuilder`, ignoring its argument.

That is it. The RNA class in Listing 25.8 implements all collection methods at their natural types. Its implementation requires a little bit of protocol. In essence, you need to know where to put the `newBuilder` factories and the `canBuildFrom` implicits. On the plus side, with relatively little code you get a large number of methods automatically defined. Also, if you don't intend to do bulk operations like `take`, `drop`, `map`, or ++ on your collection, you can choose to not go the extra length and stop at the implementation shown in Listing 25.6.

The discussion so far centered on the minimal amount of definitions needed to define new sequences with methods that obey certain types. But in practice you might also want to add new functionality to your sequences or override existing methods for better efficiency. An example of this is the overridden `foreach` method in class RNA. `foreach` is an important method in its own right because it implements loops over collections. Furthermore, many other collection methods are implemented in terms of `foreach`. So it makes sense to invest some effort optimizing the method's implementation.

The standard implementation of `foreach` in `IndexedSeq` will simply select every i'th element of the collection using `apply`, where i ranges from 0 to the collection's length minus one. So this standard implementation selects an array element and unpacks a base from it once for every element in an RNA strand. The overriding `foreach` in class RNA is smarter than that. For every selected array element it immediately applies the given function to all bases contained in it. So the effort for array selection and bit unpacking is much reduced.

### Integrating new sets and maps

As a second example you'll learn how to integrate a new kind of map into the collection framework. The idea is to implement a mutable map with `String` as the type of keys by a "Patricia trie".[2] The term *Patricia* is an abbreviation for "Practical Algorithm to Retrieve Information Coded in Alphanumeric." The idea is to store a set or a map as a tree where subsequent characters in a search key determines uniquely a descendant tree.

---

[2]Morrison, "PATRICIA—Practical Algorithm To Retrieve Information Coded in Alphanumeric" [Mor68]

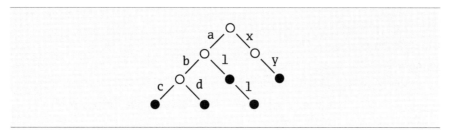

Figure 25.1 · An example Patricia trie.

For instance, a Patricia trie storing the five strings, "abc", "abd", "al", "all", "xy", would look like the tree given in Figure 25.1. To find the node corresponding to the string "abc" in this trie, simply follow the subtree labeled "a", proceed from there to the subtree labeled "b" to finally reach its subtree labeled "c". If the Patricia trie is used as a map, the value that's associated with a key is stored in the nodes that can be reached by the key. If it is a set, you simply store a marker saying that the node is present in the set.

Patricia tries support very efficient lookups and updates. Another nice feature is that they support selecting a subcollection by giving a prefix. For instance, in the tree in Figure 25.1 you can obtain the sub-collection of all keys that start with an "a" simply by following the "a" link from the root of the tree.

Based on these ideas we will now walk you through the implementation of a map that's implemented as a Patricia trie. We call the map a PrefixMap, which means that it provides a method withPrefix that selects a submap of all keys starting with a given prefix.

We'll first define a prefix map with the keys shown in Figure 25.1:

```
scala> val m = PrefixMap("abc" -> 0, "abd" -> 1, "al" -> 2,
   "all" -> 3, "xy" -> 4)
m: PrefixMap[Int] = Map((abc,0), (abd,1), (al,2), (all,3),
   (xy,4))
```

Then calling withPrefix on m will yield another prefix map:

```
scala> m withPrefix "a"
res14: PrefixMap[Int] = Map((bc,0), (bd,1), (l,2), (ll,3))
```

Listing 25.10 shows the definition of PrefixMap. This class is parameterized with the type of associated values T, and extends mutable.Map[String, T]

589

```
import collection._

class PrefixMap[T]
extends mutable.Map[String, T]
   with mutable.MapLike[String, T, PrefixMap[T]] {

  var suffixes: immutable.Map[Char, PrefixMap[T]] = Map.empty
  var value: Option[T] = None

  def get(s: String): Option[T] =
    if (s.isEmpty) value
    else suffixes get (s(0)) flatMap (_.get(s substring 1))

  def withPrefix(s: String): PrefixMap[T] =
    if (s.isEmpty) this
    else {
      val leading = s(0)
      suffixes get leading match {
        case None =>
          suffixes = suffixes + (leading -> empty)
        case _ =>
      }
      suffixes(leading) withPrefix (s substring 1)
    }

  override def update(s: String, elem: T) =
    withPrefix(s).value = Some(elem)

  override def remove(s: String): Option[T] =
    if (s.isEmpty) { val prev = value; value = None; prev }
    else suffixes get (s(0)) flatMap (_.remove(s substring 1))

  def iterator: Iterator[(String, T)] =
    (for (v <- value.iterator) yield ("", v)) ++
    (for ((chr, m) <- suffixes.iterator;
          (s, v) <- m.iterator) yield (chr +: s, v))

  def += (kv: (String, T)): this.type = { update(kv._1, kv._2); this }

  def -= (s: String): this.type  = { remove(s); this }

  override def empty = new PrefixMap[T]
}
```

Listing 25.10 · An implementation of prefix maps with Patricia tries.

and mutable.MapLike[String, T, PrefixMap[T]]. You have seen this
pattern already for sequences in the RNA strand example. Then as now
inheriting an implementation class such as MapLike serves to get the right
result type for transformations such as filter.

A prefix map node has two mutable fields: suffixes and value. The
value field contains an optional value that's associated with the node. It is
initialized to None. The suffixes field contains a map from characters to
PrefixMap values. It is initialized to the empty map. You might ask, Why
did we pick an immutable map as the implementation type for suffixes?
Would not a mutable map been more standard since PrefixMap as a whole
is also mutable? The answer is that immutable maps that contain only a few
elements are very efficient in both space and execution time.

For instance, maps that contain fewer than 5 elements are represented
as a single object. By contrast, as described in Section 17.2, the standard
mutable map is a HashMap, which typically occupies around 80 bytes, even
if it is empty. So if small collections are common, it's better to pick immutable over mutable. In the case of Patricia tries, we'd expect that most
nodes, except the ones at the very top of the tree, would contain only a few
successors. So storing these successors in an immutable map is likely to be
more efficient.

Now have a look at the first method that needs to be implemented for
a map: get. The algorithm is as follows: To get the value associated with
the empty string in a prefix map, simply select the optional value stored in
the root of the tree. Otherwise, if the key string is not empty, try to select
the submap corresponding to the first character of the string. If that yields
a map, follow up by looking up the remainder of the key string after its
first character in that map. If the selection fails, the key is not stored in the
map, so return with None. The combined selection over an option value is
elegantly expressed using flatMap. When applied to an optional value, ov,
and a closure, f, which in turn returns an optional value, ov flatMap f will
succeed if both ov and f return a defined value. Otherwise ov flatMap f
will return None.

The next two methods to implement for a mutable map are += and -=. In
the implementation of Listing 25.10, these are defined in terms of two other
methods: update and remove. The remove method is very similar to get,
except that before returning any associated value, the field containing that
value is set to None. The update method first calls withPrefix to navigate
to the tree node that needs to be updated, then sets the value field of that

node to the given value. The `withPrefix` method navigates through the tree, creating sub-maps as necessary if some prefix of characters is not yet contained as a path in the tree.

The last abstract method to implement for a mutable map is `iterator`. This method needs to produce an iterator that yields all key/value pairs stored in the map. For any given prefix map this iterator is composed of the following parts: First, if the map contains a defined value, `Some(x)`, in the `value` field at its root, then (`""`, x) is the first element returned from the iterator. Furthermore, the iterator needs to traverse the iterators of all submaps stored in the `suffixes` field, but it needs to add a character in front of every key string returned by those iterators. More precisely, if m is the submap reached from the root through a character chr, and (s, v) is an element returned from m.`iterator`, then the root's iterator will return (chr +: s, v) instead.

This logic is implemented quite concisely as a concatenation of two `for` expressions in the implementation of the `iterator` method in Listing 25.10. The first `for` expression iterates over `value.iterator`. This makes use of the fact that `Option` values define an iterator method that returns either no element, if the option value is None, or exactly one element x, if the option value is Some(x).

Note that there is no `newBuilder` method defined in `PrefixMap`. There is no need because maps and sets come with default builders, which are instances of class `MapBuilder`. For a mutable map the default builder starts with an empty map and then adds successive elements using the map's +=
method. Mutable sets behave the same. The default builders for immutable maps and sets use the non-destructive element addition method +, instead of method +=. However, in all these cases, to build the right kind of set or map, you need to start with an empty set or map of this kind. This is provided by the `empty` method, which is the last method defined in `PrefixMap`. In Listing 25.10, this method simply returns a fresh `PrefixMap`.

We'll now turn to the companion object `PrefixMap`, which is shown in Listing 25.11. In fact it is not strictly necessary to define this companion object, as class `PrefixMap` can stand well on its own. The main purpose of object `PrefixMap` is to define some convenience factory methods. It also defines a `CanBuildFrom` implicit to make typing work out better.

The two convenience methods are `empty` and `apply`. The same methods are present for all other collections in Scala's collection framework so it makes sense to define them here too. With the two methods, you can write `PrefixMap` literals like you do for any other collection:

```scala
import scala.collection.mutable.{Builder, MapBuilder}
import scala.collection.generic.CanBuildFrom

object PrefixMap {
  def empty[T] = new PrefixMap[T]

  def apply[T](kvs: (String, T)*): PrefixMap[T] = {
    val m: PrefixMap[T] = empty
    for (kv <- kvs) m += kv
    m
  }

  def newBuilder[T]: Builder[(String, T), PrefixMap[T]] =
    new MapBuilder[String, T, PrefixMap[T]](empty)

  implicit def canBuildFrom[T]
    : CanBuildFrom[PrefixMap[_], (String, T), PrefixMap[T]] =
      new CanBuildFrom[PrefixMap[_], (String, T), PrefixMap[T]] {
        def apply(from: PrefixMap[_]) = newBuilder[T]
        def apply() = newBuilder[T]
      }
}
```

Listing 25.11 · The companion object for prefix maps.

```scala
scala> PrefixMap("hello" -> 5, "hi" -> 2)
res0: PrefixMap[Int] = Map((hello,5), (hi,2))

scala> PrefixMap.empty[String]
res2: PrefixMap[String] = Map()
```

The other member in object PrefixMap is an implicit CanBuildFrom instance. It has the same purpose as the CanBuildFrom definition in the last section: to make methods like map return the best possible type. For instance, consider mapping a function over the key/value pairs of a PrefixMap. As long as that function produces pairs of strings and some second type, the result collection will again be a PrefixMap. Here's an example:

```scala
scala> res0 map { case (k, v) => (k + "!", "x" * v) }
res8: PrefixMap[String] = Map((hello!,xxxxx), (hi!,xx))
```

The given function argument takes the key/value bindings of the prefix map res0 and produces pairs of strings. The result of the map is a PrefixMap, this time with value type String instead of Int. Without the canBuildFrom implicit in PrefixMap the result would just have been a general mutable map, not a prefix map.

**Summary**

If you want to fully integrate a new collection class into the framework, you need to pay attention to the following points:

1. Decide whether the collection should be mutable or immutable.

2. Pick the right base traits for the collection.

3. Inherit from the right implementation trait to implement most collection operations.

4. If you want map and similar operations to return instances of your collection type, provide an implicit CanBuildFrom in your class's companion object.

## 25.4 Conclusion

You have now seen how Scala's collections are built and how you can build new kinds of collections. Because of Scala's rich support for abstraction, each new collection type can have a large number of methods without having to reimplement them all over again.

# Chapter 26

# Extractors

By now you have probably grown accustomed to the concise way data can be decomposed and analyzed using pattern matching. This chapter shows you how to generalize this concept further. Until now, constructor patterns were linked to case classes. For instance, Some(x) is a valid pattern because Some is a case class. Sometimes you might wish that you could write patterns like this without creating an associated case class. In fact, you might wish to be able to create your own kinds of patterns. Extractors give you a way to do so. This chapter explains what extractors are and how you can use them to define patterns that are decoupled from an object's representation.

## 26.1   An example: extracting email addresses

To illustrate the problem extractors solve, imagine that you need to analyze strings that represent email addresses. Given a string, you want to decide whether it is an email address or not, and, if it is, you want to access the user and domain parts of the address. The traditional way to do this uses three helper functions:

```
def isEMail(s: String): Boolean
def domain(s: String): String
def user(s: String): String
```

With these functions, you could parse a given string s as follows:

```
if (isEMail(s)) println(user(s) + " AT " + domain(s))
else println("not an email address")
```

This works, but is kind of clumsy. What's more, things would become more complicated if you combined several such tests. For instance you might want to find two successive strings in a list that are both email addresses with the same user. You can try this yourself with the access functions defined previously to see what would be involved.

You saw already in Chapter 15 that pattern matching is ideal for attacking problems like this. Let's assume for the moment that you could match a string with a pattern:

```
EMail(user, domain)
```

The pattern would match if the string contained an embedded at sign (@). In that case it would bind variable user to the part of the string before the @ and variable domain to the part after it. Postulating a pattern like this, the previous expression could be written more clearly like this:

```
s match {
  case EMail(user, domain) => println(user + " AT " + domain)
  case _ => println("not an email address")
}
```

The more complicated problem of finding two successive email addresses with the same user part would translate to the following pattern:

```
ss match {
  case EMail(u1, d1) :: EMail(u2, d2) :: _ if (u1 == u2) => ...
    ...
}
```

This is much more legible than anything that could be written with access functions. However, the problem is that strings are not case classes; they do not have a representation that conforms to EMail(user, domain). This is where Scala's extractors come in: they let you define new patterns for pre-existing types, where the pattern need not follow the internal representation of the type.

## 26.2  Extractors

An extractor in Scala is an object that has a method called unapply as one of its members. The purpose of that unapply method is to match a value and take it apart. Often, the extractor object also defines a dual method `apply` for building values, but this is not required. As an example, Listing 26.1 shows an extractor object for email addresses:

```
object EMail {
  // The injection method (optional)
  def apply(user: String, domain: String) = user + "@" + domain

  // The extraction method (mandatory)
  def unapply(str: String): Option[(String, String)] = {
    val parts = str split "@"
    if (parts.length == 2) Some(parts(0), parts(1)) else None
  }
}
```

Listing 26.1 · The `EMail` string extractor object.

This object defines both `apply` and `unapply` methods. The `apply` method has the same meaning as always: it turns `EMail` into an object that can be applied to arguments in parentheses in the same way a method is applied. So you can write `EMail("John", "epfl.ch")` to construct the string `"John@epfl.ch"`. To make this more explicit, you could also let `EMail` inherit from Scala's function type, like this:

```
object EMail extends ((String, String) => String) { ... }
```

> **Note**
> The "`(String, String) => String`" portion of the previous object declaration means the same as `Function2[String, String, String]`, which declares an abstract `apply` method that `EMail` implements. As a result of this declaration, you could, for example, pass `EMail` to a method expecting a `Function2[String, String, String]`.

The `unapply` method is what turns `EMail` into an extractor. In a sense, it reverses the construction process of `apply`. Where `apply` takes two strings and forms an email address string out of them, `unapply` takes an email address and returns potentially two strings: the user and the domain of the address. But `unapply` must also handle the case where the given string is not an email address. That's why `unapply` returns an `Option`-type over pairs of strings. Its result is either `Some(user, domain)` if the string `str` is an email address with the given user and domain parts,[1] or None, if str is not an email address. Here are some examples:

```
unapply("John@epfl.ch")  equals  Some("John", "epfl.ch")
unapply("John Doe")  equals  None
```

Now, whenever pattern matching encounters a pattern referring to an extractor object, it invokes the extractor's unapply method on the selector expression. For instance, executing the code:

```
selectorString match { case EMail(user, domain) => ... }
```

would lead to the call:

```
EMail.unapply(selectorString)
```

As you saw previously, this call to EMail.unapply will return either None or Some(u, d), for some values u for the user part of the address and d for the domain part. In the None case, the pattern does not match, and the system tries another pattern or fails with a MatchError exception. In the Some(u, d) case, the pattern matches and its variables are bound to the elements of the returned value. In the previous match, user would be bound to u and domain would be bound to d.

In the EMail pattern matching example, the type String of the selector expression, selectorString, conformed to unapply's argument type (which in the example was also String). This is quite common, but not necessary. It would also be possible to use the EMail extractor to match selector expressions for more general types. For instance, to find out whether an arbitrary value x was an email address string, you could write:

```
val x: Any = ...
x match { case EMail(user, domain) => ... }
```

Given this code, the pattern matcher will first check whether the given value x conforms to String, the parameter type of EMail's unapply method. If it does conform, the value is cast to String and pattern matching proceeds as before. If it does not conform, the pattern fails immediately.

---

[1]As demonstrated here, where Some is applied to the tuple, (user, domain), you can leave off one pair of parentheses when passing a tuple to a function that takes a single argument. Thus, Some(user, domain) means the same as Some((user, domain)).

In object EMail, the apply method is called an *injection*, because it takes some arguments and yields an element of a given set (in our case: the set of strings that are email addresses). The unapply method is called an *extraction*, because it takes an element of the same set and extracts some of its parts (in our case: the user and domain substrings). Injections and extractions are often grouped together in one object, because then you can use the object's name for both a constructor and a pattern, which simulates the convention for pattern matching with case classes. However, it is also possible to define an extraction in an object without a corresponding injection. The object itself is called an *extractor*, regardless of whether or not it has an apply method.

If an injection method is included, it should be the dual to the extraction method. For instance, a call of:

```
EMail.unapply(EMail.apply(user, domain))
```

should return:

```
Some(user, domain)
```

*i.e.*, the same sequence of arguments wrapped in a Some. Going in the other direction means running first the unapply and then the apply, as shown in the following code:

```
EMail.unapply(obj) match {
  case Some(u, d) => EMail.apply(u, d)
}
```

In that code, if the match on obj succeeds, you'd expect to get back that same object from the apply. These two conditions for the duality of apply and unapply are good design principles. They are not enforced by Scala, but it's recommended to keep to them when designing your extractors.

## 26.3 Patterns with zero or one variables

The unapply method of the previous example returned a pair of element values in the success case. This is easily generalized to patterns of more than two variables. To bind N variables, an unapply would return an N-element tuple, wrapped in a Some.

The case where a pattern binds just one variable is treated differently, however. There is no one-tuple in Scala. To return just one pattern element, the unapply method simply wraps the element itself in a Some. For example, the extractor object shown in Listing 26.2 defines apply and unapply for strings that consist of the same substring appearing twice in a row:

```
object Twice {
  def apply(s: String): String = s + s
  def unapply(s: String): Option[String] = {
    val length = s.length / 2
    val half = s.substring(0, length)
    if (half == s.substring(length)) Some(half) else None
  }
}
```

Listing 26.2 · The Twice string extractor object.

It's also possible that an extractor pattern does not bind any variables. In that case the corresponding unapply method returns a boolean—true for success and false for failure. For instance, the extractor object shown in Listing 26.3 characterizes strings consisting of all uppercase characters:

```
object UpperCase {
  def unapply(s: String): Boolean = s.toUpperCase == s
}
```

Listing 26.3 · The UpperCase string extractor object.

This time, the extractor only defines an unapply, but not an apply. It would make no sense to define an apply, as there's nothing to construct.

The following userTwiceUpper function applies all previously defined extractors together in its pattern matching code:

```
def userTwiceUpper(s: String) = s match {
  case EMail(Twice(x @ UpperCase()), domain) =>
    "match: " + x + " in domain " + domain
  case _ =>
    "no match"
}
```

The first pattern of this function matches strings that are email addresses whose user part consists of two occurrences of the same string in uppercase letters. For instance:

```
scala> userTwiceUpper("DIDI@hotmail.com")
res0: String = match: DI in domain hotmail.com

scala> userTwiceUpper("DIDO@hotmail.com")
res1: String = no match

scala> userTwiceUpper("didi@hotmail.com")
res2: String = no match
```

Note that UpperCase in function userTwiceUpper takes an empty parameter list. This cannot be omitted as otherwise the match would test for equality with the object UpperCase! Note also that, even though UpperCase() itself does not bind any variables, it is still possible to associate a variable with the whole pattern matched by it. To do this, you use the standard scheme of variable binding explained in Section 15.2: the form x @ UpperCase() associates the variable x with the pattern matched by UpperCase(). For instance, in the first userTwiceUpper invocation above, x was bound to "DI", because that was the value against which the UpperCase() pattern was matched.

## 26.4 Variable argument extractors

The previous extraction methods for email addresses all returned a fixed number of element values. Sometimes, this is not flexible enough. For example, you might want to match on a string representing a domain name, so that every part of the domain is kept in a different sub-pattern. This would let you express patterns such as the following:

```
dom match {
  case Domain("org", "acm") => println("acm.org")
  case Domain("com", "sun", "java") => println("java.sun.com")
  case Domain("net", _*) => println("a .net domain")
}
```

In this example things were arranged so that domains are expanded in re-verse order—from the top-level domain down to the sub-domains. This was done so that you could better profit from sequence patterns. You saw in Section 15.2 that a sequence wildcard pattern, _*, at the end of an argument list matches any remaining elements in a sequence. This feature is more useful if the top-level domain comes first, because then you can use sequence wildcards to match sub-domains of arbitrary depth.

The question remains how an extractor can support *vararg matching* as shown in the previous example, where patterns can have a varying number of sub-patterns. The unapply methods encountered so far are not sufficient, because they each return a fixed number of sub-elements in the success case. To handle this case, Scala lets you define a different extraction method specifically for vararg matching. This method is called unapplySeq. To see how it is written, have a look at the Domain extractor, shown in Listing 26.4:

```
object Domain {

  // The injection method (optional)
  def apply(parts: String*): String =
    parts.reverse.mkString(".")

  // The extraction method (mandatory)
  def unapplySeq(whole: String): Option[Seq[String]] =
    Some(whole.split("\\.").reverse)
}
```

Listing 26.4 · The Domain string extractor object.

The Domain object defines an unapplySeq method that first splits the string into parts separated by periods. This is done using Java's split method on strings, which takes a regular expression as its argument. The result of split is an array of substrings. The result of unapplySeq is then that array with all elements reversed and wrapped in a Some.

The result type of an unapplySeq must conform to Option[Seq[T]], where the element type T is arbitrary. As you saw in Section 17.1, Seq is an important class in Scala's collection hierarchy. It's a common superclass of several classes describing different kinds of sequences: Lists, Arrays, WrappedString, and several others.

For symmetry, Domain also has an apply method that builds a domain string from a variable argument parameter of domain parts starting with the top-level domain. As always, the apply method is optional.

You can use the Domain extractor to get more detailed information out of email strings. For instance, to search for an email address named "tom" in some ".com" domain, you could write the following function:

```
def isTomInDotCom(s: String): Boolean = s match {
  case EMail("tom", Domain("com", _*)) => true
  case _ => false
}
```

This gives the expected results:

```
scala> isTomInDotCom("tom@sun.com")
res3: Boolean = true

scala> isTomInDotCom("peter@sun.com")
res4: Boolean = false

scala> isTomInDotCom("tom@acm.org")
res5: Boolean = false
```

It's also possible to return some fixed elements from an unapplySeq together with the variable part. This is expressed by returning all elements in a tuple, where the variable part comes last, as usual. As an example, Listing 26.5 shows a new extractor for emails where the domain part is already expanded into a sequence:

```
object ExpandedEMail {
  def unapplySeq(email: String)
      : Option[(String, Seq[String])] = {
    val parts = email split "@"
    if (parts.length == 2)
      Some(parts(0), parts(1).split("\\.").reverse)
    else
      None
  }
}
```

Listing 26.5 · The ExpandedEMail extractor object.

The unapplySeq method in ExpandedEMail returns an optional value of a pair (a Tuple2). The first element of the pair is the user part. The second

603

element is a sequence of names representing the domain. You can match on
this as usual:

```
scala> val s = "tom@support.epfl.ch"
s: String = tom@support.epfl.ch

scala> val ExpandedEMail(name, topdom, subdoms @ _*) = s
name: String = tom
topdom: String = ch
subdoms: Seq[String] = WrappedArray(epfl, support)
```

## 26.5   Extractors and sequence patterns

You saw in Section 15.2 that you can access the elements of a list or an array
using sequence patterns such as:

```
List()
List(x, y, _*)
Array(x, 0, 0, _)
```

In fact, these sequence patterns are all implemented using extractors in the
standard Scala library. For instance, patterns of the form List(...) are
possible because the scala.List companion object is an extractor that defines an unapplySeq method. Listing 26.6 shows the relevant definitions:

```
package scala
object List {
  def apply[T](elems: T*) = elems.toList
  def unapplySeq[T](x: List[T]): Option[Seq[T]] = Some(x)

  ...
}
```

Listing 26.6 · An extractor that defines an unapplySeq method.

The List object contains an apply method that takes a variable number of
arguments. That's what lets you write expressions such as:

```
List()
List(1, 2, 3)
```

604

It also contains an `unapplySeq` method that returns all elements of the list as a sequence. That's what supports `List(...)` patterns. Very similar definitions exist in the object `scala.Array`. These support analogous injections and extractions for arrays.

## 26.6    Extractors versus case classes

Even though they are very useful, case classes have one shortcoming: they expose the concrete representation of data. This means that the name of the class in a constructor pattern corresponds to the concrete representation type of the selector object. If a match against:

```
case C(...)
```

succeeds, you know that the selector expression is an instance of class C.

Extractors break this link between data representations and patterns. You have seen in the examples in this section that they enable patterns that have nothing to do with the data type of the object that's selected on. This property is called *representation independence*. In open systems of large size, representation independence is very important because it allows you to change an implementation type used in a set of components without affecting clients of these components.

If your component had defined and exported a set of case classes, you'd be stuck with them because client code could already contain pattern matches against these case classes. Renaming some case classes or changing the class hierarchy would affect client code. Extractors do not share this problem, because they represent a layer of indirection between a data representation and the way it is viewed by clients. You could still change a concrete representation of a type, as long as you update all your extractors with it.

Representation independence is an important advantage of extractors over case classes. On the other hand, case classes also have some advantages of their own over extractors. First, they are much easier to set up and to define, and they require less code. Second, they usually lead to more efficient pattern matches than extractors, because the Scala compiler can optimize patterns over case classes much better than patterns over extractors. This is because the mechanisms of case classes are fixed, whereas an `unapply` or `unapplySeq` method in an extractor could do almost anything. Third, if your case classes inherit from a `sealed` base class, the Scala compiler will check your pattern matches for exhaustiveness and will complain if some combination of possible values is not covered by a pattern. No such exhaustiveness checks are available for extractors.

So which of the two methods should you prefer for your pattern matches? It depends. If you write code for a closed application, case classes are usually preferable because of their advantages in conciseness, speed and static checking. If you decide to change your class hierarchy later, the application needs to be refactored, but this is usually not a problem. On the other hand, if you need to expose a type to unknown clients, extractors might be preferable because they maintain representation independence.

Fortunately, you need not decide right away. You could always start with case classes and then, if the need arises, change to extractors. Because patterns over extractors and patterns over case classes look exactly the same in Scala, pattern matches in your clients will continue to work.

Of course, there are also situations where it's clear from the start that the structure of your patterns does not match the representation type of your data. The email addresses discussed in this chapter were one such example. In that case, extractors are the only possible choice.

## 26.7 Regular expressions

One particularly useful application area of extractors are regular expressions. Like Java, Scala provides regular expressions through a library, but extractors make it much nicer to interact with them.

### Forming regular expressions

Scala inherits its regular expression syntax from Java, which in turn inherits most of the features of Perl. We assume you know that syntax already; if not, there are many accessible tutorials, starting with the Javadoc documentation of class `java.util.regex.Pattern`. Here are just some examples that should be enough as refreshers:

ab?    An 'a', possibly followed by a 'b'.

\d+    A number consisting of one or more digits represented by \d.

| | |
|---|---|
| `[a-dA-D]\w*` | A word starting with a letter between a and d in lower or upper case, followed by a sequence of zero or more "word characters" denoted by \w. (A word character is a letter, digit, or underscore.) |
| `(-)?(\d+)(\.\d*)?` | A number consisting of an optional minus sign, followed by one or more digits, optionally followed by a period and zero or more digits. The number contains three *groups, i.e.*, the minus sign, the part before the decimal point, and the fractional part including the decimal point. Groups are enclosed in parentheses. |

Scala's regular expression class resides in package `scala.util.matching`.

```
scala> import scala.util.matching.Regex
```

A new regular expression value is created by passing a string to the `Regex` constructor. For instance:

```
scala> val Decimal = new Regex("(-)?(\\d+)(\\.\\d*)?")
Decimal: scala.util.matching.Regex = (-)?(\d+)(\.\d*)?
```

Note that, compared to the regular expression for decimal numbers given previously, every backslash appears twice in the string above. This is because in Java and Scala a single backslash is an escape character in a string literal, not a regular character that shows up in the string. So instead of '\' you need to write '\\' to get a single backslash in the string.

If a regular expression contains many backslashes this might be a bit painful to write and to read. Scala's raw strings provide an alternative. As you saw in Section 5.2, a raw string is a sequence of characters between triple quotes. The difference between a raw and a normal string is that all characters in a raw string appear exactly as they are typed. This includes backslashes, which are not treated as escape characters. So you could write equivalently and somewhat more legibly:

```
scala> val Decimal = new Regex("""(-)?(\d+)(\.\d*)?""")
Decimal: scala.util.matching.Regex = (-)?(\d+)(\.\d*)?
```

607

As you can see from the interpreter's output, the generated result value for
Decimal is exactly the same as before.

Another, even shorter way to write a regular expression in Scala is this:

```
scala> val Decimal = """(-)?(\d+)(\.\d*)?""".r
Decimal: scala.util.matching.Regex = (-)?(\d+)(\.\d*)?
```

In other words, simply append a .r to a string to obtain a regular expression.
This is possible because there is a method named r in class StringOps,
which converts a string to a regular expression. The method is defined as
shown in Listing 26.7:

```
package scala.runtime
import scala.util.matching.Regex

class StringOps(self: String) ... {

  ...

  def r = new Regex(self)
}
```

Listing 26.7 · How the r method is defined in StringOps.

### Searching for regular expressions

You can search for occurrences of a regular expression in a string using several different operators:

regex findFirstIn str

Finds first occurrence of regular expression regex in string str, returning the result in an Option type.

regex findAllIn str

Finds all occurrences of regular expression regex in string str, returning the results in an Iterator.

regex findPrefixOf str

Finds an occurrence of regular expression regex at the start of string
str, returning the result in an Option type.

For instance, you could define the input sequence below and then search decimal numbers in it:

```
scala> val Decimal = """(-)?(\d+)(\.\d*)?""".r
Decimal: scala.util.matching.Regex = (-)?(\d+)(\.\d*)?

scala> val input = "for -1.0 to 99 by 3"
input: String = for -1.0 to 99 by 3

scala> for (s <- Decimal findAllIn input)
         println(s)
-1.0
99
3

scala> Decimal findFirstIn input
res7: Option[String] = Some(-1.0)

scala> Decimal findPrefixOf input
res8: Option[String] = None
```

### Extracting with regular expressions

What's more, every regular expression in Scala defines an extractor. The extractor is used to identify substrings that are matched by the groups of the regular expression. For instance, you could decompose a decimal number string as follows:

```
scala> val Decimal(sign, integerpart, decimalpart) = "-1.23"
sign: String = -
integerpart: String = 1
decimalpart: String = .23
```

In this example, the pattern, Decimal(...), is used in a val definition, as described in Section 15.7. What happens here is that the Decimal regular expression value defines an unapplySeq method. That method matches every string that corresponds to the regular expression syntax for decimal numbers. If the string matches, the parts that correspond to the three groups in the regular expression (-)?(\d+)(\.\d*)? are returned as elements of the pattern and are then matched by the three pattern variables sign, integerpart, and decimalpart. If a group is missing, the element value is set to null, as can be seen in the following example:

```
scala> val Decimal(sign, integerpart, decimalpart) = "1.0"
sign: String = null
integerpart: String = 1
decimalpart: String = .0
```

It's also possible to mix extractors with regular expression searches in a `for` expression. For instance, the following expression decomposes all decimal numbers it finds in the `input` string:

```
scala> for (Decimal(s, i, d) <- Decimal findAllIn input)
         println("sign: " + s + ", integer: " +
             i + ", decimal: " + d)
sign: -, integer: 1, decimal: .0
sign: null, integer: 99, decimal: null
sign: null, integer: 3, decimal: null
```

## 26.8   Conclusion

In this chapter you saw how to generalize pattern matching with extractors. Extractors let you define your own kinds of patterns, which need not correspond to the type of the expressions you select on. This gives you more flexibility in the kinds of patterns you can use for matching. In effect it's like having different possible views on the same data. It also gives you a layer between a type's representation and the way clients view it. This lets you do pattern matching while maintaining representation independence, a property which is very useful in large software systems.

Extractors are one more element in your tool box that let you define flexible library abstractions. They are used heavily in Scala's libraries, for instance, to enable convenient regular expression matching.

Chapter 27

# Annotations

Annotations are structured information added to program source code. Like comments, they can be sprinkled throughout a program and attached to any variable, method, expression, or other program element. Unlike comments, they have structure, thus making them easier to machine process.

This chapter shows how to use annotations in Scala. It shows their general syntax and how to use several standard annotations.

This chapter does not show how to write new annotation processing tools, because it is beyond the scope of this book. Chapter 31 shows one technique, but not the only one. Instead, this chapter focuses on how to use annotations, because it is more common to use annotations than to define new annotation processors.

## 27.1 Why have annotations?

There are many things you can do with a program other than compiling and running it. Some examples are:

1. Automatic generation of documentation as with Scaladoc.

2. Pretty printing code so that it matches your preferred style.

3. Checking code for common errors such as opening a file but, on some control paths, never closing it.

4. Experimental type checking, for example to manage side effects or ensure ownership properties.

Such tools are called *meta-programming* tools, because they are programs that take other programs as input. Annotations support these tools by letting the programmer sprinkle directives to the tool throughout their source code. Such directives let the tools be more effective than if they could have no user input. For example, annotations can improve the previously listed tools as follows:

1. A documentation generator could be instructed to document certain methods as deprecated.

2. A pretty printer could be instructed to skip over parts of the program that have been carefully hand formatted.

3. A checker for non-closed files could be instructed to ignore a particular file that has been manually verified to be closed.

4. A side-effects checker could be instructed to verify that a specified method has no side effects.

In all of these cases, it would in theory be possible for the programming language to provide ways to insert the extra information. In fact, most of these are directly supported in some language or another. However, there are too many such tools for one language to directly support them all. Further, all of this information is completely ignored by the compiler, which after all just wants to make the code run.

Scala's philosophy in cases like this is to include the minimum, orthogonal support in the core language such that a wide variety of meta-programming tools can be written. In this case, that minimum support is a system of annotations. The compiler understands just one feature, annotations, but it doesn't attach any meaning to individual annotations. Each meta-programming tool can then define and use its own specific annotations.

## 27.2 Syntax of annotations

A typical use of an annotation looks like this:

```
@deprecated def bigMistake() = //...
```

The annotation is the @deprecated part, and it applies to the entirety of the bigMistake method (not shown—it's too embarrassing). In this case, the method is being marked as something the author of `bigMistake` wishes you not to use. Maybe `bigMistake` will be removed entirely from a future version of the code.

In the previous example, a method is annotated as `@deprecated`. Annotations are allowed in other places too. Annotations are allowed on any kind of declaration or definition, including `vals`, `vars`, `defs`, `classes`, `objects`, `traits`, and `types`. The annotation applies to the entirety of the declaration or definition that follows it:

```
@deprecated class QuickAndDirty {
  //...
}
```

Annotations can also be applied to an expression, as with the `@unchecked` annotation for pattern matching (see Chapter 15). To do so, place a colon (`:`) after the expression and then write the annotation. Syntactically, it looks like the annotation is being used as a type:

```
(e: @unchecked) match {
  // non-exhaustive cases...
}
```

Finally, annotations can be placed on types. Annotated types are described later in this chapter.

So far the annotations shown have been simply an at sign followed by an annotation class. Such simple annotations are common and useful, but annotations have a richer general form:

$$@annot(exp_1, exp_2, ...)$$

The *annot* specifies the class of annotation. All annotations must include that much. The *exp* parts are arguments to the annotation. For annotations like `@deprecated` that do not need any arguments, you would normally leave off the parentheses, but you can write `@deprecated()` if you like. For annotations that do have arguments, place the arguments in parentheses, for example, `@serial(1234)`.

The precise form of the arguments you may give to an annotation depends on the particular annotation class. Most annotation processors only let you supply immediate constants such as `123` or `"hello"`. The compiler itself supports arbitrary expressions, however, so long as they type check.

Some annotation classes can make use of this, for example, to let you refer to other variables that are in scope:

```
@cool val normal = "Hello"
@coolerThan(normal) val fonzy = "Heeyyy"
```

Internally, Scala represents an annotation as just a constructor call of an annotation class—replace the '@' by 'new' and you have a valid instance creation expression. This means that named and default annotation arguments are supported naturally, because Scala already has named and default arguments for method and constructor calls. One slightly tricky bit concerns annotations that conceptually take other annotations as arguments, which are required by some frameworks. You cannot write an annotation directly as an argument to an annotation, because annotations are not valid expressions. In such cases you must use 'new' instead of '@', as illustrated here:

```
scala> import annotation._
import annotation._

scala> class strategy(arg: Annotation) extends Annotation
defined class strategy

scala> class delayed extends Annotation
defined class delayed

scala> @strategy(@delayed) def f() = {}
<console>:1: error: illegal start of simple expression
       @strategy(@delayed) def f() = {}
                 ^

scala> @strategy(new delayed) def f() = {}
f: ()Unit
```

## 27.3   Standard annotations

Scala includes several standard annotations. They are for features that are used widely enough to merit putting in the language specification, but that are not fundamental enough to merit their own syntax. Over time, there should be a trickle of new annotations that are added to the standard in just the same way.

**Deprecation**

Sometimes you write a class or method that you later wish you had not. Once it is available, though, code written by other people might call the method. Thus, you cannot simply delete the method, because this would cause other people's code to stop compiling.

Deprecation lets you gracefully remove a method or class that turns out to be a mistake. You mark the method or class as deprecated, and then anyone who calls that method or class will get a deprecation warning. They had better heed this warning and update their code! The idea is that after a suitable amount of time has passed, you feel safe in assuming that all reasonable clients will have stopped accessing the deprecated class or method and thus that you can safely remove it.

You mark a method as deprecated simply by writing @deprecated before it. For example:

```
@deprecated def bigMistake() =  //...
```

Such an annotation will cause the Scala compiler to emit deprecation warnings whenever Scala code accesses the method.

If you supply a string as an argument to @deprecated, that string will be emitted along with the error message. Use this message to explain to developers what they should use instead of the deprecated method.

```
@deprecated("use newShinyMethod() instead")
def bigMistake() =  //...
```

Now any callers will get a message like this:

```
$ scalac -deprecation Deprecation2.scala
Deprecation2.scala:33: warning: method bigMistake in object
Deprecation2 is deprecated: use newShinyMethod() instead
    bigMistake()
    ^
one warning found
```

## Volatile fields

Concurrent programming does not mix well with shared mutable state. For this reason, the focus of Scala's concurrency support is message passing and a minimum of shared mutable state. See Chapter **??** for the details.

Nonetheless, sometimes programmers want to use mutable state in their concurrent programs. The @volatile annotation helps in such cases. It informs the compiler that the variable in question will be used by multiple threads. Such variables are implemented so that reads and writes to the variable are slower, but accesses from multiple threads behave more predictably.

The @volatile keyword gives different guarantees on different platforms. On the Java platform, however, you get the same behavior as if you wrote the field in Java code and marked it with the Java volatile modifier.

## Binary serialization

Many languages include a framework for binary *serialization*. A serialization framework helps you convert objects into a stream of bytes and *vice versa*. This is useful if you want to save objects to disk or send them over the network. XML can help with the same goals (see Chapter 28), but it has different trade offs regarding speed, space usage, flexibility, and portability.

Scala does not have its own serialization framework. Instead, you should use a framework from your underlying platform. What Scala does is provide three annotations that are useful for a variety of frameworks. Also, the Scala compiler for the Java platform interprets these annotations in the Java way (see Chapter 31).

The first annotation indicates whether a class is serializable at all. Most classes are serializable, but not all. A handle to a socket or GUI window, for example, cannot be serialized. By default, a class is not considered serializable. You should add a @serializable annotation to any class you would like to be serializable.

The second annotation helps deal with serializable classes changing as time goes by. You can attach a serial number to the current version of a class by adding an annotation like @SerialVersionUID(1234), where 1234 should be replaced by your serial number of choice. The framework should store this number in the generated byte stream. When you later reload that byte stream and try to convert it to an object, the framework can check that the current version of the class has the same version number as the version in the byte stream. If you want to make a serialization-incompatible change to your class, then you can change the version number. The framework will then automatically refuse to load old instances of the class.

Finally, Scala provides a `@transient` annotation for fields that should not be serialized at all. If you mark a field as `@transient`, then the framework should not save the field even when the surrounding object is serialized. When the object is loaded, the field will be restored to the default value for the type of the field annotated as `@transient`.

**Automatic get and set methods**

Scala code normally does not need explicit `get` and `set` methods for fields, because Scala blends the syntax for field access and method invocation. Some platform-specific frameworks do expect `get` and `set` methods, however. For that purpose, Scala provides the `@scala.reflect.BeanProperty` annotation. If you add this annotation to a field, the compiler will automatically generate `get` and `set` methods for you. If you annotate a field named `crazy`, the `get` method will be named `getCrazy` and the `set` method will be named `setCrazy`.

The generated `get` and `set` methods are only available after a compilation pass completes. Thus, you cannot call these `get` and `set` methods from code you compile at the same time as the annotated fields. This should not be a problem in practice, because in Scala code you can access the fields directly. This feature is intended to support frameworks that expect regular `get` and `set` methods, and typically you do not compile the framework and the code that uses it at the same time.

**Tailrec**

You would typically add the `@tailrec` annotation to a method that needs to be tail recursive, for instance because you expect that it would recurse very deeply otherwise. To make sure that the Scala compiler does perform the tail-recursion optimization described in Section 8.9 on this method, you can add `@tailrec` in front of the method definition. If the optimization cannot be performed, you will then get a warning together with an explanation of the reasons.

**Unchecked**

The @unchecked annotation is interpreted by the compiler during pattern matches. It tells the compiler not to worry if the match expression seems to leave out some cases. See Section 15.5 for details.

**Native methods**

The @native annotation informs the compiler that a method's implementation is supplied by the runtime rather than in Scala code. The compiler will toggle the appropriate flags in the output, and it will be up to the developer to supply the implementation using a mechanism such as the Java Native Interface (JNI).

When using the @native annotation, a method body must be supplied, but it will not be emitted into the output. For example, here is how to declare that method beginCountdown will be supplied by the runtime:

```
@native
def beginCountdown() = {}
```

## 27.4   Conclusion

This chapter described the platform-independent aspects of annotations that you will most commonly need to know about. First of all it covered the syntax of annotations, because using annotations is far more common than defining new ones. Second it showed how to use several annotations that are supported by the standard Scala compiler, including @deprecated, @volatile, @serializable, @BeanProperty, @tailrec, and @unchecked.

Chapter 31 gives additional, Java-specific information on annotations. It covers annotations only available when targeting Java, additional meanings of standard annotations when targeting Java, how to interoperate with Java-based annotations, and how to use Java-based mechanisms to define and process annotations in Scala.

# Chapter 28

# Working with XML

This chapter introduces Scala's support for XML. After discussing semi-structured data in general, it shows the essential functionality in Scala for manipulating XML: how to make nodes with XML literals, how to save and load XML to files, and how to take apart XML nodes using query methods and pattern matching. This chapter is just a brief introduction to what is possible with XML, but it shows enough to get you started.

## 28.1 Semi-structured data

XML is a form of *semi-structured data*. It is more structured than plain strings, because it organizes the contents of the data into a tree. Plain XML is less structured than the objects of a programming language, though, as it admits free-form text between tags and it lacks a type system.[1]

Semi-structured data is very helpful any time you need to serialize program data for saving in a file or shipping across a network. Instead of converting structured data all the way down to bytes, you convert it to and from semi-structured data. You then use pre-existing library routines to convert between semi-structured data and binary data, saving your time for more important problems.

There are many forms of semi-structured data, but XML is the most widely used on the Internet. There are XML tools on most operating systems, and most programming languages have XML libraries available. Its popularity is self-reinforcing. The more tools and libraries are developed

---

[1] There are type systems for XML, such as XML Schemas, but they are beyond the scope of this book.

in response to XML's popularity, the more likely software engineers are to choose XML as part of their formats. If you write software that communicates over the Internet, then sooner or later you will need to interact with some service that speaks XML.

For all of these reasons, Scala includes special support for processing XML. This chapter shows you Scala's support for constructing XML, processing it with regular methods, and processing it with Scala's pattern matching. In addition to these nuts and bolts, the chapter shows along the way several common idioms for using XML in Scala.

## 28.2  XML overview

XML is built out of two basic elements, text and tags.[2] Text is, as usual, any sequence of characters. Tags, written like <pod>, consist of a less-than sign, an alphanumeric label, and a greater than sign. Tags can be *start* or *end* tags. An end tag looks just like a start tag except that it has a slash just before the tag's label, like this: </pod>.

Start and end tags must match each other, just like parentheses. Any start tag must eventually be followed by an end tag with the same label. Thus the following is illegal:

```
// Illegal XML
One <pod>, two <pod>, three <pod> zoo
```

Further, the contents of any two matching tags must itself be valid XML. You cannot have two pairs of matching tags overlap each other:

```
// Also illegal
<pod>Three <peas> in the </pod></peas>
```

You could, however, write it like this:

```
<pod>Three <peas></peas> in the </pod>
```

Since tags are required to match in this way, XML is structured as nested *elements*. Each pair of matching start and end tags forms an element, and elements may be nested within each other. In the above example, the entirety of <pod>Three <peas></peas> in the </pod> is an element, and <peas></peas> is an element nested within it.

---

[2]The full story is more complicated, but this is enough to be effective with XML.

Those are the basics. Two other things you should know are, first, there is a shorthand notation for a start tag followed immediately by its matching end tag. Simply write one tag with a slash put after the tag's label. Such a tag comprises an *empty element*. Using an empty element, the previous example could just as well be written as follows:

```
<pod>Three <peas/> in the </pod>
```

Second, start tags can have *attributes* attached to them. An attribute is a name-value pair written with an equals sign in the middle. The attribute name itself is plain, unstructured text, and the value is surrounded by either double quotes ("") or single quotes (' '). Attributes look like this:

```
<pod peas="3" strings="true"/>
```

## 28.3  XML literals

Scala lets you type in XML as a literal anywhere that an expression is valid. Simply type a start tag and then continue writing XML content. The compiler will go into an XML-input mode and will read content as XML until it sees the end tag matching the start tag you began with:

```
scala> <a>
          This is some XML.
          Here is a tag: <atag/>
        </a>
res0: scala.xml.Elem =
<a>
  This is some XML.
  Here is a tag: <atag/>
</a>
```

The result of this expression is of type Elem, meaning it is an XML element with a label ("a") and children ("This is some XML...," *etc.*). Some other important XML classes are:

- Class Node is the abstract superclass of all XML node classes.

- Class Text is a node holding just text. For example, the "stuff" part of <a>stuff</a> is of class Text.

- Class NodeSeq holds a sequence of nodes. Many methods in the XML library process NodeSeqs in places you might expect them to process individual Nodes. You can still use such methods with individual nodes, however, since Node extends from NodeSeq. This may sound weird, but it works out well for XML. You can think of an individual Node as a one-element NodeSeq.

You are not restricted to writing out the exact XML you want, character for character. You can evaluate Scala code in the middle of an XML literal by using curly braces ({}) as an escape. Here is a simple example:

```
scala> <a> {"hello" + ", world"} </a>
res1: scala.xml.Elem = <a> hello, world </a>
```

A braces escape can include arbitrary Scala content, including further XML literals. Thus, as the nesting level increases, your code can switch back and forth between XML and ordinary Scala code. Here's an example:

```
scala> val yearMade = 1955
yearMade: Int = 1955

scala>   <a> { if (yearMade < 2000) <old>{yearMade}</old>
                else xml.NodeSeq.Empty }
         </a>
res2: scala.xml.Elem =
<a> <old>1955</old>
  </a>
```

If the code inside the curly braces evaluates to either an XML node or a sequence of XML nodes, those nodes are inserted directly as is. In the above example, if yearMade is less than 2000, it is wrapped in <old> tags and added to the <a> element. Otherwise, nothing is added. Note in the above example that "nothing" as an XML node is denoted with xml.NodeSeq.Empty.

An expression inside a brace escape does not have to evaluate to an XML node. It can evaluate to any Scala value. In such a case, the result is converted to a string and inserted as a text node:

```
scala> <a> {3 + 4} </a>
res3: scala.xml.Elem = <a> 7 </a>
```

Any <, >, and & characters in the text will be escaped if you print the node back out:

```
scala> <a> {"</a>potential security hole<a>"} </a>
res4: scala.xml.Elem = <a> &lt;/a&gt;potential security
hole&lt;a&gt; </a>
```

To contrast, if you create XML with low-level string operations, you will run into traps such as the following:

```
scala> "<a>" + "</a>potential security hole<a>" + "</a>"
res5: String = <a></a>potential security hole<a></a>
```

What happens here is that a user-supplied string has included XML tags of its own, in this case </a> and <a>. This behavior can allow some nasty surprises for the original programmer, because it allows the user to affect the resulting XML tree outside of the space provided for the user inside the <a> element. You can prevent this entire class of problems by always constructing XML using XML literals, not string appends.

## 28.4  Serialization

You have now seen enough of Scala's XML support to write the first part of a serializer: conversion from internal data structures to XML. All you need for this are XML literals and their brace escapes.

As an example, suppose you are implementing a database to keep track of your extensive collection of vintage Coca-Cola thermometers. You might make the following internal class to hold entries in the catalog:

```
abstract class CCTherm {
  val description: String
  val yearMade: Int
  val dateObtained: String
  val bookPrice: Int      // in US cents
  val purchasePrice: Int  // in US cents
  val condition: Int      // 1 to 10

  override def toString = description
}
```

This is a straightforward, data-heavy class that holds various pieces of information such as when the thermometer was made, when you got it, and how much you paid for it.

623

To convert instances of this class to XML, simply add a `toXML` method that uses XML literals and brace escapes, like this:

```
abstract class CCTherm {
  ...
  def toXML =
    <cctherm>
      <description>{description}</description>
      <yearMade>{yearMade}</yearMade>
      <dateObtained>{dateObtained}</dateObtained>
      <bookPrice>{bookPrice}</bookPrice>
      <purchasePrice>{purchasePrice}</purchasePrice>
      <condition>{condition}</condition>
    </cctherm>
}
```

Here is the method in action:

```
scala> val therm = new CCTherm {
         val description = "hot dog #5"
         val yearMade = 1952
         val dateObtained = "March 14, 2006"
         val bookPrice = 2199
         val purchasePrice = 500
         val condition = 9
       }
therm: CCTherm = hot dog #5

scala> therm.toXML
res6: scala.xml.Elem =
<cctherm>
                <description>hot dog #5</description>
                <yearMade>1952</yearMade>
                <dateObtained>March 14, 2006</dateObtained>
                <bookPrice>2199</bookPrice>
                <purchasePrice>500</purchasePrice>
                <condition>9</condition>
        </cctherm>
```

> **Note**
> The "new CCTherm" expression in the previous example works even
> though CCTherm is an abstract class, because this syntax actually
> instantiates an anonymous subclass of CCTherm. Anonymous classes were
> described in Section 20.5.

By the way, if you want to include a curly brace ('{' or '}') as XML text,
as opposed to using them to escape to Scala code, simply write two curly
braces in a row:

```
scala> <a> {{{{brace yourself!}}}} </a>
res7: scala.xml.Elem = <a> {{brace yourself!}} </a>
```

## 28.5 Taking XML apart

Among the many methods available for the XML classes, there are three in
particular that you should be aware of. They allow you to take apart XML
without thinking too much about the precise way XML is represented in
Scala. These methods are based on the XPath language for processing XML.
As is common in Scala, you can write them directly in Scala code instead of
needing to invoke an external tool.

**Extracting text.** By calling the text method on any XML node you retrieve all of the text within that node, minus any element tags:

```
scala> <a>Sounds <tag/> good</a>.text
res8: String = Sounds  good
```

Any encoded characters are decoded automatically:

```
scala> <a> input ---&gt; output </a>.text
res9: String = " input ---> output "
```

**Extracting sub-elements.** If you want to find a sub-element by tag name,
simply call \ with the name of the tag:

```
scala> <a><b><c>hello</c></b></a> \ "b"
res10: scala.xml.NodeSeq = NodeSeq(<b><c>hello</c></b>)
```

625

You can do a "deep search" and look through sub-sub-elements, *etc.*, by using \\ instead of the \ operator:

```
scala>   <a><b><c>hello</c></b></a> \ "c"
res11: scala.xml.NodeSeq = NodeSeq()

scala>   <a><b><c>hello</c></b></a> \\ "c"
res12: scala.xml.NodeSeq = NodeSeq(<c>hello</c>)

scala>   <a><b><c>hello</c></b></a> \ "a"
res13: scala.xml.NodeSeq = NodeSeq()

scala>   <a><b><c>hello</c></b></a> \\ "a"
res14: scala.xml.NodeSeq =
NodeSeq(<a><b><c>hello</c></b></a>)
```

> **Note**
> Scala uses \ and \\ instead of XPath's / and //. The reason is that //
> starts a comment in Scala! Thus, some other symbol has to be used, and
> using the other kind of slashes works well.

**Extracting attributes.**   You can extract tag attributes using the same \ and \\ methods. Simply put an at sign (@) before the attribute name:

```
scala> val joe = <employee
             name="Joe"
             rank="code monkey"
             serial="123"/>
joe: scala.xml.Elem = <employee name="Joe" rank="code monkey"
 serial="123"/>

scala>  joe \ "@name"
res15: scala.xml.NodeSeq = Joe

scala>  joe \ "@serial"
res16: scala.xml.NodeSeq = 123
```

## 28.6   Deserialization

Using the previous methods for taking XML apart, you can now write the dual of a serializer, a parser from XML back into your internal data structures. For example, you can parse back a CCTherm instance by using the following code:

```scala
def fromXML(node: scala.xml.Node): CCTherm =
  new CCTherm {
    val description   = (node \ "description").text
    val yearMade      = (node \ "yearMade").text.toInt
    val dateObtained  = (node \ "dateObtained").text
    val bookPrice     = (node \ "bookPrice").text.toInt
    val purchasePrice = (node \ "purchasePrice").text.toInt
    val condition     = (node \ "condition").text.toInt
  }
```

This code searches through an input XML node, named node, to find each of the six pieces of data needed to specify a CCTherm. The data that is text is extracted with .text and left as is. Here is this method in action:

```scala
scala> val node = therm.toXML
node: scala.xml.Elem =
<cctherm>
                <description>hot dog #5</description>
                <yearMade>1952</yearMade>
                <dateObtained>March 14, 2006</dateObtained>
                <bookPrice>2199</bookPrice>
                <purchasePrice>500</purchasePrice>
                <condition>9</condition>
        </cctherm>

scala> fromXML(node)
res17: CCTherm = hot dog #5
```

## 28.7  Loading and saving

There is one last part needed to write a data serializer: conversion between XML and streams of bytes. This last part is the easiest, because there are library routines that will do it all for you. You simply have to call the right routine on the right data.

To convert XML to a string, all you need is toString. The presence of a workable toString is why you can experiment with XML in the Scala shell.

627

However, it is better to use a library routine and convert all the way to bytes. That way, the resulting XML can include a directive that specifies which character encoding was used. If you encode the string to bytes yourself, then the onus is on you to keep track of the character encoding.

To convert from XML to a file of bytes, you can use the XML.save command. You must specify a file name and a node to be saved:

```
scala.xml.XML.save("therm1.xml", node)
```

After running the above command, the resulting file therm1.xml looks like the following:

```
<?xml version='1.0' encoding='UTF-8'?>
<cctherm>
             <description>hot dog #5</description>
             <yearMade>1952</yearMade>
             <dateObtained>March 14, 2006</dateObtained>
             <bookPrice>2199</bookPrice>
             <purchasePrice>500</purchasePrice>
             <condition>9</condition>
          </cctherm>
```

Loading is simpler than saving, because the file includes everything the loader needs to know. Simply call XML.loadFile on a file name:

```
scala> val loadnode = xml.XML.loadFile("therm1.xml")
loadnode: scala.xml.Elem =
<cctherm>
        <description>hot dog #5</description>
        <yearMade>1952</yearMade>
        <dateObtained>March 14, 2006</dateObtained>
        <bookPrice>2199</bookPrice>
        <purchasePrice>500</purchasePrice>
        <condition>9</condition>
     </cctherm>

scala> fromXML(loadnode)
res14: CCTherm = hot dog #5
```

628

Those are the basic methods you need. There are many variations on these loading and saving methods, including methods for reading and writing to various kinds of readers, writers, input and output streams.

## 28.8 Pattern matching on XML

So far you have seen how to dissect XML using `text` and the XPath-like methods, \ and \\. These are good when you know exactly what kind of XML structure you are taking apart. Sometimes, though, there are a few possible structures the XML could have. Maybe there are multiple kinds of records within the data, for example because you have extended your thermometer collection to include clocks and sandwich plates. Maybe you simply want to skip over any white space between tags. Whatever the reason, you can use the pattern matcher to sift through the possibilities.

An XML pattern looks just like an XML literal. The main difference is that if you insert a {} escape, then the code inside the {} is not an expression but a pattern. A pattern embedded in {} can use the full Scala pattern language, including binding new variables, performing type tests, and ignoring content using the _ and _* patterns. Here is a simple example:

```scala
def proc(node: scala.xml.Node): String =
  node match {
    case <a>{contents}</a> => "It's an a: " + contents
    case <b>{contents}</b> => "It's a b: " + contents
    case _ => "It's something else."
  }
```

This function has a pattern match with three cases. The first case looks for an <a> element whose contents consist of a single sub-node. It binds those contents to a variable named `contents` and then evaluates the code to the right of the associated right arrow (=>). The second case does the same thing but looks for a <b> instead of an <a>, and the third case matches anything not matched by any other case. Here is the function in use:

```scala
scala> proc(<a>apple</a>)
res18: String = It's an a: apple
scala> proc(<b>banana</b>)
res19: String = It's a b: banana
```

629

```
scala> proc(<c>cherry</c>)
res20: String = It's something else.
```

Most likely this function is not exactly what you want, because it looks precisely for contents consisting of a single sub-node within the <a> or <b>. Thus it will fail to match in cases like the following:

```
scala> proc(<a>a <em>red</em> apple</a>)
res21: String = It's something else.
scala> proc(<a/>)
res22: String = It's something else.
```

If you want the function to match in cases like these, you can match against a sequence of nodes instead of a single one. The pattern for "any sequence" of XML nodes is written '_*'. Visually, this sequence looks like the wildcard pattern (_) followed by a regex-style Kleene star (*). Here is the updated function that matches a sequence of sub-elements instead of a single sub-element:

```
def proc(node: scala.xml.Node): String =
  node match {
    case <a>{contents @ _*}</a> => "It's an a: " + contents
    case <b>{contents @ _*}</b> => "It's a b: " + contents
    case _ => "It's something else."
  }
```

Notice that the result of the _* is bound to the contents variable by using the @ pattern described in Section 15.2. Here is the new version in action:

```
scala> proc(<a>a <em>red</em> apple</a>)
res23: String = It's an a: ArrayBuffer(a , <em>red</em>,
apple)
scala> proc(<a/>)
res24: String = It's an a: WrappedArray()
```

As a final tip, be aware that XML patterns work very nicely with for expressions as a way to iterate through some parts of an XML tree while ignoring other parts. For example, suppose you wish to skip over the white space between records in the following XML structure:

```
val catalog =
  <catalog>
    <cctherm>
      <description>hot dog #5</description>
      <yearMade>1952</yearMade>
      <dateObtained>March 14, 2006</dateObtained>
      <bookPrice>2199</bookPrice>
      <purchasePrice>500</purchasePrice>
      <condition>9</condition>
    </cctherm>
    <cctherm>
      <description>Sprite Boy</description>
      <yearMade>1964</yearMade>
      <dateObtained>April 28, 2003</dateObtained>
      <bookPrice>1695</bookPrice>
      <purchasePrice>595</purchasePrice>
      <condition>5</condition>
    </cctherm>
  </catalog>
```

Visually, it looks like there are two nodes inside the `<catalog>` element. Actually, though, there are five. There is white space before, after, and between the two elements! If you do not consider this white space, you might incorrectly process the thermometer records as follows:

```
catalog match {
  case <catalog>{therms @ _*}</catalog> =>
    for (therm <- therms)
      println("processing: " +
              (therm \ "description").text)
}
```

```
processing:
processing: hot dog #5
processing:
processing: Sprite Boy
processing:
```

631

Notice all of the lines that try to process white space as if it were a true thermometer record. What you would really like to do is ignore the white space and process only those sub-nodes that are inside a <cctherm> element. You can describe this subset using the pattern <cctherm>{_*}</cctherm>, and you can restrict the for expression to iterating over items that match that pattern:

```
catalog match {
  case <catalog>{therms @ _*}</catalog> =>
    for (therm @ <cctherm>{_*}</cctherm>  <-  therms)
      println("processing: " +
              (therm \ "description").text)
}
```

```
processing: hot dog #5
processing: Sprite Boy
```

## 28.9 Conclusion

This chapter has only scratched the surface of what you can do with XML. There are many other extensions, libraries, and tools you could learn about, some customized for Scala, some made for Java but usable in Scala, and some language-neutral. What you should walk away from this chapter with is how to use semi-structured data for interchange, and how to access semi-structured data via Scala's XML support.

Chapter 29

# Modular Programming Using Objects

In Chapter 1, we claimed that one way Scala is a scalable language is that you can use the same techniques to construct small as well as large programs. So far in this book we've focused primarily on *programming in the small*: designing and implementing the smaller program pieces out of which you can construct a larger program.[1] The other side of the story is *programming in the large*: organizing and assembling the smaller pieces into larger programs, applications, or systems. We touched on this subject when we discussed packages and access modifiers in Chapter 13. In short, packages and access modifiers enable you to organize a large program using packages as *modules*, where a module is a "smaller program piece" with a well defined interface and a hidden implementation.

While the division of programs into packages is already quite helpful, it is limited because it provides no way to abstract. You cannot reconfigure a package two different ways within the same program, and you cannot inherit between packages. A package always includes one precise list of contents, and that list is fixed until you change the code.

In this chapter, we'll discuss how you can use Scala's object-oriented features to make a program more modular. We'll first show how a simple singleton object can be used as a module. Then we'll explain how you can use traits and classes as abstractions over modules. These abstractions can be reconfigured into multiple modules, even multiple times within the same program. Finally, we'll show a pragmatic technique for using traits to divide a module across multiple files.

---

[1]This terminology was introduced in DeRemer, *et al.*, "Programming-in-the-large versus programming-in-the-small." [DeR75]

## 29.1 The problem

As a program grows in size, it becomes increasingly important to organize it in a modular way. First, being able to compile different modules that make up the system separately helps different teams work independently. In addition, being able to unplug one implementation of a module and plug in another is useful, because it allows different configurations of a system to be used in different contexts, such as unit testing on a developer's desktop, integration testing, staging, and deployment.

For example, you may have an application that uses a database and a message service. As you write code, you may want to run unit tests on your desktop that use mock versions of both the database and message service, which simulate these services sufficiently for testing without needing to talk across the network to a shared resource. During integration testing, you may want to use a mock message service but a live developer database. During staging and certainly during deployment, your organization will likely want to use live versions of both the database and message service.

Any technique that aims to facilitate this kind of modularity needs to provide a few essentials. First, there should be a module construct that provides a good separation of interface and implementation. Second, there should be a way to replace one module with another that has the same interface without changing or recompiling the modules that depend on the replaced one. Lastly, there should be a way to wire modules together. This wiring task can by thought of as *configuring* the system.

One approach to solving this problem is *dependency injection*, a technique supported on the Java platform by frameworks such as Spring and Guice, which are popular in the enterprise Java community.[2] Spring, for example, essentially allows you to represent the interface of a module as a Java interface and implementations of the module as Java classes. You can specify dependencies between modules and "wire" an application together via external XML configuration files. Although you can use Spring with Scala and thereby use Spring's approach to achieving system-level modularity of your Scala programs, with Scala you have some alternatives enabled by the language itself. In the remainder of this chapter, we'll show how to use objects as modules to achieve the desired "in the large" modularity without using an external framework.

---

[2]Fowler, "Inversion of control containers and the dependency injection pattern." [Fow04]

## 29.2 A recipe application

Imagine you are building an enterprise web application that will allow users to manage recipes. You want to partition the software into layers, including a *domain layer* and an *application layer*. In the domain layer, you'll define *domain objects*, which will capture business concepts and rules, as well as encapsulate state that will be persisted to an external relational database. In the application layer, you'll provide an API organized in terms of the services the application offers to clients (including the user interface layer). The application layer will implement these services by coordinating tasks and delegating the work to the objects of the domain layer.[3]

You want to be able to plug in real or mock versions of certain objects in each of these layers, so that you can more easily write unit tests for your application. To achieve this goal, you can treat the objects you want to mock as modules. In Scala, there is no need for objects to be "small" things, no need to use some other kind of construct for "big" things like modules. One of the ways Scala is a scalable language is that the same constructs are used for structures both small and large.

For example, since one of the "things" you want to mock in the domain layer is the object that represents the relational database, you'll make that one of the modules. In the application layer, you'll treat a "database browser" object as a module. The database will hold all of the recipes that a person has collected. The browser will help search and browse that database, for example, to find every recipe that includes an ingredient you have on hand.

The first thing to do is to model foods and recipes. To keep things simple, a food will just have a name, as shown in Listing 29.1. A recipe will have a name, a list of ingredients, and some instructions, as shown in Listing 29.2.

```
package org.stairwaybook.recipe

abstract class Food(val name: String) {
  override def toString = name
}
```

Listing 29.1 · A simple Food entity class.

---

[3]The naming of these layers follows that of Evans, *Domain-Driven Design*. [Eva03]

```
package org.stairwaybook.recipe

class Recipe(
  val name: String,
  val ingredients: List[Food],
  val instructions: String
) {
  override def toString = name
}
```

Listing 29.2 · Simple Recipe entity class.

The Food and Recipe classes shown in Listings 29.1 and 29.2 represent *entities* that will be persisted in the database.[4] Listing 29.3 shows some singleton instances of these classes, which can be used when writing tests.

```
package org.stairwaybook.recipe

object Apple extends Food("Apple")
object Orange extends Food("Orange")
object Cream extends Food("Cream")
object Sugar extends Food("Sugar")

object FruitSalad extends Recipe(
  "fruit salad",
  List(Apple, Orange, Cream, Sugar),
  "Stir it all together."
)
```

Listing 29.3 · Food and Recipe examples for use in tests.

Scala uses objects for modules, so you can start modularizing your program by making two singleton objects to serve as mock implementations of the database and browser modules during testing. Because it is a mock,

---

[4]These entity classes are simplified to keep the example uncluttered with too much real-world detail. But transforming these classes into entities that could be persisted with Hibernate or the Java Persistence Architecture, , for example, would require only a few modifications, such as adding a private Long id field and a no-arg constructor, placing scala.reflect.BeanProperty annotations on the fields, specifying appropriate mappings via annotations or a separate XML file, and so on.

```scala
package org.stairwaybook.recipe

object SimpleDatabase {
  def allFoods = List(Apple, Orange, Cream, Sugar)

  def foodNamed(name: String): Option[Food] =
    allFoods.find(_.name == name)

  def allRecipes: List[Recipe] = List(FruitSalad)
}

object SimpleBrowser {
  def recipesUsing(food: Food) =
    SimpleDatabase.allRecipes.filter(recipe =>
      recipe.ingredients.contains(food))
}
```

Listing 29.4 · Mock database and browser modules.

the database module is backed by a simple in-memory list. Implementations of these objects are shown in Listing 29.4. You can use this database and browser as follows:

```scala
scala> val apple = SimpleDatabase.foodNamed("Apple").get
apple: Food = Apple

scala> SimpleBrowser.recipesUsing(apple)
res0: List[Recipe] = List(fruit salad)
```

To make things a little more interesting, suppose the database sorts foods into categories. To implement this, you can add a FoodCategory class and a list of all categories in the database, as shown in Listing 29.5. Notice in this example that the private keyword, so useful for implementing classes, is also useful for implementing modules. Items marked private are part of the implementation of a module, and thus are particularly easy to change without affecting other modules.

At this point, many more facilities could be added, but you get the idea. Programs can be divided into singleton objects, which you can think of as modules. This is not big news, but it becomes very useful when you consider abstraction (which we'll cover next).

```scala
package org.stairwaybook.recipe

object SimpleDatabase {
  def allFoods = List(Apple, Orange, Cream, Sugar)

  def foodNamed(name: String): Option[Food] =
    allFoods.find(_.name == name)

  def allRecipes: List[Recipe] = List(FruitSalad)

  case class FoodCategory(name: String, foods: List[Food])

  private var categories = List(
    FoodCategory("fruits", List(Apple, Orange)),
    FoodCategory("misc", List(Cream, Sugar)))

  def allCategories = categories
}

object SimpleBrowser {
  def recipesUsing(food: Food) =
    SimpleDatabase.allRecipes.filter(recipe =>
      recipe.ingredients.contains(food))

  def displayCategory(category: SimpleDatabase.FoodCategory) = {
    println(category)
  }
}
```

Listing 29.5 · Database and browser modules with categories added.

## 29.3 Abstraction

Although the examples shown so far did manage to partition your application into separate database and browser modules, the design is not yet very "modular." The problem is that there is essentially a "hard link" from the browser module to the database modules:

```scala
SimpleDatabase.allRecipes.filter(recipe => ...
```

Because the SimpleBrowser module mentions the SimpleDatabase module by name, you won't be able to plug in a different implementation of the database module without modifying and recompiling the browser module. In addition, although there's no hard link from the SimpleDatabase module

```
abstract class Browser {
  val database: Database

  def recipesUsing(food: Food) =
    database.allRecipes.filter(recipe =>
      recipe.ingredients.contains(food))

  def displayCategory(category: database.FoodCategory) = {
    println(category)
  }
}
```

Listing 29.6 · A Browser class with an abstract database val.

to the SimpleBrowser module,[5] there's no clear way to enable the user interface layer, for example, to be configured to use different implementations of the browser module.

When making these modules more pluggable, however, it is important to avoid duplicating code, because much code can likely be shared by different implementations of the same module. For example, suppose you want the same code base to support multiple recipe databases, and you want to be able to create a separate browser for each of these databases. You would like to reuse the browser code for each of the instances, because the only thing different about the browsers is which database they refer to. Except for the database implementation, the rest of the code can be reused character for character. How can the program be arranged to minimize repetitive code? How can the code be made reconfigurable, so that you can configure it using either database implementation?

The answer is a familiar one: If a module is an object, then a template for a module is a class. Just like a class describes the common parts of all its instances, a class can describe the parts of a module that are common to all of its possible configurations.

The browser definition therefore becomes a class, instead of an object, and the database to use is specified as an abstract member of the class, as shown in Listing 29.6. The database also becomes a class, including as much as possible that is common between all databases, and declaring the missing

---

[5]This is good, because each of these architectural layers should depend only on layers below them.

639

parts that a database must define. In this case, all database modules must define methods for allFoods, allRecipes, and allCategories, but since they can use an arbitrary definition, the methods must be left abstract in the Database class. The foodNamed method, by contrast, can be defined in the abstract Database class, as shown in Listing 29.7.

```scala
abstract class Database {
  def allFoods: List[Food]
  def allRecipes: List[Recipe]

  def foodNamed(name: String) =
    allFoods.find(f => f.name == name)

  case class FoodCategory(name: String, foods: List[Food])
  def allCategories: List[FoodCategory]
}
```

Listing 29.7 · A Database class with abstract methods.

The SimpleDatabase object must be updated to inherit from the abstract Database class, as shown in Listing 29.8.

```scala
object SimpleDatabase extends Database {
  def allFoods = List(Apple, Orange, Cream, Sugar)

  def allRecipes: List[Recipe] = List(FruitSalad)

  private var categories = List(
    FoodCategory("fruits", List(Apple, Orange)),
    FoodCategory("misc", List(Cream, Sugar)))

  def allCategories = categories
}
```

Listing 29.8 · The SimpleDatabase object as a Database subclass.

Then, a specific browser module is made by instantiating the Browser class and specifying which database to use, as shown in Listing 29.9.

You can use these more pluggable modules the same as before:

```scala
scala> val apple = SimpleDatabase.foodNamed("Apple").get
apple: Food = Apple
```

640

```
object SimpleBrowser extends Browser {
  val database = SimpleDatabase
}
```

Listing 29.9 · The SimpleBrowser object as a Browser subclass.

```
scala> SimpleBrowser.recipesUsing(apple)
res1: List[Recipe] = List(fruit salad)
```

Now, however, you can create a second mock database, and use the same browser class with it, as shown in Listing 29.10:

```
object StudentDatabase extends Database {
  object FrozenFood extends Food("FrozenFood")

  object HeatItUp extends Recipe(
    "heat it up",
    List(FrozenFood),
    "Microwave the 'food' for 10 minutes.")

  def allFoods = List(FrozenFood)
  def allRecipes = List(HeatItUp)
  def allCategories = List(
    FoodCategory("edible", List(FrozenFood)))
}
object StudentBrowser extends Browser {
  val database = StudentDatabase
}
```

Listing 29.10 · A student database and browser.

## 29.4   Splitting modules into traits

Often a module is too large to fit comfortably into a single file. When that happens, you can use traits to split a module into separate files. For example, suppose you wanted to move categorization code out of the main Database file and into its own. You could create a trait for the code as shown in (Listing 29.11).

641

```
trait FoodCategories {
  case class FoodCategory(name: String, foods: List[Food])
  def allCategories: List[FoodCategory]
}
```

Listing 29.11 · A trait for food categories.

Now class Database can mix in the FoodCategories trait instead of defining FoodCategory and allCategories itself, as shown in Listing 29.12:

```
abstract class Database extends FoodCategories {
  def allFoods: List[Food]
  def allRecipes: List[Recipe]
  def foodNamed(name: String) =
    allFoods.find(f => f.name == name)
}
```

Listing 29.12 · A Database class that mixes in the FoodCategories trait.

You might try and divide SimpleDatabase into two traits, one for foods and one for recipes. This would allow you to define SimpleDatabase as shown in Listing 29.13:

```
object SimpleDatabase extends Database
    with SimpleFoods with SimpleRecipes
```

Listing 29.13 · A SimpleDatabase object composed solely of mixins.

The SimpleFoods trait could look as shown in Listing 29.14:

```
trait SimpleFoods {
  object Pear extends Food("Pear")
  def allFoods = List(Apple, Pear)
  def allCategories = Nil
}
```

Listing 29.14 · A SimpleFoods trait.

So far so good, but unfortunately, a problem arises if you try to define a SimpleRecipes trait like this:

```
trait SimpleRecipes { // Does not compile
  object FruitSalad extends Recipe(
    "fruit salad",
    List(Apple, Pear),  // Uh oh
    "Mix it all together."
  )
  def allRecipes = List(FruitSalad)
}
```

The problem is that Pear is located in a different trait from the one that uses it, so it is out of scope. The compiler has no idea that SimpleRecipes is only ever mixed together with SimpleFoods.

However, there is a way you can tell this to the compiler. Scala provides the *self type* for precisely this situation. Technically, a self type is an assumed type for this whenever this is mentioned within the class. Pragmatically, a self type specifies the requirements on any concrete class the trait is mixed into. If you have a trait that is only ever used when mixed in with another trait or traits, then you can specify that those other traits should be assumed. In the present case, it is enough to specify a self type of SimpleFoods, as shown in Listing 29.15:

```
trait SimpleRecipes {
  this: SimpleFoods =>

  object FruitSalad extends Recipe(
    "fruit salad",
    List(Apple, Pear),   // Now Pear is in scope
    "Mix it all together."
  )
  def allRecipes = List(FruitSalad)
}
```

Listing 29.15 · A SimpleRecipes trait with a self type.

Given the new self type, Pear is now available. Implicitly, the reference to Pear is thought of as this.Pear. This is safe, because any *concrete* class that mixes in `SimpleRecipes` must also be a subtype of `SimpleFoods`, which means that `Pear` will be a member. Abstract subclasses and traits do not have to follow this restriction, but since they cannot be instantiated with new, there is no risk that the `this.Pear` reference will fail.

## 29.5  Runtime linking

Scala modules can be linked together at runtime, and you can decide which modules will link to which depending on runtime computations. For example, Listing 29.16 shows a small program that chooses a database at runtime and then prints out all the apple recipes in it:

```
object GotApples {
  def main(args: Array[String]) = {
    val db: Database =
      if(args(0) == "student")
        StudentDatabase
      else
        SimpleDatabase

    object browser extends Browser {
      val database = db
    }

    val apple = SimpleDatabase.foodNamed("Apple").get

    for(recipe <- browser.recipesUsing(apple))
      println(recipe)
  }
}
```

Listing 29.16 · An app that dynamically selects a module implementation.

Now, if you use the simple database, you will find a recipe for fruit salad. If you use the student database, you will find no recipes at all using apples:

```
$ scala GotApples simple
fruit salad
$ scala GotApples student
$
```

> ### Configuring with Scala code
>
> You may wonder if you are not backsliding to the hard links problem of the original examples in this chapter, because the GotApples object shown in Listing 29.16 contains hard links to both StudentDatabase and SimpleDatabase. The difference here is that the hard links are localized in one file that can be replaced.
>
> Every modular application needs some way to specify the actual module implementations to use in a particular situation. This act of "configuring" the application will by definition involve the naming of concrete module implementations. For example, in a Spring application, you configure by naming implementations in an external XML file. In Scala, you can configure via Scala code itself. One advantage to using Scala source over XML for configuration is that the process of running your configuration file through the Scala compiler should uncover any misspellings in it prior to its actual use.

## 29.6  Tracking module instances

Despite using the same code, the different browser and database modules created in the previous section really are separate modules. This means that each module has its own contents, including any nested classes. FoodCategory in SimpleDatabase, for example, is a different class from FoodCategory in StudentDatabase!

```
scala> val category = StudentDatabase.allCategories.head
category: StudentDatabase.FoodCategory =
FoodCategory(edible,List(FrozenFood))

scala> SimpleBrowser.displayCategory(category)
<console>:21: error: type mismatch;
 found    : StudentDatabase.FoodCategory
 required: SimpleBrowser.database.FoodCategory
              SimpleBrowser.displayCategory(category)
                                            ^
```

If instead you prefer all FoodCategorys to be the same, you can accomplish this by moving the definition of FoodCategory outside of any class or trait.

The choice is yours, but as it is written, each `Database` gets its own, unique `FoodCategory` class.

Since the two `FoodCategory` classes shown in this example really are different, the compiler is correct to complain. Sometimes, though, you may encounter a case where two types are the same but the compiler can't verify it. You will see the compiler complaining that two types are not the same, even though you as the programmer know they are.

In such cases you can often fix the problem using *singleton types*. For example, in the `GotApples` program, the type checker does not know that db and `browser.database` are the same. This will cause type errors if you try to pass categories between the two objects:

```scala
object GotApples {
  // same definitions...

  for (category <- db.allCategories)
    browser.displayCategory(category)

  // ...
}
```

```
GotApples2.scala:14: error: type mismatch;
 found    : db.FoodCategory
 required: browser.database.FoodCategory
         browser.displayCategory(category)
                    ^
one error found
```

To avoid this error, you need to inform the type checker that they are the same object. You can do this by changing the definition of `browser.database` as shown in Listing 29.17:

```scala
object browser extends Browser {
  val database: db.type = db
}
```

Listing 29.17 · Using a singleton type.

This definition is the same as before except that `database` has the funny-looking type `db.type`. The ".type" on the end means that this is a *singleton type*. A singleton type is extremely specific and holds only one object; in this case, whichever object is referred to by `db`. Usually such types are too specific to be useful, which is why the compiler is reluctant to insert them automatically. In this case, though, the singleton type allows the compiler to know that `db` and `browser.database` are the same object—enough information to eliminate the type error.

## 29.7 Conclusion

This chapter has shown how to use Scala's objects as modules. In addition to simple static modules, this approach gives you a variety of ways to create abstract, reconfigurable modules. There are actually even more abstraction techniques than shown, since anything that works on a class also works on a class used to implement a module. As always, how much of this power you use should be a matter of taste.

Modules are part of programming in the large, and thus are hard to experiment with. You need a large program before it really makes a difference. Nonetheless, after reading this chapter you know which Scala features to think about when you want to program in a modular style. Think about these techniques when you write your own large programs and recognize these coding patterns when you see them in other people's code.

Chapter 30

# Object Equality

Comparing two values for equality is ubiquitous in programming. It is also more tricky than it looks at first glance. This chapter looks at object equality in detail and gives some recommendations to consider when you design your own equality tests.

## 30.1 Equality in Scala

As mentioned in Section 11.2, the definition of equality is different in Scala and Java. Java has two equality comparisons: the == operator, which is the natural equality for value types and object identity for reference types, and the equals method, which is (user-defined) canonical equality for reference types. This convention is problematic because the more natural symbol, ==, does not always correspond to the natural notion of equality. When programming in Java, a common pitfall for beginners is to compare objects with == when they should be compared with equals. For instance, comparing two strings x and y using "x == y" might yield false in Java, even if x and y have exactly the same characters in the same order.

Scala also has an equality method signifying object identity, but it is not used much. That kind of equality, written "x eq y", is true if x and y reference the same object. The == equality is reserved in Scala for the "natural" equality of each type. For value types, == is value comparison, just like in Java. For reference types, == is the same as equals in Scala. You can redefine the behavior of == for new types by overriding the equals method, which is always inherited from class Any. The inherited equals, which takes effect unless overridden, is object identity, as is the case in Java. So equals

649

(and with it, ==) is by default the same as eq, but you can change its behavior by overriding the equals method in the classes you define. It is not possible to override == directly, as it is defined as a final method in class Any. That is, Scala treats == as if it were defined as follows in class Any:

```
final def == (that: Any): Boolean =
  if (null eq this) {null eq that} else {this equals that}
```

## 30.2   Writing an equality method

How should the equals method be defined? It turns out that writing a correct equality method is surprisingly difficult in object-oriented languages. In fact, after studying a large body of Java code, the authors of a 2007 paper concluded that almost all implementations of equals methods are faulty.[1] This is problematic because equality is at the basis of many other things. For one, a faulty equality method for a type C might mean that you cannot reliably put an object of type C in a collection.

For example, you might have two elements, elem1 and elem2, of type C which are equal (*i.e.*, "elem1 equals elem2" yields true). Nevertheless, with commonly occurring faulty implementations of the equals method, you could still see behavior like the following:

```
var hashSet: Set[C] = new collection.immutable.HashSet
hashSet += elem1
hashSet contains elem2    // returns false!
```

Here are four common pitfalls[2] that can cause inconsistent behavior when overriding equals:

1. Defining equals with the wrong signature.

2. Changing equals without also changing hashCode.

3. Defining equals in terms of mutable fields.

4. Failing to define equals as an equivalence relation.

These four pitfalls are discussed in the remainder of this section.

---

[1] Vaziri, *et al.*, "Declarative Object Identity Using Relation Types" [Vaz07]

[2] All but the third pitfall are described in the context of Java in the book, *Effective Java Second Edition*, by Joshua Bloch. [Blo08]

## Pitfall #1: Defining equals with the wrong signature

Consider adding an equality method to the following class of simple points:

```
class Point(val x: Int, val y: Int) { ... }
```

A seemingly obvious but wrong way would be to define it like this:

```
// An utterly wrong definition of equals
def equals(other: Point): Boolean =
  this.x == other.x && this.y == other.y
```

What's wrong with this method? At first glance, it seems to work OK:

```
scala> val p1, p2 = new Point(1, 2)
p1: Point = Point@37d7d90f
p2: Point = Point@3beb846d

scala> val q = new Point(2, 3)
q: Point = Point@e0cf182

scala> p1 equals p2
res0: Boolean = true

scala> p1 equals q
res1: Boolean = false
```

However, trouble starts once you start putting points into a collection:

```
scala> import scala.collection.mutable
import scala.collection.mutable

scala> val coll = mutable.HashSet(p1)
coll: scala.collection.mutable.HashSet[Point] =
Set(Point@37d7d90f)

scala> coll contains p2
res2: Boolean = false
```

How to explain that `coll` does not contain p2, even though p1 was added to it, and p1 and p2 are equal objects? The reason becomes clear in the following interaction, where the precise type of one of the compared points is masked. Define p2a as an alias of p2, but with type Any instead of Point:

651

```
scala> val p2a: Any = p2
p2a: Any = Point@3beb846d
```

Now, were you to repeat the first comparison, but with the alias p2a instead of p2, you would get:

```
scala> p1 equals p2a
res3: Boolean = false
```

What went wrong? The version of equals given previously does not override the standard method equals because its type is different. Here is the type of the equals method as it is defined in the root class Any:[3]

```
def equals(other: Any): Boolean
```

Because the equals method in Point takes a Point instead of an Any as an argument, it does not override equals in Any. Instead, it is just an overloaded alternative. Now, overloading in Scala and in Java is resolved by the static type of the argument, not the run-time type. So as long as the static type of the argument is Point, the equals method in Point is called. However, once the static argument is of type Any, the equals method in Any is called instead. This method has not been overridden, so it is still implemented by comparing object identity.

That's why the comparison "p1 equals p2a" yields false even though points p1 and p2a have the same x and y values. That's also why the contains method in HashSet returned false. Since that method operates on generic sets, it calls the generic equals method in Object instead of the overloaded variant in Point. Here's a better equals method:

```
// A better definition, but still not perfect
override def equals(other: Any) = other match {
  case that: Point => this.x == that.x && this.y == that.y
  case _ => false
}
```

Now equals has the correct type. It takes a value of type Any as parameter and it yields a Boolean result. The implementation of this method uses a

---

[3]If you write a lot of Java, you might expect the argument to this method to be type Object instead of type Any. Don't worry about it; it is the same equals method. The compiler simply makes it appear to have type Any.

pattern match. It first tests whether the other object is also of type Point. If it is, it compares the coordinates of the two points and returns the result. Otherwise the result is false.

A related pitfall is to define == with a wrong signature. Normally, if you try to redefine == with the correct signature, which takes an argument of type Any, the compiler will give you an error because you try to override a final method of type Any.

Newcomers to Scala sometimes make two errors at once: They try to override == *and* they give it the wrong signature. For instance:

```
def ==(other: Point): Boolean = // Don't do this!
```

In this case, the user-defined == method is treated as an overloaded variant of the same-named method class Any and the program compiles. However, the behavior of the program would be just as dubious as if you had defined equals with the wrong signature.

### Pitfall #2: Changing equals without also changing hashCode

We'll continue to use the example from pitfall #1. If you repeat the comparison of p1 and p2a with the latest definition of Point, you will get true, as expected. However, if you repeat the HashSet.contains test, you will probably still get false:

```
scala> val p1, p2 = new Point(1, 2)
p1: Point = Point@122c1533
p2: Point = Point@c23d097

scala> collection.mutable.HashSet(p1) contains p2
res4: Boolean = false
```

But this outcome is not 100% certain. You might also get true from the experiment. If you do, you can try with some other points with coordinates 1 and 2. Eventually, you'll get one that is not contained in the set. What goes wrong here is that Point redefined equals without also redefining hashCode.

Note that the collection in the example here is a HashSet. This means elements of the collection are put in "hash buckets" determined by their hash code. The contains test first determines a hash bucket to look in and then compares the given elements with all elements in that bucket. Now, the last version of class Point did redefine equals, but it did not redefine hashCode at the same time. So hashCode is still what it was in its version in class AnyRef: some transformation of the address of the allocated object.

The hash codes of p1 and p2 are almost certainly different, even though the fields of both points are the same. Different hash codes mean, with high probability, different hash buckets in the set. The contains test will look for a matching element in the bucket which corresponds to p2's hash code. In most cases, point p1 will be in another bucket, so it will never be found. p1 and p2 might also end up by chance in the same hash bucket. In that case the test would return true. The problem is that the last implementation of Point violated the contract on hashCode as defined for class Any:[4]

> *If two objects are equal according to the equals method, then*
> *calling the hashCode method on each of the two objects must*
> *produce the same integer result.*

In fact, it's well known in Java that hashCode and equals should always be redefined together. Furthermore, hashCode may only depend on fields that equals depends on. For the Point class, the following would be a suitable definition of hashCode:

```
class Point(val x: Int, val y: Int) {
  override def hashCode = (x, y).##
  override def equals(other: Any) = other match {
    case that: Point => this.x == that.x && this.y == that.y
    case _ => false
  }
}
```

This is just one of many possible implementations of hashCode. Recall that the ## method is a shorthand for computing hash codes that works for primitive values, reference types, and null. When invoked on a collection or a tuple, it computes a mixed hash that is sensitive to the hash codes of all the elements in the collection. We'll provide more guidance on writing hashCode later in this chapter.

Adding hashCode fixes the problems of equality when defining classes like Point; however, there are other trouble spots to watch out for.

---

[4]The text of Any's hashCode contract is inspired by the Javadoc documentation of class java.lang.Object.

## Pitfall #3: Defining equals in terms of mutable fields

Consider the following slight variation of class `Point`:

```
class Point(var x: Int, var y: Int) { // Problematic
  override def hashCode = (x, y).##
  override def equals(other: Any) = other match {
    case that: Point => this.x == that.x && this.y == that.y
    case _ => false
  }
}
```

The only difference is that the fields x and y are now `vars` instead of `vals`. The equals and hashCode methods are now defined in terms of these mutable fields, so their results change when the fields change. This can have strange effects once you put points in collections:

```
scala> val p = new Point(1, 2)
p: Point = Point@5428bd62

scala> val coll = collection.mutable.HashSet(p)
coll: scala.collection.mutable.HashSet[Point] =
Set(Point@5428bd62)

scala> coll contains p
res5: Boolean = true
```

Now, if you change a field in point p, does the collection still contain the point? We'll try it:

```
scala> p.x += 1

scala> coll contains p
res7: Boolean = false
```

This looks strange. Where did p go? More strangeness results if you check whether the iterator of the set contains p:

```
scala> coll.iterator contains p
res8: Boolean = true
```

So here's a set that does not contain p, yet p is among the elements of the set! What happened is that after the change to the x field, the point p ended

655

up in the wrong hash bucket of the set `coll`. That is, its original hash bucket no longer corresponded to the new value of its hash code. In a manner of speaking, point p "dropped out of sight" in the set `coll` even though it still belonged to its elements.

The lesson to be drawn from this example is that when `equals` and `hashCode` depend on mutable state, it causes problems for potential users. If you put such objects into collections, you have to be careful to never modify the depended-on state, and this is tricky. If you need a comparison that takes the current state of an object into account, you should usually name it something else, not `equals`.

Considering the last definition of `Point`, it would have been preferable to omit a redefinition of `hashCode` and name the comparison method `equalContents` or some other name different from `equals`. `Point` would then have inherited the default implementation of `equals` and `hashCode`; p would have stayed locatable in `coll` even after the modification to its x field.

## Pitfall #4: Failing to define `equals` as an equivalence relation

The contract of the `equals` method in `scala.Any` specifies that `equals` must implement an equivalence relation on non-null objects:[5]

- *It is* reflexive*: For any non-null value* x*, the expression* x.equals(x) *should return* true*.*

- *It is* symmetric*: For any non-null values* x *and* y*,* x.equals(y) *should return* true *if and only if* y.equals(x) *returns* true*.*

- *It is* transitive*: For any non-null values* x*,* y*, and* z*, if* x.equals(y) *returns* true *and* y.equals(z) *returns* true*, then* x.equals(z) *should return* true*.*

- *It is* consistent*: For any non-null values* x *and* y*, multiple invocations of* x.equals(y) *should consistently return* true *or consistently return* false*, provided no information used in equals comparisons on the objects is modified.*

- *For any non-null value* x*,* x.equals(null) *should return* false*.*

---

[5]As with `hashCode`, `Any`'s equals method contract is based on `java.lang.Object`'s equals method contract.

The definition of `equals` developed for class `Point` up to now satisfies the contract for `equals`. However, things become more complicated once subclasses are considered. Say there is a subclass `ColoredPoint` of `Point` that adds a field `color` of type `Color`. Assume `Color` is defined as an enumeration, as presented in Section 20.9:

```
object Color extends Enumeration {
  val Red, Orange, Yellow, Green, Blue, Indigo, Violet = Value
}
```

`ColoredPoint` overrides `equals` to take the new `color` field into account:

```
class ColoredPoint(x: Int, y: Int, val color: Color.Value)
    extends Point(x, y) { // Problem: equals not symmetric

  override def equals(other: Any) = other match {
    case that: ColoredPoint =>
      this.color == that.color && super.equals(that)
    case _ => false
  }
}
```

This is what many programmers would likely write. Note that in this case, class `ColoredPoint` need not override `hashCode`. Because the new definition of `equals` on `ColoredPoint` is stricter than the overridden definition in `Point` (meaning it equates fewer pairs of objects), the contract for `hashCode` stays valid. If two colored points are equal, they must have the same coordinates, so their hash codes are guaranteed to be equal as well.

Taking the class `ColoredPoint` by itself, its definition of `equals` looks OK. However, the contract for `equals` is broken once points and colored points are mixed. Consider:

```
scala> val p = new Point(1, 2)
p: Point = Point@5428bd62

scala> val cp = new ColoredPoint(1, 2, Color.Red)
cp: ColoredPoint = ColoredPoint@5428bd62

scala> p equals cp
res9: Boolean = true

scala> cp equals p
res10: Boolean = false
```

The comparison "p equals cp" invokes p's equals method, which is defined in class Point. This method only takes into account the coordinates of the two points. Consequently, the comparison yields true. On the other hand, the comparison "cp equals p" invokes cp's equals method, which is defined in class ColoredPoint. This method returns false because p is not a ColoredPoint. So the relation defined by equals is not symmetric.

The loss in symmetry can have unexpected consequences for collections. Here's an example:

```
scala> collection.mutable.HashSet[Point](p) contains cp
res11: Boolean = true

scala> collection.mutable.HashSet[Point](cp) contains p
res12: Boolean = false
```

Even though p and cp are equal, one contains test succeeds whereas the other one fails:

How can you change the definition of equals so that it becomes symmetric? Essentially there are two ways. You can either make the relation more general or more strict. Making it more general means that a pair of two objects, x and y, is taken to be equal if either comparing x with y or comparing y with x yields true. Here's code that does this:

```
class ColoredPoint(x: Int, y: Int, val color: Color.Value)
    extends Point(x, y) { // Problem: equals not transitive

  override def equals(other: Any) = other match {
    case that: ColoredPoint =>
      (this.color == that.color) && super.equals(that)
    case that: Point =>
      that equals this
    case _ =>
      false
  }
}
```

The new definition of equals in ColoredPoint has one more case than the old one: If the other object is a Point but not a ColoredPoint, the method forwards to the equals method of Point. This has the desired effect of making equals symmetric. Now, both "cp equals p" and "p equals cp" result in true. However, the contract for equals is still broken. The problem is that the new relation is no longer transitive!

Here's a sequence of statements that demonstrates this. Define a point and two colored points of different colors, all at the same position:

```
scala> val redp = new ColoredPoint(1, 2, Color.Red)
redp: ColoredPoint = ColoredPoint@5428bd62

scala> val bluep = new ColoredPoint(1, 2, Color.Blue)
bluep: ColoredPoint = ColoredPoint@5428bd62
```

Taken individually, redp is equal to p and p is equal to bluep:

```
scala> redp == p
res13: Boolean = true

scala> p == bluep
res14: Boolean = true
```

However, comparing redp and bluep yields false:

```
scala> redp == bluep
res15: Boolean = false
```

Hence, the transitivity clause of the equals's contract is violated.

Making the equals relation more general seems to lead to a dead end. We'll try to make it stricter instead. One way to make equals stricter is to always treat objects of different classes as different. This could be achieved by modifying the equals methods in classes Point and ColoredPoint. In class Point, you could add an extra comparison that checks whether the run-time class of the other Point is exactly the same as this Point's class:

```
// A technically valid, but unsatisfying, equals method
class Point(val x: Int, val y: Int) {
  override def hashCode = (x, y).##
  override def equals(other: Any) = other match {
    case that: Point =>
      this.x == that.x && this.y == that.y &&
      this.getClass == that.getClass
    case _ => false
  }
}
```

659

You can then revert class `ColoredPoint`'s implementation back to the version that previously had violated the symmetry requirement:[6]

```
class ColoredPoint(x: Int, y: Int, val color: Color.Value)
    extends Point(x, y) {

  override def equals(other: Any) = other match {
    case that: ColoredPoint =>
      (this.color == that.color) && super.equals(that)
    case _ => false
  }
}
```

Here, an instance of class `Point` is considered to be equal to some other instance of the same class, only if the objects have the same coordinates and they have the same run-time class, meaning `getClass` on either object returns the same value. The new definitions satisfy symmetry and transitivity because now every comparison between objects of different classes yields `false`. So a colored point can never be equal to a point. This convention looks reasonable, but one could argue that the new definition is too strict.

Consider the following slightly roundabout way to define a point at coordinates (1, 2):

```
scala> val pAnon = new Point(1, 1) { override val y = 2 }
pAnon: Point = $anon$1@5428bd62
```

Is pAnon equal to p? The answer is no because the `java.lang.Class` objects associated with p and pAnon are different. For p it is `Point`, whereas for pAnon it is an anonymous subclass of `Point`. But clearly, pAnon is just another point at coordinates (1, 2). It does not seem reasonable to treat it as being different from p.

So it seems we are stuck. Is there a sane way to redefine equality on several levels of the class hierarchy while keeping its contract? In fact, there is such a way, but it requires one more method to redefine together with equals and hashCode. The idea is that as soon as a class redefines equals (and hashCode), it should also explicitly state that objects of this class are never equal to objects of some superclass that implement a different equality

---

[6]Given the new implementation of equals in Point, this version of ColoredPoint no longer violates the symmetry requirement.

method. This is achieved by adding a method `canEqual` to every class that redefines equals. Here's the method's signature:

```
def canEqual(other: Any): Boolean
```

The method should return `true` if the `other` object is an instance of the class in which `canEqual` is (re)defined, `false` otherwise. It is called from `equals` to make sure that the objects are comparable both ways. Listing 30.1 shows a new (and final) implementation of class `Point` along these lines:

```scala
class Point(val x: Int, val y: Int) {
  override def hashCode = (x, y).##
  override def equals(other: Any) = other match {
    case that: Point =>
      (that canEqual this) &&
      (this.x == that.x) && (this.y == that.y)
    case _ =>
      false
  }
  def canEqual(other: Any) = other.isInstanceOf[Point]
}
```

Listing 30.1 · A superclass `equals` method that calls `canEqual`.

The `equals` method in this version of class `Point` contains the additional requirement that the other object *can equal* this one, as determined by the `canEqual` method. The implementation of `canEqual` in `Point` states that all instances of `Point` can be equal.

Listing 30.2 shows the corresponding implementation of `ColoredPoint`. It can be shown that the new definition of `Point` and `ColoredPoint` keeps the contract of `equals`. Equality is symmetric and transitive. Comparing a `Point` to a `ColoredPoint` always yields `false`. Indeed, for any point p and colored point cp, "p equals cp" will return `false` because "cp canEqual p" will return `false`. The reverse comparison, "cp equals p", will also return `false` because p is not a `ColoredPoint`, so the first pattern match in the body of equals in `ColoredPoint` will fail.

On the other hand, instances of different subclasses of `Point` can be equal, as long as none of the classes redefines the equality method. For in-

```
class ColoredPoint(x: Int, y: Int, val color: Color.Value)
    extends Point(x, y) {
  override def hashCode = (super.hashCode, color).##
  override def equals(other: Any) = other match {
    case that: ColoredPoint =>
      (that canEqual this) &&
      super.equals(that) && this.color == that.color
    case _ =>
      false
  }
  override def canEqual(other: Any) =
    other.isInstanceOf[ColoredPoint]
}
```

Listing 30.2 · A subclass equals method that calls canEqual.

stance, with the new class definitions, the comparison of p and pAnon would
yield true. Here are some examples:

```
scala> val p = new Point(1, 2)
p: Point = Point@5428bd62

scala> val cp = new ColoredPoint(1, 2, Color.Indigo)
cp: ColoredPoint = ColoredPoint@e6230d8f

scala> val pAnon = new Point(1, 1) { override val y = 2 }
pAnon: Point = $anon$1@5428bd62

scala> val coll = List(p)
coll: List[Point] = List(Point@5428bd62)

scala> coll contains p
res16: Boolean = true

scala> coll contains cp
res17: Boolean = false

scala> coll contains pAnon
res18: Boolean = true
```

These examples demonstrate that if a superclass equals implementation defines and calls canEqual, then programmers who implement subclasses can

decide whether or not their subclasses may be equal to instances of the super-class. Because `ColoredPoint` overrides `canEqual`, for example, a colored point may never be equal to a plain-old point. But because the anonymous subclass referenced from pAnon does not override `canEqual`, its instance can be equal to a `Point` instance.

One potential criticism of the `canEqual` approach is that it violates the Liskov Substitution Principle. For example, the technique of implementing `equals` by comparing run-time classes, which led to the inability to define a subclass whose instances can equal instances of the superclass, has been described as a violation of the LSP.[7] The LSP states you should be able to use (substitute) a subclass instance where a superclass instance is required.

In the previous example, however, "coll contains cp" returned `false` even though cp's x and y values matched those of the point in the collection. Thus, it may seem like a violation of the LSP because you can't use a `ColoredPoint` here where a `Point` is expected. We believe this is the wrong interpretation, because the LSP doesn't require that a subclass behaves identically to its superclass, just that it behaves in a way that fulfills the contract of its superclass.

The problem with writing an `equals` method that compares run-time classes is not that it violates the LSP, but that it doesn't give you a way to create a subclass whose instances can equal superclass instances. For example, had we used the run-time class technique in the previous example, "coll contains pAnon" would have returned `false`, and that's not what we wanted. By contrast, we really did want "coll contains cp" to return `false`, because by overriding `equals` in `ColoredPoint`, we were basically saying that an indigo-colored point at coordinates (1, 2) is *not the same thing* as an uncolored point at (1, 2). Thus, in the previous example we were able to pass two different `Point` subclass instances to the collection's `contains` method, and we got back two different answers, both correct.

## 30.3 Defining equality for parameterized types

The `equals` methods in the previous examples all started with a pattern match that tested whether the type of the operand conformed to the type of the class containing the `equals` method. When classes are parameterized, this scheme needs to be adapted a little bit.

---

[7]Bloch, *Effective Java Second Edition*, p. 39 [Blo08]

As an example, consider binary trees. The class hierarchy shown in Listing 30.3 defines an abstract class Tree for a binary tree, with two alternative implementations: an EmptyTree object and a Branch class representing non-empty trees. A non-empty tree is made up of some element elem, and a left and right child tree. The type of its element is given by a type parameter T.

```
trait Tree[+T] {
  def elem: T
  def left: Tree[T]
  def right: Tree[T]
}

object EmptyTree extends Tree[Nothing] {
  def elem =
    throw new NoSuchElementException("EmptyTree.elem")
  def left =
    throw new NoSuchElementException("EmptyTree.left")
  def right =
    throw new NoSuchElementException("EmptyTree.right")
}

class Branch[+T](
  val elem: T,
  val left: Tree[T],
  val right: Tree[T]
) extends Tree[T]
```

Listing 30.3 · Hierarchy for binary trees.

We'll now add equals and hashCode methods to these classes. For class Tree itself there's nothing to do because we assume that these methods are implemented separately for each implementation of the abstract class. For object EmptyTree, there's still nothing to do because the default implementations of equals and hashCode that EmptyTree inherits from AnyRef work just fine. After all, an EmptyTree is only equal to itself, so equality should be reference equality, which is what's inherited from AnyRef.

But adding equals and hashCode to Branch requires some work. A Branch value should only be equal to other Branch values, and only if the two values have equal elem, left and right fields. It's natural to apply the schema for equals that was developed in the previous sections of this chapter. This would give you:

```scala
class Branch[T](
  val elem: T,
  val left: Tree[T],
  val right: Tree[T]
) extends Tree[T] {

  override def equals(other: Any) = other match {
    case that: Branch[T] => this.elem == that.elem &&
                            this.left == that.left &&
                            this.right == that.right
    case _ => false
  }
}
```

Compiling this example, however, gives an indication that "unchecked" warnings occurred. Compiling again with the –unchecked option reveals the following problem:

```
$ fsc -unchecked Tree.scala
Tree.scala:14: warning: non variable type-argument T in type
pattern is unchecked since it is eliminated by erasure
      case that: Branch[T] => this.elem == that.elem &&
                ^
```

As the warning says, there is a pattern match against a Branch[T] type, yet the system can only check that the other reference is (some kind of) Branch; it cannot check that the element type of the tree is T. You encountered in Chapter 19 the reason for this: Element types of parameterized types are eliminated by the compiler's erasure phase; they are not available to be inspected at run-time.

So what can you do? Fortunately, it turns out that you need not necessarily check that two Branches have the same element types when comparing them. It's quite possible that two Branches with different element types are equal, as long as their fields are the same. A simple example of this would be the Branch that consists of a single Nil element and two empty subtrees. It's plausible to consider any two such Branches to be equal, no matter what static types they have:

```
scala> val b1 = new Branch[List[String]](Nil,
          EmptyTree, EmptyTree)
b1: Branch[List[String]] = Branch@9d5fa4f

scala> val b2 = new Branch[List[Int]](Nil,
          EmptyTree, EmptyTree)
b2: Branch[List[Int]] = Branch@56cdfc29

scala> b1 == b2
res19: Boolean = true
```

The positive result of the comparison above was obtained with the implementation of equals on Branch shown previously. This demonstrates that the element type of the Branch was not checked—if it had been checked, the result would have been false.

We can disagree on which of the two possible outcomes of the comparison would be more natural. In the end, this depends on the mental model of how classes are represented. In a model where type-parameters are present only at compile-time, it's natural to consider the two Branch values b1 and b2 to be equal. In an alternative model where a type parameter forms part of an object's value, it's equally natural to consider them different. Since Scala adopts the type erasure model, type parameters are not preserved at run time, so that b1 and b2 are naturally considered to be equal.

There's only a tiny change needed to formulate an equals method that does not produce an unchecked warning. Instead of an element type T, use a lower case letter, such as t:

```
case that: Branch[t] => this.elem == that.elem &&
                        this.left == that.left &&
                        this.right == that.right
```

Recall from Section 15.2 that a type parameter in a pattern starting with a lower-case letter represents an unknown type. Now the pattern match:

```
case that: Branch[t] =>
```

will succeed for Branch values of any type. The type parameter t represents the unknown element type of the Branch. It can also be replaced by an underscore, as in the following case, which is equivalent to the previous one:

```
case that: Branch[_] =>
```

The only thing that remains is to define for class Branch the other two methods, hashCode and canEqual, which go with equals. Here's a possible implementation of hashCode:

```
override def hashCode: Int = (elem, left, right).##
```

This is only one of many possible implementations. As shown previously, the principle is to take hashCode values of all fields and combine them. Here's an implementation of method canEqual in class Branch:

```
def canEqual(other: Any) = other match {
  case that: Branch[_] => true
  case _ => false
}
```

The implementation of the canEqual method used a typed pattern match. It would also be possible to formulate it with isInstanceOf:

```
def canEqual(other: Any) = other.isInstanceOf[Branch[_]]
```

If you feel like nit-picking—and we encourage you to do so!—you might wonder what the occurrence of the underscore in the type above signifies. After all, Branch[_] is technically a type parameter of a method, not a type pattern. So how is it possible to leave some parts of it undefined?

The answer to this question is discussed in the next chapter. Branch[_] is shorthand for a so-called *wildcard type*, which is, roughly speaking, a type with some unknown parts in it. So even though technically the underscore stands for two different things in a pattern match and in a type parameter of a method call, in essence, the meaning is the same: It lets you label something that is unknown. The final version of Branch is shown in Listing 30.4.

## 30.4   Recipes for equals and hashCode

In this section, we'll provide step-by-step recipes for creating equals and hashCode methods that should suffice for most situations. As an illustration, we'll use the methods of class Rational, shown in Listing 30.5.

To create this class, we removed the mathematical operator methods from the version of class Rational shown in Listing 6.5 on page 113. We also made a minor enhancement to toString, and modified the initializers

667

```
class Branch[T](
  val elem: T,
  val left: Tree[T],
  val right: Tree[T]
) extends Tree[T] {

  override def equals(other: Any) = other match {
    case that: Branch[_] => (that canEqual this) &&
                            this.elem == that.elem &&
                            this.left == that.left &&
                            this.right == that.right
    case _ => false
  }

  def canEqual(other: Any) = other.isInstanceOf[Branch[_]]

  override def hashCode: Int = (elem, left, right).##
}
```

Listing 30.4 · A parameterized type with equals and hashCode.

of numer and denom to normalize all fractions to have a positive denominator (*i.e.*, to transform $\frac{1}{-2}$ to $\frac{-1}{2}$).

**Recipe for equals**

Here's the recipe for overriding equals:

1. To override equals in a non-final class, create a canEqual method. If the inherited definition of equals is from AnyRef (that is, equals was not redefined higher up in the class hierarchy), the definition of canEqual should be new; otherwise, it will override a previous definition of a method with the same name. The only exception to this requirement is for final classes that redefine the equals method inherited from AnyRef. For them the subclass anomalies described in Section 30.2 cannot arise; consequently they need not define canEqual. The type of object passed to canEqual should be Any:

```
def canEqual(other: Any): Boolean =
```

2. The canEqual method should yield true if the argument object is an instance of the current class (*i.e.*, the class in which canEqual is defined), and false otherwise:

```
other.isInstanceOf[Rational]
```

3. In the equals method, make sure you declare the type of the object passed as an Any:

```
override def equals(other: Any): Boolean =
```

4. Write the body of the equals method as a single match expression. The selector of the match should be the object passed to equals:

```
other match {
  // ...
}
```

5. The match expression should have two cases. The first case should declare a typed pattern for the type of the class on which you're defining the equals method:

```
case that: Rational =>
```

6. In the body of this case, write an expression that logical-ands together the individual expressions that must be true for the objects to be equal. If the equals method you are overriding is not that of AnyRef, you will most likely want to include an invocation of the superclass's equals method:

```
super.equals(that) &&
```

If you are defining equals for a class that first introduced canEqual, you should invoke canEqual on the argument to the equality method, passing this as the argument:

```
(that canEqual this) &&
```

669

Overriding redefinitions of equals should also include the canEqual invocation, unless they contain a call to super.equals. In the latter case, the canEqual test will already be done by the superclass call. Lastly, for each field relevant to equality, verify that the field in this object is equal to the corresponding field in the passed object:

```
numer == that.numer &&
denom == that.denom
```

7. For the second case, use a wildcard pattern that yields false:

```
case _ => false
```

If you adhere to this recipe for equals, equality is guaranteed to be an equivalence relation, as is required by the equals contract.

**Recipe for hashCode**

For hashCode, you can usually achieve satisfactory results if you use the following recipe, which is similar to a recipe recommended for Java classes in *Effective Java*.[8] Include in the calculation each field in your object that is used to determine equality in the equals method (the "relevant" fields). Make a tuple containing the values of all those fields. Then, invoke ## on the resulting tuple.

For example, to implement the hash code for an object that has five relevant fields named a, b, c, d, and e, you would write:

```
override def hashCode: Int = (a, b, c, d, e).##
```

If the equals method invokes super.equals(that) as part of its calculation, you should start your hashCode calculation with an invocation of super.hashCode. For example, had Rational's equals method invoked super.equals(that), its hashCode would have been:

```
override def hashCode: Int = (super.hashCode, numer, denom).##
```

---

[8]Bloch, *Effective Java Second Edition*. [Blo08]

```
class Rational(n: Int, d: Int) {

  require(d != 0)

  private val g = gcd(n.abs, d.abs)
  val numer = (if (d < 0) -n else n) / g
  val denom = d.abs / g

  private def gcd(a: Int, b: Int): Int =
    if (b == 0) a else gcd(b, a % b)

  override def equals(other: Any): Boolean =
    other match {

      case that: Rational =>
        (that canEqual this) &&
        numer == that.numer &&
        denom == that.denom

      case _ => false
    }

  def canEqual(other: Any): Boolean =
    other.isInstanceOf[Rational]

  override def hashCode: Int = (numer, denom).##

  override def toString =
    if (denom == 1) numer.toString else numer + "/" + denom
}
```

Listing 30.5 · Class Rational with equals and hashCode.

One thing to keep in mind as you write hashCode methods using this approach is that your hash code will only be as good as the hash codes you build out of it, namely the hash codes you obtain by calling hashCode on the relevant fields of your object. Sometimes you may need to do something extra besides just calling hashCode on the field to get a useful hash code for that field. For example, if one of your fields is a collection, you probably want a hash code for that field that is based on all the elements contained in the collection. If the field is a Vector, List, Set, Map, or tuple, you can simply include it in the list of items you are hashing over, because equals and hashCode are overridden in those classes to take into account the contained elements. However the same is not true for Arrays, which do not take elements into account when calculating a hash code. Thus for an array, you should treat each element of the array like an individual field of your object, calling ## on each element explicitly or passing the array to one of the hashCode methods in singleton object java.util.Arrays.

Lastly, if you find that a particular hash code calculation is harming the performance of your program, consider caching the hash code. If the object is immutable, you can calculate the hash code when the object is created and store it in a field. You can do this by simply overriding hashCode with a val instead of a def, like this:

```
override val hashCode: Int = (numer, denom).##
```

This approach trades off memory for computation time, because each instance of the immutable class will have one more field to hold the cached hash code value.

## 30.5   Conclusion

In retrospect, defining a correct implementation of equals has been surprisingly subtle. You must be careful about the type signature; you must override hashCode; you should avoid dependencies on mutable state; and you should implement and use a canEqual method if your class is non-final.

Given how difficult it is to implement a correct equality method, you might prefer to define your classes of comparable objects as case classes. That way, the Scala compiler will add equals and hashCode methods with the right properties automatically.

# Chapter 31

# Combining Scala and Java

Scala code is often used in tandem with large Java programs and frameworks. Since Scala is highly compatible with Java, most of the time you can combine the languages without worrying very much. For example, standard frameworks, such as Swing, Servlets, and JUnit, are known to work just fine with Scala. Nonetheless, from time to time, you will run into some issue combining Java and Scala.

This chapter describes two aspects of combining Java and Scala. First, it discusses how Scala is translated to Java, which is especially important if you call Scala code from Java. Second, it discusses the use of Java annotations in Scala, an important feature if you want to use Scala with an existing Java framework.

## 31.1 Using Scala from Java

Most of the time you can think of Scala at the source code level. However, you will have a richer understanding of how the system works if you know something about its translation. Further, if you call Scala code from Java, you will need to know what Scala code looks like from a Java point of view.

### General rules

Scala is implemented as a translation to standard Java bytecodes. As much as possible, Scala features map directly onto the equivalent Java features. For example, Scala classes, methods, strings, and exceptions are all compiled to the same in Java bytecode as their Java counterparts.

To make this happen required an occasional hard choice in the design of Scala. For example, it might have been nice to resolve overloaded methods at run time, using run-time types, rather than at compile time. Such a design would break with Java's, however, making it much trickier to mesh Java and Scala. In this case, Scala stays with Java's overloading resolution, and thus Scala methods and method calls can map directly to Java methods and method calls.

Scala has its own design for other features. For example, traits have no equivalent in Java. Similarly, while both Scala and Java have generic types, the details of the two systems clash. For language features like these, Scala code cannot be mapped directly to a Java construct, so it must be encoded using some combination of the structures Java does have.

For these features that are mapped indirectly, the encoding is not fixed. There is an ongoing effort to make the translations as simple as possible so, by the time you read this, some details may be different than at the time of writing. You can find out what translation your current Scala compiler uses by examining the ".class" files with tools like `javap`.

Those are the general rules. Consider now some special cases.

**Value types**

A value type like `Int` can be translated in two different ways to Java. Whenever possible, the compiler translates a Scala `Int` to a Java `int` to get better performance. Sometimes this is not possible, though, because the compiler is not sure whether it is translating an `Int` or some other data type. For example, a particular List[Any] might hold only `Int`s, but the compiler has no way to be sure.

In such cases, where the compiler is unsure whether an object is a value type or not, the compiler uses objects and relies on wrapper classes. For example, wrapper classes such as `java.lang.Integer` allow a value type to be wrapped inside a Java object and thereby manipulated by code that needs objects.[1]

**Singleton objects**

Java has no exact equivalent to a singleton object, but it does have static methods. The Scala translation of singleton objects uses a combination of

---

[1]The implementation of value types was discussed in detail in Section 11.2.

674

static and instance methods. For every Scala singleton object, the compiler will create a Java class for the object with a dollar sign added to the end. For a singleton object named App, the compiler produces a Java class named App$. This class has all the methods and fields of the Scala singleton object. The Java class also has a single static field named MODULE$ to hold the one instance of the class that is created at run time.

As a full example, suppose you compile the following singleton object:

```scala
object App {
  def main(args: Array[String]) = {
    println("Hello, world!")
  }
}
```

Scala will generate a Java App$ class with the following fields and methods:

```
$ javap App$
public final class App$ extends java.lang.Object
implements scala.ScalaObject{
    public static final App$ MODULE$;
    public static {};
    public App$();
    public void main(java.lang.String[]);
    public int $tag();
}
```

That's the translation for the general case. An important special case is if you have a "standalone" singleton object, one which does not come with a class of the same name. For example, you might have a singleton object named App, and not have any class named App. In that case, the compiler will create a Java class named App that has a static forwarder method for each method of the Scala singleton object:

```
$ javap App
Compiled from "App.scala"
public final class App extends java.lang.Object{
    public static final int $tag();
    public static final void main(java.lang.String[]);
}
```

675

To contrast, if you did have a class named App, Scala would create a corresponding Java App class to hold the members of the App class you defined. In that case it would not add any forwarding methods for the same-named singleton object, and Java code would have to access the singleton via the MODULE$ field.

### Traits as interfaces

Compiling any trait creates a Java interface of the same name. This interface is usable as a Java type, and it lets you call methods on Scala objects through variables of that type.

Implementing a trait in Java is another story. In the general case it is not practical; however, one special case is important. If you make a Scala trait that includes only abstract methods, then that trait will be translated directly to a Java interface with no other code to worry about. Essentially this means that you can write a Java interface in Scala syntax if you like.

## 31.2 Annotations

Scala's general annotations system is discussed in Chapter 27. This section discusses Java-specific aspects of annotations.

### Additional effects from standard annotations

Several annotations cause the compiler to emit extra information when targeting the Java platform. When the compiler sees such an annotation, it first processes it according to the general Scala rules, and then it does something extra for Java.

**Deprecation** For any method or class marked @deprecated, the compiler will add Java's own deprecation annotation to the emitted code. Because of this, Java compilers can issue deprecation warnings when Java code accesses deprecated Scala methods.

**Volatile fields** Likewise, any field marked @volatile in Scala is given the Java volatile modifier in the emitted code. Thus, volatile fields in Scala behave exactly according to Java's semantics, and accesses to volatile fields are sequenced precisely according to the rules specified for volatile fields in the Java memory model.

**Serialization**   Scala's three standard serialization annotations are all translated to Java equivalents. A `@serializable` class has Java's `Serializable` interface added to it. A `@SerialVersionUID(1234L)` annotation is converted to the following Java field definition:

```
// Java serial version marker
private final static long SerialVersionUID = 1234L
```

Any variable marked `@transient` is given the Java `transient` modifier.

### Exceptions thrown

Scala does not check that thrown exceptions are caught. That is, Scala has no equivalent to Java's `throws` declarations on methods. All Scala methods are translated to Java methods that declare no thrown exceptions.[2]

The reason this feature is omitted from Scala is that the Java experience with it has not been purely positive. Because annotating methods with `throws` clauses is a heavy burden, too many developers write code that swallows and drops exceptions, just to get the code to compile without adding all those `throws` clauses. They may intend to improve the exception handling later, but experience shows that all too often time-pressed programmers will never come back and add proper exception handling. The twisted result is that this well-intentioned feature often ends up making code *less* reliable. A large amount of production Java code swallows and hides runtime exceptions, and the reason it does so is to satisfy the compiler.

Sometimes when interfacing to Java, however, you may need to write Scala code that has Java-friendly annotations describing which exceptions your methods may throw. For example, each method in an RMI remote interface is required to mention `java.io.RemoteException` in its `throws` clause. Thus, if you wish to write an RMI remote interface as a Scala trait with abstract methods, you would need to list `RemoteException` in the `throws` clauses for those methods. To accomplish this, all you have to do is mark your methods with `@throws` annotations. For example, the Scala class shown in Listing 31.1 has a method marked as throwing `IOException`.

---

[2]The reason it all works is that the Java bytecode verifier does not check the declarations anyway! The Java compiler checks, but not the verifier.

```
import java.io._
class Reader(fname: String) {
  private val in =
    new BufferedReader(new FileReader(fname))

  @throws(classOf[IOException])
  def read() = in.read()
}
```

Listing 31.1 · A Scala method that declares a Java throws clause.

Here is how it looks from Java:

```
$ javap Reader
Compiled from "Reader.scala"
public class Reader extends java.lang.Object implements
scala.ScalaObject{
    public Reader(java.lang.String);
    public int read()        throws java.io.IOException;
    public int $tag();
}
$
```

Note that the read method indicates with a Java throws clause that it may throw an IOException.

## Java annotations

Existing annotations from Java frameworks can be used directly in Scala code. Any Java framework will see the annotations you write just as if you were writing in Java.

A wide variety of Java packages use annotations. As an example, consider JUnit 4. JUnit is a framework for writing and running automated tests. The latest version, JUnit 4, uses annotations to indicate which parts of your code are tests. The idea is that you write a lot of tests for your code, and then you run those tests whenever you change the source code. That way, if your changes add a new bug, one of the tests will fail and you will find out immediately.

Writing a test is easy. You simply write a method in a top-level class that exercises your code, and you use an annotation to mark the method as a test. It looks like this:

```
import org.junit.Test
import org.junit.Assert.assertEquals

class SetTest {

  @Test
  def testMultiAdd = {
    val set = Set() + 1 + 2 + 3 + 1 + 2 + 3
    assertEquals(3, set.size)
  }
}
```

The `testMultiAdd` method is a test. This test adds multiple items to a set and makes sure that each is added only once. The `assertEquals` method, which comes as part of the JUnit API, checks that its two arguments are equal. If they are different, then the test fails. In this case, the test verifies that repeatedly adding the same numbers does not increase the size of a set.

The test is marked using the annotation `org.junit.Test`. Note that this annotation has been imported, so it can be referred to as simply `@Test` instead of the more cumbersome `@org.junit.Test`.

That's all there is to it. The test can be run using any JUnit test runner. Here it is being run with the command-line test runner:

```
$ scala -cp junit-4.3.1.jar:. org.junit.runner.JUnitCore SetTest
JUnit version 4.3.1
.
Time: 0.023

OK (1 test)
```

**Writing your own annotations**

To make an annotation that is visible to Java reflection, you must use Java notation and compile it with `javac`. For this use case, writing the annotation in Scala does not seem helpful, so the standard compiler does not support it. The reasoning is that the Scala support would inevitably fall short of the

679

full possibilities of Java annotations, and further, Scala will probably one day have its own reflection, in which case you would want to access Scala annotations with Scala reflection.

Here is an example annotation:

```
import java.lang.annotation.*; // This is Java
@Retention(RetentionPolicy.RUNTIME)
@Target(ElementType.METHOD)
public @interface Ignore { }
```

After compiling the above with javac, you can use the annotation as follows:

```
object Tests {
  @Ignore
  def testData = List(0, 1, -1, 5, -5)

  def test1 = {
    assert(testData == (testData.head :: testData.tail))
  }

  def test2 = {
    assert(testData.contains(testData.head))
  }
}
```

In this example, test1 and test2 are supposed to be test methods, but testData should be ignored even though its name starts with "test".

To see when these annotations are present, you can use the Java reflection APIs. Here is sample code to show how it works:

```
for {
  method <- Tests.getClass.getMethods
  if method.getName.startsWith("test")
  if method.getAnnotation(classOf[Ignore]) == null
} {
  println("found a test method: " + method)
}
```

Here, the reflective methods getClass and getMethods are used to inspect all the fields of the input object's class. These are normal reflection methods. The annotation-specific part is the use of method getAnnotation. Many reflection objects have a getAnnotation method for searching for annotations of a specific type. In this case, the code looks for an annotation of our new Ignore type. Since this is a Java API, success is indicated by whether the result is null or an actual annotation object.

Here is the code in action:

```
$ javac Ignore.java
$ scalac Tests.scala
$ scalac FindTests.scala
$ scala FindTests
found a test method: public void Tests$.test2()
found a test method: public void Tests$.test1()
```

As an aside, notice that the methods are in class Tests$ instead of class Tests when viewed with Java reflection. As described at the beginning of the chapter, the implementation of a Scala singleton object is placed in a Java class with a dollar sign added to the end of its name. In this case, the implementation of Tests is in the Java class Tests$.

Be aware that when you use Java annotations you have to work within their limitations. For example, you can only use constants, not expressions, in the arguments to annotations. You can support @serial(1234) but not @serial(x * 2), because x * 2 is not a constant.

## 31.3 Wildcard types

All Java types have a Scala equivalent. This is necessary so that Scala code can access any legal Java class. Most of the time the translation is straightforward. Pattern in Java is Pattern in Scala, and Iterator<Component> in Java is Iterator[Component] in Scala. For some cases, though, the Scala types you have seen so far are not enough. What can be done with Java wildcard types such as Iterator<?> or Iterator<? extends Component>? What can be done about raw types like Iterator, where the type parameter is omitted? For Java wildcard types and raw types, Scala uses an extra kind of type also called a *wildcard type*.

Wildcard types are written using *placeholder syntax*, just like the shorthand function literals described in Section 8.5. In the short hand for function literals, you can use an underscore (_) in place of an expression; for example, (_ + 1) is the same as (x => x + 1). Wildcard types use the same idea, only for types instead of expressions. If you write Iterator[_], then the underscore is replacing a type. Such a type represents an Iterator where the element type is not known.

You can also insert upper and lower bounds when using this placeholder syntax. Simply add the bound after the underscore, using the same <: syntax used with type parameters (Section 19.8 and Section 19.5). For example, the type Iterator[_ <: Component] is an iterator where the element type is not known, but whatever type it is, it must be a subtype of Component.

That's how you write a wildcard type, but how do you use it? In simple cases, you can ignore the wildcard and call methods on the base type. For example, suppose you had the following Java class:

```
// This is a Java class with wildcards
public class Wild {
  public Collection<?> contents() {
    Collection<String> stuff = new Vector<String>();
    stuff.add("a");
    stuff.add("b");
    stuff.add("see");
    return stuff;
  }
}
```

If you access this in Scala code you will see that it has a wildcard type:

```
scala> val contents = (new Wild).contents
contents: java.util.Collection[_] = [a, b, see]
```

If you want to find out how many elements are in this collection, you can simply ignore the wildcard part and call the size method as normal:

```
scala> contents.size()
res0: Int = 3
```

In more complicated cases, wildcard types can be more awkward. Since the wildcard type has no name, there is no way to use it in two separate places. For example, suppose you wanted to create a mutable Scala set and initialize it with the elements of contents:

```scala
import scala.collection.mutable
val iter = (new Wild).contents.iterator
val set = mutable.Set.empty[???]     // what type goes here?
while (iter.hasMore)
  set += iter.next()
```

A problem occurs on the third line. There is no way to name the type of elements in the Java collection, so you cannot write down a satisfactory type for set. To work around this kind of problem, here are two tricks you should consider:

1. When passing a wildcard type into a method, give a parameter to the method for the placeholder. You now have a name for the type that you can use as many times as you like.

2. Instead of returning wildcard type from a method, return an object that has abstract members for each of the placeholder types. (See Chapter 20 for information on abstract members.)

Using these two tricks together, the previous code can be written as follows:

```scala
import scala.collection.mutable
import java.util.Collection

abstract class SetAndType {
  type Elem
  val set: mutable.Set[Elem]
}

def javaSet2ScalaSet[T](jset: Collection[T]): SetAndType = {
  val sset = mutable.Set.empty[T]  // now T can be named!

  val iter = jset.iterator
  while (iter.hasNext)
    sset += iter.next()

  return new SetAndType {
    type Elem = T
    val set = sset
  }
}
```

683

You can see why Scala code normally does not use wildcard types. To do anything sophisticated with them, you tend to convert them to use abstract members. So you may as well use abstract members to begin with.

## 31.4  Compiling Scala and Java together

Usually when you compile Scala code that depends on Java code, you first build the Java code to class files. You then build the Scala code, putting the Java code's class files on the classpath. However, this approach doesn't work if the Java code has references back into the Scala code. In such a case, no matter which order you compile the code, one side or the other will have unsatisfied external references. These situations are not uncommon; all it takes is a mostly Java project where you replace one Java source file with a Scala source file.

To support such builds, Scala allows compiling against Java source code as well as Java class files. All you have to do is put the Java source files on the command line as if they were Scala files. The Scala compiler won't compile those Java files, but it will scan them to see what they contain. To use this facility, you first compile the Scala code using Java source files, and then compile the Java code using Scala class files.

Here is a typical sequence of commands:

```
$ scalac -d bin InventoryAnalysis.scala InventoryItem.java \
      Inventory.java
$ javac -cp bin -d bin Inventory.java InventoryItem.java \
      InventoryManagement.java
$ scala -cp bin InventoryManagement
Most expensive item = sprocket($4.99)
```

## 31.5  Java 8 integration in Scala 2.12

Java 8 added a few improvements to the Java language and bytecodes that Scala takes advantage of in its 2.12 release.[3] By exploiting new features of Java 8, the Scala 2.12 compiler can generate smaller class and jar files and improve the binary compatibility of traits.

---

[3]Scala 2.12 *requires* Java 8 so that it can take advantage of Java 8 features.

## Lambda expressions and "SAM" types

From the Scala programmer's perspective, the most visible Java 8-related enhancement in Scala 2.12 is that Scala function literals can be used like Java 8 *lambda expressions* as a more concise form for anonymous class instance expressions. To pass behavior into a method prior to Java 8, Java programmers often defined anonymous inner class instances, like this:

```
JButton button = new JButton(); // This is Java
button.addActionListener(
  new ActionListener() {
    public void actionPerformed(ActionEvent event) {
      System.out.println("pressed!");
    }
  }
);
```

In this example, an anonymous instance of `ActionListener` is created and passed to the `addActionListener` of a Swing `JButton`. When a user clicks on the button, Swing will invoke the `actionPerformed` method on this instance, which will print `"pressed!"`.

In Java 8, a lambda expression can be used anywhere an instance of a class or interface that contains just a single abstract method (SAM) is required. `ActionListener` is such an interface, because it contains a single abstract method, `actionPerformed`. Thus a lambda expression can be used to register an action listener on a Swing button. Here's an example:

```
JButton button = new JButton(); // This is Java 8
button.addActionListener(
  event -> System.out.println("pressed!")
);
```

In Scala, you could also use an anonymous inner class instance in the same situation, but you might prefer to use a function literal, like this:

```
val button = new JButton
button.addActionListener(
  _ => println("pressed!")
)
```

685

As you have already seen in Section 21.1, you could support such a coding style by defining an implicit conversion from the `ActionEvent => Unit` function type to `ActionListener`.

Scala 2.12 enables a function literal to be used in this case even in the absence of such an implicit conversion. As with Java 8, Scala 2.12 will allow a function type to be used where an instance of a class or trait declaring a single abstract method (SAM) is required. This will work with any SAM in Scala 2.12. For example, you might define a trait, `Increaser`, with a single abstract method, `increase`:

```
scala> trait Increaser {
          def increase(i: Int): Int
       }
defined trait Increaser
```

You could then define a method that takes an `Increaser`:

```
scala> def increaseOne(increaser: Increaser): Int =
          increaser.increase(1)
increaseOne: (increaser: Increaser)Int
```

To invoke your new method, you could pass in an anonymous instance of trait `Increaser`, like this:

```
scala> increaseOne(
          new Increaser {
            def increase(i: Int): Int = i + 7
          }
       )
res0: Int = 8
```

In Scala 2.12, however, you could alternatively just use a function literal, because `Increaser` is a SAM type:

```
scala> increaseOne(i => i + 7) // Scala 2.12
res1: Int = 8
```

## Using Java 8 Streams from Scala 2.12

Java's Stream is a functional data structure that offers a map method taking a java.util.function.IntUnaryOperator. From Scala you could invoke Stream.map to increment each element of an Array, like this:

```
scala> import java.util.function.IntUnaryOperator
import java.util.function.IntUnaryOperator

scala> import java.util.Arrays
import java.util.Arrays

scala> val stream = Arrays.stream(Array(1, 2, 3))
stream: java.util.stream.IntStream = ...

scala> stream.map(
          new IntUnaryOperator {
             def applyAsInt(i: Int): Int = i + 1
          }
       ).toArray
res3: Array[Int] = Array(2, 3, 4)
```

Because IntUnaryOperator is a SAM type, however, you could in Scala 2.12 invoke it more concisely with a function literal:

```
scala> val stream = Arrays.stream(Array(1, 2, 3))
stream: java.util.stream.IntStream = ...

scala> stream.map(i => i + 1).toArray // Scala 2.12
res4: Array[Int] = Array(2, 3, 4)
```

Note that only function *literals* will be adapted to SAM types, not arbitrary expressions that have a function type. For example, consider the following val, f, which has type Int => Int:

```
scala> val f = (i: Int) => i + 1
f: Int => Int = ...
```

Although f has the same type as the function literal passed to stream.map previously, you can't use f where an IntUnaryOperator is required:

```
scala> val stream = Arrays.stream(Array(1, 2, 3))
stream: java.util.stream.IntStream = ...
```

```
scala> stream.map(f).toArray
<console>:16: error: type mismatch;
 found    : Int => Int
 required: java.util.function.IntUnaryOperator
         stream.map(f).toArray
                 ^
```

To use f, you can explicitly call it using a function literal, like this:

```
scala> stream.map(i => f(i)).toArray
res5: Array[Int] = Array(2, 3, 4)
```

Or, you could annotate f with IntUnaryOperator, the type expected by Stream.map, when you define f:

```
scala> val f: IntUnaryOperator = i => i + 1
f: java.util.function.IntUnaryOperator = ...

scala> val stream = Arrays.stream(Array(1, 2, 3))
stream: java.util.stream.IntStream = ...

scala> stream.map(f).toArray
res6: Array[Int] = Array(2, 3, 4)
```

With Scala 2.12 and Java 8, you can also invoke methods compiled with Scala from Java, passing Scala function types using Java lambda expressions. Although Scala function types are defined as traits that include concrete methods, Scala 2.12 compiles traits to Java interfaces with *default methods*, a new feature of Java 8. As a result, Scala function types appear to Java as SAMs.

## 31.6   Conclusion

Most of the time, you can ignore how Scala is implemented, and simply write and run your code. But sometimes it is nice to "look under the hood," so this chapter has gone into three aspects of Scala's implementation on the Java platform: What the translation looks like, how Scala and Java annotations work together, and how Scala's wildcard types let you access Java wildcard types. It also covered using Java's concurrency primitives from Scala and compiling combined Scala and Java projects. These topics are important whenever you use Scala and Java together.

# Chapter 32

# Futures and Concurrency

One consequence of the proliferation of multicore processors has been an increased interest in concurrency. Java provides concurrency support built around shared memory and locking. Although this support is sufficient, this approach turns out to be quite difficult to get right in practice. Scala's standard library offers an alternative that avoids these difficulties by focusing on asynchronous transformations of immutable state: the Future.

Although Java also offers a Future, it is very different from Scala's. Both represent the result of an asynchronous computation, but Java's Future requires that you access the result via a blocking get method. Although you can call isDone to find out if a Java Future has completed before calling get, thereby avoiding any blocking, you must wait until the Java Future has completed before proceeding with any computation that uses the result.

By contrast, you can specify transformations on a Scala Future whether it has completed or not. Each transformation results in a new Future representing the asynchronous result of the original Future transformed by the function. The thread that performs the computation is determined by an implicitly provided *execution context*. This allows you to describe asynchronous computations as a series of transformations of immutable values, with no need to reason about shared memory and locks.

## 32.1   Trouble in paradise

On the Java platform, each object is associated with a logical *monitor*, which can be used to control multi-threaded access to data. To use this model, you decide what data will be shared by multiple threads and mark as "synchronized" sections of the code that access, or control access to, the shared data. The Java runtime employs a locking mechanism to ensure that only one thread at a time enters synchronized sections guarded by the same lock, thereby enabling you to orchestrate multi-threaded access to the shared data.

For compatibility's sake, Scala provides access to Java's concurrency primitives. The `wait`, `notify`, and `notifyAll` methods can be called in Scala, and they have the same meaning as in Java. Scala doesn't technically have a `synchronized` keyword, but it includes a predefined `synchronized` method that can be called as follows:

```
var counter = 0
synchronized {
  // One thread in here at a time
  counter = counter + 1
}
```

Unfortunately, programmers have found it very difficult to reliably build robust multi-threaded applications using the shared data and locks model, especially as applications grow in size and complexity. The problem is that at each point in the program, you must reason about what data you are modifying or accessing that might be modified or accessed by other threads, and what locks are being held. At each method call, you must reason about what locks it will try to hold and convince yourself that it will not deadlock while trying to obtain them. Compounding the problem, the locks you reason about are not fixed at compile time, because the program is free to create new locks at run time as it progresses.

Making things worse, testing is not reliable with multi-threaded code. Since threads are non-deterministic, you might successfully test a program one thousand times—and the program could still go wrong the first time it runs on a customer's machine. With shared data and locks, you must get the program correct through reason alone.

Moreover, you can't solve the problem by over-synchronizing either. It can be just as problematic to synchronize everything as it is to synchronize nothing. Although new lock operations may remove possibilities for race conditions, they simultaneously add possibilities for deadlocks. A correct lock-using program must have neither race conditions nor deadlocks, so you cannot play it safe by overdoing it in either direction.

The `java.util.concurrent` library provides higher level abstractions for concurrent programming. Using the concurrency utilities makes multi-threaded programming far less error prone than rolling your own abstractions with Java's low-level synchronization primitives. Nevertheless, the concurrent utilities are also based on the shared data and locks model, and as a result, do not solve the fundamental difficulties of using that model.

## 32.2  Asynchronous execution and Trys

Although not a silver bullet, Scala's Future offers one way to deal with concurrency that can reduce, and often eliminate, the need to reason about shared data and locks. When you invoke a Scala method, it performs a computation "while you wait" and returns a result. If that result is a Future, the Future represents another computation to be performed asynchronously, often by a completely different thread. As a result, many operations on Future require an implicit *execution context* that provides a strategy for executing functions asynchronously. For example, if you try to create a future via the Future.apply factory method without providing an implicit execution context, an instance of scala.concurrent.ExecutionContext, you'll get a compiler error:

```
scala> import scala.concurrent.Future
import scala.concurrent.Future

scala> val fut = Future { Thread.sleep(10000); 21 + 21 }
<console>:11: error: Cannot find an implicit ExecutionContext.
    You might pass an (implicit ec: ExecutionContext)
    parameter to your method or import
    scala.concurrent.ExecutionContext.Implicits.global.
      val fut = Future { Thread.sleep(10000); 21 + 21 }
                  ^
```

The error message gives you one way to solve the problem: importing a global execution context provided by Scala itself. On the JVM, the global execution context uses a thread pool.[1] Once you bring an implicit execution context into scope, you can create a future:

```
scala> import scala.concurrent.ExecutionContext.Implicits.global
import scala.concurrent.ExecutionContext.Implicits.global
```

---

[1] On Scala.js, the global execution context places tasks on the JavaScript event queue.

691

```
scala> val fut = Future { Thread.sleep(10000); 21 + 21 }
fut: scala.concurrent.Future[Int] = ...
```

The future created in the previous example asynchronously executes the block of code, using the global execution context, then completes with the value 42. Once it starts execution, that thread will sleep for ten seconds. Thus this future will take at least ten seconds to complete.

Two methods on Future allow you to poll: isCompleted and value. When invoked on a future that has not yet completed, isCompleted will return false and value will return None.

```
scala> fut.isCompleted
res0: Boolean = false

scala> fut.value
res1: Option[scala.util.Try[Int]] = None
```

Once the future completes (in this case, after at least ten seconds has gone by), isCompleted will return true and value will return a Some:

```
scala> fut.isCompleted
res2: Boolean = true

scala> fut.value
res3: Option[scala.util.Try[Int]] = Some(Success(42))
```

The option returned by value contains a Try. As shown in Figure 32.1, a Try is either a Success, which contains a value of type T, or a Failure, which contains an exception (an instance of java.lang.Throwable). The purpose of Try is to provide for asynchronous computations what the try expression provides for synchronous computations: It allows you to deal with the possibility that the computation will complete abruptly with an exception rather than return a result.[2]

For synchronous computations you can use try/catch to ensure that a thread that invokes a method catches and handles exceptions thrown by the method. For asynchronous computations, however, the thread that initiates the computation often moves on to other tasks. Later if that asynchronous

---

[2]Note that the Java Future also has a way to deal with the potential of an exception being thrown by the asynchronous computation: its get method will throw that exception wrapped in an ExecutionException.

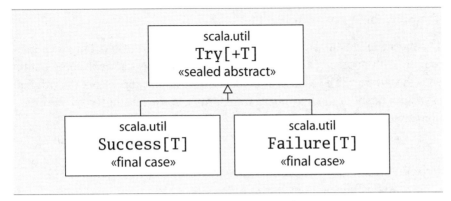

Figure 32.1 · Class hierarchy for Try.

computation fails with an exception, the original thread is no longer able to handle the exception in a catch clause. Thus when working with a Future representing an asynchronous activity, you use Try to deal with the possibility that the activity fails to yield a value and instead completes abruptly an exception. Here's an example that shows what happens when an asynchronous activity fails:

```
scala> val fut = Future { Thread.sleep(10000); 21 / 0 }
fut: scala.concurrent.Future[Int] = ...

scala> fut.value
res4: Option[scala.util.Try[Int]] = None
```

Then, after ten seconds:

```
scala> fut.value
res5: Option[scala.util.Try[Int]] =
    Some(Failure(java.lang.ArithmeticException: / by zero))
```

## 32.3   Working with Futures

Scala's Future allows you to specify transformations on Future results and obtain a *new future* that represents the composition of the two asynchronous computations: the original and the transformation.

## Transforming Futures with map

The most fundamental such operation is map. Instead of blocking then continuing with another computation, you can just map the next computation onto the future. The result will be a new future that represents the original asynchronously computed result transformed asynchronously by the function passed to map.

For example, the following future will complete after ten seconds:

```
scala> val fut = Future { Thread.sleep(10000); 21 + 21 }
fut: scala.concurrent.Future[Int] = ...
```

Mapping this future with a function that increments by one will yield another future. This new future will represent a computation consisting of the original addition followed by the subsequent increment:

```
scala> val result = fut.map(x => x + 1)
result: scala.concurrent.Future[Int] = ...

scala> result.value
res5: Option[scala.util.Try[Int]] = None
```

Once the original future completes and the function has been applied to its result, the future returned by map will complete:

```
scala> result.value
res6: Option[scala.util.Try[Int]] = Some(Success(43))
```

Note that the operations performed in this example—the future creation, the 21 + 21 sum calculation, and the 42 + 1 increment—may be performed by three different threads.

## Transforming Futures with for expressions

Because Scala's future also declares a flatMap method, you can transform futures using a for expression. For example, consider the following two futures that will, after ten seconds, produce 42 and 46:

```
scala> val fut1 = Future { Thread.sleep(10000); 21 + 21 }
fut1: scala.concurrent.Future[Int] = ...
```

```
scala> val fut2 = Future { Thread.sleep(10000); 23 + 23 }
fut2: scala.concurrent.Future[Int] = ...
```

Given these two futures, you can obtain a new future representing the asynchronous sum of their results like this:

```
scala> for {
         x <- fut1
         y <- fut2
       } yield x + y
res7: scala.concurrent.Future[Int] = ...
```

Once the original futures have completed, and the subsequent sum completes, you'll be able to see the result:

```
scala> res7.value
res8: Option[scala.util.Try[Int]] = Some(Success(88))
```

Because for expressions serialize their transformations,[3] if you don't create the futures before the for expression, they won't run in parallel. For example, although the previous for expression requires around ten seconds to complete, the following for expression requires at least twenty seconds:

```
scala> for {
         x <- Future { Thread.sleep(10000); 21 + 21 }
         y <- Future { Thread.sleep(10000); 23 + 23 }
       } yield x + y
res9: scala.concurrent.Future[Int] = ...

scala> res9.value
res27: Option[scala.util.Try[Int]] = None

scala> // Will need at least 20 seconds to complete

scala> res9.value
res28: Option[scala.util.Try[Int]] = Some(Success(88))
```

---

[3]The for expression shown in this example will be rewritten as a call to fut1.flatMap passing in a function that calls fut2.map: fut1.flatMap(x => fut2.map(y => x + y)).

**Creating the Future: Future.failed, Future.successful, Future.fromTry, and Promises**

Besides the apply method, used in earlier examples to create futures, the Future companion object also includes three factory methods for creating already-completed futures: successful, failed, and fromTry. These factory methods do not require an ExecutionContext.

The successful factory method creates a future that has already succeeded:

```
scala> Future.successful { 21 + 21 }
res2: scala.concurrent.Future[Int] = ...
```

The failed method creates a future that has already failed:

```
scala> Future.failed(new Exception("bummer!"))
res3: scala.concurrent.Future[Nothing] = ...
```

The fromTry method creates an already completed future from a Try:

```
scala> import scala.util.{Success,Failure}
import scala.util.{Success, Failure}

scala> Future.fromTry(Success { 21 + 21 })
res4: scala.concurrent.Future[Int] = ...

scala> Future.fromTry(Failure(new Exception("bummer!")))
res5: scala.concurrent.Future[Nothing] = ...
```

The most general way to create a future is to use a Promise. Given a promise you can obtain a future that is controlled by the promise. The future will complete when you complete the promise. Here's an example:

```
scala> val pro = Promise[Int]
pro: scala.concurrent.Promise[Int] = ...

scala> val fut = pro.future
fut: scala.concurrent.Future[Int] = ...

scala> fut.value
res8: Option[scala.util.Try[Int]] = None
```

696

You can complete the promise with methods named success, failure, and complete. These methods on Promise are similar to those described previously for constructing already completed futures. For example, the success method will complete the future successfully:

```
scala> pro.success(42)
res9: pro.type = ...

scala> fut.value
res10: Option[scala.util.Try[Int]] = Some(Success(42))
```

The failure method takes an exception that will cause the future to fail with that exception. The complete method takes a Try. A completeWith method, which takes a future, also exists; the promise's future will thereafter mirror the completion status of the future you passed to completeWith.

### Filtering: filter and collect

Scala's future offers two methods, filter and collect, that allow you to ensure a property holds true about a future value. The filter method validates the future result, leaving it the same if it is valid. Here's an example that ensures an Int is positive:

```
scala> val fut = Future { 42 }
fut: scala.concurrent.Future[Int] = ...

scala> val valid = fut.filter(res => res > 0)
valid: scala.concurrent.Future[Int] = ...

scala> valid.value
res0: Option[scala.util.Try[Int]] = Some(Success(42))
```

If the future value is not valid, the future returned by filter will fail with a NoSuchElementException:

```
scala> val invalid = fut.filter(res => res < 0)
invalid: scala.concurrent.Future[Int] = ...

scala> invalid.value
res1: Option[scala.util.Try[Int]] =
  Some(Failure(java.util.NoSuchElementException:
  Future.filter predicate is not satisfied))
```

697

Because Future also offers a withFilter method, you can perform the same operation with for expression filters:

```
scala> val valid = for (res <- fut if res > 0) yield res
valid: scala.concurrent.Future[Int] = ...

scala> valid.value
res2: Option[scala.util.Try[Int]] = Some(Success(42))

scala> val invalid = for (res <- fut if res < 0) yield res
invalid: scala.concurrent.Future[Int] = ...

scala> invalid.value
res3: Option[scala.util.Try[Int]] =
  Some(Failure(java.util.NoSuchElementException:
  Future.filter predicate is not satisfied))
```

Future's collect method allows you to validate the future value and transform it in one operation. If the partial function passed to collect is defined at the future result, the future returned by collect will succeed with that value transformed by the function:

```
scala> val valid =
         fut collect { case res if res > 0 => res + 46 }
valid: scala.concurrent.Future[Int] = ...

scala> valid.value
res17: Option[scala.util.Try[Int]] = Some(Success(88))
```

Otherwise, the future will fail with NoSuchElementException:

```
scala> val invalid =
         fut collect { case res if res < 0 => res + 46 }
invalid: scala.concurrent.Future[Int] = ...

scala> invalid.value
res18: Option[scala.util.Try[Int]] =
  Some(Failure(java.util.NoSuchElementException:
  Future.collect partial function is not defined at: 42))
```

**Dealing with failure: `failed`, `fallbackTo`, `recover`, and `recoverWith`**

Scala's future provides ways to work with futures that fail, including `failed`, `fallbackTo`, `recover`, and `recoverWith`. The `failed` method will transform a failed future of any type into a successful `Future[Throwable]` that holds onto the exception that caused the failure. Here's an example:

```
scala> val failure = Future { 42 / 0 }
failure: scala.concurrent.Future[Int] = ...

scala> failure.value
res23: Option[scala.util.Try[Int]] =
  Some(Failure(java.lang.ArithmeticException: / by zero))

scala> val expectedFailure = failure.failed
expectedFailure: scala.concurrent.Future[Throwable] = ...

scala> expectedFailure.value
res25: Option[scala.util.Try[Throwable]] =
  Some(Success(java.lang.ArithmeticException: / by zero))
```

If the future on which the `failed` method is called ultimately succeeds, the future returned by `failed` will itself fail with a `NoSuchElementException`. The `failed` method is appropriate, therefore, only when you expect that the future will fail. Here's an example:

```
scala> val success = Future { 42 / 1 }
success: scala.concurrent.Future[Int] = ...

scala> success.value
res21: Option[scala.util.Try[Int]] = Some(Success(42))

scala> val unexpectedSuccess = success.failed
unexpectedSuccess: scala.concurrent.Future[Throwable] = ...

scala> unexpectedSuccess.value
res26: Option[scala.util.Try[Throwable]] =
  Some(Failure(java.util.NoSuchElementException:
  Future.failed not completed with a throwable.))
```

The `fallbackTo` method allows you to provide an alternate future to use in case the future on which you invoke `fallbackTo` fails. Here's an example in which a failed future falls back to a successful future:

```
scala> val fallback = failure.fallbackTo(success)
fallback: scala.concurrent.Future[Int] = ...

scala> fallback.value
res27: Option[scala.util.Try[Int]] = Some(Success(42))
```

If the original future on which `fallbackTo` is invoked fails, a failure of the future passed to `fallbackTo` is essentially ignored. The future returned by `fallbackTo` will fail with the initial exception. Here's an example:

```
scala> val failedFallback = failure.fallbackTo(
         Future { val res = 42; require(res < 0); res }
       )
failedFallback: scala.concurrent.Future[Int] = ...

scala> failedFallback.value
res28: Option[scala.util.Try[Int]] =
  Some(Failure(java.lang.ArithmeticException: / by zero))
```

The `recover` method allows you to transform a failed future into a successful one, allowing a successful future's result to pass through unchanged. For example, on a future that fails with `ArithmeticException`, you can use the `recover` method to transform the failure into a success, like this:

```
scala> val recovered = failedFallback recover {
         case ex: ArithmeticException => -1
       }
recovered: scala.concurrent.Future[Int] = ...

scala> recovered.value
res32: Option[scala.util.Try[Int]] = Some(Success(-1))
```

If the original future doesn't fail, the future returned by `recover` will complete with the same value:

```
scala> val unrecovered = fallback recover {
         case ex: ArithmeticException => -1
       }
unrecovered: scala.concurrent.Future[Int] = ...

scala> unrecovered.value
res33: Option[scala.util.Try[Int]] = Some(Success(42))
```

Similarly, if the partial function passed to `recover` isn't defined at the exception with which the original future ultimately fails, that original failure will pass through:

```scala
scala> val alsoUnrecovered = failedFallback recover {
         case ex: IllegalArgumentException => -2
       }
alsoUnrecovered: scala.concurrent.Future[Int] = ...

scala> alsoUnrecovered.value
res34: Option[scala.util.Try[Int]] =
  Some(Failure(java.lang.ArithmeticException: / by zero))
```

The `recoverWith` method is similar to `recover`, except instead of recovering to a value like `recover`, the `recoverWith` method allows you to recover to a future value. Here's an example:

```scala
scala> val alsoRecovered = failedFallback recoverWith {
         case ex: ArithmeticException => Future { 42 + 46 }
       }
alsoRecovered: scala.concurrent.Future[Int] = ...

scala> alsoRecovered.value
res35: Option[scala.util.Try[Int]] = Some(Success(88))
```

As with `recover`, if either the original future doesn't fail, or the partial function passed to `recoverWith` isn't defined at the exception the original future ultimately fails with, the original success (or failure) will pass through to the future returned by `recoverWith`.

## Mapping both possibilities: `transform`

Future's `transform` method accepts two functions with which to transform a future: one to use in case of success and the other in case of failure:

```scala
scala> val first = success.transform(
         res => res * -1,
         ex => new Exception("see cause", ex)
       )
first: scala.concurrent.Future[Int] = ...
```

701

If the future succeeds, the first function is used:

```
scala> first.value
res42: Option[scala.util.Try[Int]] = Some(Success(-42))
```

If the future fails, the second function is used:

```
scala> val second = failure.transform(
          res => res * -1,
          ex => new Exception("see cause", ex)
       )
second: scala.concurrent.Future[Int] = ...

scala> second.value
res43: Option[scala.util.Try[Int]] =
   Some(Failure(java.lang.Exception: see cause))
```

Note that with the `transform` method shown in the previous examples, you can't change a successful future into a failed one, nor can you change a failed future into a successful one. To make this kind of transformation easier, Scala 2.12 introduced an alternate overloaded form of `transform` that takes a function from Try to Try. Here are some examples:

```
scala> val firstCase = success.transform { // Scala 2.12
          case Success(res) => Success(res * -1)
          case Failure(ex) =>
            Failure(new Exception("see cause", ex))
       }
first: scala.concurrent.Future[Int] = ...

scala> firstCase.value
res6: Option[scala.util.Try[Int]] = Some(Success(-42))

scala> val secondCase = failure.transform {
          case Success(res) => Success(res * -1)
          case Failure(ex) =>
            Failure(new Exception("see cause", ex))
       }
secondCase: scala.concurrent.Future[Int] = ...

scala> secondCase.value
res8: Option[scala.util.Try[Int]] =
   Some(Failure(java.lang.Exception: see cause))
```

Here's an example of using the new `transform` method to transform a failure into a success:

```
scala> val nonNegative = failure.transform { // Scala 2.12
         case Success(res) => Success(res.abs + 1)
         case Failure(_) => Success(0)
       }
nonNegative: scala.concurrent.Future[Int] = ...

scala> nonNegative.value
res11: Option[scala.util.Try[Int]] = Some(Success(0))
```

### Combining futures: `zip`, `Future.fold`, `Future.reduce`, `Future.sequence`, and `Future.traverse`

Future and its companion object offer methods that allow you to combine multiple futures. The `zip` method will transform two successful futures into a future tuple of both values. Here's an example:

```
scala> val zippedSuccess = success zip recovered
zippedSuccess: scala.concurrent.Future[(Int, Int)] = ...

scala> zippedSuccess.value
res46: Option[scala.util.Try[(Int, Int)]] =
       Some(Success((42,-1)))
```

If either of the futures fail, however, the future returned by `zip` will also fail with the same exception:

```
scala> val zippedFailure = success zip failure
zippedFailure: scala.concurrent.Future[(Int, Int)] = ...

scala> zippedFailure.value
res48: Option[scala.util.Try[(Int, Int)]] =
  Some(Failure(java.lang.ArithmeticException: / by zero))
```

If both futures fail, the failed future that results will contain the exception stored in the initial future, the one on which `zip` was invoked.

Future's companion object offers a `fold` method that allows you to accumulate a result across a `TraversableOnce` collection of futures, yielding a future result. If all futures in the collection succeed, the resulting future will succeed with the accumulated result. If any future in the collection fails, the resulting future will fail. If multiple futures fail, the result will fail with the same exception with which the first future (earliest in the TraversableOnce collection) fails. Here's an example:

```
scala> val fortyTwo = Future { 21 + 21 }
fortyTwo: scala.concurrent.Future[Int] = ...

scala> val fortySix = Future { 23 + 23 }
fortySix: scala.concurrent.Future[Int] = ...

scala> val futureNums = List(fortyTwo, fortySix)
futureNums: List[scala.concurrent.Future[Int]] = ...

scala> val folded =
         Future.fold(futureNums)(0) { (acc, num) =>
           acc + num
         }
folded: scala.concurrent.Future[Int] = ...

scala> folded.value
res53: Option[scala.util.Try[Int]] = Some(Success(88))
```

The Future.reduce method performs a fold without a zero, using the initial future result as the start value. Here's an example:

```
scala> val reduced =
         Future.reduce(futureNums) { (acc, num) =>
           acc + num
         }
reduced: scala.concurrent.Future[Int] = ...

scala> reduced.value
res54: Option[scala.util.Try[Int]] = Some(Success(88))
```

If you pass an empty collection to reduce, the resulting future will fail with a NoSuchElementException.

The Future.sequence method transforms a TraversableOnce collection of futures into a future TraversableOnce of values. For instance, in the following example, sequence is used to transform a List[Future[Int]] to a Future[List[Int]]:

704

```
scala> val futureList = Future.sequence(futureNums)
futureList: scala.concurrent.Future[List[Int]] = ...

scala> futureList.value
res55: Option[scala.util.Try[List[Int]]] =
  Some(Success(List(42, 46)))
```

The Future.traverse method will change a TraversableOnce of any element type into a TraversableOnce of futures and "sequence" that into a future TraversableOnce of values. For example, here a List[Int] is transformed into a Future[List[Int]] by Future.traverse:

```
scala> val traversed =
        Future.traverse(List(1, 2, 3)) { i => Future(i) }
traversed: scala.concurrent.Future[List[Int]] = ...

scala> traversed.value
res58: Option[scala.util.Try[List[Int]]] =
  Some(Success(List(1, 2, 3)))
```

## Performing side-effects: foreach, onComplete, and andThen

Sometimes you may need to perform a side effect after a future completes. Future provides several methods for this purpose. The most basic method is foreach, which will perform a side effect if a future completes successfully. For instance, in the following example a println is not executed in the case of a failed future, just a successful future:

```
scala> failure.foreach(ex => println(ex))

scala> success.foreach(res => println(res))
42
```

Since for without yield will rewrite to an invocation of foreach, you can also accomplish the same effect using for expressions:

```
scala> for (res <- failure) println(res)

scala> for (res <- success) println(res)
42
```

Future also offers two methods for registering "callback" functions. The onComplete method will be executed whether the future ultimately succeeds or fails. The function will be passed a Try—a Success holding the result if the future succeeded, else a Failure holding the exception that caused the future to fail. Here's an example:

```
scala> import scala.util.{Success, Failure}
import scala.util.{Success, Failure}

scala> success onComplete {
         case Success(res) => println(res)
         case Failure(ex) => println(ex)
       }
42

scala> failure onComplete {
         case Success(res) => println(res)
         case Failure(ex) => println(ex)
       }
java.lang.ArithmeticException: / by zero
```

Future does not guarantee any order of execution for callback functions registered with onComplete. If you want to enforce an order for callback functions, you must use andThen instead. The andThen method returns a new future that mirrors (succeeds or fails in the same way as) the original future on which you invoke andThen, but it does not complete until the callback function has been fully executed:

```
scala> val newFuture = success andThen {
         case Success(res) => println(res)
         case Failure(ex) => println(ex)
       }
42
newFuture: scala.concurrent.Future[Int] = ...

scala> newFuture.value
res76: Option[scala.util.Try[Int]] = Some(Success(42))
```

Note that if a callback function passed to andThen throws an exception when executed, that exception will not be propagated to subsequent callbacks or reported via the resulting future.

**Other methods added in 2.12: `flatten`, `zipWith`, and `transformWith`**

The `flatten` method, added in 2.12, transforms a Future nested inside another Future into a Future of the nested type. For example, `flatten` can transform a `Future[Future[Int]]` into a `Future[Int]`:

```
scala> val nestedFuture = Future { Future { 42 } }
nestedFuture: Future[Future[Int]] = ...

scala> val flattened = nestedFuture.flatten // Scala 2.12
flattened: scala.concurrent.Future[Int] = Future(Success(42))
```

The `zipWith` method, added in 2.12, essentially zips two Futures together, then performs a map on the resulting tuple. Here's an example of the two-step process, a `zip` followed by a `map`:

```
scala> val futNum = Future { 21 + 21 }
futNum: scala.concurrent.Future[Int] = ...

scala> val futStr = Future { "ans" + "wer" }
futStr: scala.concurrent.Future[String] = ...

scala> val zipped = futNum zip futStr
zipped: scala.concurrent.Future[(Int, String)] = ...

scala> val mapped = zipped map {
         case (num, str) => s"$num is the $str"
       }
mapped: scala.concurrent.Future[String] = ...

scala> mapped.value
res2: Option[scala.util.Try[String]] =
    Some(Success(42 is the answer))
```

The `zipWith` method allows you to perform the same operation in one step:

```
scala> val fut = futNum.zipWith(futStr) { // Scala 2.12
         case (num, str) => s"$num is the $str"
       }
zipWithed: scala.concurrent.Future[String] = ...

scala> fut.value
res3: Option[scala.util.Try[String]] =
    Some(Success(42 is the answer))
```

Future also gained a `transformWith` method in Scala 2.12, which allows you to transform a future using a function from Try to Future. Here's an example:

```
scala> val flipped = success.transformWith { // Scala 2.12
         case Success(res) =>
           Future { throw new Exception(res.toString) }
         case Failure(ex) => Future { 21 + 21 }
       }
flipped: scala.concurrent.Future[Int] = ...

scala> flipped.value
res5: Option[scala.util.Try[Int]] =
      Some(Failure(java.lang.Exception: 42))
```

The `transformWith` method is similar to the new, overloaded `transform` method added in Scala 2.12, except instead of yielding a Try in your passed function as in `transform`, `transformWith` allows you to yield a future.

## 32.4   Testing with Futures

One advantage of Scala's futures is that they help you avoid blocking. On most JVM implementations, after creating just a few thousand threads, the cost of context switching between threads will degrade performance to an unacceptable level. By avoiding blocking, you can keep the finite number of threads you decide to work with busy. Nevertheless, Scala does allow you to block on a future result when you need to. Scala's Await object facilitates blocking to wait for future results. Here's an example:

```
scala> import scala.concurrent.Await
import scala.concurrent.Await

scala> import scala.concurrent.duration._
import scala.concurrent.duration._

scala> val fut = Future { Thread.sleep(10000); 21 + 21 }
fut: scala.concurrent.Future[Int] = ...

scala> val x = Await.result(fut, 15.seconds) // blocks
x: Int = 42
```

`Await.result` takes a `Future` and a `Duration`. The `Duration` indicates how long `Await.result` should wait for a `Future` to complete before timing out. In this example, fifteen seconds was specified for the `Duration`. Thus the `Await.result` method should not time out before the future completes with its eventual value, 42.

One place where blocking has been generally accepted is in tests of asynchronous code. Now that the `Await.result` has returned, you can perform a computation using that result, such as an assertion in a test:

```
scala> import org.scalatest.Matchers._
import org.scalatest.Matchers._

scala> x should be (42)
res0: org.scalatest.Assertion = Succeeded
```

Alternatively, you can use blocking constructs provided by ScalaTest's trait `ScalaFutures`. For example, the `futureValue` method, implicitly added to `Future` by `ScalaFutures`, will block until the future completes. If the future fails, `futureValue` will throw a `TestFailedException` describing the problem. If the future succeeds, `futureValue` will return the successful result of the future so you can perform asssertions on that value:

```
scala> import org.scalatest.concurrent.ScalaFutures._
import org.scalatest.concurrent.ScalaFutures._

scala> val fut = Future { Thread.sleep(10000); 21 + 21 }
fut: scala.concurrent.Future[Int] = ...

scala> fut.futureValue should be (42) // futureValue blocks
res1: org.scalatest.Assertion = Succeeded
```

While blocking in tests is often fine, ScalaTest 3.0 adds "async" testing styles that allow you to test futures without blocking. Given a future, instead of blocking and performing assertions on the result, you can map assertions directly onto that future and return the resulting `Future[Assertion]` to ScalaTest. An example is shown in Listing 32.1. When the future assertion completes, ScalaTest will fire events (test succeeded, test failed, *etc.*) to the test reporter asynchronously.

The async testing use case illustrates a general principle for working with futures: Once in "future space," try to stay in future space. Don't block on a

```
import org.scalatest.AsyncFunSpec
import scala.concurrent.Future

class AddSpec extends AsyncFunSpec {

  def addSoon(addends: Int*): Future[Int] =
      Future { addends.sum }

  describe("addSoon") {
    it("will eventually compute a sum of passed Ints") {
      val futureSum: Future[Int] = addSoon(1, 2)
      // You can map assertions onto a Future, then return
      // the resulting Future[Assertion] to ScalaTest:
      futureSum map { sum => assert(sum == 3) }
    }
  }
}
```

Listing 32.1 · Returning a future assertion to ScalaTest.

future then continue the computation with the result. Stay asynchronous by performing a series of transformations, each of which returns a new future to transform. To get results out of future space, register side effects to be performed asynchronously once futures complete. This approach will help you make maximum use of your threads.

## 32.5 Conclusion

Concurrent programming gives you great power. It lets you simplify your code and take advantage of multiple processors. It's unfortunate that the most widely used concurrency primitives, threads, locks, and monitors, are such a minefield of deadlocks and race conditions. Futures provide a way out of that minefield, letting you write concurrent programs without as great a risk of deadlocks and race conditions. This chapter has introduced several fundamental constructs for working with futures in Scala, including how to create futures, how to transform them, and how to test them, among other nuts and bolts. It then showed you how to use these constructs as part of a general futures style.

# Chapter 33

# Combinator Parsing

Occasionally, you may need to process a small, special-purpose language. For example, you may need to read configuration files for your software, and you want to make them easier to modify by hand than XML. Alternatively, maybe you want to support an input language in your program, such as search terms with boolean operators (computer, find me a movie "with 'space ships' and without 'love stories'"). Whatever the reason, you are going to need a *parser*. You need a way to convert the input language into some data structure your software can process.

Essentially, you have only a few choices. One choice is to roll your own parser (and lexical analyzer). If you are not an expert, this is hard. If you are an expert, it is still time consuming.

An alternative choice is to use a parser generator. There exist quite a few of these generators. Some of the better known are Yacc and Bison for parsers written in C and ANTLR for parsers written in Java. You'll probably also need a scanner generator such as Lex, Flex, or JFlex to go with it. This might be the best solution, except for a couple of inconveniences. You need to learn new tools, including their—sometimes obscure—error messages. You also need to figure out how to connect the output of these tools to your program. This might limit the choice of your programming language, and complicate your tool chain.

This chapter presents a third alternative. Instead of using the standalone domain specific language of a parser generator, you will use an *internal domain specific language*, or internal DSL for short. The internal DSL will consist of a library of *parser combinators*—functions and operators defined in Scala that will serve as building blocks for parsers. These building blocks will map one to one to the constructions of a context-free grammar, to make them easy to understand.

This chapter introduces only one language feature that was not explained before: `this` aliasing, in Section 33.6. The chapter does, however, heavily use several other features that were explained in previous chapters. Among others, parameterized types, abstract types, functions as objects, operator overloading, by-name parameters, and implicit conversions all play important roles. The chapter shows how these language elements can be combined in the design of a very high-level library.

The concepts explained in this chapter tend to be a bit more advanced than previous chapters. If you have a good grounding in compiler construction, you'll profit from it reading this chapter, because it will help you put things better in perspective. However, the only prerequisite for understanding this chapter is that you know about regular and context-free grammars. If you don't, the material in this chapter can also safely be skipped.

## 33.1   Example: Arithmetic expressions

We'll start with an example. Say you want to construct a parser for arithmetic expressions consisting of floating-point numbers, parentheses, and the binary operators +, -, *, and /. The first step is always to write down a grammar for the language to be parsed. Here's the grammar for arithmetic expressions:

$$
\begin{aligned}
\textit{expr} \quad &::= \quad \textit{term} \ \{\texttt{"+"} \ \textit{term} \ | \ \texttt{"-"} \ \textit{term}\}. \\
\textit{term} \quad &::= \quad \textit{factor} \ \{\texttt{"*"} \ \textit{factor} \ | \ \texttt{"/"} \ \textit{factor}\}. \\
\textit{factor} \quad &::= \quad \textit{floatingPointNumber} \ | \ \texttt{"("} \ \textit{expr} \ \texttt{")"}.
\end{aligned}
$$

Here, | denotes alternative productions, and { ... } denotes repetition (zero or more times). And although there's no use of it in this example, [ ... ] denotes an optional occurrence.

This context-free grammar defines formally a language of arithmetic expressions. Every expression (represented by *expr*) is a *term*, which can be followed by a sequence of + or - operators and further *term*s. A *term* is a *factor*, possibly followed by a sequence of * or / operators and further *factor*s. A *factor* is either a numeric literal or an expression in parentheses. Note that the grammar already encodes the relative precedence of operators. For instance, * binds more tightly than +, because a * operation gives a *term*, whereas a + operation gives an *expr*, and *exprs* can contain *terms* but a *term* can contain an *expr* only when the latter is enclosed in parentheses.

Now that you have defined the grammar, what's next? If you use Scala's combinator parsers, you are basically done! You only need to perform some systematic text replacements and wrap the parser in a class, as shown in Listing 33.1:

```scala
import scala.util.parsing.combinator._

class Arith extends JavaTokenParsers {
  def expr: Parser[Any] = term~rep("+"~term | "-"~term)
  def term: Parser[Any] = factor~rep("*"~factor | "/"~factor)
  def factor: Parser[Any] = floatingPointNumber | "("~expr~")"
}
```

Listing 33.1 · An arithmetic expression parser.

The parsers for arithmetic expressions are contained in a class that inherits from the trait `JavaTokenParsers`. This trait provides the basic machinery for writing a parser and also provides some primitive parsers that recognize some word classes: identifiers, string literals and numbers. In the example in Listing 33.1 you need only the primitive `floatingPointNumber` parser, which is inherited from this trait.

The three definitions in class `Arith` represent the productions for arithmetic expressions. As you can see, they follow very closely the productions of the context-free grammar. In fact, you could generate this part automatically from the context-free grammar, by performing a number of simple text replacements:

1. Every production becomes a method, so you need to prefix it with `def`.

2. The result type of each method is `Parser[Any]`, so you need to change the ::= symbol to ": `Parser[Any]` =". You'll find out later in this chapter what the type `Parser[Any]` signifies, and also how to make it more precise.

3. In the grammar, sequential composition was implicit, but in the program it is expressed by an explicit operator: ~. So you need to insert a ~ between every two consecutive symbols of a production. In the example in Listing 33.1 we chose not to write any spaces around the ~

713

operator. That way, the parser code keeps closely to the visual appearance of the grammar—it just replaces spaces by ~ characters.

4. Repetition is expressed rep( ... ) instead of { ... }. Analogously (though not shown in the example), option is expressed opt( ... ) instead of [ ... ].

5. The period (.) at the end of each production is omitted—you can, however, write a semicolon (;) if you prefer.

That's all there is to it. The resulting class Arith defines three parsers, expr, term and factor, which can be used to parse arithmetic expressions and their parts.

## 33.2 Running your parser

You can exercise your parser with the following small program:

```
object ParseExpr extends Arith {
  def main(args: Array[String]) = {
    println("input : " + args(0))
    println(parseAll(expr, args(0)))
  }
}
```

The ParseExpr object defines a main method that parses the first command-line argument passed to it. It prints the original input argument, and then prints its parsed version. Parsing is done by the expression:

```
parseAll(expr, input)
```

This expression applies the parser, expr, to the given input. It expects that all of the input matches, i.e., that there are no characters trailing a parsed expression. There's also a method parse, which allows you to parse an input prefix, leaving some remainder unread.

You can run the arithmetic parser with the following command:

```
$ scala ParseExpr "2 * (3 + 7)"
input: 2 * (3 + 7)
[1.12] parsed: ((2~List((*~(((~((3~List())~List((+
~(7~List()))))))~)))))~List())
```

The output tells you that the parser successfully analyzed the input string up to position [1.12]. That means the first line and the twelfth column—in other words, the whole input string—was parsed. Disregard for the moment the result after "parsed:". It is not very useful, and you will find out later how to get more specific parser results.

You can also try to introduce some input string that is not a legal expression. For instance, you could write one closing parenthesis too many:

```
$ scala ParseExpr "2 * (3 + 7))"
input: 2 * (3 + 7))
[1.12] failure: `-' expected but `)' found

2 * (3 + 7))
           ^
```

Here, the expr parser parsed everything until the final closing parenthesis, which does not form part of the arithmetic expression. The parseAll method then issued an error message, which said that it expected a – operator at the point of the closing parenthesis. You'll find out later in this chapter why it produced this particular error message, and how you can improve it.

## 33.3  Basic regular expression parsers

The parser for arithmetic expressions made use of another parser, named floatingPointNumber. This parser, which was inherited from Arith's supertrait, JavaTokenParsers, recognizes a floating point number in the format of Java. But what do you do if you need to parse numbers in a format that's a bit different from Java's? In this situation, you can use a *regular expression parser*.

The idea is that you can use any regular expression as a parser. The regular expression parses all strings that it can match. Its result is the parsed string. For instance, the regular expression parser shown in Listing 33.2 describes Java's identifiers:

```
object MyParsers extends RegexParsers {
  val ident: Parser[String] = """[a-zA-Z_]\w*""".r
}
```

Listing 33.2 · A regular expression parser for Java identifiers.

The `MyParsers` object of Listing 33.2 inherits from trait `RegexParsers`, whereas `Arith` inherited from `JavaTokenParsers`. Scala's parsing combinators are arranged in a hierarchy of traits, which are all contained in package `scala.util.parsing.combinator`. The top-level trait is `Parsers`, which defines a very general parsing framework for all sorts of input. One level below is trait `RegexParsers`, which requires that the input is a sequence of characters and provides for regular expression parsing. Even more specialized is trait `JavaTokenParsers`, which implements parsers for basic classes of words (or *tokens*) as they are defined in Java.

## 33.4  Another example: JSON

JSON, the JavaScript Object Notation, is a popular data interchange format. In this section, we'll show you how to write a parser for it. Here's a grammar that describes the syntax of JSON:

$$
\begin{aligned}
value \quad &::= \quad obj \mid arr \mid stringLiteral \mid \\
&\qquad floatingPointNumber \mid \\
&\qquad \texttt{"null"} \mid \texttt{"true"} \mid \texttt{"false"}. \\
obj \quad &::= \quad \texttt{"\{"} \; [members] \; \texttt{"\}"}. \\
arr \quad &::= \quad \texttt{"["} \; [values] \; \texttt{"]"}. \\
members \quad &::= \quad member \; \{\texttt{","} \; member\}. \\
member \quad &::= \quad stringLiteral \; \texttt{":"} \; value. \\
values \quad &::= \quad value \; \{\texttt{","} \; value\}.
\end{aligned}
$$

A JSON value is an object, array, string, number, or one of the three reserved words `null`, `true`, or `false`. A JSON object is a (possibly empty) sequence of members separated by commas and enclosed in braces. Each member is a string/value pair where the string and the value are separated by a colon. Finally, a JSON array is a sequence of values separated by commas and enclosed in square brackets. As an example, Listing 33.3 contains an address-book formatted as a JSON object.

Parsing such data is straightforward when using Scala's parser combinators. The complete parser is shown in Listing 33.4. This parser follows the same structure as the arithmetic expression parser. It is again a straightforward mapping of the productions of the JSON grammar. The productions

```
{
  "address book": {
    "name": "John Smith",
    "address": {
      "street": "10 Market Street",
      "city"  : "San Francisco, CA",
      "zip"   : 94111
    },
    "phone numbers": [
      "408 338-4238",
      "408 111-6892"
    ]
  }
}
```

Listing 33.3 · Data in JSON format.

use one shortcut that simplifies the grammar: The `repsep` combinator parses a (possibly empty) sequence of terms that are separated by a given separator string. For instance, in the example in Listing 33.4, `repsep(member, ",")` parses a comma-separated sequence of `member` terms. Otherwise, the productions in the parser correspond exactly to the productions in the grammar, as was the case for the arithmetic expression parsers.

To try out the JSON parsers, we'll change the framework a bit, so that the parser operates on a file instead of on the command line:

```
import java.io.FileReader

object ParseJSON extends JSON {
  def main(args: Array[String]) = {
    val reader = new FileReader(args(0))
    println(parseAll(value, reader))
  }
}
```

The `main` method in this program first creates a `FileReader` object. It then parses the characters returned by that reader according to the `value` production of the JSON grammar. Note that `parseAll` and `parse` exist in overloaded variants: both can take a character sequence or alternatively an input reader as second argument.

If you store the "address book" object shown in Listing 33.3 into a file named address-book.json and run the ParseJSON program on it, you should get:

```
$ scala ParseJSON address-book.json
[13.4] parsed: (({~List((("address book"~:)~(({~List(((
"name"~:)~"John Smith"), (("address"~:)~(({~List(((
"street"~:)~"10 Market Street"), (("city"~:)~"San Francisco
,CA"), (("zip"~:)~94111)))~})), (("phone numbers"~:)~((([~
List("408 338-4238", "408 111-6892"))~]))))~}))))~})
```

## 33.5  Parser output

The ParseJSON program successfully parsed the JSON address book. However, the parser output looks strange. It seems to be a sequence composed of bits and pieces of the input glued together with lists and ~ combinations. This output is not very useful. It is less readable for humans than the input, but it is also too disorganized to be easily analyzable by a computer. It's time to do something about this.

```
import scala.util.parsing.combinator._

class JSON extends JavaTokenParsers {
  def value : Parser[Any] = obj | arr |
                            stringLiteral |
                            floatingPointNumber |
                            "null" | "true" | "false"

  def obj   : Parser[Any] = "{"~repsep(member, ",")~"}"

  def arr   : Parser[Any] = "["~repsep(value, ",")~"]"

  def member: Parser[Any] = stringLiteral~":"~value
}
```

Listing 33.4 · A simple JSON parser.

To figure out what to do, you need to know first what the individual parsers in the combinator frameworks return as a result (provided they succeed in parsing the input). Here are the rules:

1. Each parser written as a string (such as: "{" or ":" or "null") returns the parsed string itself.

2. Regular expression parsers such as """[a-zA-Z_]\w*""".r also return the parsed string itself. The same holds for regular expression parsers such as stringLiteral or floatingPointNumber, which are inherited from trait JavaTokenParsers.

3. A sequential composition P~Q returns the results of both P and of Q. These results are returned in an instance of a case class that is also written ~. So if P returns "true" and Q returns "?", then the sequential composition P~Q returns ~("true", "?"), which prints as (true~?).

4. An alternative composition P | Q returns the result of either P or Q, whichever one succeeds.

5. A repetition rep(P) or repsep(P, separator) returns a list of the results of all runs of P.

6. An option opt(P) returns an instance of Scala's Option type. It returns Some(R) if P succeeds with result R and None if P fails.

With these rules you can now deduce *why* the parser output appeared as it did in the previous examples. However, the output is still not very convenient. It would be much better to map a JSON object into an internal Scala representation that represents the meaning of the JSON value. A more natural representation would be as follows:

- A JSON object is represented as a Scala map of type Map[String, Any]. Every member is represented as a key/value binding in the map.

- A JSON array is represented as a Scala list of type List[Any].

- A JSON string is represented as a Scala String.

- A JSON numeric literal is represented as a Scala Double.

- The values `true`, `false`, and `null` are represented as the Scala values with the same names.

To produce this representation, you need to make use of one more combination form for parsers: `^^`.

The `^^` operator *transforms* the result of a parser. Expressions using this operator have the form P `^^` f where P is a parser and f is a function. P `^^` f parses the same sentences as just P. Whenever P returns with some result R, the result of P `^^` f is f(R).

As an example, here is a parser that parses a floating point number and converts it to a Scala value of type `Double`:

```
floatingPointNumber ^^ (_.toDouble)
```

And here is a parser that parses the string `"true"` and returns Scala's boolean true value:

```
"true" ^^ (x => true)
```

Now for more advanced transformations. Here's a new version of a parser for JSON objects that returns a Scala `Map`:

```
def obj: Parser[Map[String, Any]] = // Can be improved
  "{"~repsep(member, ",")~"}" ^^
    { case "{"~ms~"}" => Map() ++ ms }
```

Remember that the ~ operator produces as its result an instance of a case class with the same name: ~. Here's a definition of that class—it's an inner class of trait `Parsers`:

```
case class ~[+A, +B](x: A, y: B) {
  override def toString = "(" + x + "~" + y + ")"
}
```

The name of the class is intentionally the same as the name of the sequence combinator method, ~. That way, you can match parser results with patterns that follow the same structure as the parsers themselves. For instance, the pattern `"{"~ms~"}"` matches a result string `"{"` followed by a result variable ms, which is followed in turn by a result string `"}"`. This pattern corresponds exactly to what is returned by the parser on the left of the `^^`.

In its desugared versions where the ~ operator comes first, the same pattern reads ~(~("{", ms), "}"), but this is much less legible.

The purpose of the "{"~ms~"}" pattern is to strip off the braces so that you can get at the list of members resulting from the repsep(member, ",") parser. In cases like these there is also an alternative that avoids producing unnecessary parser results that are immediately discarded by the pattern match. The alternative makes use of the ~> and <~ parser combinators. Both express sequential composition like ~, but ~> keeps only the result of its right operand, whereas <~ keeps only the result of its left operand. Using these combinators, the JSON object parser can be expressed more succinctly:

```
def obj: Parser[Map[String, Any]] =
  "{"~> repsep(member, ",") <~"}" ^^ (Map() ++ _)
```

Listing 33.5 shows a full JSON parser that returns meaningful results. If you run this parser on the address-book.json file, you will get the following result (after adding some newlines and indentation):

```
$ scala JSON1Test address-book.json
[14.1] parsed: Map(
  address book -> Map(
    name -> John Smith,
    address -> Map(
      street -> 10 Market Street,
      city -> San Francisco, CA,
      zip -> 94111),
    phone numbers -> List(408 338-4238, 408 111-6892)
  )
)
```

This is all you need to know in order to get started writing your own parsers. As an aide to memory, Table 33.1 lists the parser combinators that were discussed so far.

**Symbolic versus alphanumeric names**

Many of the parser combinators in Table 33.1 use symbolic names. This has both advantages and disadvantages. On the minus side, symbolic names take time to learn. Users who are unfamiliar with Scala's combinator parsing

```
import scala.util.parsing.combinator._

class JSON1 extends JavaTokenParsers {

  def obj: Parser[Map[String, Any]] =
    "{"~> repsep(member, ",") <~"}" ^^ (Map() ++ _)

  def arr: Parser[List[Any]] =
    "["~> repsep(value, ",") <~"]"

  def member: Parser[(String, Any)] =
    stringLiteral~":"~value ^^
      { case name~":"~value => (name, value) }

  def value: Parser[Any] = (
    obj
  | arr
  | stringLiteral
  | floatingPointNumber ^^ (_.toDouble)
  | "null"  ^^ (x => null)
  | "true"  ^^ (x => true)
  | "false" ^^ (x => false)
  )
}
```

Listing 33.5 · A full JSON parser that returns meaningful results.

Table 33.1 · Summary of parser combinators

| | |
|---|---|
| "..." | literal |
| "...".r | regular expression |
| P~Q | sequential composition |
| P <~ Q, P ~> Q | sequential composition; keep left/right only |
| P \| Q | alternative |
| opt(P) | option |
| rep(P) | repetition |
| repsep(P, Q) | interleaved repetition |
| P ^^ f | result conversion |

---

### Turning off semicolon inference

Note that the body of the `value` parser in Listing 33.5 is enclosed in
parentheses. This is a little trick to disable semicolon inference in
parser expressions. You saw in Section 4.2 that Scala assumes there's
a semicolon between any two lines that can be separate statements
syntactically, unless the first line ends in an infix operator, or the two
lines are enclosed in parentheses or square brackets. Now, you could
have written the | operator at the end of the each alternative instead of
at the beginning of the following one, like this:

```
def value: Parser[Any] =
  obj |
  arr |
  stringLiteral |
  ...
```

In that case, no parentheses around the body of the `value` parser would
have been required. However, some people prefer to see the | operator
at the beginning of the second alternative rather than at the end of the
first. Normally, this would lead to an unwanted semicolon between the
two lines, like this:

```
  obj;   // semicolon implicitly inserted
| arr
```

The semicolon changes the structure of the code, causing it to fail
compilation. Putting the whole expression in parentheses avoids the
semicolon and makes the code compile correctly.

---

libraries are probably mystified what ~, ~>, or ˆˆ mean. On the plus side,
symbolic names are short, and can be chosen to have the "right" precedences
and associativities. For instance, the parser combinators ~, ˆˆ, and | are
chosen intentionally in decreasing order of precedence. A typical grammar
production is composed of alternatives that have a parsing part and a transformation part. The parsing part usually contains several sequential items
separated by ~ operators. With the chosen precedences of ~, ˆˆ, and | you
can write such a grammar production without needing any parentheses.

Furthermore, symbolic operators take less visual real estate than alphabetic ones. That's important for a parser because it lets you concentrate on the grammar at hand, instead of the combinators themselves. To see the difference, imagine for a moment that sequential composition (~) was called andThen and alternative (|) was called orElse. The arithmetic expression parsers in Listing 33.1 on page 713 would look as follows:

```
class ArithHypothetical extends JavaTokenParsers {
  def expr: Parser[Any]   =
    term andThen rep(("+" andThen term) orElse
                     ("-" andThen term))
  def term: Parser[Any]   =
    factor andThen rep(("*" andThen factor) orElse
                       ("/" andThen factor))
  def factor: Parser[Any] =
    floatingPointNumber orElse
    ("(" andThen expr andThen ")")
}
```

You notice that the code becomes much longer, and that it's hard to "see" the grammar among all those operators and parentheses. On the other hand, somebody new to combinator parsing could probably figure out better what the code is supposed to do.

## 33.6  Implementing combinator parsers

The previous sections have shown that Scala's combinator parsers provide a convenient means for constructing your own parsers. Since they are nothing more than a Scala library, they fit seamlessly into your Scala programs. So it's very easy to combine a parser with some code that processes the results it delivers, or to rig a parser so that it takes its input from some specific source (say, a file, a string, or a character array).

How is this achieved? In the rest of this chapter you'll take a look "under the hood" of the combinator parser library. You'll see what a parser is, and how the primitive parsers and parser combinators encountered in previous sections are implemented. You can safely skip these parts if all you want to do is write some simple combinator parsers. On the other hand, reading the rest of this chapter should give you a deeper understanding of combinator

---

## Choosing between symbolic and alphabetic names

As guidelines for choosing between symbolic and alphabetic names we recommend the following:

- Use symbolic names in cases where they already have a universally established meaning. For instance, nobody would recommend writing add instead of + for numeric addition.

- Otherwise, give preference to alphabetic names if you want your code to be understandable to casual readers.

- You can still choose symbolic names for domain-specific libraries, if this gives clear advantages in legibility and you do not expect anyway that a casual reader without a firm grounding in the domain would be able to understand the code immediately.

In the case of parser combinators we are looking at a highly domain-specific language, which casual readers may have trouble understanding even with alphabetic names. Furthermore, symbolic names give clear advantages in legibility for the expert. So we believe their use is warranted in this application.

---

parsers in particular, and of the design principles of a combinator domain-specific language in general.

The core of Scala's combinator parsing framework is contained in the trait scala.util.parsing.combinator.Parsers. This trait defines the Parser type as well as all fundamental combinators. Except where stated explicitly otherwise, the definitions explained in the following two subsections all reside in this trait. That is, they are assumed to be contained in a trait definition that starts as follows:

```
package scala.util.parsing.combinator
trait Parsers {
    ... // code goes here unless otherwise stated
}
```

A Parser is in essence just a function from some input type to a parse result. As a first approximation, the type could be written as follows:

```
type Parser[T] = Input => ParseResult[T]
```

## Parser input

Sometimes, a parser reads a stream of tokens instead of a raw sequence of characters. A separate lexical analyzer is then used to convert a stream of raw characters into a stream of tokens. The type of parser inputs is defined as follows:

```
type Input = Reader[Elem]
```

The class Reader comes from the package scala.util.parsing.input. It is similar to a Stream, but also keeps track of the positions of all the elements it reads. The type Elem represents individual input elements. It is an abstract type member of the Parsers trait:

```
type Elem
```

This means that subclasses and subtraits of Parsers need to instantiate class Elem to the type of input elements that are being parsed. For instance, RegexParsers and JavaTokenParsers fix Elem to be equal to Char. But it would also be possible to set Elem to some other type, such as the type of tokens returned from a separate lexer.

## Parser results

A parser might either succeed or fail on some given input. Consequently class ParseResult has two subclasses for representing success and failure:

```
sealed abstract class ParseResult[+T]
case class Success[T](result: T, in: Input)
  extends ParseResult[T]
case class Failure(msg: String, in: Input)
  extends ParseResult[Nothing]
```

The Success case carries the result returned from the parser in its result parameter. The type of parser results is arbitrary; that's why ParseResult, Success, and Parser are all parameterized with a type parameter T. The type parameter represents the kinds of results returned by a given parser. Success also takes a second parameter, in, which refers to the input immediately following the part that the parser consumed. This field is needed for chaining parsers, so that one parser can operate after another. Note that this is a purely functional approach to parsing. Input is not read as a side effect, but it is kept in a stream. A parser analyzes some part of the input stream, and then returns the remaining part in its result.

The other subclass of `ParseResult` is `Failure`. This class takes as a parameter a message that describes why the parser failed. Like `Success`, `Failure` also takes the remaining input stream as a second parameter. This is needed not for chaining (the parser won't continue after a failure), but to position the error message at the correct place in the input stream.

Note that parse results are defined to be covariant in the type parameter T. That is, a parser returning `Strings` as result, say, is compatible with a parser returning `AnyRefs`.

## The Parser class

The previous characterization of parsers as functions from inputs to parse results was a bit oversimplified. The previous examples showed that parsers also implement *methods* such as ~ for sequential composition of two parsers and | for their alternative composition. So `Parser` is in reality a class that inherits from the function type `Input => ParseResult[T]` and additionally defines these methods:

```
abstract class Parser[+T] extends (Input => ParseResult[T])
{ p =>
    // An unspecified method that defines
    // the behavior of this parser.
    def apply(in: Input): ParseResult[T]

    def ~ ...
    def | ...
    ...
}
```

Since parsers are (*i.e.*, inherit from) functions, they need to define an `apply` method. You see an abstract `apply` method in class `Parser`, but this is just for documentation, as the same method is in any case inherited from the parent type `Input => ParseResult[T]` (recall that this type is an abbreviation for `scala.Function1[Input, ParseResult[T]]`). The `apply` method still needs to be implemented in the individual parsers that inherit from the abstract `Parser` class. These parsers will be discussed after the following section on `this` aliasing.

**Aliasing this**

The body of the Parser class starts with a curious expression:

```
abstract class Parser[+T] extends ... { p =>
```

A clause such as "id =>" immediately after the opening brace of a class template defines the identifier id as an alias for this in the class. It's as if you had written:

```
val id = this
```

in the class body, except that the Scala compiler knows that id is an alias for this. For instance, you could access an object-private member m of the class using either id.m or this.m; the two are completely equivalent. The first expression would not compile if id were just defined as a val with this as its right hand side, because in that case the Scala compiler would treat id as a normal identifier.

You saw syntax like this in Section 29.4, where it was used to give a self type to a trait. Aliasing can also be a good abbreviation when you need to access the this of an outer class. Here's an example:

```
class Outer { outer =>
  class Inner {
    println(Outer.this eq outer) // prints: true
  }
}
```

The example defines two nested classes, Outer and Inner. Inside Inner the this value of the Outer class is referred to twice, using different expressions. The first expression shows the Java way of doing things: You can prefix the reserved word this with the name of an outer class and a period; such an expression then refers to the this of the outer class. The second expression shows the alternative that Scala gives you. By introducing an alias named outer for this in class Outer, you can refer to this alias directly also in inner classes. The Scala way is more concise, and can also improve clarity, if you choose the name of the alias well. You'll see examples of this in pages 729 and 730.

728

**Single-token parsers**

Trait Parsers defines a generic parser elem that can be used to parse any single token:

```
def elem(kind: String, p: Elem => Boolean) =
  new Parser[Elem] {
    def apply(in: Input) =
      if (p(in.first)) Success(in.first, in.rest)
      else Failure(kind + " expected", in)
  }
```

This parser takes two parameters: a kind string describing what kind of token should be parsed and a predicate p on Elems, which indicates whether an element fits the class of tokens to be parsed.

When applying the parser elem(kind, p) to some input in, the first element of the input stream is tested with predicate p. If p returns true, the parser succeeds. Its result is the element itself, and its remaining input is the input stream starting just after the element that was parsed. On the other hand, if p returns false, the parser fails with an error message that indicates what kind of token was expected.

**Sequential composition**

The elem parser only consumes a single element. To parse more interesting phrases, you can string parsers together with the sequential composition operator ~. As you have seen before, P~Q is a parser that applies first the P parser to a given input string. Then, if P succeeds, the Q parser is applied to the input that's left after P has done its job.

The ~ combinator is implemented as a method in class Parser. Its definition is shown in Listing 33.6. The method is a member of the Parser class. Inside this class, p is specified by the "p =>" part as an alias of this, so p designates the left operand (or: receiver) of ~. Its right operand is represented by parameter q. Now, if p~q is run on some input in, first p is run on in and the result is analyzed in a pattern match. If p succeeds, q is run on the remaining input in1. If q also succeeds, the parser as a whole succeeds. Its result is a ~ object containing both the result of p (*i.e.*, x) and the result of q (*i.e.*, y). On the other hand, if either p or q fails the result of p~q is the Failure object returned by p or q.

```
abstract class Parser[+T] ... { p =>
  ...
  def ~ [U](q: => Parser[U]) = new Parser[T~U] {
    def apply(in: Input) = p(in) match {
      case Success(x, in1) =>
        q(in1) match {
          case Success(y, in2) => Success(new ~(x, y), in2)
          case failure => failure
        }
      case failure => failure
    }
  }
}
```

Listing 33.6 · The ~ combinator method.

The result type of ~ is a parser that returns an instance of the case class
~ with elements of types T and U. The type expression T~U is just a more
legible shorthand for the parameterized type ~[T, U]. Generally, Scala always interprets a binary type operation such as A op B, as the parameterized
type op[A, B]. This is analogous to the situation for patterns, where a binary
pattern P op Q is also interpreted as an application, *i.e.*, op(P, Q).

The other two sequential composition operators, <~ and ~>, could be
defined just like ~, only with some small adjustment in how the result is
computed. A more elegant technique, though, is to define them in terms of ~
as follows:

```
def <~ [U](q: => Parser[U]): Parser[T] =
  (p~q) ^^ { case x~y => x }
def ~> [U](q: => Parser[U]): Parser[U] =
  (p~q) ^^ { case x~y => y }
```

## Alternative composition

An alternative composition P | Q applies either P or Q to a given input. It
first tries P. If P succeeds, the whole parser succeeds with the result of P.
Otherwise, if P fails, then Q is tried *on the same input* as P. The result of Q is
then the result of the whole parser.

Here is a definition of | as a method of class `Parser`:

```
def | (q: => Parser[T]) = new Parser[T] {
  def apply(in: Input) = p(in) match {
    case s1 @ Success(_, _) => s1
    case failure => q(in)
  }
}
```

Note that if P and Q both fail, then the failure message is determined by Q. This subtle choice is discussed later, in Section 33.9.

## Dealing with recursion

Note that the q parameter in methods ~ and | is by-name—its type is preceded by =>. This means that the actual parser argument will be evaluated only when q is needed, which should only be the case after p has run. This makes it possible to write recursive parsers like the following one which parses a number enclosed by arbitrarily many parentheses:

```
def parens = floatingPointNumber | "("~parens~")"
```

If | and ~ took *by-value parameters*, this definition would immediately cause a stack overflow without reading anything, because the value of `parens` occurs in the middle of its right-hand side.

## Result conversion

The last method of class `Parser` converts a parser's result. The parser P ^^ f succeeds exactly when P succeeds. In that case it returns P's result converted using the function f. Here is the implementation of this method:

```
def ^^ [U](f: T => U): Parser[U] = new Parser[U] {
  def apply(in: Input) = p(in) match {
    case Success(x, in1) => Success(f(x), in1)
    case failure => failure
  }
}
} // end Parser
```

## Parsers that don't read any input

There are also two parsers that do not consume any input: success and failure. The parser success(result) always succeeds with the given result. The parser failure(msg) always fails with error message msg. Both are implemented as methods in trait Parsers, the outer trait that also contains class Parser:

```
def success[T](v: T) = new Parser[T] {
  def apply(in: Input) = Success(v, in)
}
def failure(msg: String) = new Parser[Nothing] {
  def apply(in: Input) = Failure(msg, in)
}
```

## Option and repetition

Also defined in trait Parsers are the option and repetition combinators opt, rep, and repsep. They are all implemented in terms of sequential composition, alternative, and result conversion:

```
def opt[T](p: => Parser[T]): Parser[Option[T]] = (
  p ^^ Some(_)
| success(None)
)
def rep[T](p: => Parser[T]): Parser[List[T]] = (
  p~rep(p) ^^ { case x~xs => x :: xs }
| success(List())
)
def repsep[T](p: => Parser[T],
    q: => Parser[Any]): Parser[List[T]] = (
  p~rep(q~> p) ^^ { case r~rs => r :: rs }
| success(List())
)
} // end Parsers
```

## 33.7 String literals and regular expressions

The parsers you saw so far made use of string literals and regular expressions to parse single words. The support for these comes from `RegexParsers`, a subtrait of `Parsers`:

```
trait RegexParsers extends Parsers {
```

This trait is more specialized than trait `Parsers` in that it only works for inputs that are sequences of characters:

```
type Elem = Char
```

It defines two methods, `literal` and `regex`, with the following signatures:

```
implicit def literal(s: String): Parser[String] = ...
implicit def regex(r: Regex): Parser[String] = ...
```

Note that both methods have an `implicit` modifier, so they are automatically applied whenever a `String` or `Regex` is given but a `Parser` is expected. That's why you can write string literals and regular expressions directly in a grammar, without having to wrap them with one of these methods. For instance, the parser `"("~expr~")"` will be automatically expanded to `literal("(")~expr~literal(")")`.

The RegexParsers trait also takes care of handling white space between symbols. To do this, it calls a method named handleWhiteSpace before running a `literal` or `regex` parser. The handleWhiteSpace method skips the longest input sequence that conforms to the whiteSpace regular expression, which is defined by default as follows:

```
    protected val whiteSpace = """\s+""".r
} // end RegexParsers
```

If you prefer a different treatment of white space, you can override the whiteSpace val. For instance, if you want white space not to be skipped at all, you can override whiteSpace with the empty regular expression:

```
object MyParsers extends RegexParsers {
  override val whiteSpace = "".r

  ...
}
```

## 33.8 Lexing and parsing

The task of syntax analysis is often split into two phases. The *lexer* phase recognizes individual words in the input and classifies them into some *token* classes. This phase is also called *lexical analysis*. This is followed by a *syntactical analysis* phase that analyzes sequences of tokens. Syntactical analysis is also sometimes just called parsing, even though this is slightly imprecise, as lexical analysis can also be regarded as a parsing problem.

The `Parsers` trait as described in the previous section can be used for either phase, because its input elements are of the abstract type `Elem`. For lexical analysis, `Elem` would be instantiated to `Char`, meaning the individual characters that make up a word are being parsed. The syntactical analyzer would in turn instantiate `Elem` to the type of token returned by the lexer.

Scala's parsing combinators provide several utility classes for lexical and syntactic analysis. These are contained in two sub-packages, one for each kind of analysis:

```
scala.util.parsing.combinator.lexical
scala.util.parsing.combinator.syntactical
```

If you want to split your parser into a separate lexer and syntactical analyzer, you should consult the Scaladoc documentation for these packages. But for simple parsers, the regular expression based approach shown previously in this chapter is usually sufficient.

## 33.9 Error reporting

There's one final topic that was not covered yet: how does the parser issue an error message? Error reporting for parsers is somewhat of a black art. One problem is that when a parser rejects some input, it generally has encountered many different failures. Each alternative parse must have failed, and recursively so at each choice point. Which of the usually numerous failures should be emitted as error message to the user?

Scala's parsing library implements a simple heuristic: among all failures, the one that occurred at the latest position in the input is chosen. In other words, the parser picks the longest prefix that is still valid and issues an error message that describes why parsing the prefix could not be continued further. If there are several failure points at that latest position, the one that was visited last is chosen.

For instance, consider running the JSON parser on a faulty address book which starts with the line:

```
{ "name": John,
```

The longest legal prefix of this phrase is "{ "name": ". So the JSON parser will flag the word John as an error. The JSON parser expects a value at this point, but John is an identifier, which does not count as a value (presumably, the author of the document had forgotten to enclose the name in quotation marks). The error message issued by the parser for this document is:

```
[1.13] failure: "false" expected but identifier John found
  { "name": John,
            ^
```

The part that "false" was expected comes from the fact that "false" is the last alternative of the production for value in the JSON grammar. So this was the last failure at this point. Users who know the JSON grammar in detail can reconstruct the error message, but for non-experts this error message is probably surprising and can also be quite misleading.

A better error message can be engineered by adding a "catch-all" failure point as last alternative of a value production:

```
def value: Parser[Any] =
  obj | arr | stringLit | floatingPointNumber | "null" |
  "true" | "false" | failure("illegal start of value")
```

This addition does not change the set of inputs that are accepted as valid documents. What it does is improve the error messages, because now it will be the explicitly added failure that comes as last alternative and therefore gets reported:

```
[1.13] failure: illegal start of value
  { "name": John,
            ^
```

The implementation of the "latest possible" scheme of error reporting uses a field named lastFailure in trait Parsers to mark the failure that occurred at the latest position in the input:

735

```
var lastFailure: Option[Failure] = None
```

The field is initialized to None. It is updated in the constructor of the `Failure` class:

```
case class Failure(msg: String, in: Input)
    extends ParseResult[Nothing] {

  if (lastFailure.isDefined &&
      lastFailure.get.in.pos <= in.pos)
    lastFailure = Some(this)
}
```

The field is read by the `phrase` method, which emits the final error message if the parser failed. Here is the implementation of `phrase` in trait `Parsers`:

```
def phrase[T](p: Parser[T]) = new Parser[T] {
  lastFailure = None
  def apply(in: Input) = p(in) match {
    case s @ Success(out, in1) =>
      if (in1.atEnd) s
      else Failure("end of input expected", in1)
    case f : Failure =>
      lastFailure
  }
}
```

The `phrase` method runs its argument parser p. If p succeeds with a completely consumed input, the success result of p is returned. If p succeeds but the input is not read completely, a failure with message "end of input expected" is returned. If p fails, the failure or error stored in `lastFailure` is returned. Note that the treatment of `lastFailure` is non-functional; it is updated as a side effect by the constructor of `Failure` and by the `phrase` method itself. A functional version of the same scheme would be possible, but it would require threading the `lastFailure` value through every parser result, no matter whether this result is a `Success` or a `Failure`.

## 33.10  Backtracking versus LL(1)

The parser combinators employ *backtracking* to choose between different parsers in an alternative. In an expression P | Q, if P fails, then Q is run on the same input as P. This happens even if P has parsed some tokens before failing. In this case the same tokens will be parsed again by Q.

Backtracking imposes only a few restrictions on how to formulate a grammar so that it can be parsed. Essentially, you just need to avoid left-recursive productions. A production such as:

$$expr \quad ::= \quad expr \; "+" \; term \; | \; term.$$

will always fail because `expr` immediately calls itself and thus never progresses any further.[1] On the other hand, backtracking is potentially costly because the same input can be parsed several times. Consider for instance the production:

$$expr \quad ::= \quad term \; "+" \; expr \; | \; term.$$

What happens if the `expr` parser is applied to an input such as $(1 + 2) * 3$ which constitutes a legal term? The first alternative would be tried, and would fail when matching the + sign. Then the second alternative would be tried on the same term and this would succeed. In the end the term ended up being parsed twice.

It is often possible to modify the grammar so that backtracking can be avoided. For instance, in the case of arithmetic expressions, either one of the following productions would work:

$$expr \quad ::= \quad term \; ["+" \; expr].$$
$$expr \quad ::= \quad term \; \{"+" \; term\}.$$

Many languages admit so-called "LL(1)" grammars.[2] When a combinator parser is formed from such a grammar, it will never backtrack, *i.e.*, the input position will never be reset to an earlier value. For instance, the grammars for arithmetic expressions and JSON terms earlier in this chapter are both LL(1), so the backtracking capabilities of the parser combinator framework are never exercised for inputs from these languages.

The combinator parsing framework allows you to express the expectation that a grammar is LL(1) explicitly, using a new operator ~!. This operator is

---

[1] There are ways to avoid stack overflows even in the presence of left-recursion, but this requires a more refined parsing combinator framework, which to date has not been implemented.

[2] Aho, *et al.*, *Compilers: Principles, Techniques, and Tools.* [Aho86]

like sequential composition ~ but it will never backtrack to "un-read" input elements that have already been parsed. Using this operator, the productions in the arithmetic expression parser could alternatively be written as follows:

```
def expr  : Parser[Any] =
  term ~! rep("+" ~! term | "-" ~! term)
def term  : Parser[Any] =
  factor ~! rep("*" ~! factor | "/" ~! factor)
def factor: Parser[Any] =
  "(" ~! expr ~! ")" | floatingPointNumber
```

One advantage of an LL(1) parser is that it can use a simpler input technique. Input can be read sequentially, and input elements can be discarded once they are read. That's another reason why LL(1) parsers are usually more efficient than backtracking parsers.

## 33.11   Conclusion

You have now seen all the essential elements of Scala's combinator parsing framework. It's surprisingly little code for something that's genuinely useful. With the framework you can construct parsers for a large class of context-free grammars. The framework lets you get started quickly, but it is also customizable to new kinds of grammars and input methods. Being a Scala library, it integrates seamlessly with the rest of the language. So it's easy to integrate a combinator parser in a larger Scala program.

One downside of combinator parsers is that they are not very efficient, at least not when compared with parsers generated from special purpose tools such as Yacc or Bison. There are two reasons for this. First, the backtracking method used by combinator parsing is itself not very efficient. Depending on the grammar and the parse input, it might yield an exponential slow-down due to repeated backtracking. This can be fixed by making the grammar LL(1) and by using the committed sequential composition operator, ~!.

The second problem affecting the performance of combinator parsers is that they mix parser construction and input analysis in the same set of operations. In effect, a parser is generated anew for each input that's parsed.

This problem can be overcome, but it requires a different implementation of the parser combinator framework. In an optimizing framework, a parser would no longer be represented as a function from inputs to parse results.

Instead, it would be represented as a tree, where every construction step was represented as a case class. For instance, sequential composition could be represented by a case class Seq, alternative by Alt, and so on. The "outermost" parser method, phrase, could then take this symbolic representation of a parser and convert it to highly efficient parsing tables, using standard parser generator algorithms.

What's nice about all this is that from a user perspective nothing changes compared to plain combinator parsers. Users still write parsers in terms of ident, floatingPointNumber, ~, |, and so on. They need not be aware that these methods generate a symbolic representation of a parser instead of a parser function. Since the phrase combinator converts these representations into real parsers, everything works as before.

The advantage of this scheme with respect to performance is two-fold. First, you can now factor out parser construction from input analysis. If you were to write:

```
val jsonParser = phrase(value)
```

and then apply jsonParser to several different inputs, the jsonParser would be constructed only once, not every time an input is read.

Second, the parser generation can use efficient parsing algorithms such as LALR(1).[3] These algorithms usually lead to much faster parsers than parsers that operate with backtracking.

At present, such an optimizing parser generator has not yet been written for Scala. But it would be perfectly possible to do so. If someone contributes such a generator, it will be easy to integrate into the standard Scala library. Even postulating that such a generator will exist at some point in the future, however, there are reasons for keeping the current parser combinator framework around. It is much easier to understand and to adapt than a parser generator, and the difference in speed would often not matter in practice, unless you want to parse very large inputs.

---

[3] Aho, *et al.*, *Compilers: Principles, Techniques, and Tools.* [Aho86]

# Chapter 34

# GUI Programming

In this chapter you'll learn how to develop in Scala applications that use a graphical user interface (GUI). The applications we'll develop are based on a Scala library that provides access to Java's Swing framework of GUI classes. Conceptually, the Scala library resembles the underlying Swing classes, but hides much of their complexity. You'll find out that developing GUI applications using the framework is actually quite easy.

Even with Scala's simplifications, a framework like Swing is quite rich, with many different classes and many methods in each class. To find your way in such a rich library, it helps to use an IDE such as Scala's Eclipse plugin. The advantage is that the IDE can show you interactively with its command completion which classes are available in a package and which methods are available for objects you reference. This speeds up your learning considerably when you first explore an unknown library space.

## 34.1 A first Swing application

As a first Swing application, we'll start with a window containing a single button. To program with Swing, you need to import various classes from Scala's Swing API package:

```
import scala.swing._
```

Listing 34.1 shows the code of your first Swing application in Scala. If you compile and run that file, you should see a window as shown on the left of Figure 34.1. The window can be resized to a larger size as shown on the right of Figure 34.1.

Figure 34.1 · A simple Swing application: initial (left) and resized (right).

```
import scala.swing._

object FirstSwingApp extends SimpleSwingApplication {
  def top = new MainFrame {
    title = "First Swing App"
    contents = new Button {
      text = "Click me"
    }
  }
}
```

Listing 34.1 · A simple Swing application in Scala.

If you analyze the code in Listing 34.1 line by line, you'll notice the following elements:

```
object FirstSwingApp extends SimpleSwingApplication {
```

In the first line after the import, the FirstSwingApp object inherits from scala.swing.SimpleSwingApplication. This application differs from traditional command-line applications, which may inherit from scala.App. The SimpleSwingApplication class already defines a main method that contains some setup code for Java's Swing framework. The main method then proceeds to call the top method, which you supply:

```
def top = new MainFrame {
```

The next line implements the top method. This method contains the code that defines your top-level GUI component. This is usually some kind of Frame—*i.e.*, a window that can contain arbitrary data. In Listing 34.1, we chose a `MainFrame` as the top-level component. A `MainFrame` is like a normal Swing `Frame` except that closing it will also close the whole GUI application.

```
title = "First Swing App"
```

Frames have a number of attributes. Two of the most important are the frame's `title`, which will be written in the title bar, and its `contents`, which will be displayed in the window itself. In Scala's Swing API, such attributes are modeled as properties. You know from Section 18.2 that properties are encoded in Scala as pairs of getter and setter methods. For instance, the `title` property of a `Frame` object is modeled as a getter method:

```
def title: String
```

and a setter method:

```
def title_=(s: String)
```

It is this setter method that gets invoked by the above assignment to `title`. The effect of the assignment is that the chosen title is shown in the header of the window. If you leave it out, the window will have an empty title.

```
contents = new Button {
```

The `top` frame is the root component of the Swing application. It is a `Container`, which means that further components can be defined in it. Every Swing container has a `contents` property, which allows you to get and set the components it contains. The getter `contents` of this property has type `Seq[Component]`, indicating that a component can in general have several objects as its contents. Frames, however, always have just a single component as their `contents`. This component is set and potentially changed using the setter `contents_=`. For example, in Listing 34.1 a single `Button` constitutes the `contents` of the `top` frame.

```
text = "Click me"
```

The button also gets a title, in this case "Click me."

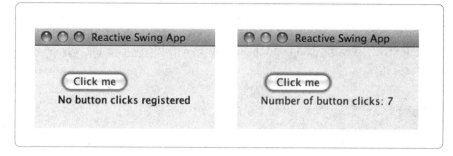

Figure 34.2 · A reactive Swing application: initial (left) after clicks (right).

```scala
import scala.swing._

object SecondSwingApp extends SimpleSwingApplication {
  def top = new MainFrame {
    title = "Second Swing App"
    val button = new Button {
      text = "Click me"
    }
    val label = new Label {
      text = "No button clicks registered"
    }
    contents = new BoxPanel(Orientation.Vertical) {
      contents += button
      contents += label
      border = Swing.EmptyBorder(30, 30, 10, 30)
    }
  }
}
```

Listing 34.2 · Component assembly on a panel.

## 34.2 Panels and layouts

As next step, we'll add some text as a second content element to the top frame of the application. The left part of Figure 34.2 shows what the application should look like.

You saw in the last section that a frame contains exactly one child component. Hence, to make a frame with both a button and a label, you need to create a different container component that holds both. That's what *panels* are used for. A Panel is a container that displays all the components it contains according to some fixed layout rules. There are a number of different possible layouts that are implemented by various subclasses of class Panel, ranging from simple to quite intricate. In fact, one of the hardest parts of a complex GUI application can be getting the layouts right—it's not easy to come up with something that displays reasonably well on all sorts of devices and for all window sizes.

Listing 34.2 shows a complete implementation. In this class, the two sub-components of the top frame are named button and label. The button is defined as before. The label is a displayed text field that can't be edited:

```
val label = new Label {
  text = "No button clicks registered"
}
```

The code in Listing 34.2 picks a simple vertical layout where components are stacked on top of each other in a BoxPanel:

```
contents = new BoxPanel(Orientation.Vertical) {
```

The contents property of the BoxPanel is an (initially empty) buffer, to which the button and label elements are added with the += operator:

```
contents += button
contents += label
```

We also add a border around the two objects by assigning to the border property of the panel:

```
border = Swing.EmptyBorder(30, 30, 10, 30)
```

As is the case with other GUI components, borders are represented as objects. EmptyBorder is a factory method in object Swing that takes four parameters indicating the width of the borders on the top, right, bottom, and left sides of the objects to be drawn.

Simple as it is, the example has already shown the basic way to structure a GUI application. It is built from components, which are instances of scala.swing classes such as Frame, Panel, Label or Button. Components

745

have properties, which can be customized by the application. `Panel` components can contain several other components in their `contents` property, so that in the end a GUI application consists of a tree of components.

## 34.3   Handling events

On the other hand, the application still misses an essential property. If you run the code in Listing 34.2 and click on the displayed button, nothing happens. In fact, the application is completely static; it does not react in any way to user events except for the close button of the `top` frame, which terminates the application. So as a next step, we'll refine the application so that it displays together with the button a label that indicates how often the button was clicked. The right part of Figure 34.2 contains a snapshot of what the application should look like after a few button clicks.

To achieve this behavior, you need to connect a user-input event (the button was clicked) with an action (the displayed label is updated). Java and Scala have fundamentally the same "publish/subscribe" approach to event handling: Components may be publishers and/or subscribers. A publisher publishes events. A subscriber subscribes with a publisher to be notified of any published events. Publishers are also called "event sources," and subscribers are also called "event listeners". For instance a `Button` is an event source, which publishes an event, `ButtonClicked`, indicating that the button was clicked.

In Scala, subscribing to an event source `source` is done by the call `listenTo(source)`. There's also a way to unsubscribe from an event source using `deafTo(source)`. In the current example application, the first thing to do is to get the `top` frame to listen to its button, so that it gets notified of any events that the button issues. To do that you need to add the following call to the body of the `top` frame:

```
listenTo(button)
```

Being notified of events is only half the story; the other half is handling them. It is here that the Scala Swing framework is most different from (and radically simpler than) the Java Swing API's. In Java, signaling an event means calling a "notify" method in an object that has to implement some `Listener` interfaces. Usually, this involves a fair amount of indirection and boilerplate code, which makes event-handling applications hard to write and read.

By contrast, in Scala, an event is a real object that gets sent to subscribing components much like messages are sent to actors. For instance, pressing a button will create an event which is an instance of the following case class:

```
case class ButtonClicked(source: Button)
```

The parameter of the case class refers to the button that was clicked. As with all other Scala Swing events, this event class is contained in a package named scala.swing.event.

To have your component react to incoming events you need to add a handler to a property called reactions. Here's the code that does this:

```
var nClicks = 0
reactions += {
  case ButtonClicked(b) =>
    nClicks += 1
    label.text = "Number of button clicks: " + nClicks
}
```

The first line above defines a variable, nClicks, which holds the number of times a button was clicked. The remaining lines add the code between braces as a *handler* to the reactions property of the top frame. Handlers are functions defined by pattern matching on events, much like Akka actor receive methods are defined by pattern matching on messages. The handler above matches events of the form ButtonClicked(b), *i.e.*, any event which is an instance of class ButtonClicked. The pattern variable b refers to the actual button that was clicked. The action that corresponds to this event in the code above increments nClicks and updates the text of the label.

Generally, a handler is a PartialFunction that matches on events and performs some actions. It is also possible to match on more than one kind of event in a single handler by using multiple cases.

The reactions property implements a collection, just like the contents property does. Some components come with predefined reactions. For instance, a Frame has a predefined reaction that it will close if the user presses the close button on the upper right. If you install your own reactions by adding them with += to the reactions property, the reactions you define will be considered in addition to the standard ones. Conceptually, the handlers installed in reactions form a stack. In the current example, if the top frame receives an event, the first handler tried will be the one that matches

```scala
import scala.swing._
import scala.swing.event._

object ReactiveSwingApp extends SimpleSwingApplication {
  def top = new MainFrame {
    title = "Reactive Swing App"
    val button = new Button {
      text = "Click me"
    }
    val label = new Label {
      text = "No button clicks registered"
    }
    contents = new BoxPanel(Orientation.Vertical) {
      contents += button
      contents += label
      border = Swing.EmptyBorder(30, 30, 10, 30)
    }
    listenTo(button)
    var nClicks = 0
    reactions += {
      case ButtonClicked(b) =>
        nClicks += 1
        label.text = "Number of button clicks: " + nClicks
    }
  }
}
```

Listing 34.3 · Implementing a reactive Swing application.

on ButtonClicked, because it was the last handler installed for the frame. If the received event is of type ButtonClicked, the code associated with the pattern will be invoked. After that code has completed, the system will search for further handlers in the event stack that might also be applicable. If the received event is not of type ButtonClicked, the event is immediately propagated to the rest of the installed handler stack. It's also possible to remove handlers from the reaction property, using the -= operator.

Listing 34.3 shows the completed application, including reactions. The code illustrates the essential elements of a GUI application in Scala's Swing

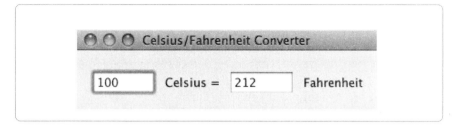

Figure 34.3 · A converter between degrees Celsius and Fahrenheit.

framework: The application consists of a tree of components, starting with the top frame. The components shown in the code are Frame, BoxPanel, Button, and Label, but there are many other kinds of components defined in the Swing libraries. Each component is customized by setting attributes. Two important attributes are contents, which fixes the children of a component in the tree, and reactions, which determines how the component reacts to events.

## 34.4 Example: Celsius/Fahrenheit converter

As another example, we'll write a GUI program that converts between temperature degrees in Celsius and Fahrenheit. The user interface of the program is shown in Figure 34.3. It consists of two text fields (shown in white) with a label following each. One text field shows temperatures in degrees Celsius, the other in degrees Fahrenheit. Each of the two fields can be edited by the user of the application. Once the user has changed the temperature in either field, the temperature in the other field should automatically update.

Listing 34.4 shows the complete code that implements this application. The imports at the top of the code use a short-hand:

```
import swing._
import event._
```

This is in fact equivalent to the imports used before:

```
import scala.swing._
import scala.swing.event._
```

749

```scala
import swing._
import event._

object TempConverter extends SimpleSwingApplication {
  def top = new MainFrame {
    title = "Celsius/Fahrenheit Converter"
    object celsius extends TextField { columns = 5 }
    object fahrenheit extends TextField { columns = 5 }
    contents = new FlowPanel {
      contents += celsius
      contents += new Label(" Celsius  =  ")
      contents += fahrenheit
      contents += new Label(" Fahrenheit")
      border = Swing.EmptyBorder(15, 10, 10, 10)
    }
    listenTo(celsius, fahrenheit)
    reactions += {
      case EditDone(`fahrenheit`) =>
        val f = fahrenheit.text.toInt
        val c = (f - 32) * 5 / 9
        celsius.text = c.toString
      case EditDone(`celsius`) =>
        val c = celsius.text.toInt
        val f = c * 9 / 5 + 32
        fahrenheit.text = f.toString
    }
  }
}
```

Listing 34.4 · An implementation of the temperature converter.

The reason you can use the shorthand is that packages nest in Scala. Because package scala.swing is contained in package scala, and everything in that package imported automatically, you can write just swing to refer to the package. Likewise, package scala.swing.event, is contained as subpackage event in package scala.swing. Because you have imported everything in scala.swing in the first import, you can refer to the event package with just event thereafter.

750

The two components `celsius` and `fahrenheit` in `TempConverter` are objects of class `TextField`. A `TextField` in Swing is a component that lets you edit a single line of text. It has a default width, which is given in the columns property measured in characters (set to 5 for both fields).

The `contents` of `TempConverter` are assembled into a panel, which includes the two text fields and two labels that explain what the fields are. The panel is of class `FlowPanel`, which means it displays all its elements one after another, in one or more rows, depending on the width of the frame.

The `reactions` of `TempConverter` are defined by a handler that contains two cases. Each case matches an `EditDone` event for one of the two text fields. Such an event gets issued when a text field has been edited by the user. Note the form of the patterns, which include back ticks around the element names:

```
case EditDone(`celsius`)
```

As was explained in Section 15.2, the back ticks around `celsius` ensure that the pattern matches only if the source of the event was the `celsius` object. If you had omitted the back ticks and just written `case EditDone(celsius)`, the pattern would have matched every event of class `EditDone`. The changed field would then be stored in the pattern variable `celsius`. Obviously, this is not what you want. Alternatively, you could have defined the two `TextField` objects starting with upper case characters, *i.e.*, `Celsius` and `Fahrenheit`. In that case you could have matched them directly without back ticks, as in `case EditDone(Celsius)`.

The two actions of the `EditDone` events convert one quantity to another. Each starts by reading out the contents of the modified field and converting it to an `Int`. It then applies the formula for converting one temperature degree to the other, and stores the result back as a string in the other text field.

## 34.5 Conclusion

This chapter has given you a first taste of GUI programming, using Scala's wrappers for the Swing framework. It has shown how to assemble GUI components, how to customize their properties, and how to handle events. For space reasons, we could discuss only a small number of simple components. There are many more kinds of components. You can find out about them by consulting the Scala documentation of the package `scala.swing`. The next section will develop an example of a more complicated Swing application.

There are also many tutorials on the original Java Swing framework, on which the Scala wrapper is based.[1] The Scala wrappers resemble the underlying Swing classes, but try to simplify concepts where possible and make them more uniform. The simplification makes extensive use of the properties of the Scala language. For instance, Scala's emulation of properties and its operator overloading allow convenient property definitions using assignments and += operations. Its "everything is an object" philosophy makes it possible to inherit the main method of a GUI application. The method can thus be hidden from user applications, including the boilerplate code for setting things up that comes with it. Finally, and most importantly, Scala's first-class functions and pattern matching make it possible to formulate event handling as the `reactions` component property, which greatly simplifies life for the application developer.

---

[1]See, for instance, *The Java Tutorials*. [Jav]

Chapter 35

# The `SCells` Spreadsheet

In the previous chapters you saw many different constructs of the Scala programming language. In this chapter you'll see how these constructs play together in the implementation of a sizable application. The task is to write a spreadsheet application, which will be named `SCells`.

There are several reasons why this task is interesting. First, everybody knows spreadsheets, so it is easy to understand what the application should do. Second, spreadsheets are programs that exercise a large range of different computing tasks. There's the visual aspect, where a spreadsheet is seen as a rich GUI application. There's the symbolic aspect, having to do with formulas and how to parse and interpret them. There's the calculational aspect, dealing with how to update possibly large tables incrementally. There's the reactive aspect, where spreadsheets are seen as programs that react in intricate ways to events. Finally, there's the component aspect where the application is constructed as a set of reusable components. All these aspects will be treated in depth in this chapter.

## 35.1 The visual framework

We'll start by writing the basic visual framework of the application. Figure 35.1 shows the first iteration of the user interface. You can see that a spreadsheet is a scrollable table. It has rows going from 0 to 99 and columns going from A to Z. You express this in Swing by defining a spreadsheet as a `ScrollPane` containing a `Table`. Listing 35.1 shows the code.

The spreadsheet component shown in Listing 35.1 is defined in package `org.stairwaybook.scells`, which will contain all classes, traits, and

753

Figure 35.1 · A simple spreadsheet table.

objects needed for the application. It imports from package scala.swing essential elements of Scala's Swing wrapper. Spreadsheet itself is a class that takes height and width (in numbers of cells) as parameters. The class extends ScrollPane, which gives it the scroll-bars at the bottom and right in Figure 35.1. It contains two sub-components named table and rowHeader.

The table component is an instance of an anonymous subclass of class scala.swing.Table. The four lines in its body set some of its attributes: rowHeight for the height of a table row in points, autoResizeMode to turn auto-sizing the table off, showGrid to show a grid of lines between cells, and gridColor to set the color of the grid to a dark gray.

The rowHeader component, which contains the row-number headers at the left of the spreadsheet in Figure 35.1, is a ListView that displays in its

```
package org.stairwaybook.scells
import swing._

class Spreadsheet(val height: Int, val width: Int)
    extends ScrollPane {

  val table = new Table(height, width) {
    rowHeight = 25
    autoResizeMode = Table.AutoResizeMode.Off
    showGrid = true
    gridColor = new java.awt.Color(150, 150, 150)
  }

  val rowHeader =
    new ListView((0 until height) map (_.toString)) {
      fixedCellWidth = 30
      fixedCellHeight = table.rowHeight
    }
  viewportView = table
  rowHeaderView = rowHeader
}
```

Listing 35.1 · Code for spreadsheet in Figure 35.1.

elements the strings 0 through 99. The two lines in its body fix the width of a cell to be 30 points and the height to be the same as the table's rowHeight.

The whole spreadsheet is assembled by setting two fields in ScrollPane. The field viewportView is set to the table, and the field rowHeaderView is set to the rowHeader list. The difference between the two views is that a view port of a scroll pane is the area that scrolls with the two bars, whereas the row header on the left stays fixed when you move the horizontal scroll bar. By some quirk, Swing already supplies by default a column header at the top of the table, so there's no need to define one explicitly.

To try out the rudimentary spreadsheet shown in Listing 35.1, you just need to define a main program that creates the Spreadsheet component. Such a program is shown in Listing 35.2.

Program Main inherits from SimpleSwingApplication, which takes care of all the low-level details that need to be set up before a Swing application can be run. You only need to define the top-level window of the appli-

```
package org.stairwaybook.scells
import swing._

object Main extends SimpleSwingApplication {
  def top = new MainFrame {
    title = "ScalaSheet"
    contents = new Spreadsheet(100, 26)
  }
}
```

Listing 35.2 · The main program for the spreadsheet application.

cation in the top method. In our example, top is a MainFrame that has two elements defined: its title, set to "ScalaSheet," and its contents, set to an instance of class Spreadsheet with 100 rows and 26 columns. That's all. If you launch this application with scala org.stairwaybook.scells.Main, you should see the spreadsheet in Figure 35.1.

## 35.2 Disconnecting data entry and display

If you play a bit with the spreadsheet written so far, you'll quickly notice that the output that's displayed in a cell is always exactly what you entered in the cell. A real spreadsheet does not behave like that. In a real spreadsheet, you would enter a formula and you'd see its value. So what is entered into a cell is different from what is displayed.

As a first step to a real spreadsheet application, you should concentrate on disentangling data entry and display. The basic mechanism for display is contained in the rendererComponent method of class Table. By default, rendererComponent always displays what's entered. If you want to change that, you need to override rendererComponent to do something different. Listing 35.3 shows a new version of Spreadsheet with a rendererComponent method.

The rendererComponent method overrides a default method in class Table. It takes four parameters. The isSelected and hasFocus parameters are Booleans that indicate whether the cell has been selected and whether it has focus, meaning that keyboard events will go into the cell. The remaining two parameters, row and column, give the cell's coordinates.

756

```
package org.stairwaybook.scells
import swing._

class Spreadsheet(val height: Int, val width: Int)
    extends ScrollPane {

  val cellModel = new Model(height, width)
  import cellModel._

  val table = new Table(height, width) {

    // settings as before...

    override def rendererComponent(isSelected: Boolean,
        hasFocus: Boolean, row: Int, column: Int): Component =

      if (hasFocus) new TextField(userData(row, column))
      else
        new Label(cells(row)(column).toString) {
          xAlignment = Alignment.Right
        }

    def userData(row: Int, column: Int): String = {
      val v = this(row, column)
      if (v == null) "" else v.toString
    }
  }
  // rest as before...
}
```

Listing 35.3 · A spreadsheet with a rendererComponent method.

The new rendererComponent method checks whether the cell has input focus. If hasFocus is true, the cell is used for editing. In this case you want to display an editable TextField that contains the data the user has entered so far. This data is returned by the helper method userData, which displays the contents of the table at a given row and column. The contents are retrieved by the call this(row, column).[1] The userData method also takes care to display a null element as the empty string instead of "null."

---

[1]Although "this(row, column)" may look similar to a constructor invocation, it is in this case an invocation of the apply method on the current Table instance.

```
package org.stairwaybook.scells

class Model(val height: Int, val width: Int) {

  case class Cell(row: Int, column: Int)

  val cells = Array.ofDim[Cell](height, width)

  for (i <- 0 until height; j <- 0 until width)
    cells(i)(j) = new Cell(i, j)
}
```

Listing 35.4 · First version of the Model class.

So far so good. But what should be displayed if the cell does not have focus? In a real spreadsheet this would be the value of a cell. Thus, there are really two tables at work. The first table, named table contains what the user entered. A second "shadow" table contains the internal representation of cells and what should be displayed. In the spreadsheet example, this table is a two-dimensional array called cells. If a cell at a given row and column does not have editing focus, the rendererComponent method will display the element cells(row)(column). The element cannot be edited, so it should be displayed in a Label instead of in an editable TextField.

It remains to define the internal array of cells. You could do this directly in the Spreadsheet class, but it's generally preferable to separate the view of a GUI component from its internal model. That's why in the example above the cells array is defined in a separate class named Model. The model is integrated into the Spreadsheet by defining a value cellModel of type Model. The import clause that follows this val definition makes the members of cellModel available inside Spreadsheet without having to prefix them. Listing 35.4 shows a first simplified version of a Model class. The class defines an inner class, Cell, and a two-dimensional array, cells, of Cell elements. Each element is initialized to be a fresh Cell.

That's it. If you compile the modified Spreadsheet class with the Model class and run the Main application you should see a window as in Figure 35.2.

The objective of this section was to arrive at a design where the displayed value of a cell is different from the string that was entered into it. This objective has clearly been met, albeit in a very crude way. In the new spreadsheet you can enter anything you want into a cell, but it will always display just its coordinates once it loses focus. Clearly, we are not done yet.

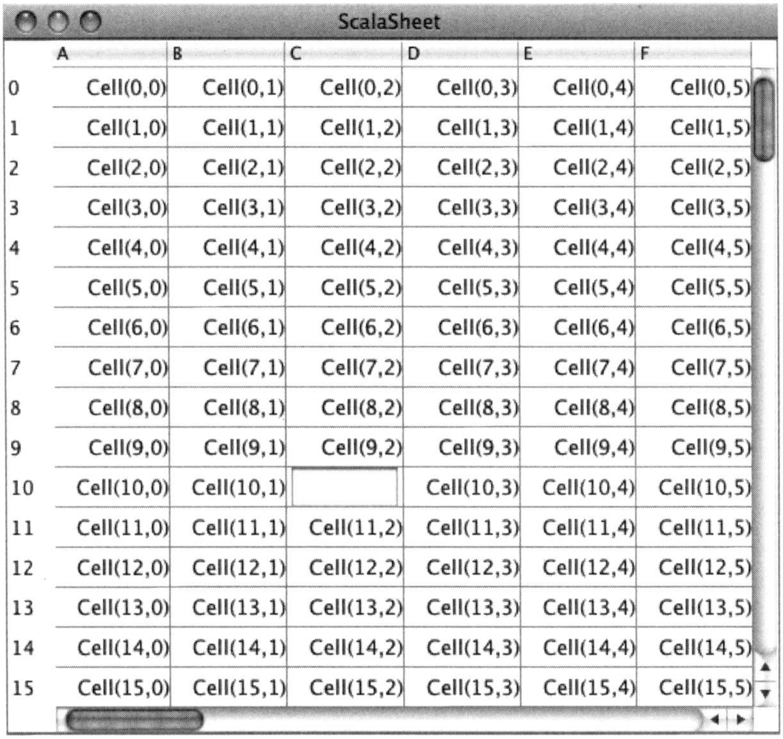

ScalaSheet

| | A | B | C | D | E | F |
|---|---|---|---|---|---|---|
| 0 | Cell(0,0) | Cell(0,1) | Cell(0,2) | Cell(0,3) | Cell(0,4) | Cell(0,5) |
| 1 | Cell(1,0) | Cell(1,1) | Cell(1,2) | Cell(1,3) | Cell(1,4) | Cell(1,5) |
| 2 | Cell(2,0) | Cell(2,1) | Cell(2,2) | Cell(2,3) | Cell(2,4) | Cell(2,5) |
| 3 | Cell(3,0) | Cell(3,1) | Cell(3,2) | Cell(3,3) | Cell(3,4) | Cell(3,5) |
| 4 | Cell(4,0) | Cell(4,1) | Cell(4,2) | Cell(4,3) | Cell(4,4) | Cell(4,5) |
| 5 | Cell(5,0) | Cell(5,1) | Cell(5,2) | Cell(5,3) | Cell(5,4) | Cell(5,5) |
| 6 | Cell(6,0) | Cell(6,1) | Cell(6,2) | Cell(6,3) | Cell(6,4) | Cell(6,5) |
| 7 | Cell(7,0) | Cell(7,1) | Cell(7,2) | Cell(7,3) | Cell(7,4) | Cell(7,5) |
| 8 | Cell(8,0) | Cell(8,1) | Cell(8,2) | Cell(8,3) | Cell(8,4) | Cell(8,5) |
| 9 | Cell(9,0) | Cell(9,1) | Cell(9,2) | Cell(9,3) | Cell(9,4) | Cell(9,5) |
| 10 | Cell(10,0) | Cell(10,1) | | Cell(10,3) | Cell(10,4) | Cell(10,5) |
| 11 | Cell(11,0) | Cell(11,1) | Cell(11,2) | Cell(11,3) | Cell(11,4) | Cell(11,5) |
| 12 | Cell(12,0) | Cell(12,1) | Cell(12,2) | Cell(12,3) | Cell(12,4) | Cell(12,5) |
| 13 | Cell(13,0) | Cell(13,1) | Cell(13,2) | Cell(13,3) | Cell(13,4) | Cell(13,5) |
| 14 | Cell(14,0) | Cell(14,1) | Cell(14,2) | Cell(14,3) | Cell(14,4) | Cell(14,5) |
| 15 | Cell(15,0) | Cell(15,1) | Cell(15,2) | Cell(15,3) | Cell(15,4) | Cell(15,5) |

Figure 35.2 · Cells displaying themselves.

## 35.3 Formulas

In reality, a spreadsheet cell holds two things: An actual *value* and a *formula* to compute this value. There are three types of formulas in a spreadsheet:

1. Numeric values such as `1.22`, `-3`, or `0`.

2. Textual labels such as `Annual sales`, `Deprecation`, or `total`.

3. Formulas that compute a new value from the contents of cells, such as "`=add(A1,B2)`", or "`=sum(mul(2, A2), C1:D16)`"

A formula that computes a value always starts with an equals sign and is followed by an arithmetic expression. The `SCells` spreadsheet has a particularly simple and uniform convention for arithmetic expressions: every expression is an application of some function to a list of arguments. The function name is an identifier such as add for binary addition, or sum for summation of an arbitrary number of operands. A function argument can be a number, a reference to a cell, a reference to a range of cells such as C1:D16, or another function application. You'll see later that SCells has an open architecture that makes it easy to install your own functions via mixin composition.

The first step to handling formulas is writing down the types that represent them. As you might expect, the different kinds of formulas are represented by case classes. Listing 35.5 shows the contents of a file named Formulas.scala, where these case classes are defined:

```
package org.stairwaybook.scells

trait Formula

case class Coord(row: Int, column: Int) extends Formula {
  override def toString = ('A' + column).toChar.toString + row
}
case class Range(c1: Coord, c2: Coord) extends Formula {
  override def toString = c1.toString + ":" + c2.toString
}
case class Number(value: Double) extends Formula {
  override def toString = value.toString
}
case class Textual(value: String) extends Formula {
  override def toString = value
}
case class Application(function: String,
    arguments: List[Formula]) extends Formula {

  override def toString =
    function + arguments.mkString("(", ",", ")")
}
object Empty extends Textual("")
```

Listing 35.5 · Classes representing formulas.

Trait Formula, shown in Listing 35.5, has five case classes as children:

| | |
|---|---|
| Coord | for cell coordinates such as A3, |
| Range | for cell ranges such as A3:B17, |
| Number | for floating-point numbers such as 3.1415, |
| Textual | for textual labels such as Deprecation, |
| Application | for function applications such as sum(A1,A2). |

Each case class overrides the `toString` method so that it displays its kind of formula in the standard way shown above. For convenience there's also an `Empty` object that represents the contents of an empty cell. The `Empty` object is an instance of the `Textual` class with an empty string argument.

## 35.4  Parsing formulas

In the previous section you saw the different kinds of formulas and how they display as strings. In this section you'll see how to reverse the process: to transform a user input string into a `Formula` tree. The rest of this section explains one by one the different elements of a class `FormulaParsers`, which contains the parsers that do the transformation. The class builds on the combinator framework given in Chapter 33. Specifically, formula parsers are an instance of the `RegexParsers` class explained in that chapter:

```
package org.stairwaybook.scells
import scala.util.parsing.combinator._

object FormulaParsers extends RegexParsers {
```

The first two elements of object `FormulaParsers` are auxiliary parsers for identifiers and decimal numbers:

```
def ident: Parser[String] = """[a-zA-Z_]\w*""".r
def decimal: Parser[String] = """-?\d+(\.\d*)?""".r
```

As you can see from the first regular expression above, an identifier starts with a letter or underscore. This is followed by an arbitrary number of "word" characters represented by the regular expression code \w, which recognizes letters, digits or underscores. The second regular expression describes decimal numbers, which consist of an optional minus sign, one or more digits that are represented by regular expression code \d, and an optional decimal part consisting of a period followed by zero or more digits.

The next element of object `FormulaParsers` is the `cell` parser, which recognizes the coordinates of a cell, such as C11 or B2. It first calls a regular expression parser that determines the form of a coordinate: a single letter followed by one or more digits. The string returned from that parser is then converted to a cell coordinate by separating the letter from the numerical part and converting the two parts to indices for the cell's column and row:

```
def cell: Parser[Coord] =
  """[A-Za-z]\d+""".r ^^ { s =>
    val column = s.charAt(0).toUpper - 'A'
    val row = s.substring(1).toInt
    Coord(row, column)
  }
```

Note that the `cell` parser is a bit restrictive in that it allows only column coordinates consisting of a single letter. Hence the number of spreadsheet columns is in effect restricted to be at most 26, because further columns cannot be parsed. It's a good idea to generalize the parser so that it accepts cells with several leading letters. This is left as an exercise to you.

The range parser recognizes a range of cells. Such a range is composed of two cell coordinates with a colon between them:

```
def range: Parser[Range] =
  cell~":"~cell ^^ {
    case c1~":"~c2 => Range(c1, c2)
  }
```

The number parser recognizes a decimal number, which is converted to a `Double` and wrapped in an instance of the `Number` class:

```
def number: Parser[Number] =
  decimal ^^ (d => Number(d.toDouble))
```

The `application` parser recognizes a function application. Such an application is composed of an identifier followed by a list of argument expressions in parentheses:

```
def application: Parser[Application] =
  ident~"("~repsep(expr, ",")~")" ^^ {
    case f~"("~ps~")" => Application(f, ps)
  }
```

The expr parser recognizes a formula expression—either a top-level formula following an '=', or an argument to a function. Such a formula expression is defined to be a cell, a range of cells, a number, or an application:

```
def expr: Parser[Formula] =
  range | cell | number | application
```

This definition of the expr parser contains a slight oversimplification because ranges of cells should only appear as function arguments; they should not be allowed as top-level formulas. You could change the formula grammar so that the two uses of expressions are separated, and ranges are excluded syntactically from top-level formulas. In the spreadsheet presented here such an error is instead detected once an expression is evaluated.

The textual parser recognizes an arbitrary input string, as long as it does not start with an equals sign (recall that strings that start with '=' are considered to be formulas):

```
def textual: Parser[Textual] =
  """[^=].*""".r ^^ Textual
```

The formula parser recognizes all kinds of legal inputs into a cell. A formula is either a number, or a textual entry, or a formula starting with an equals sign:

```
def formula: Parser[Formula] =
  number | textual | "="~>expr
```

This concludes the grammar for spreadsheet cells. The final method parse uses this grammar in a method that converts an input string into a Formula tree:

```
def parse(input: String): Formula =
  parseAll(formula, input) match {
    case Success(e, _) => e
    case f: NoSuccess => Textual("[" + f.msg + "]")
  }
} //end FormulaParsers
```

The parse method parses all of the input with the formula parser. If that succeeds, the resulting formula is returned. If it fails, a Textual object with an error message is returned instead.

```
package org.stairwaybook.scells
import swing._
import event._

class Spreadsheet(val height: Int, val width: Int) ... {
  val table = new Table(height, width) {

    ...

    reactions += {
      case TableUpdated(table, rows, column) =>
        for (row <- rows)
          cells(row)(column).formula =
            FormulaParsers.parse(userData(row, column))
    }
  }
}
```

Listing 35.6 · A spreadsheet that parses formulas.

That's everything there is to parsing formulas. The only thing that remains is to integrate the parser into the spreadsheet. To do this, you can enrich the Cell class in class Model by a formula field:

```
case class Cell(row: Int, column: Int) {
  var formula: Formula = Empty
  override def toString = formula.toString
}
```

In the new version of the Cell class, the toString method is defined to display the cell's formula. That way you can check whether formulas have been correctly parsed.

The last step in this section is to integrate the parser into the spreadsheet. Parsing a formula happens as a reaction to the user's input into a cell. A completed cell input is modeled in the Swing library by a TableUpdated event. The TableUpdated class is contained in package scala.swing.event. The event is of the form:

```
TableUpdated(table, rows, column)
```

It contains the table that was changed, as well as a set of coordinates of affected cells given by rows and column. The rows parameter is a range

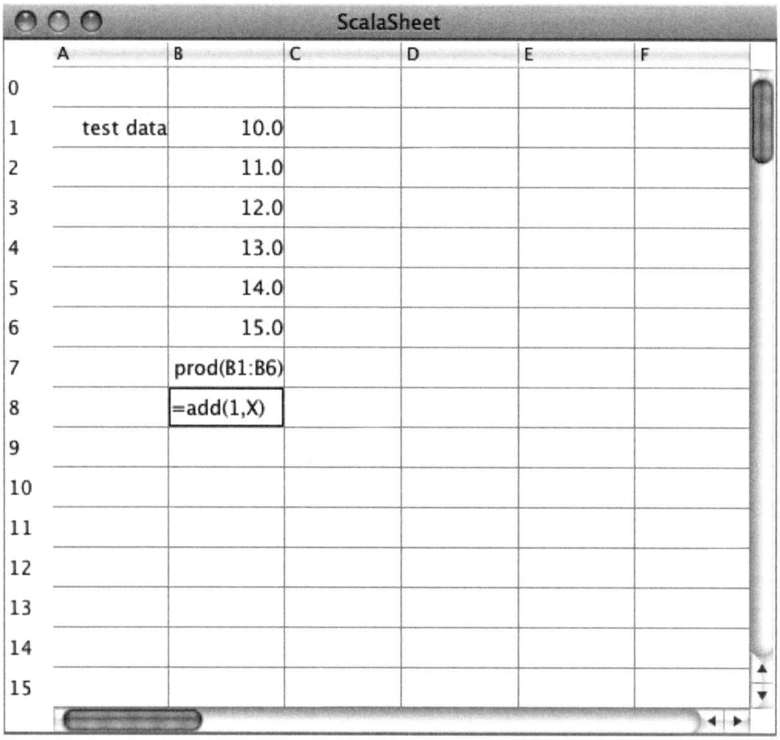

Figure 35.3 · Cells displaying their formulas.

value of type `Range[Int]`.[2] The `column` parameter is an integer. So in general a `TableUpdated` event can refer to several affected cells, but they would be on a consecutive range of rows and share the same column.

Once a table is changed, the affected cells need to be re-parsed. To react to a `TableUpdated` event, you add a case to the `reactions` value of the `table` component, as is shown in Listing 35.6. Now, whenever the table is edited the formulas of all affected cells will be updated by parsing the corresponding user data. When compiling the classes discussed so far and launching the `scells.Main` application you should see a spreadsheet application like the one shown in Figure 35.3. You can edit cells by typing into

---

[2]`Range[Int]` is also the type of a Scala expression such as "`1 to N`".

them. After editing is done, a cell displays the formula it contains. You can also try to type some illegal input such as the one reading =add(1, X) in the field that has the editing focus in Figure 35.3. Illegal input will show up as an error message. For instance, once you'd leave the edited field in Figure 35.3 you should see the error message [`(' expected] in the cell (to see all of the error message you might need to widen the column by dragging the separation between the column headers to the right).

## 35.5  Evaluation

Of course, in the end a spreadsheet should evaluate formulas, not just display them. In this section, we'll add the necessary components to achieve this.

What's needed is a method, evaluate, which takes a formula and returns the value of that formula in the current spreadsheet, represented as a Double. We'll place this method in a new trait, Evaluator. The method needs to access the cells field in class Model to find out about the current values of cells that are referenced in a formula. On the other hand, the Model class needs to call evaluate. Hence, there's a mutual dependency between the Model and the Evaluator. A good way to express such mutual dependencies between classes was shown in Chapter 29: you use inheritance in one direction and self types in the other.

In the spreadsheet example, class Model inherits from Evaluator and thus gains access to its evaluation method. To go the other way, class Evaluator defines its self type to be Model, like this:

```
package org.stairwaybook.scells
trait Evaluator { this: Model => ...
```

That way, the this value inside class Evaluator is assumed to be Model and the cells array is accessible by writing either cells or this.cells.

Now that the wiring is done, we'll concentrate on defining the contents of class Evaluator. Listing 35.7 shows the implementation of the evaluate method. As you might expect, the method contains a pattern match over the different types of formulas. For a coordinate Coord(row, column), it returns the value of the cells array at that coordinate. For a number Number(v), it returns the value v. For a textual label Textual(s), it returns zero. Finally, for an application Application(function, arguments), it

```scala
def evaluate(e: Formula): Double = try {
  e match {
    case Coord(row, column) =>
      cells(row)(column).value
    case Number(v) =>
      v
    case Textual(_) =>
      0
    case Application(function, arguments) =>
      val argvals = arguments flatMap evalList
      operations(function)(argvals)
  }
} catch {
  case ex: Exception => Double.NaN
}
```

Listing 35.7 · The evaluate method of trait Evaluator.

computes the values of all arguments, retrieves a function object corresponding to the function name from an operations table and applies that function to all argument values.

The operations table maps function names to function objects. It is defined as follows:

```scala
type Op = List[Double] => Double
val operations = new collection.mutable.HashMap[String, Op]
```

As you can see from this definition, operations are modeled as functions from lists of values to values. The Op type introduces a convenient alias for the type of an operation.

The computation in evaluate is wrapped in a try-catch to guard against input errors. There are actually quite a few things that can go wrong when evaluating a cell formula: coordinates might be out of range; function names might be undefined; functions might have the wrong number of arguments; arithmetic operations might be illegal or overflow. The reaction to any of these errors is the same: a "not-a-number" value is returned. The returned value, Double.NaN, is the IEEE representation for a computation that does not have a representable floating-point value. This might happen because of an overflow or a division by zero, for example. The evaluate method of Listing 35.7 chooses to return the same value also for all other kinds of errors. The advantage of this scheme is that it's simple to understand and doesn't require much code to implement. Its disadvantage is that all kinds of errors are lumped together, so a spreadsheet user does not get any detailed feedback on what went wrong. If you wish you can experiment with more refined ways of representing errors in the SCells application.

The evaluation of arguments is different from the evaluation of top-level formulas. Arguments may be lists whereas top-level functions may not. For instance, the argument expression A1:A3 in sum(A1:A3) returns the values of cells A1, A2, A3 in a list. This list is then passed to the sum operation. It's also possible to mix lists and single values in argument expressions, for instance the operation sum(A1:A3, 1.0, C7), which would sum up five values. To handle arguments that might evaluate to lists, there's another evaluation function, called evalList, which takes a formula and returns a list of values:

```
private def evalList(e: Formula): List[Double] = e match {
  case Range(_, _) => references(e) map (_.value)
  case _ => List(evaluate(e))
}
```

If the formula argument passed to evalList is a Range, the returned value is a list consisting of the values of all cells referenced by the range. For every other formula the result is a list consisting of the single result value of that formula. The cells referenced by a formula are computed by a third function, references. Here is its definition:

```
def references(e: Formula): List[Cell] = e match {
  case Coord(row, column) =>
    List(cells(row)(column))
  case Range(Coord(r1, c1), Coord(r2, c2)) =>
    for (row <- (r1 to r2).toList; column <- c1 to c2)
    yield cells(row)(column)
  case Application(function, arguments) =>
    arguments flatMap references
  case _ =>
    List()
  }
} // end Evaluator
```

The references method is actually more general than needed right now in that it computes the list of cells referenced by any sort of formula, not just a Range formula. It will turn out later that the added functionality is needed to compute the sets of cells that need updating. The body of the method is a straightforward pattern match on kinds of formulas. For a coordinate Coord(row, column), it returns a single-element list containing the cell at that coordinate. For a range expression Range(coord1, coord2), it returns all cells between the two coordinates, computed by a for expression. For a function application Application(function, arguments), it returns the cells referenced by each argument expression, concatenated via flatMap into a single list. For the other two types of formulas, Textual and Number, it returns an empty list.

## 35.6 Operation libraries

The class Evaluator itself defines no operations that can be performed on cells: its operations table is initially empty. The idea is to define such operations in other traits, which are then mixed into the Model class. Listing 35.8 shows an example trait that implements common arithmetic operations:

```
package org.stairwaybook.scells
trait Arithmetic { this: Evaluator =>
  operations += (
    "add" -> { case List(x, y) => x + y },
    "sub" -> { case List(x, y) => x - y },
    "div" -> { case List(x, y) => x / y },
    "mul" -> { case List(x, y) => x * y },
    "mod" -> { case List(x, y) => x % y },
    "sum" -> { xs => (0.0 /: xs)(_ + _) },
    "prod" -> { xs => (1.0 /: xs)(_ * _) }
  )
}
```

Listing 35.8 · A library for arithmetic operations.

Interestingly, this trait has no exported members. The only thing it does is populate the operations table during its initialization. It gets access to that table by using a self type Evaluator, *i.e.*, by the same technique the Arithmetic class uses to get access to the model.

Of the seven operations that are defined by the Arithmetic trait, five are binary operations and two take an arbitrary number of arguments. The binary operations all follow the same schema. For instance, the addition operation add is defined by the expression:

```
{ case List(x, y) => x + y }
```

That is, it expects an argument list consisting of two elements x and y and returns the sum of x and y. If the argument list contains a number of elements different from two, a MatchError is thrown. This corresponds to the general "let it crash" philosophy of SCell's evaluation model, where incorrect input is expected to lead to a runtime exception that then gets caught by the try-catch inside the evaluation method.

The last two operations, sum and prod, take a list of arguments of arbitrary length and insert a binary operation between successive elements. So they are instances of the "fold left" schema that's expressed in class List by the /: operation. For instance, to sum a list of numbers List(x, y, z), the operation computes 0 + x + y + z. The first operand, 0, is the result if the list is empty.

You can integrate this operation library into the spreadsheet application by mixing the Arithmetic trait into the Model class, like this:

```
package org.stairwaybook.scells

class Model(val height: Int, val width: Int)
    extends Evaluator with Arithmetic {

  case class Cell(row: Int, column: Int) {
    var formula: Formula = Empty
    def value = evaluate(formula)

    override def toString = formula match {
      case Textual(s) => s
      case _ => value.toString
    }
  }

  ... // rest as before
}
```

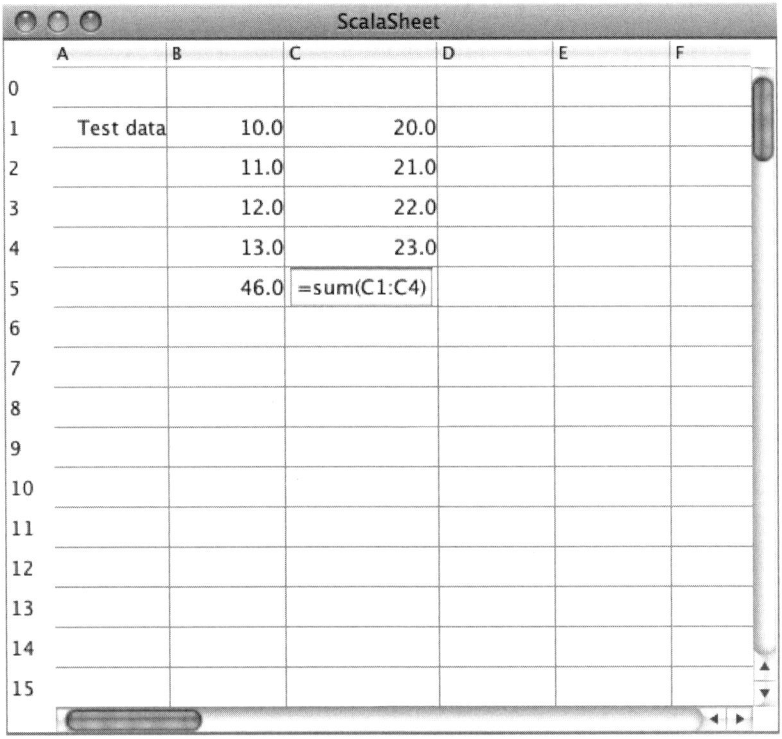

Figure 35.4 · Cells that evaluate.

Another change in the Model class concerns the way cells display themselves. In the new version, the displayed value of a cell depends on its formula. If the formula is a Textual field, the contents of the field are displayed literally. In all other cases, the formula is evaluated and the result value of that evaluation is displayed.

If you compile the changed traits and classes and relaunch the Main program you get something that starts to resemble a real spreadsheet. Figure 35.4 shows an example. You can enter formulas into cells and get them to evaluate themselves. For instance, once you close the editing focus on cell C5 in Figure 35.4, you should see 86.0, the result of evaluating the formula sum(C1:C4).

However, there's a crucial element still missing. If you change the value

of cell C1 in Figure 35.4 from 20 to 100, the sum in cell C5 will not be
automatically updated to 166. You'll have to click on C5 manually to see a
change in its value. What's still missing is a way to have cells recompute
their values automatically after a change.

## 35.7 Change propagation

If a cell's value has changed, all cells that depend on that value should have
their results recomputed and redisplayed. The simplest way to achieve this
would be to recompute the value of every cell in the spreadsheet after each
change. However such an approach does not scale well as the spreadsheet
grows in size.

A better approach is to recompute the values of only those cells that refer to a changed cell in their formula. The idea is to use an event-based
publish/subscribe framework for change propagation: once a cell gets assigned a formula, it will subscribe to be notified of all value changes in cells
to which the formula refers. A value change in one of these cells will trigger
a re-evaluation of the subscriber cell. If such a re-evaluation causes a change
in the value of the cell, it will in turn notify all cells that depend on it. The
process continues until all cell values have stabilized, *i.e.*, there are no more
changes in the values of any cell.[3]

The publish/subscribe framework is implemented in class Model using
the standard event mechanism of Scala's Swing framework. Here's a new
(and final) version of this class:

```
package org.stairwaybook.scells
import swing._

class Model(val height: Int, val width: Int)
extends Evaluator with Arithmetic {
```

Compared to the previous version of Model, this version adds a new import
of swing._, which makes Swing's event abstractions directly available.

The main modifications of class Model concern the nested class Cell.
Class Cell now inherits from Publisher, so that it can publish events. The
event-handling logic is completely contained in the setters of two properties:
value and formula. Here is Cell's new version:

---

[3]This assumes that there are no cyclic dependencies between cells. We discuss dropping
this assumption at the end of this chapter.

```
case class Cell(row: Int, column: Int) extends Publisher {
  override def toString = formula match {
    case Textual(s) => s
    case _ => value.toString
  }
}
```

To the outside, it looks like value and formula are two variables in class Cell. Their actual implementation is in terms of two private fields that are equipped with public getters, value and formula, and setters, value_= and formula_=. Here are the definitions implementing the value property:

```
private var v: Double = 0
def value: Double = v
def value_=(w: Double) = {
  if (!(v == w || v.isNaN && w.isNaN)) {
    v = w
    publish(ValueChanged(this))
  }
}
```

The value_= setter assigns a new value w to the private field v. If the new value is different from the old one, it also publishes a ValueChanged event with the cell itself as argument. Note that the test whether the value has changed is a bit tricky because it involves the value NaN. The Java spec says that NaN is different from every other value, including itself! Therefore, a test whether two values are the same has to treat NaN specially: two values v, w are the same if they are equal with respect to ==, or they are both the value NaN, *i.e.*, v.isNaN and w.isNaN both yield true.

Whereas the value_= setter does the publishing in the publish/subscribe framework, the formula_= setter does the subscribing:

```
private var f: Formula = Empty
def formula: Formula = f
def formula_=(f: Formula) = {
  for (c <- references(formula)) deafTo(c)
  this.f = f
  for (c <- references(formula)) listenTo(c)
  value = evaluate(f)
}
```

If a cell is assigned a new formula, it first unsubscribes with deafTo from all cells referenced by the previous formula value. It then stores the new formula in the private variable f and subscribes with listenTo to all cells referenced by it. Finally, it recomputes its value using the new formula.

The last piece of code in the revised class Cell specifies how to react to a ValueChanged event:

```
reactions += {
  case ValueChanged(_) => value = evaluate(formula)
}
} // end class Cell
```

The ValueChanged class is also contained in class Model:

```
case class ValueChanged(cell: Cell) extends event.Event
```

The rest of class Model is as before:

```
val cells = Array.ofDim[Cell](height, width)
for (i <- 0 until height; j <- 0 until width)
  cells(i)(j) = new Cell(i, j)
} // end class Model
```

The spreadsheet code is now almost complete. The final piece missing is the re-display of modified cells. So far, all value propagation concerned the internal Cell values only; the visible table was not affected. One way to change this would be to add a redraw command to the value_= setter. However, this would undermine the strict separation between model and view that you have seen so far. A more modular solution is to notify the table of all ValueChanged events and let it do the redrawing itself. Listing 35.9 shows the final spreadsheet component, which implements this scheme.

Class Spreadsheet of Listing 35.9 has only two new revisions. First, the table component now subscribes with listenTo to all cells in the model. Second, there's a new case in the table's reactions: if it is notified of a ValueChanged(cell) event, it demands a redraw of the corresponding cell with a call of updateCell(cell.row, cell.column).

```
package org.stairwaybook.scells
import swing._, event._

class Spreadsheet(val height: Int, val width: Int)
    extends ScrollPane {

  val cellModel = new Model(height, width)
  import cellModel._

  val table = new Table(height, width) {
    ... // settings as in Listing 35.1

    override def rendererComponent(
        isSelected: Boolean, hasFocus: Boolean,
        row: Int, column: Int) =
      ... // as in Listing 35.3

    def userData(row: Int, column: Int): String =
      ... // as in Listing 35.3

    reactions += {
      case TableUpdated(table, rows, column) =>
        for (row <- rows)
          cells(row)(column).formula =
            FormulaParsers.parse(userData(row, column))
      case ValueChanged(cell) =>
        updateCell(cell.row, cell.column)
    }

    for (row <- cells; cell <- row) listenTo(cell)
  }

  val rowHeader = new ListView(0 until height) {
    fixedCellWidth = 30
    fixedCellHeight = table.rowHeight
  }

  viewportView = table
  rowHeaderView = rowHeader
}
```

Listing 35.9 · The finished spreadsheet component.

## 35.8 Conclusion

The spreadsheet developed in this chapter is fully functional, even though at some points it adopts the simplest solution to implement rather than the most convenient one for the user. That way, it could be written in just under 200 lines of code. Nevertheless, the architecture of the spreadsheet makes modifications and extensions easy. In case you would like to experiment with the code a bit further, here are some suggestions of what you could change or add:

1. You could make the spreadsheet resizable, so that the number of rows and columns can be changed interactively.

2. You could add new kinds of formulas, for instance binary operations, or other functions.

3. You might think about what to do when cells refer recursively to themselves. For instance, if cell A1 holds the formula add(B1, 1) and cell B1 holds the formula mul(A1, 2), a re-evaluation of either cell will trigger a stack-overflow. Clearly, that's not a very good solution. As alternatives, you could either disallow such a situation, or just compute one iteration each time one of the cells is touched.

4. You could enhance error handling, giving more detailed messages describing what went wrong.

5. You could add a formula entry field at the top of the spreadsheet, so that long formulas could be entered more conveniently.

At the beginning of this book we stressed the scalability aspect of Scala. We claimed that the combination of Scala's object-oriented and functional constructs makes it suitable for programs ranging from small scripts to very large systems. The spreadsheet presented here is clearly still a small system, even though it would probably take up much more than 200 lines in most other languages. Nevertheless, you can see many of the details that make Scala scalable at play in this application.

The spreadsheet uses Scala's classes and traits with their mixin composition to combine its components in flexible ways. Recursive dependencies between components are expressed using self types. The need for static state is completely eliminated—the only top-level components that are not classes are formula trees and formula parsers, and both of these are purely functional. The application also uses higher-order functions and pattern matching extensively, both for accessing formulas and for event handling. So it is a good showcase of how functional and object-oriented programming can be combined smoothly.

One important reason why the spreadsheet application is so concise is that it can base itself on powerful libraries. The parser combinator library provides in effect an internal domain-specific language for writing parsers. Without it, parsing formulas would have been much more difficult. The event handling in Scala's Swing libraries is a good example of the power of control abstractions. If you know Java's Swing libraries, you probably appreciate the conciseness of Scala's reactions concept, particularly when compared to the tedium of writing notify methods and implementing listener interfaces in the classical publish/subscribe design pattern. So the spreadsheet demonstrates the benefits of extensibility, where high-level libraries can be made to look just like language extensions.

# Appendix A

# Scala Scripts on Unix and Windows

If you're on some flavor of Unix, you can run a Scala script as a shell script by prepending a "pound bang" directive at the top of the file. For example, type the following into a file named `helloarg`:

```
#!/bin/sh
exec scala "$0" "$@"
!#
// Say hello to the first argument
println("Hello, " + args(0) + "!")
```

The initial `#!/bin/sh` must be the very first line in the file. Once you set its execute permission:

```
$ chmod +x helloarg
```

You can run the Scala script as a shell script by simply saying:

```
$ ./helloarg globe
```

If you're on Windows, you can achieve the same effect by naming the file `helloarg.bat` and placing this at the top of your script:

```
::#!
@echo off
call scala %0 %*
goto :eof
::!#
```

# Glossary

**algebraic data type** A type defined by providing several alternatives, each of which comes with its own constructor. It usually comes with a way to decompose the type through pattern matching. The concept is found in specification languages and functional programming languages. Algebraic data types can be emulated in Scala with case classes.

**alternative** A branch of a `match` expression. It has the form "`case` *pattern* => *expression*." Another name for alternative is *case*.

**annotation** An *annotation* appears in source code and is attached to some part of the syntax. Annotations are computer processable, so you can use them to effectively add an extension to Scala.

**anonymous class** An anonymous class is a synthetic subclass generated by the Scala compiler from a new expression in which the class or trait name is followed by curly braces. The curly braces contains the body of the anonymous subclass, which may be empty. However, if the name following `new` refers to a trait or class that contains abstract members, these must be made concrete inside the curly braces that define the body of the anonymous subclass.

**anonymous function** Another name for function literal.

**apply** You can *apply* a method, function, or closure *to* arguments, which means you invoke it on those arguments.

**argument** When a function is invoked, an *argument* is passed for each parameter of that function. The parameter is the variable that refers to the argument. The argument is the object passed at invocation time. In addition, applications can take (command line) arguments that show up in the `Array[String]` passed to `main` methods of singleton objects.

**assign** You can *assign* an object *to* a variable. Afterwards, the variable will refer to the object.

**auxiliary constructor** Extra constructors defined inside the curly braces of the class definition, which look like method definitions named `this`, but with no result type.

**block** One or more expressions and declarations surrounded by curly braces. When the block evaluates, all of its expressions and declarations are processed in order, and then the block returns the value of the last expression as its own value. Blocks are commonly used as the bodies of functions, `for` expressions, `while` loops, and any other place where you want to group a number of statements together. More formally, a block is an encapsulation construct for which you can only see side effects and a result value. The curly braces in which you define a class or object do not, therefore, form a block, because fields and methods (which are defined inside those curly braces) are visible from the outside. Such curly braces form a *template*.

**bound variable** A *bound variable* of an expression is a variable that's both used and defined inside the expression. For instance, in the function literal expression (x: Int) => (x, y), both variables x and y are used, but only x is bound, because it is defined in the expression as an `Int` and the sole argument to the function described by the expression.

**by-name parameter** A parameter that is marked with a => in front of the parameter type, *e.g.*, (x: => Int). The argument corresponding to a by-name parameter is evaluated not before the method is invoked, but each time the parameter is referenced *by name* inside the method. If a parameter is not by-name, it is *by-value*.

**by-value parameter** A parameter that is *not* marked with a => in front of the parameter type, *e.g.*, (x: Int). The argument corresponding to a by-value parameter is evaluated before the method is invoked. By-value parameters contrast with *by-name* parameters.

**class** Defined with the `class` keyword, a *class* may either be abstract or concrete, and may be parameterized with types and values when instantiated. In "new Array[String](2)", the class being instantiated is Array and the type of the value that results is Array[String]. A class that takes type parameters is called a *type constructor*. A type can be said to have a class as well, as in: the class of type Array[String] is Array.

**closure** A function object that captures free variables, and is said to be "closed" over the variables visible at the time it is created.

**companion class** A class that shares the same name with a singleton object defined in the same source file. The class is the singleton object's companion class.

**companion object** A singleton object that shares the same name with a class defined in the same source file. Companion objects and classes have access to each other's private members. In addition, any implicit conversions defined in the companion object will be in scope anywhere the class is used.

**contravariant** A *contravariant* annotation can be applied to a type parameter of a class or trait by putting a minus sign (–) before the type parameter. The class or trait then subtypes contravariantly with—in the opposite direction as—the type annotated parameter. For example, Function1 is contravariant in its first type parameter, and so Function1[Any, Any] is a subtype of Function1[String, Any].

**covariant** A *covariant* annotation can be applied to a type parameter of a class or trait by putting a plus sign (+) before the type parameter. The class or trait then subtypes covariantly with—in the same direction as—the type annotated parameter. For example, List is covariant in its type parameter, so List[String] is a subtype of List[Any].

**currying** A way to write functions with multiple parameter lists. For instance def f(x: Int)(y: Int) is a curried function with two parameter lists. A curried function is applied by passing several arguments lists, as in: f(3)(4). However, it is also possible to write a *partial application* of a curried function, such as f(3).

**declare** You can *declare* an abstract field, method, or type, which gives an entity a name but not an implementation. The key difference between

783

declarations and definitions is that definitions establish an implementation for the named entity, declarations do not.

**define** To *define* something in a Scala program is to give it a name and an implementation. You can define classes, traits, singleton objects, fields, methods, local functions, local variables, *etc.* Because definitions always involve some kind of implementation, abstract members are *declared* not defined.

**direct subclass** A class is a *direct subclass* of its direct superclass.

**direct superclass** The class from which a class or trait is immediately derived, the nearest class above it in its inheritance hierarchy. If a class Parent is mentioned in a class Child's optional extends clause, then Parent is the direct superclass of Child. If a trait is mentioned in Child's extends clause, the trait's direct superclass is the Child's direct superclass. If Child has no extends clause, then AnyRef is the direct superclass of Child. If a class's direct superclass takes type parameters, for example class Child extends Parent[String], the direct superclass of Child is still Parent, not Parent[String]. On the other hand, Parent[String] would be the direct *supertype* of Child. See *supertype* for more discussion of the distinction between class and type.

**equality** When used without qualification, *equality* is the relation between values expressed by '=='. See also *reference equality*.

**expression** Any bit of Scala code that yields a result. You can also say that an expression *evaluates to* a result or *results in* a value.

**filter** An if followed by a boolean expression in a for expression. In for(i <- 1 to 10; if i % 2 == 0), the filter is "if i % 2 == 0". The value to the right of the if is the *filter expression*.

**filter expression** A *filter expression* is the boolean expression following an if in a for expression. In for(i <- 1 to 10; if i % 2 == 0), the filter expression is "i % 2 == 0".

**first-class function** Scala supports *first-class functions*, which means you can express functions in *function literal* syntax, *i.e.*, (x: Int) => x + 1, and that functions can be represented by objects, which are called *function values*.

**for comprehension**  Another name for `for expression`.

**free variable**  A *free variable* of an expression is a variable that's used inside the expression but not defined inside the expression. For instance, in the function literal expression (x: Int) => (x, y), both variables x and y are used, but only y is a free variable, because it is not defined inside the expression.

**function**  A *function* can be *invoked* with a list of arguments to produce a result. A function has a parameter list, a body, and a result type. Functions that are members of a class, trait, or singleton object are called *methods*. Functions defined inside other functions are called *local functions*. Functions with the result type of Unit are called *procedures*. Anonymous functions in source code are called *function literals*. At run time, function literals are instantiated into objects called *function values*.

**function literal**  A function with no name in Scala source code, specified with function literal syntax. For example, (x: Int, y: Int) => x + y.

**function value**  A function object that can be invoked just like any other function. A function value's class extends one of the FunctionN traits (*e.g.*, Function0, Function1) from package scala, and is usually expressed in source code via *function literal* syntax. A function value is "invoked" when its apply method is called. A function value that captures free variables is a *closure*.

**functional style**  The *functional style* of programming emphasizes functions and evaluation results and deemphasizes the order in which operations occur. The style is characterized by passing function values into looping methods, immutable data, methods with no side effects. It is the dominant paradigm of languages such as Haskell and Erlang, and contrasts with the *imperative style*.

**generator**  A *generator* defines a named val and assigns to it a series of values in a for expression. For example, in for(i <- 1 to 10), the generator is "i <- 1 to 10". The value to the right of the <- is the *generator expression*.

**generator expression** A *generator expression* generates a series of values in a for expression. For example, in for(i <- 1 to 10), the generator expression is "1 to 10".

**generic class** A class that takes type parameters. For example, because scala.List takes a type parameter, scala.List is a generic class.

**generic trait** A trait that takes type parameters. For example, because trait scala.collection.Set takes a type parameter, it is a generic trait.

**helper function** A function whose purpose is to provide a service to one or more other functions nearby. Helper functions are often implemented as local functions.

**helper method** A helper function that's a member of a class. Helper methods are often private.

**immutable** An object is *immutable* if its value cannot be changed after it is created in any way visible to clients. Objects may or may not be immutable.

**imperative style** The *imperative style* of programming emphasizes careful sequencing of operations so that their effects happen in the right order. The style is characterized by iteration with loops, mutating data in place, and methods with side effects. It is the dominant paradigm of languages such as C, C++, C# and Java, and contrasts with the *functional style*.

**initialize** When a variable is defined in Scala source code, you must *initialize* it with an object.

**instance** An *instance*, or class instance, is an object, a concept that exists only at run time.

**instantiate** To *instantiate* a class is to make a new object from the class, an action that happens only at run time.

786

**invariant** *Invariant* is used in two ways. It can mean a property that always holds true when a data structure is well-formed. For example, it is an invariant of a sorted binary tree that each node is ordered before its right subnode, if it has a right subnode. *Invariant* is also sometimes used as a synonym for nonvariant: "class `Array` is invariant in its type parameter."

**invoke** You can *invoke* a method, function, or closure *on* arguments, meaning its body will be executed with the specified arguments.

**JVM** The *JVM* is the Java Virtual Machine, or *runtime*, that hosts a running Scala program.

**literal** 1, "One", and (x: Int) => x + 1 are examples of *literals*. A literal is a shorthand way to describe an object, where the shorthand exactly mirrors the structure of the created object.

**local function** A *local function* is a `def` defined inside a block. To contrast, a `def` defined as a member of a class, trait, or singleton object is called a *method*.

**local variable** A *local variable* is a `val` or `var` defined inside a block. Although similar to local variables, parameters to functions are not referred to as local variables, but simply as parameters or "variables" without the "local."

**member** A *member* is any named element of the template of a class, trait, or singleton object. A member may be accessed with the name of its owner, a dot, and its simple name. For example, top-level fields and methods defined in a class are members of that class. A trait defined inside a class is a member of its enclosing class. A type defined with the `type` keyword in a class is a member of that class. A class is a member of the package in which is it defined. By contrast, a local variable or local function is not a member of its surrounding block.

**message** Actors communicate with each other by sending each other *messages*. Sending a message does not interrupt what the receiver is doing. The receiver can wait until it has finished its current activity and its invariants have been reestablished.

**meta-programming** Meta-programming software is software whose input is itself software. Compilers are meta-programs, as are tools like scaladoc. Meta-programming software is required in order to do anything with an *annotation*.

**method** A *method* is a function that is a member of some class, trait, or singleton object.

**mixin** *Mixin* is what a trait is called when it is being used in a mixin composition. In other words, in "trait Hat," Hat is just a trait, but in "new Cat extends AnyRef with Hat," Hat can be called a mixin. When used as a verb, "mix in" is two words. For example, you can *mix* traits *in*to classes or other traits.

**mixin composition** The process of mixing traits into classes or other traits. *Mixin composition* differs from traditional multiple inheritance in that the type of the super reference is not known at the point the trait is defined, but rather is determined anew each time the trait is mixed into a class or other trait.

**modifier** A keyword that qualifies a class, trait, field, or method definition in some way. For example, the private modifier indicates that a class, trait, field, or method being defined is private.

**multiple definitions** The same expression can be assigned in *multiple definitions* if you use the syntax val *v1*, *v2*, *v3* = *exp*.

**nonvariant** A type parameter of a class or trait is by default *nonvariant*. The class or trait then does not subtype when that parameter changes. For example, because class Array is nonvariant in its type parameter, Array[String] is neither a subtype nor a supertype of Array[Any].

**operation** In Scala, every *operation* is a method call. Methods may be invoked in *operator notation*, such as b + 2, and when in that notation, + is an *operator*.

**parameter** Functions may take zero to many *parameters*. Each parameter has a name and a type. The distinction between parameters and arguments is that arguments refer to the actual objects passed when a function is invoked. Parameters are the variables that refer to those passed arguments.

**parameterless function** A function that takes no parameters, which is defined without any empty parentheses. Invocations of parameterless functions may not supply parentheses. This supports the *uniform access principle*, which enables the def to be changed into a val without requiring a change to client code.

**parameterless method** A *parameterless method* is a parameterless function that is a member of a class, trait, or singleton object.

**parametric field** A field defined as a class parameter.

**partially applied function** A function that's used in an expression and that misses some of its arguments. For instance, if function f has type Int => Int => Int, then f and f(1) are *partially applied functions*.

**path-dependent type** A type like swiss.cow.Food. The swiss.cow part is a *path* that forms a reference to an object. The meaning of the type is sensitive to the path you use to access it. The types swiss.cow.Food and fish.Food, for example, are different types.

**pattern** In a match expression alternative, a *pattern* follows each case keyword and precedes either a *pattern guard* or the => symbol.

**pattern guard** In a match expression alternative, a *pattern guard* can follow a *pattern*. For example, in "case x if x % 2 == 0 => x + 1", the pattern guard is "if x % 2 == 0". A case with a pattern guard will only be selected if the pattern matches and the pattern guard yields true.

**predicate** A *predicate* is a function with a Boolean result type.

**primary constructor** The main constructor of a class, which invokes a superclass constructor, if necessary, initializes fields to passed values, and executes any top-level code defined between the curly braces of the class. Fields are initialized only for value parameters not passed to the superclass constructor, except for any that are not used in the body of the class and can therefore be optimized away.

**procedure** A *procedure* is a function with result type of Unit, which is therefore executed solely for its side effects.

**reassignable**  A variable may or may not be *reassignable*. A `var` is reassignable while a `val` is not.

**recursive**  A function is *recursive* if it calls itself. If the only place the function calls itself is the last expression of the function, then the function is *tail recursive*.

**reference**  A *reference* is the Java abstraction of a pointer, which uniquely identifies an object that resides on the JVM's heap. Reference type variables hold references to objects, because reference types (instances of `AnyRef`) are implemented as Java objects that reside on the JVM's heap. Value type variables, by contrast, may sometimes hold a reference (to a boxed wrapper type) and sometimes not (when the object is being represented as a primitive value). Speaking generally, a Scala variable *refers* to an object. The term "refers" is more abstract than "holds a reference." If a variable of type `scala.Int` is currently represented as a primitive Java `int` value, then that variable still refers to the `Int` object, but no reference is involved.

**reference equality**  *Reference equality* means that two references identify the very same Java object. Reference equality can be determined, for reference types only, by calling `eq` in `AnyRef`. (In Java programs, reference equality can be determined using `==` on Java reference types.)

**reference type**  A *reference type* is a subclass of `AnyRef`. Instances of reference types always reside on the JVM's heap at run time.

**referential transparency**  A property of functions that are independent of temporal context and have no side effects. For a particular input, an invocation of a referentially transparent function can be replaced by its result without changing the program semantics.

**refers**  A variable in a running Scala program always *refers* to some object. Even if that variable is assigned to `null`, it conceptually refers to the `Null` object. At runtime, an object may be implemented by a Java object or a value of a primitive type, but Scala allows programmers to think at a higher level of abstraction about their code as they imagine it running. See also *reference*.

**refinement type**  A type formed by supplying a base type a number of members inside curly braces. The members in the curly braces refine the types that are present in the base type. For example, the type of "animal that eats grass" is `Animal { type SuitableFood = Grass }`.

**result** An expression in a Scala program yields a *result*. The result of every expression in Scala is an object.

**result type** A method's *result type* is the type of the value that results from calling the method. (In Java, this concept is called the return type.)

**return** A function in a Scala program *returns* a value. You can call this value the *result* of the function. You can also say the function *results in* the value. The result of every function in Scala is an object.

**runtime** The Java Virtual Machine, or JVM, that hosts a running Scala program. *Runtime* encompasses both the virtual machine, as defined by the Java Virtual Machine Specification, and the runtime libraries of the Java API and the standard Scala API. The phrase *at run time* (with a space between *run* and *time*) means when the program is running, and contrasts with compile time.

**runtime type** The type of an object at run time. To contrast, a *static type* is the type of an expression at compile time. Most runtime types are simply bare classes with no type parameters. For example, the runtime type of `"Hi"` is `String`, and the runtime type of `(x: Int) => x + 1` is `Function1`. Runtime types can be tested with `isInstanceOf`.

**script** A file containing top level definitions and statements, which can be run directly with `scala` without explicitly compiling. A script must end in an expression, not a definition.

**selector** The value being matched on in a `match` expression. For example, in "`s match { case _ => }`", the selector is s.

**self type** A *self type* of a trait is the assumed type of `this`, the receiver, to be used within the trait. Any concrete class that mixes in the trait must ensure that its type conforms to the trait's self type. The most common use of self types is for dividing a large class into several traits as described in Chapter 29.

**semi-structured data** XML data is semi-structured. It is more structured than a flat binary file or text file, but it does not have the full structure of a programming language's data structures.

**serialization** You can *serialize* an object into a byte stream which can then be saved to files or transmitted over the network. You can later *deserialize* the byte stream, even on different computer, and obtain an object that is the same as the original serialized object.

**shadow** A new declaration of a local variable *shadows* one of the same name in an enclosing scope.

**signature** *Signature* is short for *type signature*.

**singleton object** An object defined with the `object` keyword. Each singleton object has one and only one instance. A singleton object that shares its name with a class, and is defined in the same source file as that class, is that class's *companion object*. The class is its *companion class*. A singleton object that doesn't have a companion class is a *standalone object*.

**standalone object** A singleton object that has no companion class.

**statement** An expression, definition, or import, *i.e.*, things that can go into a template or a block in Scala source code.

**static type** See *type*.

**subclass** A class is a *subclass* of all of its superclasses and supertraits.

**subtrait** A trait is a *subtrait* of all of its supertraits.

**subtype** The Scala compiler will allow any of a type's *subtypes* to be used as a substitute wherever that type is required. For classes and traits that take no type parameters, the subtype relationship mirrors the subclass relationship. For example, if class `Cat` is a subclass of abstract class `Animal`, and neither takes type parameters, type `Cat` is a subtype of type `Animal`. Likewise, if trait `Apple` is a subtrait of trait `Fruit`, and neither takes type parameters, type `Apple` is a subtype of type `Fruit`. For classes and traits that take type parameters, however, variance comes into play. For example, because abstract class `List` is declared to be covariant in its lone type parameter (*i.e.*, `List` is declared `List[+A]`), `List[Cat]` is a subtype of `List[Animal]`, and `List[Apple]` a subtype of `List[Fruit]`. These subtype relationships exist even though the class of each of these types is `List`. By contrast, because `Set` is not declared to be covariant in its type parameter (*i.e.*, `Set` is declared `Set[A]` with no plus sign), `Set[Cat]` is *not* a subtype of `Set[Animal]`. A subtype should correctly implement the contracts of its supertypes, so that the Liskov Substitution Principle applies, but the compiler only verifies this property at the level of type checking.

**superclass** A class's *superclasses* include its direct superclass, its direct superclass's direct superclass, and so on, all the way up to Any.

**supertrait** A class's or trait's *supertraits*, if any, include all traits directly mixed into the class or trait or any of its superclasses, plus any supertraits of those traits.

**supertype** A type is a *supertype* of all of its subtypes.

**synthetic class** A *synthetic class* is generated automatically by the compiler rather than being written by hand by the programmer.

**tail recursive** A function is *tail recursive* if the only place the function calls itself is the last operation of the function.

**target typing** *Target typing* is a form of type inference that takes into account the type that's expected. In nums.filter((x) => x > 0), for example, the Scala compiler infers type of x to be the element type of nums, because the `filter` method invokes the function on each element of nums.

**template** A *template* is the body of a class, trait, or singleton object definition. It defines the type signature, behavior, and initial state of the class, trait, or object.

**trait** A *trait*, which is defined with the `trait` keyword, is like an abstract class that cannot take any value parameters and can be "mixed into" classes or other traits via the process known as *mixin composition*. When a trait is being mixed into a class or trait, it is called a *mixin*. A trait may be parameterized with one or more types. When parameterized with types, the trait constructs a type. For example, `Set` is a trait that takes a single type parameter, whereas `Set[Int]` is a type. Also, `Set` is said to be "the trait of" type `Set[Int]`.

**type** Every variable and expression in a Scala program has a *type* that is known at compile time. A type restricts the possible values to which a variable can refer, or an expression can produce, at run time. A variable or expression's type can also be referred to as a *static type* if necessary to differentiate it from an object's *runtime type*. In other words, "type" by itself means static type. Type is distinct from class because a class that takes type parameters can construct many types. For example, `List` is a class, but not a type. `List[T]` is a type with a free type parameter. `List[Int]` and `List[String]` are also types (called *ground types* because they have no free type parameters). A type can have a "class" or "trait." For example, the class of type `List[Int]` is `List`. The trait of type `Set[String]` is `Set`.

**type constraint** Some annotations are *type constraints*, meaning that they add additional limits, or constraints, on what values the type includes. For example, `@positive` could be a type constraint on the type `Int`, limiting the type of 32-bit integers down to those that are positive. Type constraints are not checked by the standard Scala compiler, but must instead be checked by an extra tool or by a compiler plugin.

**type constructor** A class or trait that takes type parameters.

**type parameter** A parameter to a generic class or generic method that must be filled in by a type. For example, class `List` is defined as "`class List[T] { ... }`", and method `identity`, a member of object `Predef`, is defined as "`def identity[T](x:T) = x`". The T in both cases is a type parameter.

**type signature** A method's *type signature* comprises its name, the number, order, and types of its parameters, if any, and its result type. The type signature of a class, trait, or singleton object comprises its name, the type signatures of all of its members and constructors, and its declared inheritance and mixin relations.

**uniform access principle**  The *uniform access principle* states that variables and parameterless functions should be accessed using the same syntax. Scala supports this principle by not allowing parentheses to be placed at call sites of parameterless functions. As a result, a parameterless function definition can be changed to a `val`, or *vice versa*, without affecting client code.

**unreachable**  At the Scala level, objects can become *unreachable*, at which point the memory they occupy may be reclaimed by the runtime. Unreachable does not necessarily mean unreferenced. Reference types (instances of `AnyRef`) are implemented as objects that reside on the JVM's heap. When an instance of a reference type becomes unreachable, it indeed becomes unreferenced, and is available for garbage collection. Value types (instances of `AnyVal`) are implemented as both primitive type values and as instances of Java wrapper types (such as `java.lang.Integer`), which reside on the heap. Value type instances can be boxed (converted from a primitive value to a wrapper object) and unboxed (converted from a wrapper object to a primitive value) throughout the lifetime of the variables that refer to them. If a value type instance currently represented as a wrapper object on the JVM's heap becomes unreachable, it indeed becomes unreferenced, and is available for garbage collection. But if a value type currently represented as a primitive value becomes unreachable, then it does not become unreferenced, because it does not exist as an object on the JVM's heap at that point in time. The runtime may reclaim memory occupied by unreachable objects, but if an `Int`, for example, is implemented at run time by a primitive Java `int` that occupies some memory in the stack frame of an executing method, then the memory for that object is "reclaimed" when the stack frame is popped as the method completes. Memory for reference types, such as `Strings`, may be reclaimed by the JVM's garbage collector after they become unreachable.

**unreferenced**  See *unreachable*.

**value**  The result of any computation or expression in Scala is a *value*, and in Scala, every value is an object. The term value essentially means the image of an object in memory (on the JVM's heap or stack).

**value type** A *value type* is any subclass of AnyVal, such as Int, Double, or Unit. This term has meaning at the level of Scala source code. At runtime, instances of value types that correspond to Java primitive types may be implemented in terms of primitive type values or instances of wrapper types, such as java.lang.Integer. Over the lifetime of a value type instance, the runtime may transform it back and forth between primitive and wrapper types (*i.e.*, to box and unbox it).

**variable** A named entity that refers to an object. A variable is either a val or a var. Both vals and vars must be initialized when defined, but only vars can be later reassigned to refer to a different object.

**variance** A type parameter of a class or trait can be marked with a *variance* annotation, either *covariant* (+) or *contravariant* (−). Such variance annotations indicate how subtyping works for a generic class or trait. For example, the generic class List is covariant in its type parameter, and thus List[String] is a subtype of List[Any]. By default, *i.e.*, absent a + or − annotation, type parameters are *nonvariant*.

**wildcard type** A wildcard type includes references to type variables that are unknown. For example, Array[_] is a wildcard type. It is an array where the element type is completely unknown.

**yield** An expression can *yield* a result. The yield keyword designates the result of a for expression.

# Bibliography

[Abe96] Abelson, Harold and Gerald Jay Sussman. *Structure and Interpretation of Computer Programs*. The MIT Press, second edition, 1996.

[Aho86] Aho, Alfred V., Ravi Sethi, and Jeffrey D. Ullman. *Compilers: Principles, Techniques, and Tools*. Addison-Wesley Longman Publishing Co., Inc., Boston, MA, USA, 1986. ISBN 0-201-10088-6.

[Bay72] Bayer, Rudolf. "Symmetric binary B-Trees: Data structure and maintenance algorithms." *Acta Informatica*, 1(4):290–306, 1972.

[Blo08] Bloch, Joshua. *Effective Java Second Edition*. Addison-Wesley, 2008.

[DeR75] DeRemer, Frank and Hans Kron. "Programming-in-the large versus programming-in-the-small." In *Proceedings of the international conference on Reliable software*, pages 114–121. ACM, New York, NY, USA, 1975. doi:http://doi.acm.org/10.1145/800027.808431.

[Dij70] Dijkstra, Edsger W. "Notes on Structured Programming.", April 1970. Circulated privately. Available at http://www.cs.utexas.edu/users/EWD/ewd02xx/EWD249.PDF as EWD249 (accessed June 6, 2008).

[Eck98] Eckel, Bruce. *Thinking in Java*. Prentice Hall, 1998.

[Emi07] Emir, Burak, Martin Odersky, and John Williams. "Matching Objects With Patterns." In *Proc. ECOOP*, Springer LNCS, pages 273–295. July 2007.

[Eva03] Evans, Eric. *Domain-Driven Design: Tackling Complexity in the Heart of Software*. Addison-Wesley Professional, 2003.

797

[Fow04]  Fowler, Martin. "Inversion of Control Containers and the Dependency Injection pattern." January 2004. Available on the web at http://martinfowler.com/articles/injection.html (accesssed August 6, 2008).

[Gam95]  Gamma, Erich, Richard Helm, Ralph Johnson, and John Vlissides. *Design Patterns : Elements of Reusable Object-Oriented Software*. Addison-Wesley, 1995.

[Goe06]  Goetz, Brian, Tim Peierls, Joshua Bloch, Joseph Bowbeer, David Homes, and Doug Lea. *Java Concurrency in Practice*. Addison Wesley, 2006.

[Jav]  *The Java Tutorials: Creating a GUI with JFC/Swing*. Available on the web at http://java.sun.com/docs/books/tutorial/uiswing.

[Kay96]  Kay, Alan C. "The Early History of Smalltalk." In *History of programming languages—II*, pages 511–598. ACM, New York, NY, USA, 1996. ISBN 0-201-89502-1. doi:http://doi.acm.org/10.1145/234286.1057828.

[Kay03]  Kay, Alan C. An email to Stefan Ram on the meaning of the term "object-oriented programming", July 2003. The email is published on the web at http://www.purl.org/stefan_ram/pub/doc_kay_oop_en (accesssed June 6, 2008).

[Lan66]  Landin, Peter J. "The Next 700 Programming Languages." *Communications of the ACM*, 9(3):157–166, 1966.

[Mey91]  Meyers, Scott. *Effective C++*. Addison-Wesley, 1991.

[Mey00]  Meyer, Bertrand. *Object-Oriented Software Construction*. Prentice Hall, 2000.

[Mor68]  Morrison, Donald R. "PATRICIA—Practical Algorithm To Retrieve Information Coded in Alphanumeric." *J. ACM*, 15(4):514–534, 1968. ISSN 0004-5411. doi:http://doi.acm.org/10.1145/321479.321481.

[Ode03]  Odersky, Martin, Vincent Cremet, Christine Röckl, and Matthias Zenger. "A Nominal Theory of Objects with Dependent Types." In *Proc. ECOOP'03*, Springer LNCS, pages 201–225. July 2003.

[Ode05] Odersky, Martin and Matthias Zenger. "Scalable Component Abstractions." In *Proceedings of OOPSLA*, pages 41–58. October 2005.

[Ode11] Odersky, Martin. *The Scala Language Specification, Version 2.9.* EPFL, May 2011. Available on the web at http://www.scala-lang.org/docu/manuals.html (accessed April 20, 2014).

[Ray99] Raymond, Eric. *The Cathedral & the Bazaar: Musings on Linux and Open Source by an Accidental Revolutionary.* O'Reilly, 1999.

[Rum04] Rumbaugh, James, Ivar Jacobson, and Grady Booch. *The Unified Modeling Language Reference Manual (2nd Edition).* Addison-Wesley, 2004.

[SPJ02] Simon Peyton Jones, et.al. "Haskell 98 Language and Libraries, Revised Report." Technical report, http://www.haskell.org/onlinereport, 2002.

[Ste99] Steele, Jr., Guy L. "Growing a Language." *Higher-Order and Symbolic Computation*, 12:221–223, 1999. Transcript of a talk given at OOPSLA 1998.

[Vaz07] Vaziri, Mandana, Frank Tip, Stephen Fink, and Julian Dolby. "Declarative Object Identity Using Relation Types." In *Proc. ECOOP 2007*, pages 54–78. 2007.

# About the Authors

**Martin Odersky** is the creator of the Scala language. He is a professor at EPFL in Lausanne, Switzerland, and a founder of Typesafe, Inc. He works on programming languages and systems, more specifically on the topic of how to combine object-oriented and functional programming. Since 2001 he has concentrated on designing, implementing, and refining Scala. Previously, he has influenced the development of Java as a co-designer of Java generics and as the original author of the current javac reference compiler. He is a fellow of the ACM.

    **Lex Spoon** is a software engineer at Semmle, Ltd. He worked on Scala for two years as a post-doc at EPFL. He has a Ph.D. from Georgia Tech, where he worked on static analysis of dynamic languages. In addition to Scala, he has helped develop a wide variety of programming languages, including the dynamic language Smalltalk, the scientific language X10, and the logic language that powers Semmle. He and his wife live in Atlanta with two cats and a chihuahua.

    **Bill Venners** is president of Artima, Inc., publisher of the Artima Developer website (www.artima.com), and cofounder of Escalate Software, LLC. He is author of the book, *Inside the Java Virtual Machine*, a programmer-oriented survey of the Java platform's architecture and internals. His popular columns in JavaWorld magazine covered Java internals, object-oriented design, and Jini. Active in the Jini Community since its inception, Bill led the Jini Community's ServiceUI project, whose ServiceUI API became the de facto standard way to associate user interfaces to Jini services. Bill is also the lead developer and designer of ScalaTest, an open source testing tool for Scala and Java developers.

# Index

# Other titles from Artima Press

Property-based testing allows you to express your tests in terms of functions called "properties" that describe your code's behavior, leaving the task of test case generation and evaluation to the testing tool. *ScalaCheck: The Definitive Guide* is the authoritative guide to ScalaCheck, a property-based testing tool for Scala. Written by the creator of ScalaCheck, this book will help you learn how to take advantage of this valuable complement to traditional testing techniques.

## ScalaCheck: The Definitive Guide
by Rickard Nilsson
ISBN: 978-0-9815316-9-4
$24.95 paper book / $15.00 PDF eBook
Order it now at: `http://www.artima.com/shop/scalacheck`

*Akka Concurrency* is the authoritative guide to concurrent programming with Akka. Written as a practical guide, it will teach you not just the "what" and "how" of Akka, but also the "why." Akka isn't just a toolkit you can use to write your concurrent applications: it embodies a set of paradigms you can use to *reason* about those applications. This book will give you a whole new perspective on how you design, build, and think about concurrent applications on the JVM.

## Akka Concurrency
by Derek Wyatt
ISBN: 978-0-9815316-6-3
$45.95 paper book / $26.95 PDF eBook
Order it now at: `http://www.artima.com/shop/akka_concurrency`

If you liked this book, you'll like this book...
in electronic form!

A comprehensive step-by-step guide

Programming in

# Scala

Third Edition

Updated for Scala 2.12

Martin Odersky
Lex Spoon
Bill Venners

artima

The PDF eBook is a great companion to the paper book, because it is easy to carry around on your laptop, shows code with color syntax highlighting, lets you copy and paste Scala code directly into the interpreter, is quickly searchable and extensively hyperlinked. The book is also available for the Kindle, with more eBook formats coming soon.

Get your PDF or Kindle version today at:

http://www.artima.com/shop/programming_in_scala_3ed

0 4 MAY 2018